Probability Theory and Stochastic Modelling

Volume 75

The Probability Theory and Stochastic Modelling series is a merger and continuation of Springer's two well established series Stochastic Modelling and Applied Probability and Probability and Its Applications series. It publishes research monographs that make a significant contribution to probability theory or an applications domain in which advanced probability methods are fundamental. Books in this series are expected to follow rigorous mathematical standards, while also displaying the expository quality necessary to make them useful and accessible to advanced students as well as researchers. The series covers all aspects of modern probability theory including

- Gaussian processes
- Markov processes
- Random Fields, point processes and random sets
- Random matrices
- Statistical mechanics and random media
- Stochastic analysis

as well as applications that include (but are not restricted to):

- Branching processes and other models of population growth
- Communications and processing networks
- Computational methods in probability and stochastic processes, including simulation
- Genetics and other stochastic models in biology and the life sciences
- Information theory, signal processing, and image synthesis
- Mathematical economics and finance
- Statistical methods (e.g. empirical processes, MCMC)
- Statistics for stochastic processes
- Stochastic control
- Stochastic models in operations research and stochastic optimization
- Stochastic models in the physical sciences

More information about this series at http://www.springer.com/series/13205

Pierre Carpentier · Jean-Philippe Chancelier
Guy Cohen · Michel De Lara

Stochastic Multi-Stage Optimization

At the Crossroads between Discrete Time Stochastic Control and Stochastic Programming

Springer

Pierre Carpentier
ENSTA ParisTech
Palaiseau
France

Jean-Philippe Chancelier
CERMICS
École Nationale des Ponts
et Chaussées ParisTech
Marne la Vallée
France

Guy Cohen
CERMICS
École Nationale des Ponts
et Chaussées ParisTech
Marne la Vallée
France

Michel De Lara
CERMICS
École Nationale des Ponts
et Chaussées ParisTech
Marne la Vallée
France

ISSN 2199-3130 ISSN 2199-3149 (electronic)
Probability Theory and Stochastic Modelling
ISBN 978-3-319-18137-0 ISBN 978-3-319-18138-7 (eBook)
DOI 10.1007/978-3-319-18138-7

Library of Congress Control Number: 2015937212

Springer Cham Heidelberg New York Dordrecht London

Printed on acid-free paper

Springer International Publishing AG Switzerland is part of Springer Science+Business Media (www.springer.com)

Preface

This book can be considered as the result of a ten-year cooperation (starting in 2000) of the four authors within the so-called Stochastic Optimization Working Group (SOWG), a research team of the CERMICS (Applied Mathematics Laboratory) of École Nationale des Ponts et Chaussées (ENPC-ParisTech). Among the topics addressed in this working group, a major concern was to devise numerical methods to effectively solve stochastic optimization problems, particularly in a dynamic context, as this was the context of most real-life applications also tackled by the group.

The background of the four authors is system theory and control but the 2000s have seen the emergence of the Stochastic Programming stream, a stochastic expansion of Mathematical Programming, so the group was interested in bridging the gap between these two communities.

Of course, several Ph.D. students took part in the activities of this group, and among them were Kengy Barty, Laetitia Andrieu, Babacar Seck, Cyrille Strugarek, Anes Dallagi, Pierre Girardeau. Their contributions are gratefully acknowledged. We hope this book can help future students to get familiar with the field.

The book comprises five parts and two appendices. The first part provides an introduction to the main issues discussed later in the book, plus a chapter on the stochastic gradient algorithm which addresses the so-called open-loop optimization problems in which on-line information is absent. Part Two introduces the theoretical tools and notions needed to mathematically formalize and handle the topic of information which plays a major part in stochastic dynamic problems. It also discusses optimality conditions for such problems, such as the dynamic programming equation, and a variational approach which will lead to numerical methods in the next part. Part Three is precisely about discretization and numerical approaches. A simple benchmark illustrates the contribution of the particle method proposed in Chap. 7. Convergence issues of all those techniques are discussed in Part Four. Part Five is devoted to more advanced topics that are more or less out of reach of the numerical methods previously discussed, namely multi-agent problems and the presence of the so-called dual effect. Appendix A recalls some basic facts on

Optimization, while Appendix B provides a brief description of essential tools of Probability theory.

Although the four authors share the responsibility of the whole book contents, the reader may be interested in knowing who was the primary writer of each chapter. Here is the list:

Pierre Carpentier: Chapter 2, Appendix A;
Jean-Philippe Chancelier: Chapter 8, Appendix B;
Guy Cohen: Notation (in preliminary pages), Chapters 1, 5, 6, 7;
Michel De Lara: Chapters 3, 4, 9, 10.

Contents

Notation

Here we explain some notation and typographical conventions that we have used throughout this book. We conclude with a short list of symbols, abbreviations and acronyms to which the reader may refer. In this discussion about notation, we raise a tricky point that stems from some divergence between conventional mathematical concepts on the one hand, and a long-standing practice and terminology used in Probability Theory on the other. Most of the time, this divergence causes no problem in understanding what is meant, but we point out a few circumstances when some confusion may arise.

Some General Principles

This book is about stochastic optimization. As such, random variables are among the main mathematical notions involved. Unless specific reasons prevent us from doing so, we denote random variables by *capital bold* letters, e.g. $\boldsymbol{U}$. As taught in any elementary course in Probability Theory (see Appendix B in this book), random *variables* are indeed *functions* or *mappings* from a set generally called Ω to some other set, say $\mathbb{U}$.[1]

The space in which random variables, and more generally functions, take their values are denoted with the $\mathbb{BLACKBOARD}$ font. However, as is expected, symbols such as $\mathbb{R}$ and $\mathbb{N}$ have a special meaning, namely the set of real and integer numbers, respectively (they are included in the list below with additional variations such as $\overline{\mathbb{R}}$). Also, $\mathbb{P}$ denotes a probability measure and $\mathbb{E}$ denotes mathematical expectation (or conditional mathematical expectation). Functional spaces are generally denoted with the *calligraphic* font; for example, a mapping $\boldsymbol{U} : \Omega \to \mathbb{U}$ belongs to the set $\mathcal{U}$.

[1]Additional ingredients are also required (σ-fields over Ω and $\mathbb{U}$, a measurability requirement about the mapping, probability measure $\mathbb{P}$, etc.) but it is not our purpose to dwell on that here.

The *script* font is generally used to denote σ-fields (e.g. $\mathcal{F}$). We now refer the reader to the list of symbols and abbreviations at the end of this introduction.

A Tricky Point

Here we make a few remarks about the effects of calling (random) "variables" objects which are indeed "functions", and the consequences of this abuse of language on notation. This abuse of language is customary in the world of Probability Theory but may cause substantial confusion for less aware readers. We discuss this issue by referring to several puzzling situations that arise in this book.

For example, consider the expression $\mathbb{E}(f(\boldsymbol{U}))$. With no hesitation, one understands that f is a mapping from $\mathbb{U}$ to some other set (say $\mathbb{R}$ to fix ideas), that $f(\boldsymbol{U})$ must be interpreted as a new random variable, namely $f \circ \boldsymbol{U} : \Omega \to \mathbb{R}$, and that its expectation—that is, the integral of this function over Ω against the probability measure $\mathbb{P}$—is then evaluated. Hence, while in $f(\boldsymbol{U})$, $\boldsymbol{U}$ seems to play the part of a "variable", namely an argument of function f according to its position within parentheses, it must indeed be remembered that this is a mapping to be composed with f in order to produce a new mapping of the argument $\omega \in \Omega$ whose integral is then to be evaluated. Thus, there is no real difficulty.

The evaluation of the considered expression would change if $\boldsymbol{U}$ was replaced by another random variable $\boldsymbol{V}$. Because of the dependence of this expression upon this random variable, one would naturally consider the result as a function of the dummy argument $\boldsymbol{U}$. If g denotes this function, we may write

$$g(\boldsymbol{U}) = \mathbb{E}(f(\boldsymbol{U})) = \int_{\Omega} f \circ \boldsymbol{U}(\omega)\,\mathbb{P}(\mathrm{d}\omega)\,,$$

and we may even replace the first sign $=$ by $:=$ (which means that the left-hand side is defined by the expression on the right-hand side). Observe that the parts played by $\boldsymbol{U}$ in $g(\boldsymbol{U})$ and in $f(\boldsymbol{U})$ are quite different, despite the similarity in notation. Strictly speaking, $g(\boldsymbol{U})$ is a correct mathematical expression since g is indeed a function of the random variable $\boldsymbol{U}$, whereas $f(\boldsymbol{U})$ is an ambiguous shortcut that experienced readers are able to interpret. However, a problem may arise when both expressions appear on both sides of an equality as in the first of the two equalities above. Notice that if the intermediate expression in these two equalities is cancelled and only the two extreme members of the equalities are kept, no question arises since, now, everywhere $\boldsymbol{U}$ is interpreted as a *function* (and g is generally called a "*functional*" as a function of a function).

Therefore, the correct notation would be $g(\boldsymbol{U})$, whereas $f(\boldsymbol{U})$ is a shortcut that requires some appropriate interpretation, but in order to conform with a long-standing tradition in Probability Theory, we sometimes change $g(\boldsymbol{U})$ to $g([\boldsymbol{U}])$ in order to emphasize the fact that the "argument" $\boldsymbol{U}$ must rigorously be interpreted as the "global function" object and not only be used for the collection of its values $U(\omega)$ as in $f(U)$.

Let us give other instances when such a distinction is necessary. In this book, stochastic optimization problems of the following generic form are considered:

$$\min_{U} \mathbb{E}(j(U, W)) ,$$

in which

- U is a random variable taking values in $\mathbb{U}$ and plays the part of the decision variable;
- W is another random variable taking values in $\mathbb{W}$ and plays the part of the "noise";
- j is a real-valued mapping defined over $\mathbb{U} \times \mathbb{W}$ playing the part of the cost function.

A decision U is thus a random variable possibly subject to various constraints that are described in this book, and whose performance is evaluated by computing the expectation of the cost function which also involves an exogenous disturbance W. The expression behind the min operator in the above formulation must be interpreted as we did for $f(U)$ in the previous discussion. Namely, a real-valued random variable $j(U(\cdot), W(\cdot))$ must be considered and its expectation must be evaluated. On the contrary, the minimization operation involves the random variable U "as a whole"; in particular, as we shall see later in this book, some constraints (so-called informational or measurability constraints) may prevent independent consideration of the individual values $U(\omega)$ and force us to globally consider the whole function U in this minimization operation. Thus, according to our notational convention, we should instead write the previous stochastic optimization problem as

$$\min_{[U]} \mathbb{E}(j(U, W)) .$$

Nevertheless, for the sake of simplicity, we keep the former notation since the particular position of the decision in the min should prevent any ambiguity.

Finally, a third instance when this notation $[X]$ proves useful is the following. The reader may refer to Appendix B to find definitions of conditional expectations $\mathbb{E}(X|Y))$ where X and Y are two random variables with values in $\mathbb{X}$ and $\mathbb{Y}$, respectively. This conditional expectation is also a random variable with values in $\mathbb{X}$. Sometimes, we are also led to manipulate the function $\Psi : \mathbb{Y} \to \mathbb{X}$ which, whenever the event $\{Y = y\}$ (that is the subset $Y^{-1}(y) = \{\omega \mid Y(\omega) = y\}$) has a positive probability for a given value $y \in \mathbb{Y}$, may be interpreted as the "expectation of X conditioned by the event $\{Y = y\}$". It is explained in the appendix that Ψ is a function of $y \in \mathbb{Y}$, that is, of the values taken by the random variable Y, but that this function also depends on the "whole" function Y (and of course also on the function X as does the expectation $\mathbb{E}(X)$ itself). To emphasize this fact, we could write

$\Psi_{[Y]}(y)$ instead of merely $\Psi(y)$. In the former expression, both the values y taken by $\boldsymbol{Y}$ and the "global" random variable $\boldsymbol{Y}$ appear to play a part.

We will occasionally refer back to this discussion in the rest of this book.

Symbols and Abbreviations

$\mathbb{N}$	Set of integer (natural) numbers
$\mathbb{R}$	Set of real numbers
$\overline{\mathbb{R}}$	$\mathbb{R} \cup \{-\infty\} \cup \{+\infty\}$
$\mathbb{E}$	Mathematical expectation
$\mathbb{P}$	Probability measure
Var	Variance (of a random variable)
$\mathrm{I}_{\mathbb{A}}$	Identity function over set $\mathbb{A}$
χ_A	Characteristic function of subset A
1_A	Indicator function of subset A
$\lvert \cdot \rvert$	Absolute value
$\langle \cdot , \cdot \rangle$	Scalar product
$\lVert \cdot \rVert$	Norm
∇	Gradient
∇_x	Partial gradient (with respect to x)
∂	Subdifferential
$\partial \cdot / \partial x$	Partial derivative (with respect to x)
proj_A	Projection onto subset A
$\boldsymbol{U} \preceq \boldsymbol{V}$	Random variable $\boldsymbol{U}$ measurable with respect to $\boldsymbol{V}$ (same as $\boldsymbol{V} \succeq \boldsymbol{U}$; used also with functions, σ-fields, partitions, etc.)
x^{T}	Transposition of vector x
dom	Domain (of a function)
$\mathrm{co}\, A$	Convex hull of subset A
$\overline{\mathrm{co}}\, A$	Closed convex hull of subset A
$\xrightarrow{\mathcal{D}}$	Convergence in distribution
$\xrightarrow{\mathbb{P}}$	Convergence in probability
$\xrightarrow{\text{a.s.}}$	Almost sure convergence
l.s.c.	Lower semicontinuous
u.s.c.	Upper semicontinuous
i.i.d.	Independently identically distributed
iff	If and only if
w.r.t.	With respect to
s.t.	Subject to
a.s.	Almost surely (or almost sure)
$\mathbb{P}$-a.s.	Almost surely (or almost sure) w.r.t. to the probability measure $\mathbb{P}$

Acronyms

ADP	Approximate Dynamic Programming
APP	Auxiliary Problem Principle
DIS	Dynamic Information Structure
DP	Dynamic Programming
LBG	Linearly Bounded Gradient
LQG	Linear-Quadratic-Gaussian
MASIOS	Multi-Agent Stochastic Input–Output System
MQE	Mean Quadratic Error
NOLDE	No Open-Loop Dual Effect
SA	Stochastic Approximation
SAA	Sample Average Approximation
SDDP	Stochastic Dual Dynamic Programming
SIS	Static Information Structure
SOC	Stochastic Optimal Control
SP	Stochastic Programming

Part I
Preliminaries

Chapter 1
Issues and Problems in Decision Making Under Uncertainty

1.1 Introduction

The future cannot be predicted exactly, but one may learn from past observations. Past decisions can also improve future predictability. This is the context in which decisions are generally made. Herein, we discuss some mathematical issues pertaining to this topic.

1.1.1 Decision Making as Constrained Optimization Problems

Making decisions in a rational way is a problem which can be mathematically formulated as an *optimization* problem. Generally, several conflicting goals must be taken into account simultaneously. A choice must be made about which goals are formulated as constraints to be satisfied at a certain "level" (apart from constraints which are imposed by physical limitations), and which goals are reflected by (and aggregated within) a *cost function*.[1] Duality theory for *constrained optimization* problems provides a way to analyze, afterwards, the sensitivity of the best achievable cost as a function of constraint levels which were fixed a priori, and, possibly, to tune those levels to achieve a better trade-off between conflicting goals.

Problems that involve systems evolving in time enter the realm of *Optimal Control*. In a deterministic setting, Optimal Control has a long history dating back to the fifties with famous names such as Pontryagin [124] and Bellman [15]. The former, with his *Maximum Principle*, was more in the line of a *variational* approach of such problems, whereas the latter introduced the *Dynamic Programming* (DP) technique in connection with the state space approach.

---

[1] Throughout this book, without loss of generality, optimization problems are formulated as *minimization* problems, hence the objective function to be minimized is called a *cost*.

P. Carpentier et al., *Stochastic Multi-Stage Optimization*, Probability Theory and Stochastic Modelling 75, DOI 10.1007/978-3-319-18138-7_1

1.1.2 Facing Uncertainty

In general, when making decisions, one is faced with *uncertainties* which affect the cost function and, generally, the constraints. There are several possible attitudes associated with uncertainties, and consequently, several possible mathematical formulations of decision making problems under uncertainty. Let us mention two main possibilities.

Worst Case Design
The assumption here is that uncertainties lie in particular bounded subsets and, that one must consider the *worst situation* to be faced and try to make it as good as possible. In more mathematical terms, and considering the cost only for the time being (see hereafter for constraints), since one would like to minimize that cost, one must minimize the *maximal* possible value Nature can give to that cost by playing with uncertainties within the assumed bounded subsets. That is, a *min-max* (game like) problem is formulated and a *guaranteed* performance can be evaluated (as long as assumptions on uncertainties hold true).

The treatment of constraints in such an approach should normally follow the same lines of thought (one must fight against the worst possible uncertainty outcomes from the point of view of constraint satisfaction). Sometimes the terminology of *robust* decision making (or control) is used for approaches along those lines [16].

Stochastic Programming or Stochastic Control
Here, uncertainties are viewed as random variables following *a priori* probability laws. We shall call them "primitive" random variables as opposed to other "secondary" random variables involved in the problem and which are derived from the primitive ones by applying functions such as dynamic equations, feedback laws (see hereafter), etc. Then the cost to be minimized is the mathematical expectation of some performance index depending on those random variables and on decisions.

For this mathematical expectation to make sense, the decisions must also become random variables defined on the same underlying probability space. A trivial case is when those decisions are indeed *deterministic*: we shall call them *open-loop* decisions or "controls" later on. But they may also be true random variables because they are produced by applying functions to either primitive or secondary random variables. Here, we enter the domain of *feedback* or *closed-loop* control which plays a prominent part in decision making under uncertainty.

Let us now say a few words about constraint satisfaction. Constraints may be imposed as *almost sure* (a.s.) constraints. This is generally the case of equality or inequality constraints expressing physical laws or limitations. Other constraints may be formulated with mathematical expectations, although it is generally difficult to give a sound practical meaning to this approach. If a.s. requirements may sometimes be either unfeasible or not economically viable, one may appeal to "constraints in probability": the satisfaction of those constraints is required only "sufficiently often", that is, with a certain prescribed probability. We do not pursue this discussion here, as we mostly consider a.s. constraints in this book.

In the title of this section, we have used the words "Stochastic Programming" and "Stochastic Control". Stochastic Control, or rather Stochastic Optimal Control (SOC), is the extension of the theory of Deterministic Optimal Control to the situation when uncertainties are present and modeled by random variables, or stochastic processes since control theory mostly addresses dynamic problems. SOC problems were introduced not long after their deterministic counterparts, and the DP approach has been readily extended (under specific assumptions) to the stochastic framework. "Pontryagin like" or "variational" approaches appeared much later in the literature [25] and we shall come back to explanations for this fact. SOC is used to deal with *dynamic* problems. The notion of *feedback*, as naturally delivered by the DP approach, plays a central part in this area.

Stochastic Programming (SP), which can be traced back to such early contributors as Dantzig [50], is the extension of Mathematical Programming to the stochastic framework. As such, the initial emphasis is on optimization, possibly in a *static* setting, and numerical resolution methods are based on variational techniques; randomness is generally addressed by appealing to the Monte Carlo technique which, roughly speaking, amounts to representing this uncertainty through the consideration of several "samples" or "scenarios". This is why, historically, the notions of *feedback* and *information* were less present in SP than they were in SOC.

However, the SP community[2] has progressively considered two-stage, and then multi-stage problems. Inevitably, the question of *information structures* popped up in the field, at least to handle the elementary constraint of *nonanticipativeness*: one should not assume that the exact realizations of random variables at and after stage $t+1$ are known when making decisions at stage t; only a probabilistic description of future occurrences can be taken into account.

It is therefore natural that the two communities of SOC and SP tend to merge and borrow ideas from each other. The concepts of information and feedback are more developed in the former, and the variational and Monte Carlo approaches are more widespread in the latter. Getting closer to each other for the two communities should perhaps begin with unifying the terminology: as far as we understand, *recourse* in the SP community is used as a substitute for *feedback*. This book is an attempt to close the gap. The comparison between SOC and SP approaches is already addressed by Varaiya and Wets in this interesting paper [148].

1.1.3 The Role of Information in the Presence of Uncertainty

In Deterministic Optimal Control, as mentioned previously, there are two main approaches in connection with Pontryagin's and Bellman's contributions. The former

[2]The official web page of the SP community http://www.stoprog.org/ offers links to several tutorials and examples of applications of SP.

focuses on open-loop controls, whereas the latter provides closed-loop solutions. By open-loop controls, we mean that the decisions are given as a function of *time* only, whereas closed-loop strategies compute the control to be implemented at each time instant as a function of both *time and observations*; the observations may be the state itself.

In fact, there are no discrepancies in the performance achieved by both approaches because, in a deterministic situation, everything is uniquely determined by the decision maker. Therefore, if closed-loop strategies are implemented, one can simulate the closed-loop dynamic system, record the trajectories of state, control and observations variables, substitute those trajectories in the control strategy, and compute an open-loop control history that would generate exactly the same trajectories.

The situation is quite different in an uncertain environment, since trajectories are not predictable in advance (off-line) because they depend on on-line realizations of random variables. Available observations reveal some information about those realizations, at least on *past* realizations (because of *causality*). By using this on-line information, one can do better than simply apply a *blind* open-loop control which has been determined only on the basis of a priori probability laws followed by the random "noises".

This means that the achievable performance is dependent on what we call the *information pattern* or *information structure* of the problem: a decision making problem under uncertainty is not well-posed until the exact amount of information available prior to making every decision has been defined. Open-loop problems are problems in which no actual realization can be observed, and thus, the optimal decisions solely depend on a priori probability laws. In dynamic situations, every decision may depend on certain on-line observations that must be specified. Of course, the optimal decisions also depend on a priori probability laws since, generally, not all random realizations can be observed prior to making decisions, if only because of causality or nonanticipativeness.

Because of these considerations, one must keep in mind that solving stochastic optimization problems, especially in dynamic situations when on-line observations are made available, is not just a matter of optimization, of dealing with conventional constraints, or even of computing or evaluating mathematical expectations (which is generally a difficult task by itself); it is also the question of properly handling specific constraints that we shall call *informational constraints*. Indeed, as this book illustrates, there are essentially two ways of dealing with such constraints. That used by the DP approach is a *functional* way: decisions are searched for as *functions* of observations (feedback laws). But another way, which is more adapted to variational approaches in stochastic optimization, may also be considered: all variables of the problem, including decisions, are considered as random variables or stochastic processes; then the dependency of decisions upon observations must go through notions of *measurability* as used by Measure Theory. We shall call this alternative approach an *algebraic* handling of informational constraints (this terminology stems from the fact that information may be mathematically captured by σ-algebras, also called σ-fields, another important notion introduced by Measure Theory). A difficult aspect of numerical resolution schemes is precisely the practical translation of those measurability or algebraic constraints into the numerical problem.

An even more difficult aspect of *dynamic* information patterns is that future information may be affected by past decisions. Such situations are called situations with *dual effect*, a terminology which tries to convey the idea that present decisions have two, very often conflicting, effects or objectives: directly contributing to optimizing the cost function on the one hand, modifying the informational constraints to which future decisions are subject, on the other. Problems with dual effect are generally among the most difficult decision making problems (see again [148] about this topic).

1.2 Problem Formulations and Information Structures

In this section, two formulations of stochastic optimization problems are proposed: they pertain to the two schools of SOC and SP alluded to above. The important issue of *information structures* is also discussed.

1.2.1 Stochastic Optimal Control (SOC)

General Formulation

We consider the following formulation of a stochastic optimal control (SOC) problem in discrete time: for every time instant t, $\boldsymbol{X}_t$ ("state"[3]), $\boldsymbol{U}_t$ (control) and $\boldsymbol{W}_t$ (noise) are all random variables over a probability space $(\Omega, \mathcal{A}, \mathbb{P})$. They are related to each other by the *dynamics*

$$\boldsymbol{X}_{t+1} = f_t(\boldsymbol{X}_t, \boldsymbol{U}_t, \boldsymbol{W}_{t+1}) \tag{1.1a}$$

which is satisfied $\mathbb{P}$-almost surely for $t = 0, \ldots, T-1$. Here, to keep things simple, T, the *time horizon*, should be a given deterministic integer value, but it may be a random variable in more general formulations. The variable $\boldsymbol{X}_0$ is a given random variable. It is convenient to view $\boldsymbol{X}_0$ as a given function of some other random variable called $\boldsymbol{W}_0$, in such a way that all primitive random variables are denoted $\boldsymbol{W}_s$, $s = 0, \ldots, T$, whereas $\boldsymbol{W}$ denotes the corresponding stochastic process $\{\boldsymbol{W}_s\}_{s=0,\ldots,T}$. The purpose is to minimize a cost function

$$\mathbb{E}\left(\sum_{t=0}^{T-1} L_t(\boldsymbol{X}_t, \boldsymbol{U}_t, \boldsymbol{W}_{t+1}) + K(\boldsymbol{X}_T)\right) \tag{1.1b}$$

[3] Those quotes around the word *state* become clearer when discussing the *Markovian case* by the end of this subsection.

in which K is the *final* cost whereas L_t is called the *instantaneous* cost. The symbol $\mathbb{E}(\cdot)$ denotes *expectation* w.r.t. $\mathbb{P}$ (assuming of course that the functions involved are measurable and integrable). The minimization is achieved by choosing the control variable $\boldsymbol{U}_t$ at each time instant t, but as previously mentioned, this is done after some *on-line* information has been collected (in addition to the *off-line* information composed of the model—dynamics and cost—and the a priori distribution of $\{\boldsymbol{W}_s\}_{s=0,\dots,T}$). This on-line information is supposed to be at least *causal* or *nonanticipative*, that is, the largest possible amount of information available at time instant t is equivalent to the observation of the realizations of the random variables $\boldsymbol{W}_s$ for $s = 0, \dots, t$ (but not beyond t). In the language of Probability Theory, this amounts to saying that $\boldsymbol{U}_t$, as a random variable, is *measurable* w.r.t. the σ-field generated by $\{\boldsymbol{W}_s\}_{s=0,\dots,t}$ which is denoted $\mathcal{F}_t$:

$$\mathcal{F}_t = \sigma\big(\{\boldsymbol{W}_s\}_{s=0,\dots,t}\big) \tag{1.1c}$$

(the reader may refer to Appendix B for all those standard notions.) Of course, this σ-field increases as time passes, that is, $\mathcal{F}_t \subset \mathcal{F}_{t+1}$: it is then called a *filtration*.

Remark 1.1 Observe that in the right-hand side of (1.1a), $\boldsymbol{U}_t$ must be chosen *before* $\boldsymbol{W}_{t+1}$ is observed: this is called the *decision-hazard* framework, as opposed to the *hazard-decision* framework in which the decision maker plays after "nature" at each time stage. This is why we put $\boldsymbol{W}_{t+1}$ rather than $\boldsymbol{W}_t$ in the right-hand side of (1.1a). ◊

It may be that $\boldsymbol{U}_t$ is constrained to be measurable w.r.t. some σ-field $\mathcal{G}_t$ smaller than $\mathcal{F}_t$:

$$\boldsymbol{U}_t \text{ is } \mathcal{G}_t\text{-measurable}, \ \mathcal{G}_t \subset \mathcal{F}_t, \quad t = 0, \dots, T-1. \tag{1.1d}$$

Unlike $\mathcal{F}_t$, the σ-field $\mathcal{G}_t$ is not necessarily increasing with t (see hereafter).

Information Structure

Very often, $\mathcal{G}_t$ itself is a σ-field generated by some random variable $\boldsymbol{Y}_t$ called *observation*. Actually, $\boldsymbol{Y}_t$ should be considered as the collection of *all* observations available at t. That is, if $\boldsymbol{Z}_t$ denotes a new observation made available at t, but if the decision maker has *perfect memory* of all observations made so far, then $\boldsymbol{Y}_t = \{\boldsymbol{Z}_s\}_{s=0,\dots,t}$. In this case, as for $\mathcal{F}_t$, the σ-field $\mathcal{G}_t$ is increasing with t, but this is not necessarily always true.

The σ-fields $\mathcal{F}_t$, generated by $\{\boldsymbol{W}_s\}_{s=0,\dots,t}$, are of course only dependent upon the data of the problem, and this is also the case of the $\mathcal{G}_t$ if the observations $\boldsymbol{Y}_t$ are solely dependent on the primitive random variables $\boldsymbol{W}_s$. But if the observations depend also on the controls $\boldsymbol{U}_s$ (for example, if $\boldsymbol{Z}_t$ is a function of the "state" $\boldsymbol{X}_t$, possibly a function corrupted by noise), it is likely that the σ-field $\mathcal{G}_t$ depends on controls too, and therefore, the measurability constraint (1.1d) is an implicit constraint in that control is subject to constraints depending on controls! Fortunately, thanks to causality, this implicit character is only apparent, that is, the constraint on $\boldsymbol{U}_t$ depends on controls $\boldsymbol{U}_s$ with s strictly less than t.

Nevertheless, this is generally a source of huge complexity in SOC problems which is known under the name of the *dual effect* of control. This terminology tries to convey the fact that when making decisions at every time instant s, the decision maker has to take care of the following double effect: on the one hand, his decision affects cost (directly, at the same time instant, and in the future time instants, through the "state" variables); but, on the other hand, it makes the next decisions $\boldsymbol{U}_t$, $t > s$ more or less constrained through (1.1d).

Example 1.2 Let us give an example of this double or dual effect in the real life: the decision of investing in research in any industrial activity. On the one hand, investing in research costs money. On the other hand, an improved knowledge of the field of activity may help save money in the future by allowing better decisions to be made. This example shows that this future effect is very often contradictory with immediate cost considerations and thus the matter of a trade-off to be achieved. △

We now return to our general discussion of information structure in SOC problems. Even if the observations $\boldsymbol{Y}_t$ depend on past controls, it may happen than the σ-fields $\mathcal{G}_t$ they generate *do not* depend on those controls. This tricky phenomenon is discussed in Chap. 10. Apart from this rather exceptional situation, there are other circumstances when things turn out to be less complex than it may have seemed a priori.

The most classical such case is the *Markovian case*. Suppose the stochastic process $\boldsymbol{W}$ is a "white noise", that is, the random variables $\{\boldsymbol{W}_s\}_{s=0,\dots,T}$, are all mutually independent. Then, $\boldsymbol{X}_t$ truly deserves the name of the *state* variable at time t (this is why, until now, we put the word "state" between quotes—see Footnote 3). Indeed, because of this assumption of white noise, the past realizations of the noise process $\boldsymbol{W}$ provide no additional information about the likelihood of future realizations. Hence, remembering $\boldsymbol{X}_t$ is sufficient information to keep to predict the future evolution of the system after t. That is, $\boldsymbol{X}_t$ "summarizes" the past and additional observations are therefore useless. The *Markovian case* is defined as the situation when $\boldsymbol{W}$ is a white noise stochastic process and $\mathcal{G}_t$ is generated at each time t by the variable $\boldsymbol{X}_t$. Otherwise stated, the available observation $\boldsymbol{Y}_t$ at time t is simply $\boldsymbol{X}_t$. This is a *perfect (noiseless) and full size* observation of the state vector. If the observation is *partial* (a non injective function of $\boldsymbol{X}_t$) and/or a *noisy* such function, then the Markovian situation is broken.

In the Markovian case, $\mathcal{G}_t$ *does depend*, in general, upon past controls $\boldsymbol{U}_s$, $s < t$, but we *would not do better* with $\mathcal{F}_t$ replacing $\mathcal{G}_t$. This is why the Markovian case, although potentially falling into the most difficult category of problems with a dual effect, is not so complex as more general problems in this category. The Markovian feature is exploited by the Dynamic Programming (DP) approach (see Sect. 4.4) which is conceptually simple, but quickly becomes numerically difficult, and, indeed, impossible when the dimension of the state vector $\boldsymbol{X}_t$ becomes large.

1.2.2 Stochastic Programming (SP)

Formulation

Here we consider another formulation of stochastic optimization problems which ignores "intermediate" variables (such as the "state" $\boldsymbol{X}$ in the previous SOC formulation) and which concentrates on the essential items, namely, the

control or **decision** $\boldsymbol{U}$: a random variable over a probability space $(\Omega, \mathcal{A}, \mathbb{P})$ with values in a measurable space $(\mathbb{U}, \mathcal{U})$;

noise $\boldsymbol{W}$: another random variable with values in a measurable space $(\mathbb{W}, \mathcal{W})$;

cost function: a measurable mapping $j : \mathbb{U} \times \mathbb{W} \to \mathbb{R}$;

σ**-fields**: $\mathcal{F}$ denotes the σ-field generated by $\boldsymbol{W}$ whereas $\mathcal{G}$ denotes the one w.r.t. which $\boldsymbol{U}$ is constrained to be measurable; generally, $\mathcal{G}$ is generated by an

observation $\boldsymbol{Y}$: another random variable with values in a measurable space $(\mathbb{Y}, \mathcal{Y})$; in this case, we use the notation

$$\boldsymbol{U} \preceq \boldsymbol{Y} \tag{1.2}$$

to mean that $\boldsymbol{U}$ is measurable w.r.t. (the σ-field generated by) $\boldsymbol{Y}$. As we see in Chap. 3, this relation between random variables corresponds to an order relation. We also use this notation in constraints as $\boldsymbol{U} \preceq \mathcal{G}$ to mean that the random variable $\boldsymbol{U}$ is measurable w.r.t. the σ-field $\mathcal{G}$.

With these ingredients at hand, the problem under consideration is set as follows:

$$\min_{\boldsymbol{U} \preceq \mathcal{G}} \mathbb{E}\big(j(\boldsymbol{U}, \boldsymbol{W})\big) \quad \text{or} \quad \min_{\boldsymbol{U} \preceq \boldsymbol{Y}} \mathbb{E}\big(j(\boldsymbol{U}, \boldsymbol{W})\big). \tag{1.3}$$

Without going into detailed technical assumptions, we assume that expectations do exist, and that infima are reached (hence the use of the min symbol).

Typology of Information Structures

According to the nature of $\mathcal{G}$ or $\boldsymbol{Y}$, we distinguish the following three cases.

Open-loop optimization: this is the case when $\mathcal{G}$ is the trivial σ-field $\{\emptyset, \Omega\}$, or equivalently, $\boldsymbol{Y}$ is any deterministic variable (that is, a constant map over Ω). In this case, an optimal decision is based solely on the a priori (off-line) knowledge of the model, and not on any on-line observation. Therefore, the decision itself is a deterministic variable $u \in \mathbb{U}$ which must minimize a cost function $J(u)$ defined as an expectation of $j(u, \boldsymbol{W})$. The numerical resolution of such problems is considered in Chap. 2.

Static Information Structure (SIS): this is the case when $\mathcal{G}$ or $\boldsymbol{Y}$ are non trivial but *fixed*, that is, a priori given, independently of the decision $\boldsymbol{U}$. The terminology "static" does not imply that no dynamics such as (1.1a) are involved in the problem formulation. It just expresses that the σ-field $\mathcal{G}$ constraining the decision is a priori given at the problem formulation stage. If time t is involved, one must rewrite the measurability constraint as prescribed at each time stage t as "$\boldsymbol{U}_t$ is $\mathcal{G}_t$-measurable" as in (1.1d), and this *does* leave room for information made available on-line as time evolves. "Static" just says that this on-line information cannot be manipulated by past controls.

Remark 1.3 When the collection $\{\boldsymbol{U}_s\}_{s=0,\dots,T-1}$ of random variables is interpreted as a random vector over the probability space $(\Omega, \mathcal{A}, \mathbb{P})$, then its measurability is characterized by the σ-field $\sigma(\{\boldsymbol{U}_s\}_{s=0,\dots,T-1})$ on $(\Omega, \mathcal{A})$. However, with this interpretation, the collection of constraints (1.1d) cannot in general be reduced to a single "vector" constraint $\boldsymbol{U} \preceq \mathcal{G}$ where $\boldsymbol{U}$ would be the "vector" $\{\boldsymbol{U}_s\}_{s=0,\dots,T-1}$ and $\mathcal{G}$ a σ-field on $(\Omega, \mathcal{A})$, like $\sigma(\{\boldsymbol{U}_s\}_{s=0,\dots,T-1})$ is. For example, over a probability space $(\Omega, \mathcal{A}, \mathbb{P})$, with $T = 2$, $\mathcal{G}_0 = \{\emptyset, \Omega\}$ and $\mathcal{G}_1 = \mathcal{A}$, consider a random variable $\boldsymbol{U}_1$ such that $\sigma(\boldsymbol{U}_1) = \mathcal{A}$. Writing $\boldsymbol{U} \preceq \mathcal{G}$ implies that $\mathcal{G}$ would be the σ-field $\mathcal{A}$, which does not translate that $\boldsymbol{U}_0$ must be a *constant* (deterministic) variable as implied by $\boldsymbol{U}_0 \preceq \mathcal{G}_0$. ◊

Remark 1.4 If $\mathcal{G}$ is generated by an observation $\boldsymbol{Y}$, either $\boldsymbol{Y}$ does not depend on $\boldsymbol{U}$, or the σ-field it generates is fixed despite $\boldsymbol{Y}$ does depend on $\boldsymbol{U}$ (as already mentioned, this may also happen in some special situations addressed in Chap. 10). One may also wonder whether $\boldsymbol{Y}$ has any relation with $\boldsymbol{W}$, for example, whether $\boldsymbol{Y}$ is given as a function $h(\boldsymbol{W})$, in which case $\mathcal{G}$ would be a sub-σ-field of $\mathcal{F}$, the σ-field generated by $\boldsymbol{W}$. For example, in the SOC problem (1.1), $\boldsymbol{Y}_t$ may be the complete or partial observation of past noises $\boldsymbol{W}_s$, $s = 0, \dots, t$, so that $\mathcal{G}_t \subseteq \mathcal{F}_t \subset \mathcal{F}_T$. Nevertheless, the fact that $\boldsymbol{Y}$ does or does not have a connection with $\boldsymbol{W}$ is not fundamental. Indeed, by manipulating notation, one can consider that this connection does exist. As a matter of fact, one can redefine the noise variable as the couple $\boldsymbol{W}' = (\boldsymbol{W}, \boldsymbol{Y})$ so that $\boldsymbol{Y}$ *is* a function of $\boldsymbol{W}'$. That the cost function j does not depend on the "full" $\boldsymbol{W}'$ does not matter. ◊

Dynamic Information Structure (DIS): this is the situation when $\mathcal{G}$ or $\boldsymbol{Y}$ depends on $\boldsymbol{U}$, which yields a seemingly implicit measurability constraint. Actually, it is difficult to imagine such problems without explicitly introducing several stages at which decisions must be taken based on observations which may depend on decisions at other stages.

Those stages may be a priori ordered, and the order may be a total order. This is the case of SOC problems (1.1); but other examples are considered hereafter in which those stages are not directly interpreted as "time instants" but rather as "agents" acting one after the other. As soon as such a total order of stages can be defined a priori, the notion of *causality* (who is "upstream" and who is "downstream") is natural and helps untangling the implicit character of the measurability constraint. Nevertheless, the difficulty of such problems with DIS still remains sometimes tremendous as it is shown with help of an example in Sect. 1.3.3.

More general problems may arise in which the order of stages or agent actions is only partial, and the situation may be even worse if this order itself depend on outcomes of the decisions and/or of hazard. At least in the case of a fixed but partial order, it turns out that two notions are paramount for the level of difficulty of the problem resolution:

- Who influences the available observations of whom?
- Who knows more than whom?

We shall not pursue the discussion of this difficult topic here. It is more thoroughly examined in Chap. 9. The forthcoming examples help us scratch the surface.

1.3 Examples

This section introduces a few simple examples in order to illustrate the impact of information structures on the formulation of stochastic optimization problems. The stress is more on this aspect than on being fussy about mathematical details (in particular, we assume that all expectations make sense without going into more precise assumptions).

1.3.1 A Basic Example in Static Information

Consider two given scalar random variables, $\boldsymbol{W}$ and $\boldsymbol{Y}$, plus the decision $\boldsymbol{U}$, and finally the following problem of type (1.3):

$$\min_{\boldsymbol{U} \preceq \boldsymbol{Y}} \mathbb{E}\big((\boldsymbol{W} - \boldsymbol{U})^2\big). \tag{1.4}$$

It is well known that the solution of this problem, which consists in finding the best approximation of $\boldsymbol{W}$ which is $\boldsymbol{Y}$-measurable (that is, the projection of $\boldsymbol{W}$ onto the subspace of $\boldsymbol{Y}$-measurable random variables), is given by $\boldsymbol{U}^\sharp = \mathbb{E}(\boldsymbol{W} \mid \boldsymbol{Y})$, that is, the conditional expectation of $\boldsymbol{W}$ knowing $\boldsymbol{Y}$ (see Sect. 3.5.3 and Definition B.5).

Generally speaking, as we see it later on in Sects. 3.5.2 and 8.3.5, Problem (1.3) can be reformulated as follows:

$$\mathbb{E}\Big(\min_{u \in \mathbb{U}} \mathbb{E}\big(j(u, \boldsymbol{W}) \mid \boldsymbol{Y}\big)\Big). \tag{1.5}$$

In this form, since the conditional expectation subject to minimization is indeed a $\boldsymbol{Y}$-measurable random variable, it should be understood that the minimization operates parametrically for every realization driven by ω and this yields an arg min also parametrized by ω, that is, in fact, a random variable which is also $\boldsymbol{Y}$-measurable. When using this new formulation for Problem (1.4), the solution is readily derived (Hint: expand the square in the cost function and observe that $\boldsymbol{Y}$-measurable random variables "get out" of the inner conditional expectation).

1.3.2 The Communication Channel

Description of the Problem

This is the story of two agents trying to communicate through a noisy channel. This story is depicted in Fig. 1.1. The first agent (called the "encoder") gets a "message", here simply a random variable $\boldsymbol{W}_0$ supposed to be centered ($\mathbb{E}(\boldsymbol{W}_0) = 0$), and he wants to communicate it to the other agent. We may consider that the encoder's observation $\boldsymbol{Y}_0$ is precisely this $\boldsymbol{W}_0$. He knows that the channel adds a noise, say a centered random variable $\boldsymbol{W}_1$, to the message he sends, and so he must choose which "best" message to send. He has to "encode" the original signal $\boldsymbol{Y}_0$ into another variable $\boldsymbol{U}_0$ (what he decides to send through the channel), but the other agent (the "decoder") receives a noisy message $\boldsymbol{U}_0 + \boldsymbol{W}_1$. Finally, the decoder has to make his decision $\boldsymbol{U}_1$ about what was the original message $\boldsymbol{W}_0$, based on his observation, namely $\boldsymbol{Y}_1 = \boldsymbol{U}_0 + \boldsymbol{W}_1$, the message he received. That is, he has to "decode", in an "optimal" manner, the signal $\boldsymbol{Y}_1$ which is his observation.

This game is *cooperative* in that the encoder and the decoder try to help each other so as to reduce the error of communication as much as possible (a problem in "team theory" [104], which deals with decision problems involving several agents or decision makers with a common objective function but possibly different observations). Mathematically, this can be expressed by saying that they seek to minimize the expected square error $\mathbb{E}\big((\boldsymbol{U}_1 - \boldsymbol{W}_0)^2\big)$. However, without any other limitation or penalty, such a problem turns out to be rather trivial. For example, if the encoder sends an amplified signal $\boldsymbol{U}_0 = k\boldsymbol{Y}_0$ where k is an arbitrarily large constant, then the noise $\boldsymbol{W}_1$ added by the channel is negligible in front of this very large signal, and the decoder can then decode it by dividing it by the same constant k. For the game to be interesting and realistic, one must put a penalty on the "power" $\mathbb{E}(\boldsymbol{U}_0^2)$ sent over the channel, either with help of a constraint limiting this power to a maximum level, or by introducing an additional term proportional to this power into the cost. To stay closer to the generic formulation (1.3), we choose the latter option. Finally, the problem under consideration is the following:

$$\min_{\boldsymbol{U}_0, \boldsymbol{U}_1} \mathbb{E}\big(\alpha \boldsymbol{U}_0^2 + (\boldsymbol{U}_1 - \boldsymbol{W}_0)^2\big) \tag{1.6a}$$

$$\text{s.t.} \quad \boldsymbol{U}_0 \preceq \boldsymbol{Y}_0 , \quad \boldsymbol{U}_1 \preceq \boldsymbol{Y}_1 . \tag{1.6b}$$

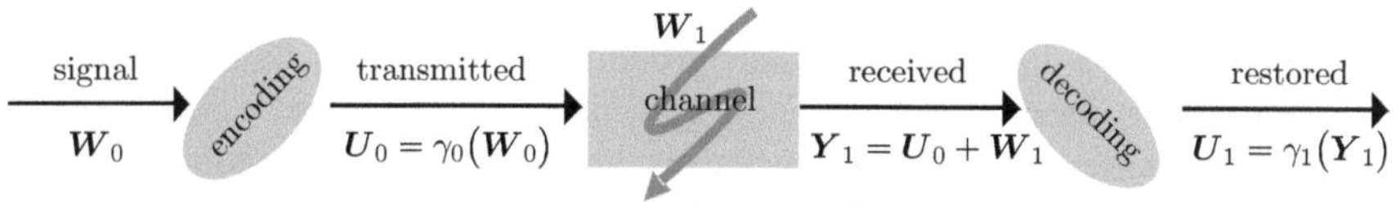

Fig. 1.1 Communication through a noisy channel

The positive parameter α is the unit cost for the power transmitted over the channel. The measurability constraints (1.6b) reflect what each agent knows before making his decision.

Discussion

There are a few remarks to make at this point:

- there is no time index t explicitly involved in this formulation, but still there is a natural order of the agents: the encoder acts first in that his action has an influence on what the decoder observes;
- there is no inclusion (in either direction) between the information available to the encoder and to the decoder although, as just highlighted, the decoder is "downstream" the encoder; if we interpret agents as time stages, it means that, at the second time stage, not all the information available at the first time stage has been retained, a fact referred to as "no perfect memory".

The fact that the encoder can influence what the decoder observes, whereas the decoder does not know as much as the encoder knows, is a source of tremendous difficulties. We are actually here in the heart of what we called "dual effect" earlier: the encoder, when making his decision, should not spend too much money according to the cost function (in particular, he should limit the power send over the channel) but, at the same time, he should be aware of the fact that his encoding impacts the information revealed to the decoder. To make this consideration more concrete, we discuss it further in a simplified setting in the next paragraph.

At this stage, let us say what is known about the resolution of Problem (1.6) [84, 154].

- The exact solution is yet unknown in the general case (see hereafter).
- There are particular cases when the solution is known, namely when the dimension of the message to be transmitted is exactly the same as the dimension of the encoded message, that is, when $\dim \mathbb{W}_0 = \dim \mathbb{U}_0$ (with certain additional assumptions, in particular Gaussian noises). Then, the encoder simply sends the original message ($\boldsymbol{U}_0 = \boldsymbol{W}_0$) and the decoder computes the conditional expectation $\boldsymbol{U}_1 = \mathbb{E}(\boldsymbol{W}_0 \mid \boldsymbol{Y}_1)$, which is a linear function of the observation $\boldsymbol{Y}_1$ when assuming that all primitive random variables are Gaussian. But what is important to notice is that the solution is proved to be optimal not because it satisfies some optimality condition (that, at present, nobody knows how to write), but because it achieves the lower bound of the expected square error provided by the Information Theory of Shannon [11].
- When $\dim \mathbb{W}_0 < \dim \mathbb{U}_0$ (redundancy in coding) or $\dim \mathbb{W}_0 > \dim \mathbb{U}_0$ (compression in coding), the exact solution is not known yet, but it is known to be a *nonlinear* function of observations. Indeed, on the one hand, the best linear feedback strategy satisfying (1.6b) can easily be obtained, and, on the other hand, clever nonlinear feedback strategies have been proposed which outperform the best linear strategy (although they are not claimed to be optimal). This appearance of nonlinear strategies in a Linear-Quadratic-Gaussian (LQG) stochastic optimization problem is an

illustration of what is known under the name of *signaling*: by using tricky nonlinear strategies, the encoder tries to provide to the decoder as much information about his observation as possible (here, the message to communicate) at the cheapest cost, using the system "dynamics" itself as the medium of this information transmission. Note that these signaling strategies would be impossible if the encoder could not influence the decoder's observation. In addition, it would be useless if the decoder knew at least as much information as the encoder knows (this would be the case of "perfect memory" in SOC problem (1.1)).

How Signaling Works?
We try to give the feeling of how signaling works, assuming that the encoder uses only linear strategies. Thus, let $\boldsymbol{U}_0 = k\boldsymbol{W}_0$. Of course, the decoder knows k because the strategy is elaborated (off line) jointly by the two decision makers. On line, the decoder observes the value of $\boldsymbol{Y}_1 = k\boldsymbol{W}_0 + \boldsymbol{W}_1$ from which he must guess the value realized by $\boldsymbol{W}_0$.

The primitive random variables of the problem are the couple $(\boldsymbol{W}_0, \boldsymbol{W}_1)$. For the purpose of graphical representation, we assume that this couple lies in the square $[0, 1] \times [0, 1]$. Figure 1.2 represents this square and the parallel lines corresponding to equations $w_1 = -kw_0 + y_1$ (with slope $-k$ and value-at-zero y_1). Therefore, after the realized value of $\boldsymbol{Y}_1$ has been observed, the decoder knows on which particular line the true realization of the noises is located. Given that his purpose is to determine the realization of $\boldsymbol{W}_0$, it is graphically intuitive that the uncertainty about this value decreases as $|k|$ (that is, the slope, be it negative or positive) increases. In terms of Communication Theory, this means that the ratio signal/noise improves as $|k|$ increases. This shows how the encoder can make the problem of the decoder more

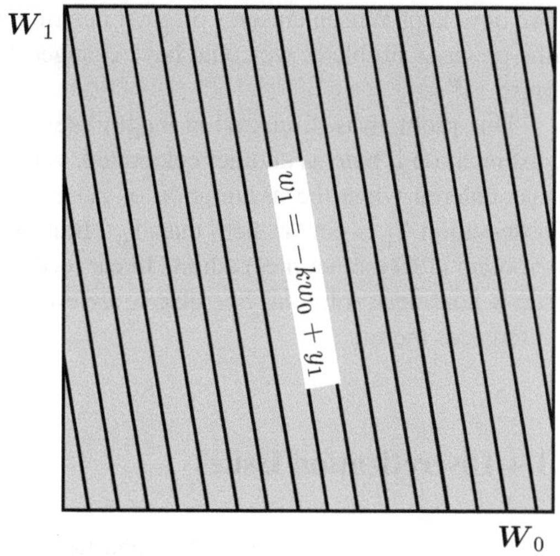

Fig. 1.2 Partition generated by $\boldsymbol{Y}_1$

or less tractable by choosing his own strategy. But remember that large values of $|k|$, and hence of $\mathbb{E}\big((\boldsymbol{U}_0)^2\big)$, cause a large cost (see (1.6a)).

1.3.3 Witsenhausen's Celebrated Counterexample

The following problem was proposed by Hans Witsenhausen in 1968 [155] as evidence that LQG problems may lead to nonlinear feedback solutions whenever the information structure is not "classical" (say, here, when it does not reflect *perfect memory*). This information feature is similar to that of the previous problem (Sect. 1.3.2) and several other features are similar (linear dynamics, dimensions, etc.). The main difference lies in the fact that Witsenhausen's problem belongs to the SOC class (1.1); therefore its cost function is additive in time as (1.1b), whereas (1.6a) is not so because of the cross-product $\boldsymbol{U}_1 \boldsymbol{W}_0$. The statement of this problem is as follows:

$$\min_{\boldsymbol{U}_0, \boldsymbol{U}_1} \mathbb{E}\big(k^2 \boldsymbol{U}_0^2 + \boldsymbol{X}_2^2\big) \tag{1.7a}$$

$$\text{s.t.} \quad \boldsymbol{U}_0 \preceq \boldsymbol{Y}_0, \quad \boldsymbol{U}_1 \preceq \boldsymbol{Y}_1, \tag{1.7b}$$

$$\boldsymbol{X}_1 = \boldsymbol{X}_0 + \boldsymbol{U}_0, \tag{1.7c}$$

$$\boldsymbol{X}_2 = \boldsymbol{X}_1 - \boldsymbol{U}_1, \tag{1.7d}$$

$$\boldsymbol{Y}_0 = \boldsymbol{X}_0, \tag{1.7e}$$

$$\boldsymbol{Y}_1 = \boldsymbol{X}_1 + \boldsymbol{W}. \tag{1.7f}$$

We have kept Witsenhausen's original notation, but to enhance the parallelism with the previous problem, we could have changed k^2 into α and $\boldsymbol{X}_0$ (resp. $\boldsymbol{W}$) into $\boldsymbol{W}_0$ (resp. $\boldsymbol{W}_1$).

This problem is discussed at length later on in this book (see Sect. 4.2), so we just mention it here as another celebrated, yet simple, example of all the difficulties encountered when the assumption of *perfect memory* is dropped (here again, the observation $\boldsymbol{Y}_1$ is not "richer" than $\boldsymbol{Y}_0$). Bansal and Basar [11] discuss the fact that Problem (1.6) (sometimes) admits linear feedback solutions whereas Problem (1.7) has a nonlinear solution. See also a review of this problem by Y.C. Ho [80] and references therein.

1.4 Discretization Issues

So far, several formulations of stochastic optimization problems have been considered, and the role and importance of their information structure have been discussed. Those problems involve random variables and measurability or informational constraints, and they are *infinite-dimensional* problems for which closed-form solutions are scarcely obtainable. Therefore, a numerical resolution goes through some discretization process to make them amenable to a finite-dimensional approximation. However, due to the particular nature of informational constraints, this discretization process requires special care.

1.4.1 Problems with Static Information Structure (SIS)

Most problems with DIS are presently out of reach from the numerical point of view, sometimes even at the early stage of writing down optimality conditions. An exception is provided by problems which are amenable to a Markovian formulation with a very moderate state space dimension. This book mainly concentrates on problems with SIS (nevertheless, problems with no dual effect are also in principle amenable to a SIS formulation).

Accordingly, we may consider problems under the SOC formulation (1.1) or under the more compact SP formulation (1.3).

The subclass of open-loop problems are simpler in that their solution is deterministic (the solution is an element of the control space $\mathbb{U}$ and not an application from Ω to $\mathbb{U}$). However, the cost function involves computing an expectation, a task that cannot generally be achieved analytically. Thus, one must appeal to some sort of *Monte Carlo sampling* one way or another. Chapter 2 considers different ways of exploiting this idea and combining it with numerical optimization itself.

The more general SP or SOC problems with SIS involve the same issue of computing expectations, if not even *conditional expectations*, but their solution, unlike in open-loop problems, are random variables. In addition, this solution is subject to informational or measurability constraints. Such constraints must be reflected, one way or another, in a discretized version of the problem, since, in general, some discretization technique must be used to come up with a numerical problem that can be solved with a computer. It turns out that this twofold aspect of discretization, namely,

- Monte Carlo like sampling for estimating expectations or conditional expectations;
- finite dimensional representation of random variables with mutual measurability constraints;

is a rather subtle issue that must be handled very carefully for, otherwise, a completely irrelevant discrete problem may result. An example is given hereafter.

As already mentioned at the end of Sect. 1.1.3, there are two different ways of translating informational constraints: one called *functional* (essentially, some random variables are represented as *functions* of other random variables), and the other one called *algebraic* (some random variables must be *measurable* with respect to other random variables). This translates into different numerical requirements, but in any case the interaction of the informational constraint representation with the Monte Carlo sampling in order to come up with a meaningful discrete problem is a tricky point as illustrated now by an example.

1.4.2 Working Out an Example

The Problem

Consider two independent random variables $\boldsymbol{W}_0$ and $\boldsymbol{W}_1$, each with a uniform probability distribution over $[-1, 1]$ (zero mean, variance $1/3$). The unique decision variable $\boldsymbol{U}$ may only use the observation of $\boldsymbol{W}_0$ (which we view as the initial state $\boldsymbol{X}_0$). The final state $\boldsymbol{X}_1$ is equal to $\boldsymbol{W}_0 + \boldsymbol{U} + \boldsymbol{W}_1$. The goal is to minimize $\mathbb{E}(\varepsilon \boldsymbol{U}^2 + \boldsymbol{X}_1^2)$, where ε is a given "small" positive number ("cheap control"). The statement is thus

$$\min_{\boldsymbol{U} \preceq \boldsymbol{W}_0} \mathbb{E}\big(\varepsilon \boldsymbol{U}^2 + (\boldsymbol{W}_0 + \boldsymbol{U} + \boldsymbol{W}_1)^2\big). \tag{1.8}$$

Exact Solution

We have that

$$\mathbb{E}\big(\varepsilon \boldsymbol{U}^2 + (\boldsymbol{W}_0 + \boldsymbol{U} + \boldsymbol{W}_1)^2\big) = \mathbb{E}\big(\boldsymbol{W}_0^2 + \boldsymbol{W}_1^2 + (1+\varepsilon)\boldsymbol{U}^2 + 2\boldsymbol{U}\,\boldsymbol{W}_0 + 2\boldsymbol{U}\,\boldsymbol{W}_1 + 2\boldsymbol{W}_0\boldsymbol{W}_1\big).$$

The last two terms on the right-hand side yield zero in expectation since $\boldsymbol{W}_0$ and $\boldsymbol{W}_1$ are centered independent random variables and since $\boldsymbol{U}$ is measurable with respect to $\boldsymbol{W}_0$. The first two terms yield twice the variance $1/3$ of the noises. Therefore, we remain with the problem of minimizing

$$\frac{2}{3} + \mathbb{E}\big((1+\varepsilon)\boldsymbol{U}^2 + 2\boldsymbol{U}\,\boldsymbol{W}_0\big) \tag{1.9}$$

by choosing $\boldsymbol{U}$ as a measurable function of $\boldsymbol{W}_0$. Using (1.5), one can prove that the solution is given by the feedback rule

$$\boldsymbol{U} = -\frac{\boldsymbol{W}_0}{1+\varepsilon},$$

and the corresponding optimal cost is readily calculated to be

$$\frac{1}{3}\,\frac{1+2\varepsilon}{1+\varepsilon} \approx \frac{1}{3}. \tag{1.10}$$

Monte Carlo Discretization

We now proceed to some discretization of this problem. To that purpose, we first consider N noise trajectories (w_0^i, w_1^i), $i = 1, \ldots, N$, which are N sample realizations of a two-dimensional vector $(\boldsymbol{W}_0, \boldsymbol{W}_1)$ with uniform probability distribution over $[-1, 1]^2$. Those samples serve to approximate the cost expectation by a usual Monte Carlo averaging.[4]

However, in this process, we must also consider N corresponding realizations $\{u^i\}_{i=1,\ldots,N}$ of the random decision variable $\boldsymbol{U}$. But, we must keep in mind that this random variable should be measurable with respect to the first component $\boldsymbol{W}_0$ of the previous vector.

To that purpose, we impose the constraint

$$\forall i, j, \; u^i = u^j \quad \text{whenever} \quad w_0^i = w_0^j, \tag{1.11}$$

which prevents $\boldsymbol{U}$ from taking different values whenever $\boldsymbol{W}_0$ assumes the same value in any two sample trajectories. For each sample i, the cost is

$$\varepsilon(u^i)^2 + (w_0^i + u^i + w_1^i)^2 = (\varepsilon + 1)(u^i)^2 + 2(w_0^i + w_1^i)u^i + (w_0^i + w_1^i)^2. \tag{1.12}$$

This expression must be minimized in u^i for every $i = 1, \ldots, N$, under the constraint (1.11). Indeed, if the N sample trajectories are produced by a random drawing with the uniform probability distribution over the square $[-1, 1]^2$, then, with probability 1, w_0^i is different from w_0^j for any couple (i, j) with $i \neq j$. Therefore, with probability 1, the constraint (1.11) is not binding, that is, (1.12) can be minimized for each value of i independently. This yields the optimal value

$$u^i = -\frac{w_0^i + w_1^i}{1 + \varepsilon} \tag{1.13}$$

and the corresponding contribution to the cost $\varepsilon(w_0^i + w_1^i)^2/(1 + \varepsilon)$. This is of order ε, and so is the average over N samples

$$\frac{1}{N(1+\varepsilon)} \sum_{i=1}^{N} \varepsilon(w_0^i + w_1^i)^2 \tag{1.14}$$

even when N goes to infinity. This is far from the actual optimal cost given by (1.10).

What Is the Real Value of this "Solution"?

However, any admissible solution (any $\boldsymbol{U}$ such that $\boldsymbol{U} \preceq \boldsymbol{W}_0$) cannot achieve a cost better than the optimal cost (1.10). The value (1.14) is just a "fake" cost estimation. The resolution of the discretized problem derived from the previous Monte Carlo

[4]What we call here "*N samples* or *sample realizations*" may be referred elsewhere in this book as a *N-sample*, whereas N is referred to as the *number of samples* or as the *size* of the N-sample.

procedure yielded an optimal value u^i (see (1.13)) associated with each sample noise trajectory represented by a point (w_0^i, w_1^i) in the square $[-1, 1]^2$. Hence, before trying to evaluate the cost associated with this "solution", we must first derive from it an *admissible* solution for the original problem, that is, a random variable $\boldsymbol{U}$ over $\Omega = [-1, 1]^2$, but with constant value along every vertical line of this square (since the abscissa represents the first component $\boldsymbol{W}_0$ of the 2-dimensional noise $(\boldsymbol{W}_0, \boldsymbol{W}_1)$).

A natural choice is as follows:

- we first renumber the N sample points so that the first component w_0^i is increasing with i;
- then, we divide the square into N vertical strips by drawing vertical lines in the middle of segments $[w_0^i, w_0^{i+1}]$ (see Fig. 1.3), that is, the i-th strip is $[a^{i-1}, a^i] \times [-1, 1]$ with $a^i = (w_0^i + w_0^{i+1})/2$ for $i = 2, \ldots, N-1$, $a^0 = -1$, and $a^N = 1$;[5]
- then, we define the solution $\boldsymbol{U}$ as the function of (w_0, w_1) which is piecewise constant over the square divided into those N strips, using of course the optimal value u^i given by (1.13) in strip i; that is, we consider

$$\boldsymbol{U}(w) = \sum_{i=1}^{N} u^i \mathbf{1}_{[a^{i-1}, a^i]\times[-1,1]}(w), \tag{1.15}$$

where w ranges in the square $[-1, 1]^2$ and $\mathbf{1}_A(\cdot)$ is the indicator function which takes the value 1 in A and 0 elsewhere.

Since this is an admissible solution for the original (continuous) problem, the corresponding cost value $\mathbb{E}(\varepsilon \boldsymbol{U}^2 + \boldsymbol{X}_1^2)$ can be evaluated. Here, the expectation is

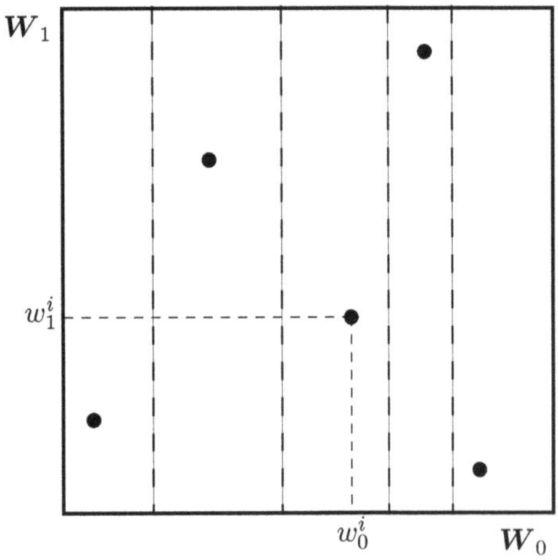

Fig. 1.3 Building an admissible solution for problem (1.8)

[5] Later on in this book (see Sect. 6.1), we discuss the concept of *Voronoi cells*: here we are defining the N Voronoi cells of the segment $[-1, 1]$ which are based on the "centroids" w_0^i.

over the argument w considered as a random variable over the square with uniform distribution.

According to (1.9), this expected cost is easily evaluated analytically as

$$\frac{2}{3}+\sum_{i=1}^{N}\left((1+\varepsilon)(u^i)^2\int_{a^{i-1}}^{a^i}\frac{1}{2}\,\mathrm{d}w_0+2u^i\int_{a^{i-1}}^{a^i}\frac{w_0}{2}\,\mathrm{d}w_0\right)$$

$$=\frac{2}{3}+\sum_{i=1}^{N}\left((1+\varepsilon)(u^i)^2\frac{a^i-a^{i-1}}{2}+u^i\frac{(a^i)^2-(a^{i-1})^2}{2}\right). \quad (1.16)$$

Although this is an "expected" cost, it is still a random variable since u^i and a^i are functions of the w_0^i's which result from random drawings (u^i also depends upon the w_1^j's). Indeed, (1.16) should be considered as an *estimation* of the optimal cost resulting from the (random) estimation (1.15) of the true solution.

In order to assess the value of this estimate, and first of all of its possible bias (not to speak of its variance), we must compute the expectation of (1.16) when considering that the w_0^i's are realizations of N independent random variables $\boldsymbol{W}_0^i$, each uniformly distributed over $[-1, 1]$. This calculation is not straightforward. The expression of the a^i's as functions of the w_0^i's is meaningful as long as the w_0^i's have been reordered into an increasing sequence. Therefore, although those N random numbers are the result of independent drawings, the calculation of expectations is made somewhat tricky by this reordering. We therefore skip it here. But, we have used a simple computer program using a pseudo-random number generator to evaluate the mean and standard deviation of this estimated cost as functions of the number N of used samples (for each value of N, the program uses 1,000 series of N drawings in order to evaluate those statistics). Figure 1.4 shows the results: the averaged cost $\pm$ the standard deviation are depicted as functions of N (here ε is taken equal to 1/100).

By observing Fig. 1.4, as N goes to infinity, the expected value of (1.16) goes to 2/3. Remember that the true optimal cost (see (1.10)) was close to 1/3! Moreover, it is readily checked that the optimal *open-loop* solution, that is the optimal $\boldsymbol{U}$ which is measurable w.r.t. the trivial σ-field $\{\emptyset, \Omega\}$, is equal to 0 and that the corresponding cost is also 2/3. Hence the solution we have produced with our naive Monte-Carlo approach (and especially the naive way (1.11) of handling the information structure of the problem) is not better than the open-loop solution!

How to Improve the Monte Carlo Approach? The Idea of Scenario Trees
Reviewing the previous procedure to provide an estimate of the solution of the original problem, one realizes that a crucial step, after the somewhat classical one of Monte Carlo sampling, is to translate the informational constraint $\boldsymbol{U} \preceq \boldsymbol{W}_0$ into the discretized version of the problem. The constraint (1.11) is rather ineffective, and it leads to the fact that the optimal value u^i (see (1.13)) found for sample i is "anticipative": u^i depends on w_1^i, which should not be the case. This explains why the apparent cost (that evaluated by averaging over the N samples) is very optimistic (of order ε whereas the true optimal cost is 1/3).

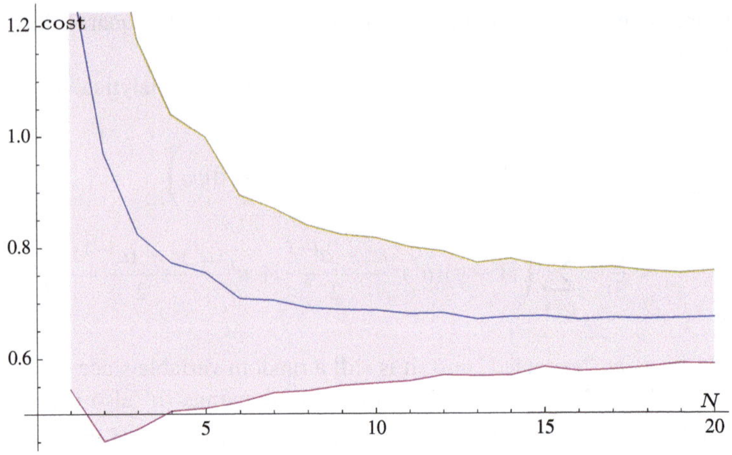

Fig. 1.4 Cost provided by the naive Monte Carlo method as a function of the number N of samples

On the other hand, when one is required to propose an *admissible* solution for the continuous problem, (namely (1.15) which satisfies the measurability constraint), this avoids the drawback of anticipativity, but then we have seen that the corresponding cost is as bad as that of the open-loop solution.

The question is thus: how to make another constraint translating the informational constraint in the discretized problem more effective than (1.11)? An obvious answer is that, in our collection of sample trajectories used in the discrete optimization problem, there should *really* be *distinct* samples with the *same* value of component w_0. This can be viewed as the origin of the idea of *"scenario trees"*. Here "scenario" is another terminology for "sample" and "tree"[6] refers to the shape depicted in Fig. 1.5. In this figure, one must imagine that a certain sample value w_0^j is attached to each node j of the first stage in the tree and that sample values w_1^k are likewise attached to nodes k at the second stage. Therefore, since distinct scenarios correspond to distinct "leaves" of the tree (they are still numbered with i ranging from 1 to N), the tree shape implies that several scenarios (couples (w_0^i, w_1^i)) share common values w_0^i. For ease of notation, we assume that all nodes of the first level (numbered with $j = 1, \ldots, N_0$) have the same number N_1 of "sons" (successors at the second stage, numbered with $k = 1, \ldots, N_1$ for each j). Hence $N = N_0 \times N_1$.

Admittedly, if the scenarios are produced randomly (according to the joint uniform probability law of $(\boldsymbol{W}_0, \boldsymbol{W}_1)$ over the square $[-1, 1] \times [-1, 1]$), or if they have been recorded from real life observations, there is a probability *zero* that a tree shape pops up spontaneously, for any arbitrary large, but finite, N. The question of how a scenario tree can be derived from real recorded data is considered in Chap. 6. The

[6]Actually, in Fig. 1.5, a "forest", that is, a collection of trees, rather than a "tree", is depicted since there are several "root nodes" which are the nodes at the first level. But we keep on speaking of "trees" to match the traditional terminology of "scenario tree".

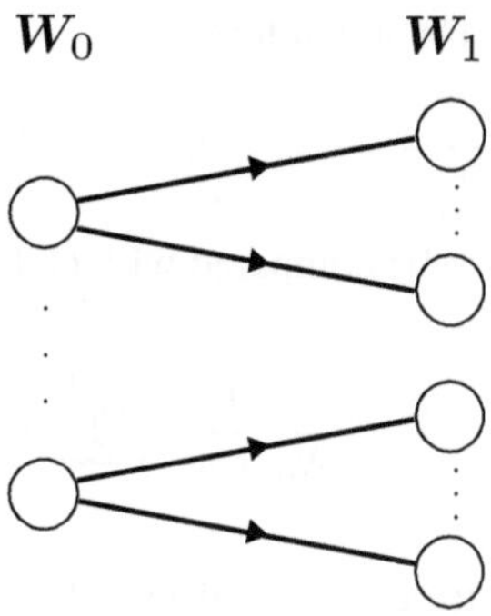

Fig. 1.5 A scenario tree on two stages

situation is easier if one knows the underlying probability law. In our example, since $\boldsymbol{W}_0$ and $\boldsymbol{W}_1$ are known to be independent (the white noise case), any element in a set of N_0 samples of $\boldsymbol{W}_0$ can be combined with the same, or N_0 distinct, sets of N_1 samples of $\boldsymbol{W}_1$ to produce such a tree. Even if $\boldsymbol{W}_0$ and $\boldsymbol{W}_1$ were not independent, one could first generate N_0 samples of $\boldsymbol{W}_0$ using the marginal probability law of this variable, and then, using each sample w_0^j and the *conditional* probability law of $\boldsymbol{W}_1$ knowing that $\boldsymbol{W}_0$ assumes the value w_0^j, one could generate N_1 associated samples w_1^k of $\boldsymbol{W}_1$ ("sons" of that w_0^j).

It is not our purpose now to discuss the production of "good" scenario trees. We just assume that such a scenario tree has been obtained, and that it reflects good statistical properties w.r.t. the underlying probability law of the noises when N_0 and N_1 go to infinity, in a sense that we leave to the reader's intuition at this stage. Our purpose is to revisit the resolution of the discretized problem formulated with this scenario tree and to examine its asymptotic behavior when the number of samples becomes very large. To fix notations, we consider scenarios $\{(w_0^j, w_1^{jk})\}_{j=1,\dots,N_0}^{k=1,\dots,N_1}$ and we introduce the following additional symbols:

$$\overline{w}_1^j = \frac{1}{N_1}\sum_{k=1}^{N_1} w_1^{jk}, \quad (\overline{\sigma}_1^j)^2 = \frac{1}{N_1}\sum_{k=1}^{N_1}(w_1^{jk})^2. \tag{1.17}$$

Notice that $\overline{w}_1^j$ can be interpreted as an estimate of the *conditional expectation* of $\boldsymbol{W}_1$ knowing that $\boldsymbol{W}_0 = w_0^j$. Likewise, $(\overline{\sigma}_1^j)^2$ can be interpreted as an estimate of the *conditional* second order moment.

To each node of the first level of the tree is attached a control variable u^j. The cost of the discretized problem is

$$\frac{1}{N_0}\sum_{j=1}^{N_0}\Big(\varepsilon(u^j)^2 + \frac{1}{N_1}\sum_{k=1}^{N_1}(u^j + w_0^j + w_1^{jk})^2\Big).$$

The arg min is

$$u^j = -\frac{w_0^j + \overline{w}_1^j}{1+\varepsilon}, \quad j = 1, \ldots, N_0, \tag{1.18}$$

to be compared with (1.13). This yields the optimal cost

$$\frac{1}{N_0(1+\varepsilon)} \sum_{j=1}^{N_0} \left(\varepsilon (w_0^j)^2 + 2\varepsilon w_0^j \overline{w}_1^j - (\overline{w}_1^j)^2 + (1+\varepsilon)(\overline{\sigma}_1^j)^2 \right), \tag{1.19}$$

to be compared with (1.14) and (1.10). If we assume that the estimates (1.17) converge towards their right values (respectively, 0 and 1/3) as N_1 goes to infinity, then (1.19) gets close to

$$\frac{1}{N_0(1+\varepsilon)} \sum_{j=1}^{N_0} \left(\varepsilon (w_0^j)^2 + \frac{1+\varepsilon}{3} \right).$$

Now, the expression $(1/N_0) \sum_{j=1}^{N_0} (w_0^j)^2$ can also be viewed as an estimate of the second order moment of $\boldsymbol{W}_0$ and, if we assume that it converges to the true value 1/3 when N_0 goes to infinity, then we recover, in the limit, the true optimal cost (1.10). Therefore, unlike with the previous naive Monte Carlo method (see (1.14)), here the optimal cost obtained in the discrete problem appears to converge to the right value.

As seen earlier (see (1.16)), it is also interesting to evaluate the real cost associated with an admissible solution derived from the collection of "optimal" values (1.18) by plugging those values into the formula (1.15) (with N replaced by N_0). Again, we have appealed to a computer program using 1,000 experiments, each consisting in:

- drawing N_0 values w_0^j at random;
- associated with each of those values, drawing a set of N_1 values w_1^{jk} at random;
- computing the $\overline{w}_1^j$'s (see (1.17)), the u^j's (see (1.18)) and forming the admissible solution (1.15) (N replaced by N_0) with those values after reordering the indices j so that w_0^j is increasing with j;
- evaluating the true cost $\mathbb{E}(\varepsilon \boldsymbol{U}^2 + \boldsymbol{X}_1^2)$ by analytic integration w.r.t. the couple $w = (w_0, w_1)$ with uniform probability distribution over the square $[-1, 1]^2$.

Remember that this integral w.r.t. argument w appearing in (1.15) is done for random values u^j depending on the random drawings w_0^j and w_1^{jk}. The 1,000 experiments are used to evaluate the mean and standard deviation of the random cost so obtained. In those experiments, we took $N_0 = N_1$, that is, $N_0 = \sqrt{N}$.

Figure 1.6 depicts the mean $\pm$ the standard deviation of the cost as a function of $N_0 = \sqrt{N}$ (still with $\varepsilon = 1/100$). The limit as N goes to infinity seems to be the correct value of the optimal cost given by (1.10), namely 0.3366, but the convergence appears to be asymptotically very slow, a fact on which we comment further in Chap. 6.

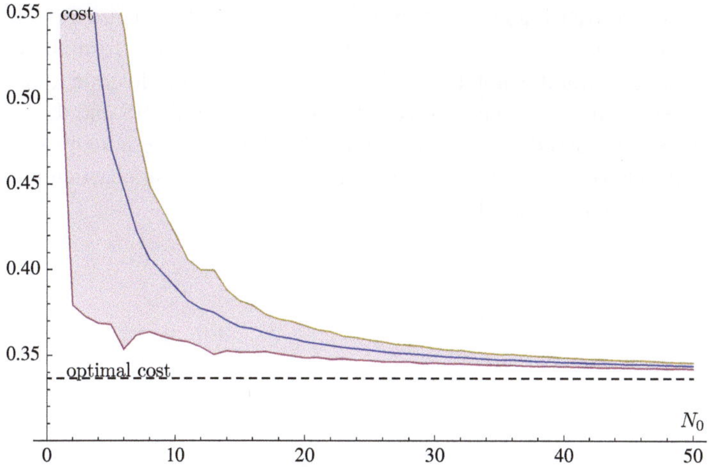

Fig. 1.6 Cost provided by the use of a stochastic tree as a function of the number N_0 of pieces of the piecewise constant $\boldsymbol{U}(\cdot)$ (N_0^2 scenarios)

Observe that in the comparison with Fig. 1.4, while the abscissa does represent the number of pieces uses to approximate the random variable $\boldsymbol{U}(\cdot)$ in both plots, in Fig. 1.4, this abscissa represents also the number of samples used to achieve the Monte Carlo approximation whereas in Fig. 1.6, this number of samples is the square of the number of pieces.

By the way, an interesting question is how to choose N_0 and N_1, for a given N with $N = N_0 \times N_1$, so as to get the minimum standard deviation of the cost estimate (or of the estimate of the true solution $\boldsymbol{U}$). This is a question that can be generalized to the question of choosing the best tree topology in a multi-stage problem (here the problem was 2-stage), given the number N of leaves of the tree.

1.5 Conclusion

When moving from *deterministic* to *stochastic* optimization, one must handle the evaluation of mathematical expectations, which typically involves the use of Monte Carlo sampling. However, when considering *dynamic* stochastic optimization, another important aspect of the formulation is the specification of the *information structure*, which amounts to defining what one knows each time a decision has to be made.

In this introductory chapter, we described various information structures and the difficulties—which are sometimes tremendous even for seemingly rather simple problems (see Witsenhausen's counterexample at Sect. 1.3.3)—that may result from those informational constraints.

Even if we restrict ourselves to problems with SIS (see Sect. 1.2.2), obtaining a sound discretized version of the problem with a consistent formulation of the informational constraint is not as trivial a task as we tried to illustrate it in Sect. 1.4.

In the rest of this book, the most complex phenomena of DIS and the associated *dual effect* are discussed (see Chaps. 4 and 10). However, the attempt to give systematic methodologies to obtain sound discrete versions of stochastic optimization problems is restricted to problems with SIS (Chap. 6).

Chapter 2
Open-Loop Control: The Stochastic Gradient Method

2.1 Introduction

The stochastic gradient method has a rather long history. The method foundations were given by Robbins and Monro [129] on the one hand, and by Kiefer and Wolfowitz [93] on the other. Later on, Polyak [120, 123] gave results about the convergence rate. Based on this work, Dodu et al. [57] studied the optimality of the stochastic gradient algorithm, that is, the asymptotic efficiency of the associated estimator. An important contribution by Polyak [121, 122] has been to combine stochastic gradient method and averaging techniques in order to reach the optimal efficiency.

Such methods have also been developed in the framework of Stochastic Approximation (SA) (see [98] for a review paper). The reference book by Kushner and Clark [96] presents the Ordinary Differential Equation method (ODE) in the nonconvex case, which makes it possible to perform a local convergence analysis for general stochastic algorithms. Other reference books are those of Duflo [59, 60] and again Kushner and Yin [97], including important topics as asymptotic normality or ways to deal with constraints. The reader is also referred to lecture notes by Delyon [54] giving a clear and detailed presentation of the subject.

The aim of this chapter is to detail the main methods available in order to analyze the behavior of stochastic gradient algorithms. After a brief discussion about open-loop optimization problems in Sect. 2.2, we present

- the general idea of stochastic gradient methods, the associated probabilistic framework, as well as "classical" theorems about almost-sure convergence (Robbins-Monro) and rate of convergence (Central Limit Theorem) in Sect. 2.3,
- a convergence result of the stochastic gradient algorithm in the framework of the Auxiliary Problem Principle in Sect. 2.4,
- the optimality analysis of the rate of convergence, that is, the optimal efficiency provided by the use of a matrix gain, and also by the averaging technique in Sect. 2.5,
- practical considerations about the numerical implementation of stochastic gradient algorithms in Sect. 2.6.

P. Carpentier et al., *Stochastic Multi-Stage Optimization*, Probability Theory and Stochastic Modelling 75, DOI 10.1007/978-3-319-18138-7_2

In this chapter, we often make use of several notions and terms specific to the optimization framework (proper function, lower semicontinuity, Lipschitz continuity, differentiability, gradient, strong convexity, strong monotonicity, coercivity, optimality conditions…). The reader is referred to Appendix A for the associated definitions and related properties.

2.2 Open-Loop Optimization Problems

We first discuss the notion of *open-loop optimization*, that is, the situation in which the decision maker is only aware of the a priori probability distribution of the random variables involved in the problem as mentioned in Sect. 1.2.2.

2.2.1 Problem Statement

Let $(\Omega, \mathcal{A}, \mathbb{P})$ be a probability space and let $\boldsymbol{W}$ be a random variable defined on Ω and taking its values in a measurable space $(\mathbb{W}, \mathcal{W})$. The probability distribution $\mathbb{P} \circ \boldsymbol{W}^{-1}$ of $\boldsymbol{W}$ is denoted by μ. Let $\mathbb{U}$ be a Hilbert space (with scalar product $\langle \cdot\,, \cdot \rangle$ and norm $\|\cdot\|$), and let U^{ad} be a non empty closed convex subset of $\mathbb{U}$. We consider a real-valued measurable function j defined on $\mathbb{U} \times \mathbb{W}$. We denote by $J(u)$ the expectation of the random variable $j(u, \boldsymbol{W})$ (we assume that the expectation exists for all $u \in U^{\mathrm{ad}}$):

$$J(u) = \mathbb{E}\big(j(u, \boldsymbol{W})\big) = \int_{\Omega} j\big(u, \boldsymbol{W}(\omega)\big)\, \mathrm{d}\mathbb{P}(\omega) = \int_{\mathbb{W}} j(u, w)\, \mathrm{d}\mu(w).$$

We assume that j is differentiable w.r.t. u, and that conditions for differentiating under the integral sign hold true. This classical issue is addressed by Integration Theory and can be found in [137, Sect. 3, Théorème 6.3.5] (see also [134] for a similar result about subdifferentiation). Then J is differentiable, its gradient is denoted by $\nabla J(u)$ and we have that

$$\nabla J(u) = \mathbb{E}\big(\nabla_u j(u, \boldsymbol{W})\big), \tag{2.1}$$

where $\nabla_u j$ is the gradient of j w.r.t. u. We are interested in the following optimization problem:

$$\min_{u \in U^{\mathrm{ad}}} J(u). \tag{2.2}$$

We consider here *open-loop* optimization problems, that is, problems in which the decision variable u is chosen without further information about $\boldsymbol{W}$ than its probability distribution.

Under standard convexity and differentiability assumptions, provided that we are able to compute the gradient of J for each $u \in U^{\mathrm{ad}}$, we may use a gradient-like algorithm (such as steepest descent, conjugate gradient, quasi-Newton, etc.) in order to compute the solution of Problem (2.2). The simplest is the projected gradient algorithm which reads

$$u^{(k+1)} = \mathrm{proj}_{U^{\mathrm{ad}}}\left(u^{(k)} - \epsilon \nabla J(u^{(k)})\right),$$

where ϵ is the gradient step size. Actually, this algorithm directly tackles the *deterministic* optimization problem (2.2) whereas the stochastic aspect is fully handled by the computation of the expectation involved in the expression (2.1) of $\nabla J(u^{(k)})$. However, this operation may be exceedingly costly if not impossible when the dimension of the space $\mathbb{W}$ is large.

Consider Problem (2.2), and replace $J(u)$ by its expression:

$$\min_{u \in U^{\mathrm{ad}}} \mathbb{E}\big(j(u, \boldsymbol{W})\big). \tag{2.3}$$

A standard way to get around the difficulty of computing an expectation is to use the Monte Carlo approach (see Sect. B.7). Using this idea in our optimization framework leads to replace Problem (2.3) by the following approximation

$$\min_{u \in U^{\mathrm{ad}}} \frac{1}{k} \sum_{l=1}^{k} j(u, w^l), \tag{2.4}$$

where $(w^1, \ldots, w^k)$ is a realization of a k-sample of $\boldsymbol{W}$.[1] Note that the gradient of the cost function of Problem (2.4), namely

$$\frac{1}{k} \sum_{l=1}^{k} \nabla_u j(u, w^l),$$

corresponds to a Monte Carlo approximation of the "true" gradient $\nabla J(u)$. This approach is known as the *Sample Average Approximation* (SAA), which is briefly presented in Sect. 2.5.3 (see [141, Chap. 5] for a detailed presentation). A drawback of the formulation (2.4) is that the sample size k is fixed prior to the resolution: one needs to solve a new optimization problem when enriching the initial sample with new realizations.

The stochastic gradient method aims to overcome the two difficulties mentioned above (that is, computing the true expectation or choosing the size of the sample prior to the resolution). In the manner of Sample Average Approximation, it uses an "easily computable" approximation of the gradient ∇J based on a sampling of $\boldsymbol{W}$.

[1] Recall that a k-sample of $\boldsymbol{W}$ is a sequence $(\boldsymbol{W}^1, \ldots, \boldsymbol{W}^k)$ of independent random variables with the same probability distribution as $\boldsymbol{W}$. See Sect. B.7.2 for further details.

Moreover, the samples are incorporated successively into the algorithm in order to produce a sequence of estimators converging towards the solution of Problem (2.3). In a sense, iterations of the gradient algorithm are also used to refine the Monte Carlo sampling process. Because of this sequential point of view in the introduction of new samples, the superscript l is now denoted (l) as an iteration index. The stochastic gradient method is presented in Sect. 2.3.

2.2.2 Sample Approximation in Stochastic Optimization

Suppose that we have built an approximation of Problem (2.3) using a k-tuple $(w^{(1)}, \ldots, w^{(k)})$ of elements of $\mathbb{W}$ related to the random variable $\boldsymbol{W}$ (see (2.4) for an example). The solution $u^{(k)}$ of the approximated problem can be viewed as a (measurable) function of that sequence:

$$u^{(k)} = \varphi^{(k)}(w^{(1)}, \ldots, w^{(k)}).$$

The performance $\mathbb{E}\big(j(\varphi^{(k)}(w^{(1)}, \ldots, w^{(k)}), \boldsymbol{W})\big)$ of the approximated solution $u^{(k)}$ can also be viewed as a (measurable) function $\psi^{(k)}$ of the sequence $(w^{(1)}, \ldots, w^{(k)})$. To alleviate the notation, we set:

$$J^{(k)} = \psi^{(k)}(w^{(1)}, \ldots, w^{(k)}) = \mathbb{E}\big(j(\varphi^{(k)}(w^{(1)}, \ldots, w^{(k)}), \boldsymbol{W})\big). \tag{2.5}$$

In the computation of $J^{(k)}$, it should be clear that the expectation operates on the random variable $\boldsymbol{W}$ whereas the $w^{(k)}$'s are considered as parameters (and therefore, the result of this calculation is also a function of those parameters). Suppose that those parameters are the result of random drawings: then $J^{(k)}$ is the realization of a random variable defined on another probability space that we are going to introduce now.

To be more specific about the approximation, suppose that the k-tuple $(w^{(1)}, \ldots, w^{(k)})$ is a realization of a k-sample $(\boldsymbol{W}^{(1)}, \ldots, \boldsymbol{W}^{(k)})$ of $\boldsymbol{W}$. As explained in Sect. B.7.2, we have to deal with two different probability spaces: the random variable $\boldsymbol{W}$ is defined on the canonical probability space $(\Omega, \mathcal{A}, \mathbb{P})$ whereas the k-tuple $(\boldsymbol{W}^{(1)}, \ldots, \boldsymbol{W}^{(k)})$ is defined on $(\widetilde{\Omega}, \widetilde{\mathcal{A}}, \widetilde{\mathbb{P}})$, the infinite-dimensional product of the probability spaces $(\mathbb{W}, \mathcal{W}, \mu)$:

$$(\widetilde{\Omega}, \widetilde{\mathcal{A}}, \widetilde{\mathbb{P}}) = (\mathbb{W}^{\mathbb{N}}, \mathcal{W}^{\otimes \mathbb{N}}, \mu^{\otimes \mathbb{N}}).$$

Of course, $(\boldsymbol{W}, \boldsymbol{W}^{(1)}, \ldots, \boldsymbol{W}^{(k)})$ can be identified with a $(k+1)$-sample, so that all random variables can be considered as living in the same probability space $(\widetilde{\Omega}, \widetilde{\mathcal{A}}, \widetilde{\mathbb{P}})$. In such a setting, $u^{(k)}$ and $J^{(k)}$ are realizations of the two random variables $\boldsymbol{U}^{(k)} = \varphi^{(k)}(\boldsymbol{W}^{(1)}, \ldots, \boldsymbol{W}^{(k)})$ and $\boldsymbol{J}^{(k)} = \psi^{(k)}(\boldsymbol{W}^{(1)}, \ldots, \boldsymbol{W}^{(k)})$. Using Theorem B.22, we deduce from (2.5) that the random variable $\boldsymbol{J}^{(k)}$ may be written as a conditional expectation:

$$J^{(k)} = \mathbb{E}\Big(j\big(\varphi^{(k)}(\boldsymbol{W}^{(1)}, \dots, \boldsymbol{W}^{(k)}), \boldsymbol{W}\big) \;\Big|\; \boldsymbol{W}^{(1)}, \dots, \boldsymbol{W}^{(k)}\Big).$$

In this text, we simplify our notation and denote the space $(\widetilde{\Omega}, \widetilde{\mathcal{A}}, \widetilde{\mathbb{P}})$ by $(\Omega, \mathcal{A}, \mathbb{P})$. Remember that such a space has to be sufficiently big to contain an infinite-dimensional sample of $\boldsymbol{W}$.

In order to assess the quality of the approximated problem, we need to study the statistical properties of the estimators $\boldsymbol{U}^{(k)}$ and $\boldsymbol{J}^{(k)}$. For example, the bias of the approximated optimal cost is evaluated by computing $\mathbb{E}(\boldsymbol{J}^{(k)})$ and comparing it to the true optimal cost $J^\sharp$ of Problem (2.3). It is important to realize that the point we are interested in is the dependency of the solution w.r.t. the sampling. In this chapter, we mainly focus on the asymptotic properties of the sequence $\{\boldsymbol{U}^{(k)}\}_{k\in\mathbb{N}}$ (convergence and convergence rate).

2.3 Stochastic Gradient Method Overview

We now present the general method of the stochastic gradient algorithm, as well as convergence results related to the method.

2.3.1 Stochastic Gradient Algorithm

Algorithm

The *stochastic gradient algorithm* applies to Problem (2.3) and consists in devising a method where the optimization variable u evolves over the iterations using the gradient of j evaluated at successive realizations of the random variable $\boldsymbol{W}$, rather than using the gradient of J. Otherwise stated, one uses gradient iterations to perform the optimization task and, in the same process, to visit successive realizations of $\boldsymbol{W}$ with the purpose of evaluating the expectation as in a Monte Carlo technique.

Algorithm 2.1 (Stochastic Gradient Algorithm).

1. Pick up some $u^{(0)} \in U^{\mathrm{ad}}$ and choose a positive real sequence $\{\epsilon^{(k)}\}_{k\in\mathbb{N}}$.
2. At iteration k, draw a realization $w^{(k+1)}$ of the random variable $\boldsymbol{W}$.
3. Compute the gradient of j w.r.t. u at point $(u^{(k)}, w^{(k+1)})$ and update $u^{(k+1)}$ by the formula: $u^{(k+1)} = \mathrm{proj}_{U^{\mathrm{ad}}}\Big(u^{(k)} - \epsilon^{(k)}\nabla_u j(u^{(k)}, w^{(k+1)})\Big)$.
4. Set $k = k + 1$ and go to step 2.

Algorithm 2.1 corresponds to the *numerical implementation* of the stochastic gradient method with a computer. The values $w^{(k)}$ involved in Algorithm 2.1 are drawn in such a way that the sequence $(w^{(1)}, \dots, w^{(k)})$ is a realization of a k–sample of $\boldsymbol{W}$ (the reader is referred to Sect. B.7.4 for further details). This assumption is of paramount importance in order to ensure that Algorithm 2.1 converges towards the solution of

Problem (2.3). Note that we did not set a stopping test in the previous algorithm. This point is discussed in Sect. 2.6.

In order to study the convergence properties of such an algorithm, it is necessary to cast it in the adequate *probabilistic framework*. We thus consider a infinite-dimensional sample $\{\boldsymbol{W}^{(k)}\}_{k\in\mathbb{N}}$ of $\boldsymbol{W}$ (as defined in Sect. B.7.2). Step 3 of Algorithm 2.1 can be interpreted as an iterative relation involving random variables, namely

$$\boldsymbol{U}^{(k+1)} = \mathrm{proj}_{U^{\mathrm{ad}}}\left(\boldsymbol{U}^{(k)} - \epsilon^{(k)}\nabla_u j(\boldsymbol{U}^{(k)}, \boldsymbol{W}^{(k+1)})\right). \tag{2.6}$$

Each value $u^{(k)}$ computed by Algorithm 2.1 corresponds to a realization of the random variable $\boldsymbol{U}^{(k)}$. The projection in (2.6) is to be understood ω per ω.

Example: Estimation of an Expectation

Let us illustrate Algorithm 2.1 in the framework of statistical estimation, more precisely as an application of the Monte Carlo method. Let $\boldsymbol{W}$ be a real-valued integrable random variable defined on $(\Omega, \mathcal{A}, \mathbb{P})$, and suppose we want to compute an estimate of its expectation

$$\mathbb{E}(\boldsymbol{W}) = \int_{\Omega} \boldsymbol{W}(\omega)\, \mathrm{d}\mathbb{P}(\omega).$$

A way to do that is to draw a realization of a k-sample $(\boldsymbol{W}^{(1)}, \ldots, \boldsymbol{W}^{(k)})$ of $\boldsymbol{W}$ and to compute the associated arithmetic mean. In terms of random variables, the estimator of the expectation associated with the k-sample is

$$\boldsymbol{U}^{(k)} = \frac{1}{k}\sum_{l=1}^{k} \boldsymbol{W}^{(l)}. \tag{2.7}$$

By the strong law of large numbers (Sect. B.7, Theorem B.27), the sequence of random variables $\{\boldsymbol{U}^{(k)}\}_{k\in\mathbb{N}}$ almost surely converges to $\mathbb{E}(\boldsymbol{W})$. From (2.7), we have that

$$\begin{aligned}
\boldsymbol{U}^{(k+1)} &= \frac{1}{k+1}\sum_{l=1}^{k} \boldsymbol{W}^{(l)} + \frac{\boldsymbol{W}^{(k+1)}}{k+1} \\
&= \frac{1}{k}\sum_{l=1}^{k} \boldsymbol{W}^{(l)} - \frac{1}{k+1}\left(\frac{1}{k}\sum_{l=1}^{k} \boldsymbol{W}^{(l)} - \boldsymbol{W}^{(k+1)}\right) \\
&= \boldsymbol{U}^{(k)} - \frac{1}{k+1}\left(\boldsymbol{U}^{(k)} - \boldsymbol{W}^{(k+1)}\right).
\end{aligned}$$

Using the notations $\epsilon^{(k)} = 1/(k+1)$ and $j(u, w) = \left(u - w\right)^2/2$, the last expression of $\boldsymbol{U}^{(k+1)}$ writes

$$\boldsymbol{U}^{(k+1)} = \boldsymbol{U}^{(k)} - \epsilon^{(k)}\nabla_u j(\boldsymbol{U}^{(k)}, \boldsymbol{W}^{(k+1)}). \tag{2.8}$$

Recalling that the expectation of $\boldsymbol{W}$ may be interpreted as the value which minimizes the dispersion of the random variable, namely

$$\mathbb{E}(\boldsymbol{W}) = \arg\min_{u \in \mathbb{R}} \frac{1}{2} \mathbb{E}\big((u - \boldsymbol{W})^2\big), \tag{2.9}$$

we conclude that the recursive form (2.8) of the Monte Carlo method exactly matches the stochastic gradient algorithm applied to the optimization problem (2.9). In the present case, U^{ad} is the whole space $\mathbb{R}$ so that $\mathrm{proj}_{U^{\mathrm{ad}}}(\cdot)$ is the identity function on $\mathbb{R}$.

This basic example makes it possible to enlighten some salient features of the stochastic gradient method.

- The step size $\epsilon^{(k)} = 1/(k+1)$ goes to zero as k goes to infinity, whereas the step size may be constant for deterministic optimization algorithms. Note however that $\epsilon^{(k)}$ goes to zero "not too fast", that is,

$$\sum_{k \in \mathbb{N}} \epsilon^{(k)} = +\infty.$$

Of c ourse, it would be awkward for the series $\{\epsilon^{(k)}\}_{k \in \mathbb{N}}$ to be convergent, because it should be clear that the algorithm would converge to a limit which depends on the initial point $u^{(0)}$ and on the sequence $\{\epsilon^{(k)}\}_{k \in \mathbb{N}}$ itself. For example, consider the case $U^{\mathrm{ad}} = \mathbb{R}$ and $j(u, w) = |u|$ (hence $\nabla_u j(u, w) = -1$ for $u < 0$). Starting from $u^{(0)} < -1$ with step sizes $\epsilon^{(k)} = 1/2^{k+1}$, Algorithm 2.1 leads to

$$u^{(k+1)} = u^{(0)} + \sum_{l=1}^{k+1} \frac{1}{2^l}, \quad \text{so that} \quad \lim_{k \to +\infty} u^{(k)} = u^{(0)} + 1 < 0,$$

whereas the solution of the optimization problem $\min_{u \in \mathbb{R}} |u|$ is $u^\sharp = 0$.
- The underlying convergence notion in this example is the one of the strong law of large numbers, that is, almost sure convergence. It is thus reasonable to expect such a convergence for the stochastic gradient algorithm (rather than a weaker notion as convergence in distribution or convergence in probability).
- As the central limit theorem applies to this example (Theorem B.28), we can expect a similar result for the rate of convergence of the sequence $\{\boldsymbol{U}^{(k)}\}_{k \in \mathbb{N}}$ generated by the stochastic gradient algorithm.

Probabilistic Considerations

Iteration k of the stochastic gradient method (2.6) can be represented by the general relation

$$\boldsymbol{U}^{(k+1)} = \mathcal{R}^{(k)}\big(\boldsymbol{U}^{(k)}, \boldsymbol{W}^{(k+1)}\big). \tag{2.10}$$

We assume that the random variable $\boldsymbol{U}^{(0)}$ is constant, equal to $u^{(0)} \in U^{\mathrm{ad}}$, and that the mappings $\mathcal{R}^{(k)}$ are measurable.

- Let $\mathcal{F}^{(k)}$ be the subfield generated by the k-sample $(\boldsymbol{W}^{(1)}, \ldots, \boldsymbol{W}^{(k)})$:

$$\mathcal{F}^{(0)} = \{\emptyset, \Omega\}, \ \mathcal{F}^{(k)} = \sigma\big(\boldsymbol{W}^{(1)}, \ldots, \boldsymbol{W}^{(k)}\big).$$

  The sequence $\{\mathcal{F}^{(k)}\}_{k\in\mathbb{N}}$ is a filtration, that is, $\mathcal{F}^{(k)} \subset \mathcal{F}^{(k+1)}$.
- By induction on (2.10), $\boldsymbol{U}^{(k)}$ is driven by $(\boldsymbol{W}^{(1)}, \ldots, \boldsymbol{W}^{(k)})$. The random variable $\boldsymbol{U}^{(k)}$ is thus $\mathcal{F}^{(k)}$-measurable for all k.
- Defining the function $\varphi^{(k)}$ as

$$\varphi^{(k)}(u) = \mathbb{E}\big(\mathcal{R}^{(k)}(u, \boldsymbol{W})\big),$$

  using the fact that the random variables $\boldsymbol{W}^{(k)}$ are independent and that $\boldsymbol{U}^{(k)}$ is $\mathcal{F}^{(k)}$-measurable, one obtains from Theorem B.22 that

$$\begin{aligned}\mathbb{E}\big(\boldsymbol{U}^{(k+1)} \mid \mathcal{F}^{(k)}\big) &= \mathbb{E}\big(\mathcal{R}^{(k)}(\boldsymbol{U}^{(k)}, \boldsymbol{W}^{(k+1)}) \mid \mathcal{F}^{(k)}\big) \\ &= \varphi^{(k)}\big(\boldsymbol{U}^{(k)}\big),\end{aligned}$$

  that is, for almost every $\omega \in \Omega$,

$$\mathbb{E}\big(\boldsymbol{U}^{(k+1)} \mid \mathcal{F}^{(k)}\big)(\omega) = \int_{\Omega} \mathcal{R}^{(k)}\big(\boldsymbol{U}^{(k)}(\omega), \boldsymbol{W}(\omega')\big)\, \mathrm{d}\mathbb{P}(\omega').$$

  The conditional expectation of $\boldsymbol{U}^{(k+1)}$ given $\mathcal{F}^{(k)}$ thus consists merely of a standard expectation.
- As observed in the previous example, the candidate convergence notion for studying (2.10) is the almost sure convergence. Note that the almost sure convergence of the sequence $\{\boldsymbol{U}^{(k)}\}_{k\in\mathbb{N}}$ towards a constant $u^\sharp$ has the following intuitive meaning: almost every run of Algorithm 2.1 produces a sequence $\{u^{(k)}\}_{k\in\mathbb{N}}$ converging to $u^\sharp$.

2.3.2 Connection with Stochastic Approximation

A classical problem considered in the Stochastic Approximation (SA) framework is to determine the zero of a function h using noisy evaluations of this function. Let $\mathbb{U}$ be the finite-dimensional Hilbert space $\mathbb{R}^n$. We consider a mapping $h : \mathbb{U} \to \mathbb{U}$, and we assume that the observation of $h(u)$ is perturbed by an additive random variable $\boldsymbol{\xi}$. The standard Stochastic Approximation algorithm consists in determining the zero of h by the following recursive formula:[2]

[2]The positive sign in front of $\epsilon^{(k)}$ in the update formula (2.11) is explained later on.

$$\boldsymbol{U}^{(k+1)} = \boldsymbol{U}^{(k)} + \epsilon^{(k)}\Big(h(\boldsymbol{U}^{(k)}) + \boldsymbol{\xi}^{(k+1)}\Big). \tag{2.11}$$

This algorithm is strongly related to the stochastic gradient algorithm. Indeed, consider the minimization problem (2.3) and assume that the admissible set U^{ad} is equal to $\mathbb{U}$. The projection onto U^{ad} is, accordingly, the identity operator, and the k-th iteration of the stochastic gradient algorithm writes

$$\boldsymbol{U}^{(k+1)} = \boldsymbol{U}^{(k)} - \epsilon^{(k)}\nabla_u j(\boldsymbol{U}^{(k)}, \boldsymbol{W}^{(k+1)}). \tag{2.12}$$

Defining the mapping h and the random variables $\boldsymbol{\xi}^{(k+1)}$ as

$$h(u) = -\nabla J(u)\ , \tag{2.13a}$$
$$\boldsymbol{\xi}^{(k+1)} = \nabla J(\boldsymbol{U}^{(k)}) - \nabla_u j(\boldsymbol{U}^{(k)}, \boldsymbol{W}^{(k+1)}), \tag{2.13b}$$

the stochastic gradient recursion (2.12) is identical to (2.11). Note that the problem of finding a point $u^\sharp \in \mathbb{U}$ such that $h(u^\sharp) = 0$ is equivalent to solving $\nabla J(u^\sharp) = 0$, a necessary condition for $u^\sharp$ to be a solution of Problem (2.2).

In the next two paragraphs, we deal with the Stochastic Approximation formulation and we present two important results about the sequence $\{\boldsymbol{U}^{(k)}\}_{k\in\mathbb{N}}$ generated by (2.11). In such a setting, a filtration $\{\mathcal{F}^{(k)}\}_{k\in\mathbb{N}}$ is given, and $\{\boldsymbol{\xi}^{(k)}\}_{k\in\mathbb{N}}$ is a sequence of $\mathbb{U}$-valued random variables. The random variable $\boldsymbol{U}^{(0)}$ is used to initiate the recursion (2.11).

Robbins-Monro Theorem

Here we focus on the convergence of the sequence $\{\boldsymbol{U}^{(k)}\}_{k\in\mathbb{N}}$ of random variables generated by (2.11). According to the observations made about the example considered in Sect. 2.3.1, the step sizes $\epsilon^{(k)}$ should be positive and should go to zero "not too fast". We first specify such a behavior.

Definition 2.2 A positive real sequence $\{\epsilon^{(k)}\}_{k\in\mathbb{N}}$ is a *σ-sequence* if it satisfies the two properties

$$\sum_{k\in\mathbb{N}} \epsilon^{(k)} = +\infty,\ \sum_{k\in\mathbb{N}} \big(\epsilon^{(k)}\big)^2 < +\infty.$$

We make the following assumptions on the different components involved in (2.11).

Assumptions 2.3

1. The random variable $\boldsymbol{U}^{(0)}$ is $\mathcal{F}^{(0)}$-measurable.
2. The mapping $h : \mathbb{U} \longrightarrow \mathbb{U}$ is continuous, such that
   - $\exists\, u^\sharp \in \mathbb{R}^n,\ h(u^\sharp) = 0$ and $\big\langle h(u)\,, u - u^\sharp\big\rangle < 0, \forall u \neq u^\sharp$;
   - $\exists\, a > 0,\ \forall u \in \mathbb{R}^n,\ \|h(u)\|^2 \leq a\big(1 + \|u\|^2\big)$.
3. The random variable $\boldsymbol{\xi}^{(k)}$ is $\mathcal{F}^{(k)}$-measurable for all k, and

- $\mathbb{E}(\boldsymbol{\xi}^{(k+1)} \mid \mathcal{F}^{(k)}) = 0$,
- $\exists\, d > 0,\ \mathbb{E}(\|\boldsymbol{\xi}^{(k+1)}\|^2 \mid \mathcal{F}^{(k)}) \le d(1 + \|\boldsymbol{U}^{(k)}\|^2)$.

4. The sequence $\{\epsilon^{(k)}\}_{k\in\mathbb{N}}$ is a *σ-sequence.*

Remark 2.4 Assumption 2.3-2 implies that $u^\sharp$ is the unique zero of h. ◊

Remark 2.5 The stepsize $\epsilon^{(k)}$ could be considered as the realization of a random variable $\boldsymbol{\epsilon}^{(k)}$ satisfying Definition 2.2 $\mathbb{P}$-a.s.. It would then be necessary to add the assumption that $\boldsymbol{\epsilon}^{(k)}$ is measurable with respect to $\mathcal{F}^{(k)}$. ◊

Theorem 2.6 below is a particular case of the standard Robbins-Monro theorem presented in [129] or in [60].

Theorem 2.6 *Under Assumptions 2.3, the sequence $\{\boldsymbol{U}^{(k)}\}_{k\in\mathbb{N}}$ of random variables generated by (2.11) almost surely converges to $u^\sharp$.*

For a proof, see [60, Sect. 1.4].

Let us detail the connection between the assumptions we may formulate about the initial problem (2.3) and the assumptions of Theorem 2.6. We assume that the σ-field $\mathcal{F}^{(k)}$ is generated by $(\boldsymbol{W}^{(0)}, \dots, \boldsymbol{W}^{(k)})$, so that we deduce from (2.13) that $\boldsymbol{\xi}^{(k)}$ is $\mathcal{F}^{(k)}$-measurable. We assume that the function j is strictly convex, coercive, continuously differentiable w.r.t. u and measurable w.r.t. w. Then J is strictly convex, coercive and continuously differentiable. The first part of Assumption 2.3-2 is related to these assumptions which ensure the existence and uniqueness of the solution of Problem (2.3), whereas the first part of Assumption 2.3-3 is an immediate consequence of (2.13). As for the second parts of Assumptions 2.3-2 and 2.3-3, they may be connected with a *linearly bounded gradient* (LBG) assumption on j, that is,

$$\exists c_1 > 0,\ c_2 > 0,\ \forall u \in \mathbb{R}^n,\ \forall w \in \mathbb{W},\ \|\nabla_u j(u, w)\| \le c_1 \|u\| + c_2,$$

which implies that (hint: use $(a + b)^2 \le 2(a^2 + b^2)$)

$$\begin{aligned}\exists\, c_3 > 0,\ c_4 > 0,\ \forall u \in \mathbb{R}^n,\ \forall w \in \mathbb{W},\ &\|\nabla_u j(u, w)\|^2 \le c_3 \|u\|^2 + c_4,\\ &\|\nabla J(u)\|^2 \le c_3 \|u\|^2 + c_4.\end{aligned}$$

These assumptions about the cost function j are natural in the convex optimization context. In Sect. 2.4, we give a more general convergence result concerning the stochastic gradient algorithm.

Remark 2.7 Theorem 2.6 can be extended to more general situations.

- As in Algorithm 2.1, a projection operator can be added to (2.11):

$$\boldsymbol{U}^{(k+1)} = \mathrm{proj}_{U^{\mathrm{ad}}}\Big(\boldsymbol{U}^{(k)} + \epsilon^{(k)}\big(h(\boldsymbol{U}^{(k)}) + \boldsymbol{\xi}^{(k+1)}\big)\Big).$$

  Here U^{ad} is a non empty closed convex subset of $\mathbb{U}$.

- A "small" additional term $\boldsymbol{R}^{(k+1)}$ can be added to (2.11):

$$\boldsymbol{U}^{(k+1)} = \boldsymbol{U}^{(k)} + \epsilon^{(k)}\big(h(\boldsymbol{U}^{(k)}) + \boldsymbol{\xi}^{(k+1)} + \boldsymbol{R}^{(k+1)}\big).$$

  Such a term may be interpreted as a bias on $h(u)$ which vanishes asymptotically, as considered in the Kiefer-Wolfowitz algorithm [93].

The reader is referred to [54, 59] for further details. ◊

Rate of Convergence

We now recall a central limit type theorem for the stochastic approximation method, that is, a result about the asymptotic normality of the random variables $\boldsymbol{U}^{(k)}$ generated by (2.11), together with an estimation of the rate of convergence of such an algorithm. Here we need to be more specific about the notion of σ-sequence and we give the following definition.

Definition 2.8 A positive real sequence $\{\epsilon^{(k)}\}_{k\in\mathbb{N}}$ is a *$\sigma(\alpha, \beta, \gamma)$-sequence* if it is such that

$$\epsilon^{(k)} = \frac{\alpha}{k^{\gamma} + \beta},$$

with $\alpha > 0$, $\beta \geq 0$ and $1/2 < \gamma \leq 1$.

An immediate consequence of this definition is that a $\sigma(\alpha, \beta, \gamma)$-sequence is also a σ-sequence.

We retain Assumptions 2.3 to ensure that the sequence $\{\boldsymbol{U}^{(k)}\}_{k\in\mathbb{N}}$ almost surely converges to $u^{\sharp}$, and we make the following additional assumptions.

Assumptions 2.9

1. The mapping h is continuously differentiable and has the following expression in a neighborhood of $u^{\sharp}$

$$h(u) = -H(u - u^{\sharp}) + \mathrm{O}(\|u - u^{\sharp}\|^2),$$

   where H is a symmetric positive-definite matrix.[3]
2. The sequence $\big\{\mathbb{E}\big(\boldsymbol{\xi}^{(k+1)}(\boldsymbol{\xi}^{(k+1)})^{\top} \mid \mathcal{F}^{(k)}\big)\big\}_{k\in\mathbb{N}}$ of conditional covariance matrices almost surely converges to a symmetric positive-definite matrix Γ.
3. There exists $\delta > 0$ such that $\sup_{k\in\mathbb{N}} \mathbb{E}\big(\|\boldsymbol{\xi}^{(k+1)}\|^{2+\delta} \mid \mathcal{F}^{(k)}\big) < +\infty$.
4. The sequence $\{\epsilon^{(k)}\}_{k\in\mathbb{N}}$ is a $\sigma(\alpha, \beta, \gamma)$-sequence.
5. The square matrix $(H - \lambda I)$ is positive-definite, λ being defined as

$$\lambda = \begin{cases} 0 & \text{if } \gamma < 1, \\ \dfrac{1}{2\alpha} & \text{if } \gamma = 1. \end{cases}$$

[3]The symbol O corresponds to the "Big-O" notation: $f(x) = \mathrm{O}\big(g(x)\big)$ as $x \to x_0$ if and only if there exist a positive constant α and a neighborhood V of x_0 such that $|f(x)| \leq \alpha\,|g(x)|$, $\forall x \in V$.

Remark 2.10 If we refer back to the initial problem (2.3) where $h = -\nabla J$, we notice that H is the Hessian matrix of J at $u^\sharp$

$$H = \nabla^2 J(u^\sharp).$$

Moreover, since $\mathbb{E}\big(\nabla_u j(u^\sharp, \boldsymbol{W})\big) = 0$, the matrix Γ introduced in Assumption 2.9-2 is equal to the covariance matrix of $\nabla_u j$ evaluated at $u^\sharp$

$$\Gamma = \mathbb{E}\Big(\nabla_u j(u^\sharp, \boldsymbol{W})\big(\nabla_u j(u^\sharp, \boldsymbol{W})\big)^\top\Big). \qquad \Diamond$$

The rate of convergence of the random variables $\boldsymbol{U}^{(k)}$ generated by (2.11) is given by Theorem 2.11. This theorem is a particular case of the one presented in [59].

Theorem 2.11 *Under Assumptions 2.3 and 2.9, the sequence of random variables* $\big\{(1/\sqrt{\epsilon^{(k)}})(\boldsymbol{U}^{(k)} - u^\sharp)\big\}_{k\in\mathbb{N}}$ *converges in law*[4] *to a centered gaussian distribution with covariance matrix* Σ, *that is,*

$$\frac{1}{\sqrt{\epsilon^{(k)}}}\Big(\boldsymbol{U}^{(k)} - u^\sharp\Big) \xrightarrow{\mathcal{D}} \mathcal{N}(0, \Sigma), \tag{2.14}$$

in which Σ *is the solution of the so-called Lyapunov equation*

$$\big(H - \lambda I\big)\Sigma + \Sigma\big(H - \lambda I\big) = \Gamma. \tag{2.15}$$

For a proof, see [59, Chap. 4]; see also [54] for a detailed step-by-step proof.

Remark 2.12 As already mentioned in Remark 2.7, the Robbins-Monro Theorem 2.6 remains valid when one adds a projection operator to (2.11). This is not true for Theorem 2.11 which only deals with unconstrained problems ($U^{\mathrm{ad}} = \mathbb{U}$), or at least with problems such that $u^\sharp$ belongs to the interior of the set U^{ad}. $\Diamond$

For the sake of completeness, we recall the characterization of solutions of Lyapunov equations. The following theorem can be found in [92, Theorem 4.6].

Proposition 2.13 *Let* H *be a positive-definite matrix and* Γ *be a symmetric positive-definite matrix. Then, the Lyapunov equation*

$$H\Sigma + \Sigma H^\top = \Gamma \tag{2.16}$$

admits a unique symmetric positive-definite solution Σ *given by:*

$$\Sigma = \int_0^{+\infty} e^{-tH}\,\Gamma\, e^{-tH^\top}\,\mathrm{d}t. \tag{2.17}$$

[4] See Sect. B.3.4 for this convergence notion and for the associated notation $\xrightarrow{\mathcal{D}}$.

Remark 2.14 This result remains true if Γ is a nonnegative-definite matrix: then, the matrix Σ given by (2.17) is a nonnegative-definite matrix, and is the solution of Eq. (2.16). $\Diamond$

In order to be more accurate about the convergence rate given by Theorem 2.11, let us examine the respective influence of the coefficients α, β and γ entering the expression of step sizes $\epsilon^{(k)}$ defined in Assumption 2.9-4.

- The convergence result of Theorem 2.11 can be rephrased as

$$k^{\frac{\gamma}{2}}\left(\boldsymbol{U}^{(k)} - u^{\sharp}\right) \xrightarrow{\mathcal{D}} \mathcal{N}\left(0, \alpha\Sigma\right), \tag{2.18}$$

  so that the coefficient β has in fact no influence on the convergence rate. The way in which β alters the transient behavior of the algorithm is explained in Sect. 2.6.2.
- It follows from (2.18) that the optimal choice for γ, that is, the value achieving the greatest convergence rate in (2.14), is $\gamma = 1$. We recover the "classical" rate $1/\sqrt{k}$ provided by a Monte Carlo estimator.

The next question is: which choice of α induces a covariance matrix $\alpha\Sigma$ in (2.18) as small as possible (in the cone of positive-definite matrices)? This problem is addressed in Sect. 2.5. Observe, for the time being, that the simplistic reasoning which consists in taking α as small as possible in order to minimize the covariance in (2.18) does not hold. Indeed, using the optimal value $\gamma = 1$, the solution Σ of the Lyapunov equation (2.15) depends on λ and hence on α, so that the covariance matrix $\alpha\Sigma$ is not a *linear* nor a *monotonic* function of α. For example, in the scalar case ($n = 1$), H and Γ are real numbers and the solution of (2.15) is

$$\Sigma = \frac{\alpha\Gamma}{2\alpha H - 1}.$$

Minimizing $\alpha\Sigma$ w.r.t. α leads to the optimal value $\alpha^{\sharp} = 1/H$, which is compatible with the condition $\alpha > 1/2H$ imposed by Assumption 2.9-5.

2.4 Convergence Analysis

We now consider a generalization of the stochastic gradient Algorithm 2.1 derived from the so-called Auxiliary Problem Principle, and we give a convergence result for this generalized algorithm.

2.4.1 Auxiliary Problem Principle

Consider the following optimization problem

$$\min_{u \in U^{\mathrm{ad}}} J(u). \tag{2.19}$$

Let $u^\sharp \in U^{\mathrm{ad}}$ be a solution of this problem. We recall (see Theorem A.10) that the associated optimality condition writes

$$\left\langle \nabla J(u^\sharp) \, , u - u^\sharp \right\rangle \geq 0, \;\; \forall u \in U^{\mathrm{ad}}. \tag{2.20}$$

In the deterministic framework, the *Auxiliary Problem Principle*[5] (**APP**) consists in replacing Problem (2.19) by a sequence of auxiliary problems indexed by $k \in \mathbb{N}$. Let K be a real-valued differentiable function defined on $\mathbb{U}$ and let ϵ be a positive constant. At iteration k, knowing $u^{(k)} \in U^{\mathrm{ad}}$, consider the auxiliary problem

$$\min_{u \in U^{\mathrm{ad}}} K(u) + \left\langle \epsilon \nabla J(u^{(k)}) - \nabla K(u^{(k)}) \, , u \right\rangle. \tag{2.21}$$

Its solution $u^{(k+1)}$ is used to formulate the $(k+1)$-th auxiliary problem.

The interest of such a principle lies in the fact that the resolution of the auxiliary problem (2.21) may be much easier to obtain than the solution of the initial problem (2.19). Namely, the function K appearing in (2.21) is part of the algorithm design (K is called a *core*). The choice of K being subject to rather mild conditions, one can take advantage of a proper choice in order to obtain many special features for Problem (2.21). The main properties of the Auxiliary Principle Problem are examined hereafter.

- **APP** is consistent. Assuming that the sequence of solutions $\{u^{(k)}\}_{k \in \mathbb{N}}$ converges to some $u^\sharp$ and taking the limit in the optimality condition of Problem (2.21)

$$\left\langle \nabla K(u^{(k+1)}) + \epsilon \nabla J(u^{(k)}) - \nabla K(u^{(k)}) \, , u - u^{(k+1)} \right\rangle \geq 0, \quad \forall u \in U^{\mathrm{ad}},$$

  we obtain the optimality conditions (2.20), up to a factor ϵ, by cancellation of the gradients of K (we assume that ∇K is continuous at $u^\sharp$). This shows that $u^\sharp$ is a solution of Problem (2.19) at least in the convex case.
- **APP** encompasses numerous classical optimization algorithms. For example, using a quadratic core $K(u) = (1/2) \left\| u \right\|^2$, Problem (2.21) writes

$$\min_{u \in U^{\mathrm{ad}}} \frac{1}{2} \left\| u \right\|^2 + \left\langle \epsilon \nabla J(u^{(k)}) - u^{(k)} \, , u \right\rangle,$$

[5] See [39] for a reference about the Auxiliary Problem Principle.

and its solution has the following closed-form expression

$$u^{(k+1)} = \mathrm{proj}_{U^{\mathrm{ad}}}\left(u^{(k)} - \epsilon \nabla J(u^{(k)})\right).$$

We thus obtain the well-known projected gradient algorithm.

- **APP** allows for decomposition. Assume that the space $\mathbb{U}$ is the Cartesian product of N spaces:

$$\mathbb{U} = \prod_{i=1}^{N} \mathbb{U}_i.$$

Assume, moreover, that the admissible set U^{ad} is the Cartesian product of N sets $(U_1^{\mathrm{ad}}, \ldots, U_N^{\mathrm{ad}})$, with $U_i^{\mathrm{ad}} \subset \mathbb{U}_i$. That is, the constraint $u \in U^{\mathrm{ad}}$ is equivalent to the set of N independent constraints $u_i \in U_i^{\mathrm{ad}}$ for the components u_i of u. If we choose a core function K additive according to that decomposition of u, namely $K(u) = \sum_{i=1}^{N} K_i(u_i)$, Problem (2.21) becomes

$$\min_{(u_1,\ldots,u_N)\in U_1^{\mathrm{ad}}\times\cdots\times U_N^{\mathrm{ad}}} \quad \sum_{i=1}^{N} \left(K_i(u_i) + \left\langle \epsilon \nabla_{u_i} J(u^{(k)}) - \nabla K_i(u_i^{(k)}) \, , u_i \right\rangle \right).$$

This problem splits up into N independent optimization subproblems, the ith subproblem being

$$\min_{u_i \in U_i^{\mathrm{ad}}} K_i(u_i) + \left\langle \epsilon \nabla_{u_i} J(u^{(k)}) - \nabla K_i(u_i^{(k)}) \, , u_i \right\rangle.$$

The reader is referred to [39–41] for a detailed description of the **APP** (see also the more recent lecture notes [38]).

2.4.2 Stochastic Auxiliary Problem Principle Algorithm

Let us consider the optimization problem (2.3), that we repeat here for convenience

$$\min_{u \in U^{\mathrm{ad}}} J(u), \tag{2.22}$$

with $J(u) = \mathbb{E}\big(j(u, \mathbf{W})\big)$. In order to mix the ideas of the Auxiliary Problem Principle and of the Stochastic Gradient Method, we first replace Problem (2.22) by the associated sequence of auxiliary problems, namely

$$\min_{u \in U^{\mathrm{ad}}} K(u) + \left\langle \epsilon \nabla J(u^{(k)}) - \nabla K(u^{(k)}) \, , u \right\rangle.$$

Then, in each auxiliary problem, we replace the gradient of J by the gradient of j evaluated at sampled realizations of $\boldsymbol{W}$; moreover, the "large" (constant) step size ϵ must be replaced by "small" (going to zero as index k goes to infinity) step sizes $\epsilon^{(k)}$. The k-th instance of the stochastic auxiliary problem is thus

$$\min_{u \in U^{\mathrm{ad}}} K(u) + \left\langle \epsilon^{(k)} \nabla_u j(u^{(k)}, w^{(k+1)}) - \nabla K(u^{(k)}) , u \right\rangle , \tag{2.23}$$

$w^{(k+1)}$ being a realization of the random variable $\boldsymbol{W}$. This results in the following algorithm.

Algorithm 2.15 (Stochastic APP Algorithm).

1. Pick up some $u^{(0)} \in U^{\mathrm{ad}}$ and choose a positive real sequence $\{\epsilon^{(k)}\}_{k \in \mathbb{N}}$.
2. At iteration k, draw a realization $w^{(k+1)}$ of the random variable $\boldsymbol{W}$.
3. Update $u^{(k+1)}$ by solving the auxiliary problem (2.23):
   $$u^{(k+1)} \in \arg\min_{u \in U^{\mathrm{ad}}} K(u) + \left\langle \epsilon^{(k)} \nabla_u j(u^{(k)}, w^{(k+1)}) - \nabla K(u^{(k)}) , u \right\rangle .$$
4. Set $k = k + 1$ and go to step 2.

As already pointed out when devising Algorithm 2.1, the values $w^{(k)}$ involved in Algorithm 2.15 are drawn in such a way that the sequence $(w^{(1)}, \dots, w^{(k)})$ is a realization of a k–sample of $\boldsymbol{W}$.

Remark 2.16 With the choice $K(u) = \|u\|^2/2$, the auxiliary problem (2.23) becomes

$$\min_{u \in U^{\mathrm{ad}}} \frac{1}{2} \|u\|^2 + \left\langle \epsilon^{(k)} \nabla_u j(u^{(k)}, w^{(k+1)}) - u^{(k)} , u \right\rangle .$$

Its unique solution $u^{(k+1)}$ is given by

$$u^{(k+1)} = \mathrm{proj}_{U^{\mathrm{ad}}} \left(u^{(k)} - \epsilon^{(k)} \nabla_u j(u^{(k)}, w^{(k+1)}) \right) .$$

This relation precisely corresponds to the stochastic gradient iteration of Algorithm 2.1. $\diamond$

We now focus on the convergence analysis of the stochastic APP Algorithm 2.15. We restrict ourselves to the differentiable case, but everything remains valid for subdifferentiable functions (see [45, 47] for further details).

2.4.3 Convergence Theorem

As in Sect. 2.3, we consider the stochastic APP Algorithm 2.15 in terms of random variables. Let $\{\boldsymbol{W}^{(k)}\}_{k \in \mathbb{N}}$ be an infinite dimensional sample of $\boldsymbol{W}$. The auxiliary problem at iteration k is

$$\min_{u \in U^{\mathrm{ad}}} K(u) + \left\langle \epsilon^{(k)} \nabla_u j(\boldsymbol{U}^{(k)}, \boldsymbol{W}^{(k+1)}) - \nabla K(\boldsymbol{U}^{(k)}) \, , u \right\rangle , \tag{2.24}$$

and the minimization in (2.24) is to be understood ω per ω. Assume that the set-valued random mapping corresponding to the arg min of Problem (2.24) admits a measurable selection $\boldsymbol{U}^{(k+1)}$ (this is justified in the proof of the following theorem). The convergence properties of the sequence of random variables $\{\boldsymbol{U}^{(k)}\}_{k \in \mathbb{N}}$ generated by (2.24) and the connection with the initial problem (2.22) are stated in the following theorem.

Theorem 2.17 *We make the following assumptions.*

1. *U^{ad} is a non empty closed convex subset of a Hilbert space $\mathbb{U}$.*
2. *The function $j : \mathbb{U} \times \mathbb{W} \to \mathbb{R}$ is a normal integrand,[6] and $\mathbb{E}\big(j(u, \boldsymbol{W})\big)$ exists for all $u \in U^{\mathrm{ad}}$.*
3. *The function $j(\cdot, w) : \mathbb{U} \to \mathbb{R}$ is proper, convex, lower semi-continuous and differentiable on an open subset containing U^{ad}, for all $w \in \mathbb{W}$.[7]*
4. *The function $j(\cdot, w)$ has linearly bounded gradients (LBG), uniformly in w:*

$$\exists c_1 > 0, \ \exists c_2 > 0, \ \forall w \in \mathbb{W}, \ \forall u \in U^{\mathrm{ad}}, \ \ \|\nabla_u j(u, w)\| \leq c_1 \|u\| + c_2 .$$

5. *The function J is coercive on U^{ad}.[8]*
6. *The core function K is proper, strongly convex with modulus b, lower semi-continuous and differentiable on an open subset containing U^{ad}.*
7. *The sequence $\{\epsilon^{(k)}\}_{k \in \mathbb{N}}$ is a σ-sequence.*

Then the following conclusions hold true.

1. *Problem (2.22) has a non empty set of solutions $U^\sharp$.*
2. *Problem (2.24) has a unique solution $\boldsymbol{U}^{(k+1)}$.*
3. *The sequence of random variables $\{J(\boldsymbol{U}^{(k)})\}_{k \in \mathbb{N}}$ almost surely converges to $\min_{u \in U^{\mathrm{ad}}} J(u)$.*
4. *The sequence of random variables $\{\boldsymbol{U}^{(k)}\}_{k \in \mathbb{N}}$ is almost surely bounded, and every cluster point of a realization of this sequence belongs to the optimal set $U^\sharp$.*

At last, if J is strongly convex, $U^\sharp$ is a singleton $\{u^\sharp\}$ and the sequence $\{\boldsymbol{U}^{(k)}\}_{k \in \mathbb{N}}$ almost surely converges to the unique solution $u^\sharp$ of Problem (2.22).

Proof The proof of Theorem 2.4.3 is rather long and technical. This is the reason why it has been postponed to the end of the present chapter, and we just give here a sketch of the proof. The proof of the first two statements is based on classical theorems in the field of convex optimization. The property that the solution $\boldsymbol{U}^{(k+1)}$ of Problem (2.24) is a random variable (hence, measurable) is a consequence of the

---

[6] See Definition 8.22. This implies that $j(u, \boldsymbol{W}) : \Omega \to \mathbb{R}$ is measurable $\forall u \in U^{\mathrm{ad}}$.

[7] Note that the semi-continuity of $j(\cdot, w)$ stems from the fact that j is a normal integrand.

[8] See (A.5) for the meaning of this term.

fact that the criterion j is a normal integrand. The proof of the last two statements involves four steps.

1. **Select a Lyapunov function** Λ. Let $u^\sharp \in U^\sharp$ be a solution of (2.22) and consider the function
$$\Lambda(u) = K(u^\sharp) - K(u) - \langle \nabla K(u) \, , u^\sharp - u \rangle .$$
From the strong convexity of K, we have that
$$\| \boldsymbol{U} - u^\sharp \|^2 \le \frac{2}{b} \Lambda(\boldsymbol{U}), \ \mathbb{P}\text{-a.s.}. \tag{2.25}$$
2. **Bound from above the variation of** Λ. The optimality conditions for the auxiliary problem (2.24) evaluated at $\boldsymbol{U} = \boldsymbol{U}^{(k)}$ together with the strong convexity of K imply that
$$\| \boldsymbol{U}^{(k+1)} - \boldsymbol{U}^{(k)} \| \le \frac{\epsilon^{(k)}}{b} \| \nabla_u j(\boldsymbol{U}^{(k)}, \boldsymbol{W}^{(k+1)}) \|, \ \mathbb{P}\text{-a.s.}. \tag{2.26}$$
From the LBG assumption and using (2.25), we obtain that there exist positive constants α and β such that
$$\| \nabla_u j(\boldsymbol{U}^{(k)}, \boldsymbol{W}^{(k+1)}) \|^2 \le \alpha \Lambda(\boldsymbol{U}^{(k)}) + \beta, \ \mathbb{P}\text{-a.s.}. \tag{2.27}$$
All these inequalities are combined to obtain the following inequality:
$$\mathbb{E}\big(\Lambda(\boldsymbol{U}^{(k+1)}) \mid \mathcal{F}^{(k)}\big) \le (1+\alpha^{(k)}) \Lambda(\boldsymbol{U}^{(k)}) + \beta^{(k)} - \epsilon^{(k)} \big(J(\boldsymbol{U}^{(k)}) - J(u^\sharp)\big), \ \mathbb{P}\text{-a.s.}. \tag{2.28}$$
with $\alpha^{(k)} = (\alpha/b)(\epsilon^{(k)})^2$ and $\beta^{(k)} = (\beta/b)(\epsilon^{(k)})^2$.
3. **Prove the convergence**. A straightforward application of the Robbins-Siegmund Theorem 2.27 shows that the sequence $\{\Lambda(\boldsymbol{U}^{(k)})\}_{k \in \mathbb{N}}$ almost surely converges to a finite random variable $\boldsymbol{\Lambda}^\infty$, and that the series $\sum \epsilon^{(k)} \big(J(\boldsymbol{U}^{(k)}) - J(u^\sharp)\big)$ almost surely converges.
4. **Characterize the sequence limits**. The convergence of $\{\Lambda(\boldsymbol{U}^{(k)})\}_{k \in \mathbb{N}}$ together with (2.27) imply that the sequence $\{\nabla_u j(\boldsymbol{U}^{(k)}, \boldsymbol{W}^{(k+1)})\}_{k \in \mathbb{N}}$ is almost surely finite. Thank to (2.26), Lemma 2.28 applies, so that the sequence $\{J(\boldsymbol{U}^{(k)})\}_{k \in \mathbb{N}}$ almost surely converges to $J(u^\sharp)$. From (2.25), we obtain that the sequence $\{\boldsymbol{U}^{(k)}\}_{k \in \mathbb{N}}$ is also almost surely finite: by a compactness argument, there exist subsequences converging to elements belonging to the set $U^\sharp$. □

2.4.4 Conclusions

We have given a general convergence theorem for the stochastic Auxiliary Problem Principle method. This theorem encompasses the standard stochastic gradient algorithm (obtained using the core function $K(u) = \|u\|^2/2$), as well as the so-called matrix-gain algorithm (the core function being in this case $K(u) = \langle u \, , Au \rangle /2$, A being a positive definite matrix).

From a theoretical point of view, Theorem 2.17 has been proved under natural assumptions. As a matter of fact, the convexity and differentiability assumptions are standard in the framework of convex optimization. Note moreover that, even if an explicit convexity property is not required in the Robbins-Monro Theorem 2.6, Assumption 2.3-2 plays in fact a very similar role.

As far as decomposition is concerned, the Auxiliary Problem Principle opens this possibility as a way to solve large stochastic optimization problems of the type (2.3).

2.5 Efficiency and Averaging

In this section we focus on the convergence rate of the stochastic gradient method. We use the setting considered in Sect. 2.3.2 for a *non constrained* stochastic optimization problem, that is,

$$\min_{u \in \mathbb{R}^n} J(u), \tag{2.29}$$

with $J(u) = \mathbb{E}\big(j(u, \boldsymbol{W})\big)$. Using a $\sigma(\alpha, \beta, \gamma)$-sequence $\{\epsilon^{(k)}\}_{k \in \mathbb{N}}$, that is, step sizes $\epsilon^{(k)}$ of the form $\alpha/(k^\gamma + \beta)$, we know from Theorem 2.11 that

$$k^{\frac{\gamma}{2}} \Big(\boldsymbol{U}^{(k)} - u^\sharp \Big) \xrightarrow{\mathcal{D}} \mathcal{N}\big(0, \alpha \Sigma\big).$$

It has already been noted that the choice $\gamma = 1$ leads to the largest convergence rate. We want now to improve the convergence speed by minimizing the covariance matrix $\alpha \Sigma$ w.r.t. the symmetric positive-definite matrix cone.

2.5.1 Stochastic Newton Algorithm

In deterministic optimization, it is well-known that pre-multiplying the gradient of the function to be optimized by a (cleverly chosen) matrix can significantly improve the algorithm behavior. For example, using the inverse of the Hessian matrix leads to the Newton algorithm, which yields a (local) quadratic convergence rate whereas the convergence rate of the gradient method is only linear. It is of course unrealistic to expect such a nice result in the field of stochastic approximation because the step size $\epsilon^{(k)}$ goes to zero as k goes to infinity, but we can expect some improvement of the method by a proper preconditioning of the gradient.

In order to apply this idea to the stochastic gradient method, we choose a symmetric positive-definite matrix A of dimension n. The step sizes $\epsilon^{(k)}$ are then built using the optimal choice $\gamma = 1$ and replacing the *scalar* gain α by the *matrix* gain A. Using these choices, the stochastic gradient iteration (2.12) becomes

$$\boldsymbol{U}^{(k+1)} = \boldsymbol{U}^{(k)} - \frac{1}{k+\beta} A \nabla_u j(\boldsymbol{U}^{(k)}, \boldsymbol{W}^{(k+1)}),$$

which in the Stochastic Approximation setting (2.11)—(2.13) writes

$$\boldsymbol{U}^{(k+1)} = \boldsymbol{U}^{(k)} + \frac{1}{k+\beta}\Big(Ah(\boldsymbol{U}^{(k)}) + A\boldsymbol{\xi}^{(k+1)}\Big). \tag{2.30}$$

The results stated in Sect. 2.3.2 are thus available, provided that we make use of modified data, namely a mapping Ah, noises $A\boldsymbol{\xi}^{(k)}$ and step sizes $1/(k+\beta)$. In the context of (2.30), Assumption 2.9-5 reads: $AH - I/2$ is a positive-definite matrix. Theorem 2.11 applies, so that the sequence $\{\boldsymbol{U}^{(k)}\}_{k\in\mathbb{N}}$ generated by (2.30) is such that

$$\sqrt{k}\Big(\boldsymbol{U}^{(k)} - u^\sharp\Big) \xrightarrow{\mathcal{D}} \mathcal{N}\big(0, \Sigma_A\big). \tag{2.31}$$

The asymptotic covariance matrix Σ_A is the unique solution of

$$\Big(AH - \frac{I}{2}\Big)\Sigma_A + \Sigma_A\Big(HA - \frac{I}{2}\Big) = A\Gamma A, \tag{2.32}$$

H and Γ being respectively the Hessian matrix of J and the covariance matrix of j, both evaluated at $u^\sharp$. Let $\mathcal{C}_H$ be the set of symmetric positive-definite matrices A, such that $AH - I/2$ is a positive-definite matrix. The next theorem characterizes the optimal choice for the gain matrix A over the set $\mathcal{C}_H$.

Theorem 2.18 *The choice $A^\sharp = H^{-1}$ for the matrix gain A in (2.30) minimizes the asymptotic covariance matrix Σ_A defined by (2.32) over the set $\mathcal{C}_H$, that is, $(\Sigma_A - \Sigma_{A^\sharp})$ is a nonnegative-definite matrix for all $A \in \mathcal{C}_H$. The expression of the minimal asymptotic covariance matrix is*

$$\Sigma_{A^\sharp} = H^{-1}\Gamma H^{-1}.$$

Proof We look for the asymptotic covariance matrix Σ_A appearing in the Lyapunov equation (2.32) in the equivalent form

$$\Sigma_A = H^{-1}\Gamma H^{-1} + \Delta_A.$$

Plugging this expression in (2.32) yields

$$\left(AH - \frac{I}{2}\right)\Delta_A + \Delta_A\left(HA - \frac{I}{2}\right) = (A - H^{-1})\Gamma(A - H^{-1}).$$

The matrix Δ_A thus satisfies another Lyapunov equation, the right-hand side of which is a nonnegative-definite matrix whatever the choice of A. According to Proposition 2.13 and Remark 2.14, the solution Δ_A is a nonnegative-definite matrix, with $\Delta_A = 0$ if $A = H^{-1}$. We deduce that the inequality $\Sigma_A \geq H^{-1}\Gamma H^{-1}$ (in the sense of symmetric nonnegative-definite matrices) is valid for any matrix $A \in \mathcal{C}_H$, the equality being obtained for the optimal value $A^\sharp = H^{-1} \in \mathcal{C}_H$. □

Remark 2.19 The gain H^{-1} corresponds to the inverse of the Hessian matrix of J evaluated at $u^\sharp$, hence the name "Stochastic Newton Algorithm" given to (2.30) with the optimal gain choice. Note, however, that the step sizes associated with the stochastic algorithm have a length $1/k$, whereas the length is equal to 1 in the deterministic Newton algorithm. This is the reason why the convergence speeds are essentially different:

- in the deterministic case, the use of the Newton algorithm leads to a quadratic convergence speed (that is a^{2k}, with $|a| < 1$),
- whereas in the stochastic case, the convergence speed of both the scalar and the matrix gain algorithms is $a/\sqrt{k}$.

In the stochastic case, the improvement provided by using a matrix gain arises from a better multiplicative constant[9] and not from the speed $\sqrt{k}$. ◊

We give the following definition, characterizing algorithms providing the same asymptotic convergence rate as the stochastic Newton algorithm.

Definition 2.20 A stochastic gradient algorithm is *Newton-efficient* if the sequence $\{\boldsymbol{U}^{(k)}\}_{k\in\mathbb{N}}$ it generates has the same asymptotic convergence rate as the stochastic Newton algorithm, namely

$$\sqrt{k}\left(\boldsymbol{U}^{(k)} - u^\sharp\right) \xrightarrow{\mathcal{D}} \mathcal{N}(0, H^{-1}\Gamma H^{-1}).$$

According to this terminology, the iterates $\boldsymbol{U}^{(k)}$ generated by such an algorithm are asymptotically unbiased Newton-efficient estimators of $u^\sharp$.

We have seen that Newton-efficient algorithms are in some sense optimal in the stochastic gradient algorithms class. A natural question then arises. *How to implement a Newton-efficient stochastic algorithm?* The problem we have to tackle is the following: the implementation of the stochastic Newton algorithm requires the prior knowledge of the optimal gain H^{-1}, that is, the Hessian matrix of J at the solution $u^\sharp$ we are looking for! Rather than approximating H^{-1} as the algorithm runs, we now introduce an averaging method leading to a Newton-efficient algorithm.

[9] In fact a better covariance matrix.

2.5.2 Stochastic Gradient Algorithm with Averaging

In order to overcome the difficulty of implementing a Newton-efficient stochastic algorithm, in [121, 122], Polyak proposed a modification of the standard stochastic gradient method which consists in adding an averaging stage in the algorithm. More precisely, assuming that the admissible set U^{ad} is equal to the whole space $\mathbb{U}$, the standard stochastic iteration

$$\boldsymbol{U}^{(k+1)} = \boldsymbol{U}^{(k)} - \epsilon^{(k)} \nabla_u j(\boldsymbol{U}^{(k)}, \boldsymbol{W}^{(k+1)}), \tag{2.33}$$

is replaced by

$$\boldsymbol{U}^{(k+1)} = \boldsymbol{U}^{(k)} - \epsilon^{(k)} \nabla_u j(\boldsymbol{U}^{(k)}, \boldsymbol{W}^{(k+1)}), \tag{2.34a}$$

$$\boldsymbol{U}_{\mathrm{M}}^{(k+1)} = \frac{1}{k+1} \sum_{l=1}^{k+1} \boldsymbol{U}^{(l)}. \tag{2.34b}$$

The first stage (2.34a) is identical to (2.33), whereas the aim of the second stage (2.34b) is to compute the arithmetic mean of the iterates $\boldsymbol{U}^{(k)}$ obtained at the first stage. An equivalent recursive form for stage (2.34b) is

$$\boldsymbol{U}_{\mathrm{M}}^{(k+1)} = \boldsymbol{U}_{\mathrm{M}}^{(k)} + \frac{1}{k+1} \big(\boldsymbol{U}^{(k+1)} - \boldsymbol{U}_{\mathrm{M}}^{(k)} \big). \tag{2.34c}$$

The algorithm associated with this averaging idea is summarized as follows.

Algorithm 2.21 (Stochastic Gradient Algorithm with Averaging).

1. Select some $u^{(0)} \in U^{\mathrm{ad}}$ and choose a positive real sequence $\{\epsilon^{(k)}\}_{k \in \mathbb{N}}$.
2. At iteration k, draw a realization $w^{(k+1)}$ of the random variable $\boldsymbol{W}$.
3. Compute the gradient of j w.r.t. u at point $(u^{(k)}, w^{(k+1)})$, and update $u^{(k+1)}$ by formula: $u^{(k+1)} = u^{(k)} - \epsilon^{(k)} \nabla_u j(u^{(k)}, w^{(k+1)})$.
4. Update $u_{\mathrm{M}}^{(k+1)}$ by formula: $u_{\mathrm{M}}^{(k+1)} = u_{\mathrm{M}}^{(k)} + \frac{1}{k+1} \big(u^{(k+1)} - u_{\mathrm{M}}^{(k)} \big)$.
5. Set $k = k + 1$ and go to step 2.

As before, the value $w^{(k)}$ involved in Algorithm 2.21 is such that the sequence $(w^{(1)}, \ldots, w^{(k)})$ is a realization of a k-sample of $\boldsymbol{W}$.

Remark 2.22 Observe that $u_{\mathrm{M}}^{(k)}$ is not recycled in the algorithm, that is, the stochastic gradient is evaluated at $u^{(k)}$ and not at $u_{\mathrm{M}}^{(k)}$. This $u_{\mathrm{M}}^{(k)}$ is just an additional output of the algorithm which does not influence its dynamics. ◊

By Cesàro's lemma, the almost sure convergence of the sequence $\{\boldsymbol{U}^{(k)}\}_{k \in \mathbb{N}}$ implies the almost sure convergence of the averaged sequence $\{\boldsymbol{U}_{\mathrm{M}}^{(k)}\}_{k \in \mathbb{N}}$. But the salient feature of the averaged recurrence (2.34) is its asymptotic convergence speed. We use here similar assumptions as those made for Theorem 2.11, but we now suppose that the exponent γ is *strictly* smaller that 1, replacing Assumption 2.9-4 by

Assumption 2.23 The sequence $\{\epsilon^{(k)}\}_{k\in\mathbb{N}}$ is a $\sigma(\alpha,\beta,\gamma)$-sequence, with $1/2<\gamma<1$.

According to Theorem 2.11, with $\gamma<1$, the convergence speed achieved by the sequence $\{\boldsymbol{U}^{(k)}\}_{k\in\mathbb{N}}$ is strictly smaller than $1/\sqrt{k}$, so that the associated convergence rate is not optimal. Better convergence properties are, however, obtained regarding the averaged sequence $\{\boldsymbol{U}_{\mathrm{M}}^{(k)}\}_{k\in\mathbb{N}}$, as shown by the following theorem.

Theorem 2.24 *Under Assumptions 2.3 and 2.9, where Item 2.9-4 is replaced by Assumption 2.23, the averaged stochastic gradient algorithm is Newton-efficient:*

$$\sqrt{k}\left(\boldsymbol{U}_{\mathrm{M}}^{(k)}-u^{\sharp}\right)\xrightarrow{\mathcal{D}}\mathcal{N}\left(0,H^{-1}\Gamma H^{-1}\right).$$

For a proof, see [59, Chap. 4].

We are thus able to easily implement a Newton-efficient stochastic gradient algorithm. The averaged stochastic gradient algorithm is also referred to as the *robust approach* in Stochastic Approximation. Such a terminology is justified in Sect. 2.6.

2.5.3 Sample Average Approximation

As illustrated by Eqs. (2.33) or (2.34a), the random variables $\boldsymbol{W}^{(k)}$ are incorporated one at a time in the different versions of the stochastic gradient algorithm. Such iterative methods belong to the *Stochastic Approximation* approach (SA). There is another method, called the *Sample Average Approximation* (SAA), which makes use of all the $\boldsymbol{W}^{(k)}$ at once. As already mentioned in Sect. 2.2.1, the Sample Average Approximation method consists of replacing the expectation to be minimized by a Monte Carlo approximation. This approach is widely used in stochastic optimization for large classes of one-stage and multi-stage problems, and there is an extensive literature on Sample Average Approximation. For references on the issue of convergence[10] treated in the framework of epi-convergence, see, for example, [5, 62]. The issue of epi-convergence of the Sample Average Approximation method is also discussed in Sect. 8.4 of this book. Central Limit Theorem-like results under regularity conditions are also available ([62] and [138]), as well as results based on large deviations theory [140]. See also [141, Chap. 5] for an overview of the method, and [108] for a comparison between the Sample Average Approximation method and the Stochastic Approximation approach.

Consider Problem (2.2), and replace $J(u)$ by its Monte Carlo approximation $\boldsymbol{J}^{(k)}(u)$ obtained using a k-sample $(\boldsymbol{W}^{(1)},\ldots,\boldsymbol{W}^{(k)})$ of $\boldsymbol{W}$:

$$\boldsymbol{J}^{(k)}(u)=\frac{1}{k}\sum_{l=1}^{k}j(u,\boldsymbol{W}^{(l)}).$$

[10] Consistency in the terminology of Statistics.

The Sample Average Approximation method consists of minimizing $\boldsymbol{J}^{(k)}(u)$ for some $\omega \in \Omega$:

$$\min_{u \in U^{\mathrm{ad}}} \frac{1}{k} \sum_{l=1}^{k} j\big(u, \boldsymbol{W}^{(l)}(\omega)\big). \tag{2.35}$$

The set of minimizers of Problem (2.35) is denoted by

$$\boldsymbol{\Upsilon}^{(k)}(\omega) = \arg\min_{u \in U^{\mathrm{ad}}} \frac{1}{k} \sum_{l=1}^{k} j\big(u, \boldsymbol{W}^{(l)}(\omega)\big).$$

The properties of measurability, convergence and convergence rate of sequences $\{\boldsymbol{U}^{(k)}\}_{k \in \mathbb{N}}$ such that $\boldsymbol{U}^{(k)}(\omega) \in \boldsymbol{\Upsilon}^{(k)}(\omega)$ are given in [62]. Here, we just recall the main result concerning the convergence rate of such sequences [62, Theorem 4.8]. Among various technical assumptions,[11] it is assumed that

- the solution $u^\sharp$ of Problem (2.2) is unique and belongs to the interior of U^{ad},
- the function J is twice continuously differentiable with nonsingular Hessian H at $u^\sharp$,
- the sequence of random variables $\big\{\sqrt{k}\,\nabla_u \boldsymbol{J}^{(k)}(u^\sharp)\big\}_{k \in \mathbb{N}}$ converges in law to a centered gaussian distribution with covariance matrix Γ.

Then, there exists a sequence $\big\{\boldsymbol{U}^{(k)}\big\}_{k \in \mathbb{N}}$ of minimizers of (2.35) such that

$$\sqrt{k}\Big(\boldsymbol{U}^{(k)} - u^\sharp\Big) \xrightarrow{\mathcal{D}} \mathcal{N}\big(0, H^{-1}\Gamma H^{-1}\big).$$

Under mild technical assumptions, the matrix Γ is the covariance matrix of j evaluated at $u^\sharp$ (recall that $\mathbb{E}\left(\nabla_u j(u^\sharp, \boldsymbol{W})\right) = 0$) :

$$\Gamma = \mathbb{E}\left(\nabla_u j(u^\sharp, \boldsymbol{W})\big(\nabla_u j(u^\sharp, \boldsymbol{W})\big)^\top\right).$$

The asymptotic covariance matrix obtained in that case is thus equal to the optimal covariance matrix obtained when using the stochastic Newton algorithm described in Sect. 2.5.1: the sequence $\{\boldsymbol{U}^{(k)}\}_{k \in \mathbb{N}}$ generated by the Sample Average Approximation (2.35) is Newton-efficient.

2.6 Practical Considerations

In order to successfully implement a stochastic gradient algorithm, one has to keep in mind some typical difficulties that we comment upon now.

[11] See [62, Sect. 4] for further details.

2.6.1 Stopping Criterion

A first question is related to the convergence assessment of the stochastic gradient algorithm. Of course, a stopping test based on the difference norm $\|u^{(k+1)} - u^{(k)}\|$ cannot be used, since this difference is forced to zero because of the assumptions on the step sizes $\epsilon^{(k)}$. Moreover, the norm of the "descent" direction $\|\nabla_u j(u^{(k)}, w^{(k+1)})\|$ does not give any information about convergence since what is minimized is J.

However, the expectation of the random variable $\nabla_u j(\boldsymbol{U}^{(k)}, \boldsymbol{W}^{(k+1)})$ converges towards the true gradient $\nabla J(u^\sharp)$ at the optimum, and is accordingly usable to test the convergence. An estimation of $\nabla J(u^\sharp)$ being given by

$$\left(\sum_{l=1}^{k} \epsilon^{(l)}\right)^{-1} \left(\sum_{l=1}^{k} \epsilon^{(l)} \nabla_u j(u^{(l)}, w^{(l+1)})\right)$$

it would be possible to test whether a certain degree of convergence has been reached.

A common practice consists of fixing a given—sufficiently large—number of iterations, and to check (through plots representing the evolution of quantities related to the problem: components or norm of the variables, of the gradient…) whether convergence is achieved. This is a major difference with the deterministic case for which stopping criteria are usually available.

2.6.2 Tuning the Standard Algorithm

A fundamental issue pertains to the choice of the step sizes $\epsilon^{(k)}$. In order to satisfy the assumptions of the convergence Theorem 2.6, it seems reasonable to take $\epsilon^{(k)}$ shaped as $1/k^\gamma$, with $1/2 < \gamma \leq 1$. This is why taking a $\sigma(\alpha, \beta, \gamma)$-sequence is quite natural. The three coefficients α, β and γ, entering the choice of $\epsilon^{(k)}$ are determined according to the following guidelines.

- From Theorem 2.11, the optimal convergence rate is reached for $\gamma = 1$, leading to the well-known $1/\sqrt{k}$ rate of the Monte Carlo approximation.
- According to (2.18), the *multiplicative* coefficient α also plays a role in the asymptotic behavior. From Eq. (2.15), with $\lambda = 1/(2\alpha)$, it is easy to figure out that the covariance matrix $\alpha\Sigma$ asymptotically grows as α goes to infinity. On the other hand, using a too small value of α generates small gradient steps, which may exceedingly slow down the convergence.[12] The choice of α thus corresponds to a trade-off between stability and precision.
- Ultimately, the coefficient β makes it possible to regulate the transient behavior of the algorithm. During the first iterations, the term k^γ may be ignored w.r.t. β

[12] From Assumption 2.9-5, the condition $\alpha > 1/(2c)$ is required, c being the strong convexity modulus of J. It is easy to produce a simple problem with extremely slow convergence in the

if this is chosen large enough. The coefficient $\epsilon^{(k)}$ is approximately equal to α/β, which thus corresponds to the initial gradient step size. If α/β is too small, the transient phase may be slow. On the contrary, taking a too large ratio may lead to a numerical burst during the first iterations. Note that a first guess for the ratio α/β is given by the step size to be used by the gradient method applied to the underlying deterministic problem.

Let us illustrate the influence of parameter α with the help of a quadratic Gaussian example. The optimization problem under consideration is

$$\min_{u\in\mathbb{R}^{10}} \mathbb{E}\Big(\frac{1}{2}u^\top Au + \boldsymbol{W}^\top u\Big),$$

where A is a symmetric positive definite matrix, $\boldsymbol{W}$ being a $\mathbb{R}^{10}$-valued Gaussian random variable with expectation m and covariance matrix Γ. The solution of this problem is obviously $u^\sharp = -A^{-1}m$. The classical Monte Carlo estimator $\widehat{\boldsymbol{U}}^{(k)}$ of $u^\sharp$, namely

$$\widehat{\boldsymbol{U}}^{(k)} = -\frac{1}{k}\sum_{l=1}^{k} A^{-1}\boldsymbol{W}^{(l)}, \tag{2.36}$$

is an efficient estimator of $u^\sharp$, that is, its normalized variance reaches the Cramer-Rao lower bound (see e.g. [90] for details):

$$k\mathrm{Var}\big(\widehat{\boldsymbol{U}}^{(k)}\big) = A^{-1}\Gamma A^{-1}. \tag{2.37}$$

Using step sizes $\epsilon^{(k)} = \alpha/(k+\beta)$, the stochastic gradient iteration writes

$$\boldsymbol{U}^{(k+1)} = \boldsymbol{U}^{(k)} - \frac{\alpha}{k+\beta}\big(A\boldsymbol{U}^{(k)} + \boldsymbol{W}^{(k+1)}\big). \tag{2.38}$$

Figure 2.1 displays four runs of the algorithm for different values of α (namely $\alpha =$ 0.3, 1.0, 5.0 and 10.0), the ratio α/β being constant and equal to 0.1. For each run, we have plotted the Monte Carlo estimator ($k \mapsto \big\|\widehat{u}^{(k)} - u^\sharp\big\|$—black curve) and the stochastic gradient algorithm estimator ($k \mapsto \big\|u^{(k)} - u^\sharp\big\|$—light gray curve), where $\widehat{u}^{(k)}$ and $u^{(k)}$ correspond to realizations of the random variables $\widehat{\boldsymbol{U}}^{(k)}$ and $\boldsymbol{U}^{(k)}$ respectively. Obviously, a "small" value of $\alpha = 0.3$ (upper left-hand side plot) prevents the algorithm from converging in a reasonable time, whereas "large" values $\alpha = 5.0$ and 10.0 (lower plots) lead to excessive oscillations. In this particular example, the choice $\alpha = 1$ (upper right-hand side plot) may be considered as optimal.

(Footnote 12 continued)
case when this condition is not satisfied. For example, with $j(u, w) = (1/2)u^2$ (deterministic cost function such that $c = 1$), with $\epsilon^{(k)} = 1/(5k)$ and starting from $u^{(0)} = 1$, the solution obtained after one billion iterations is about 0.015, hence relatively far from the optimal solution $u^\sharp = 0$ (see [108] for details).

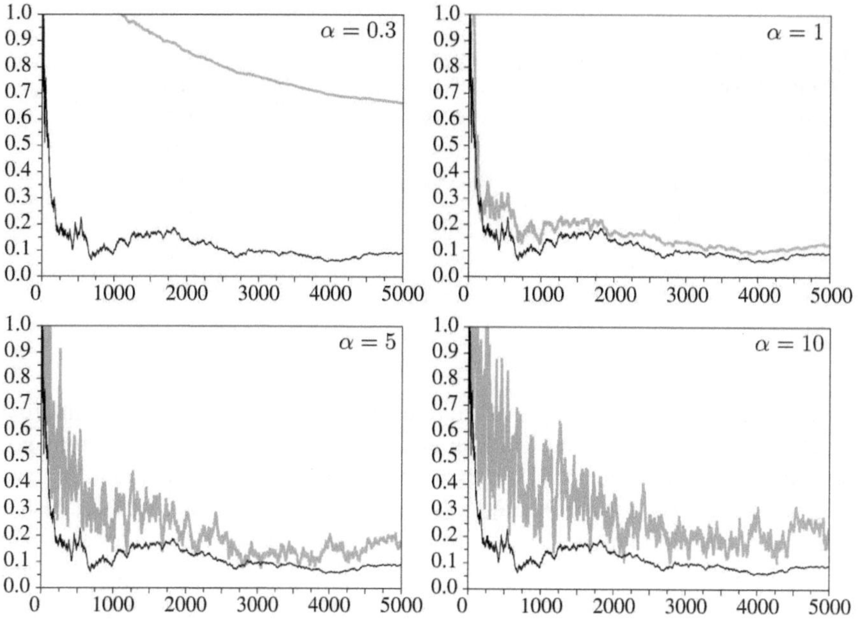

Fig. 2.1 Standard stochastic gradient runs for $\alpha = 0.3,\ 1.0,\ 5.0$ and 10.0

In order to go further into the asymptotic analysis, let us compute the covariance matrix of the iterates $\boldsymbol{U}^{(k)}$. From Eq. (2.38), denoting the identity matrix by I, we obtain that

$$\begin{aligned}\mathrm{Var}\big(\boldsymbol{U}^{(k+1)}\big) &= \mathrm{Var}\Big(\big(I - \epsilon^{(k)}A\big)\boldsymbol{U}^{(k)} - \epsilon^{(k)}\boldsymbol{W}^{(k+1)}\Big)\\ &= \big(I - \epsilon^{(k)}A\big)\mathrm{Var}\big(\boldsymbol{U}^{(k)}\big)\big(I - \epsilon^{(k)}A\big) + \big(\epsilon^{(k)}\big)^2\Gamma.\end{aligned}$$

The limit of the sequence of the normalized covariance matrices $k\mathrm{Var}\big(\boldsymbol{U}^{(k)}\big)$ induced by this relation is then compared to the Cramer-Rao bound (2.37). The lowest and greatest eigenvalues $\lambda_{\min}$ and $\lambda_{\max}$ of these matrices are reported in Table 2.1 for different values of (α, β). We notice that the greatest eigenvalue of the Cramer-Rao bound and of the "best" covariance matrix (obtained using $\alpha = 1$) are nearly identical.

This remark enlightens a result given in [57], asserting that the greatest eigenvalue of the "optimal" covariance matrix is about $(M/c)^2$, c being the strong convexity modulus of j and M being an upper bound of the norm of the gradient of j.

As a conclusion, the implementation of the stochastic gradient algorithm is not straightforward and often requires several experiments. A common error is to consider that convergence has occurred when in fact the sequence $\{\epsilon^{(k)}\}_{k\in\mathbb{N}}$ is just badly scaled.

Table 2.1 Extreme eigenvalues of the covariance matrix for different values of (α, β)

Standard stochastic gradient algorithm	λ_{min}	λ_{max}
Cramer-Rao bound	**0.108**	**11.258**
$\alpha = 0.3$—$\beta = 3.0$	0.192	6170.542
$\alpha = 0.6$—$\beta = 6.0$	0.347	24.523
$\boldsymbol{\alpha = 1.0}$—$\boldsymbol{\beta = 10.0}$	0.556	**11.286**
$\alpha = 2.0$—$\beta = 20.0$	1.083	15.244
$\alpha = 5.0$—$\beta = 50.0$	2.664	32.056
$\alpha = 10.0$—$\beta = 100.0$	5.299	60.936

Remark 2.25 Many other adaptation rules have been developed in order to improve the efficiency of the stochastic gradient algorithm. For example, Chen's projection method [35]—a theoretical tool which alleviates the assumptions required for convergence in Stochastic Approximation (see [54] for further details)—also makes it possible to prevent numerical bursts in the transient phase of the algorithm. The idea is to project the iterates $\boldsymbol{U}^{(k)}$ on compact subsets of $\mathbb{U}$ forming an increasing sequence. Another approach, namely Kesten's algorithm [91], is precisely described in [55]. There, the underlying idea is to decrease the step size $\epsilon^{(k)}$ only when the directions of two consecutive gradients are opposite. More precisely, we define a (random) sequence of integers $\boldsymbol{N}_k$ by

$$\boldsymbol{N}^{(k+1)} = \boldsymbol{N}^{(k)} + \mathbf{1}_{\left\{\left\langle \nabla_u j(\boldsymbol{U}^{(k-1)}, \boldsymbol{W}^{(k)}), \nabla_u j(\boldsymbol{U}^{(k)}, \boldsymbol{W}^{(k+1)})\right\rangle < 0\right\}},$$

$\mathbf{1}_{\Omega_0}$ being the indicator function of the set $\Omega_0 \subset \Omega$. The step size is then given by

$$\epsilon^{(k)} = \frac{\alpha}{\left(\boldsymbol{N}^{(k)}\right)^{\gamma} + \beta}.$$

Let us mention that there exist multiplicative rules [119] for the adaptation of the step size, which allow for a faster convergence towards an approximate solution of the original problem, and that numerous references deal with stochastic algorithms using constant step sizes (see e.g. [17]). ◊

2.6.3 Robustness of the Averaged Algorithm

From a theoretical point of view, the averaged stochastic gradient is, in some sense, optimal because it has the same asymptotic convergence rate as the stochastic Newton algorithm (see Theorem 2.24). From the practical point of view, the implementation of the averaged algorithm is feasible because it does not require the knowledge of the optimal matrix gain H^{-1}. The step sizes $\epsilon^{(k)}$ form a $\sigma(\alpha, \beta, \gamma)$-sequence, with $1/2 < k < 1$. The following considerations are relevant when choosing the parameters α, β and γ.

- The value $\gamma = 2/3$ is considered as a good choice by some authors (see [54] for further details).
- The tuning of parameters α and β is much easier than for the standard algorithm. Indeed, the problem of "too small" step sizes arising from a bad choice of α is not so critical because the term k^γ goes down more slowly towards zero. Of course, the ratio α/β must always be chosen in such a way that numerical bursts do not occur during the first iterations of the algorithm.

Remark 2.26 It seems wise not to start the averaging process from the very first iteration, because the whole transient phase of the algorithm is then taken into account in the averaged values $\boldsymbol{U}_{\mathrm{M}}^{(k)}$. It would be preferable to start the averaging process once the iterates $\boldsymbol{U}^{(k)}$ given by (2.34a) are oscillating near the convergence zone, but it is usually difficult to detect such a starting point. Another possibility is to average the stochastic gradient algorithm iterates $\boldsymbol{U}^{(k)}$ on a *sliding window*, with leads to the same asymptotic properties (see [99] for details). ◊

We now apply the averaged stochastic gradient algorithm to the example used in Sect. 2.6.2, namely

$$\boldsymbol{U}^{(k+1)} = \boldsymbol{U}^{(k)} - \frac{\alpha}{k^\gamma + \beta}\big(A\boldsymbol{U}^{(k)} + \boldsymbol{W}^{(k+1)}\big),$$

$$\boldsymbol{U}_{\mathrm{M}}^{(k+1)} = \frac{1}{k+1}\sum_{l=1}^{k+1} \boldsymbol{U}^{(l)}.$$

We use the same values of α and β as for the standard stochastic algorithm, γ being now equal to 2/3. The four runs of the averaged algorithm are plotted in Fig. 2.2. For each run, we have again plotted the Monte Carlo estimator given by (2.36) ($k \mapsto \|\widehat{u}^{\,(k)} - u^\sharp\|$—black curve), the stochastic gradient algorithm estimator ($k \mapsto \|u^{(k)} - u^\sharp\|$—light gray curve), and finally the averaged stochastic gradient algorithm estimator ($k \mapsto \|u_{\mathrm{M}}^{(k)} - u^\sharp\|$—dark gray curve). The changes of parameter α (from 0.3 to 10.0) affect the behavior of the stochastic gradient algorithm estimator, the oscillations of which increase with α. Nevertheless, the behavior of the averaged stochastic gradient algorithm estimator remains remarkably stable, hence the term "robust" given to the averaged algorithm.

It is again possible to iteratively compute the covariance matrices of the iterates $\boldsymbol{U}_{\mathrm{M}}^{(k)}$. The lowest and greatest eigenvalues of these matrices are given in Table 2.2 for the different values of α. We observe that the full spectrum of the Cramer-Rao bound is obtained whatever the value of α.

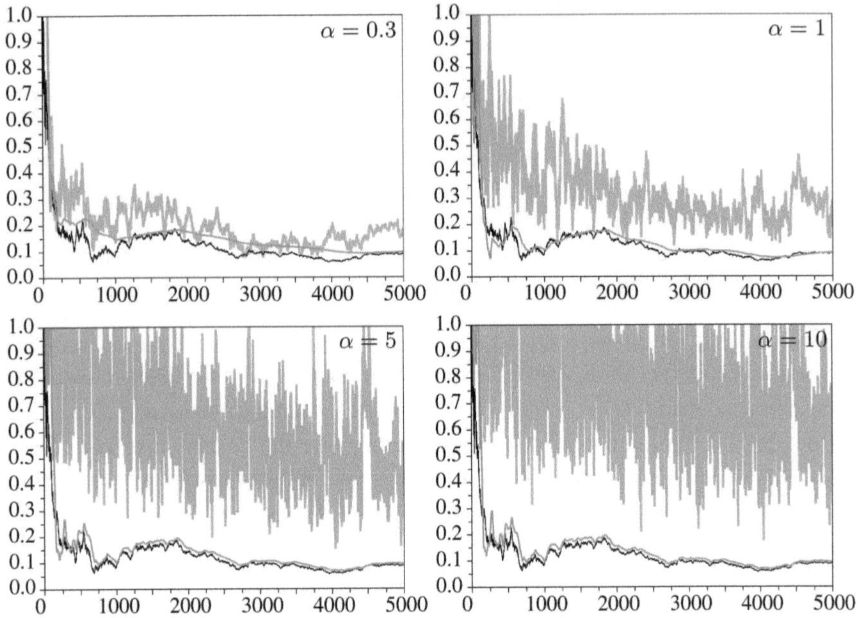

Fig. 2.2 Averaged stochastic gradient runs for $\alpha = 0.3,\ 1.0,\ 5.0$ and 10.0

Table 2.2 Extreme eigenvalues of the covariance matrix for different values of (α, β)

Averaged stochastic gradient algorithm	$\lambda_{\min}$	$\lambda_{\max}$
Cramer-Rao bound	**0.108**	**11.258**
$\alpha = 0.3$—$\beta = 3.0$	0.108	11.360
$\alpha = 0.5$—$\beta = 5.0$	0.108	11.318
$\alpha = 1.0$—$\beta = 10.0$	0.108	11.288
$\alpha = 2.0$—$\beta = 20.0$	0.108	11.273
$\alpha = 5.0$—$\beta = 50.0$	0.108	11.264
$\alpha = 10.0$—$\beta = 100.0$	0.108	11.262

2.7 Conclusion

In this chapter, we have tried to give a broad (of course non exhaustive) overview of the stochastic gradient method. After recalling some classical results from Stochastic Approximation, we have presented an algorithm based on both the Stochastic Gradient Method and on the Auxiliary Problem Principle, for which we provided a detailed convergence analysis. We then presented some issues related to the efficiency of the stochastic gradient algorithm. Finally, we have made some practical considerations about the algorithm implementation. Note that this domain is still very active, as demonstrated by the recent paper [162] providing new adaptive step length schemes in order to improve the performance of stochastic gradient algorithms, and by the paper [108] comparing the Sample Average Approximation method with a properly modified Stochastic Approximation approach. About the last paper, it is interesting to remark the strong connections between the Mirror Descent Stochastic Approximation method and the Auxiliary Problem Principle. Although restricted to the computation of open-loop solutions,[13] the stochastic gradient method is a basic component of stochastic optimization which can be embedded in many dynamic situations, when some control variables have to be decided upon once and for all or some static parameters have to be tuned. It is the case for two-stage stochastic optimization problems, for which the first time step decisions are open-loop decisions. It is also the case for multistage stochastic optimization problems when it is possible to restrict the admissible feedback laws to a particular class of functions which can be characterized in terms of a finite number of parameters, e.g., (s, S)-policies, impulse control, etc. See [148], and also [145] for a more recent application.

Throughout this book, in addition to the challenge of dealing with expectations (which was the main purpose of this chapter), we will deal with the additional difficulty related to the issue of information, that is, the measurability constraints.

2.8 Appendix

This last section is devoted to the proof of the main convergence Theorem 2.17. The proof is based on two results, namely the Robbins-Siegmund theorem and a technical lemma, that are beforehand recalled.

2.8.1 Robbins-Siegmund Theorem

The following theorem is one of the keys to Stochastic Approximation.

Theorem 2.27 *Let* $\{\boldsymbol{\Lambda}^{(k)}\}_{k\in\mathbb{N}}$, $\{\boldsymbol{\alpha}^{(k)}\}_{k\in\mathbb{N}}$, $\{\boldsymbol{\beta}^{(k)}\}_{k\in\mathbb{N}}$ *and* $\{\boldsymbol{\eta}^{(k)}\}_{k\in\mathbb{N}}$ *be four positive sequences of real-valued random variables adapted to the filtration* $\{\mathcal{F}^{(k)}\}_{k\in\mathbb{N}}$. *Assume that*

$$\mathbb{E}\big(\boldsymbol{\Lambda}^{(k+1)} \mid \mathcal{F}^{(k)}\big) \le \big(1 + \boldsymbol{\alpha}^{(k)}\big)\boldsymbol{\Lambda}^{(k)} + \boldsymbol{\beta}^{(k)} - \boldsymbol{\eta}^{(k)}, \ \forall k \in \mathbb{N},$$

and that

$$\sum_{k\in\mathbb{N}} \boldsymbol{\alpha}^{(k)} < +\infty \ \text{ and } \ \sum_{k\in\mathbb{N}} \boldsymbol{\beta}^{(k)} < +\infty, \ \mathbb{P}\text{-a.s.}.$$

[13] There however exist extensions of the stochastic gradient method to closed-loop optimization problem: see [14] for further details.

Then, the sequence $\{\boldsymbol{\Lambda}^{(k)}\}_{k\in\mathbb{N}}$ *almost surely converges to a finite*[14] *random variable* $\boldsymbol{\Lambda}^{\infty}$, *and* $\sum_{k\in\mathbb{N}} \boldsymbol{\eta}^{(k)} < +\infty$, $\mathbb{P}$*-a.s.*.

A proof can be found e.g. in [60, Theorem 1.3.12].

2.8.2 A Technical Lemma

The following lemma is also used in order to prove the convergence of the stochastic APP algorithm.

Lemma 2.28 *Let* J *be a real-valued function defined on a Hilbert space* $\mathbb{U}$. *We assume that* J *is Lipschitz continuous with constant* L. *Let* $\{u^{(k)}\}_{k\in\mathbb{N}}$ *be a sequence of elements of* $\mathbb{U}$ *and let* $\{\epsilon^{(k)}\}_{k\in\mathbb{N}}$ *be a sequence of positive real numbers such that*

(a) $\displaystyle\sum_{k\in\mathbb{N}} \epsilon^{(k)} = +\infty$,

(b) $\displaystyle\exists\, \mu \in \mathbb{R},\ \sum_{k\in\mathbb{N}} \epsilon^{(k)} \left|J(u^{(k)}) - \mu\right| < +\infty$,

(c) $\exists\, \delta > 0,\ \forall k \in \mathbb{N},\ \left\|u^{(k+1)} - u^{(k)}\right\| \leq \delta\epsilon^{(k)}$.

Then the sequence $\left\{J(u^{(k)})\right\}_{k\in\mathbb{N}}$ *converges to* μ.

Proof Let α be a given positive real number. We define the subset N_α of $\mathbb{N}$ and its complementary N_α^{c} as follows:

$$N_\alpha = \left\{k \in \mathbb{N},\ \left|J(u^{(k)}) - \mu\right| \leq \alpha\right\} \quad \text{and} \quad N_\alpha^{\mathrm{c}} = \mathbb{N} \setminus N_\alpha .$$

From the definition of N_α^{c}, we have that

$$\sum_{k\in N_\alpha^{\mathrm{c}}} \epsilon^{(k)} \left|J(u^{(k)}) - \mu\right| \geq \alpha \sum_{k\in N_\alpha^{\mathrm{c}}} \epsilon^{(k)},$$

and Property (b) implies that

$$\sum_{k\in N_\alpha^{\mathrm{c}}} \epsilon^{(k)} \left|J(u^{(k)}) - \mu\right| \leq \sum_{k\in\mathbb{N}} \epsilon^{(k)} \left|J(u^{(k)}) - \mu\right| < +\infty.$$

We thus deduce that the series $\sum_{k\in N_\alpha^{\mathrm{c}}} \epsilon^{(k)}$ converges, that is,

$$\forall \beta > 0,\ \exists n_\beta \in \mathbb{N}, \quad \sum_{k\in N_\alpha^{\mathrm{c}},\, k\geq n_\beta} \epsilon^{(k)} \leq \beta. \tag{2.39}$$

[14] A random variable $\boldsymbol{X}$ is finite if $\mathbb{P}\big(\left\{\omega \in \Omega \mid \boldsymbol{X}(\omega) = +\infty\right\}\big) = 0$.

Then, from (2.39) and Property (a), we obtain that N_α is not a finite set.

For each $\epsilon > 0$, we choose $\alpha = \epsilon/2$ and $\beta = \epsilon/(2L\delta)$. Let n_β be the integer defined by (2.39). For any $k \geq n_\beta$,

- either $k \in N_\alpha$ and, we have, by definition

$$\left|J(u^{(k)}) - \mu\right| \leq \alpha < \epsilon,$$

- or $k \in N_\alpha^c$; then let m be the smallest element of N_α such that $m > k$ (such an element exists because N_α is not a finite set); using the Lipschitz assumption on J and Property (c), we obtain

$$\begin{aligned}\left|J(u^{(k)}) - \mu\right| &\leq \left|J(u^{(k)}) - J(u^{(m)})\right| + \left|J(u^{(m)}) - \mu\right| \leq L \left\|u^{(k)} - u^{(m)}\right\| + \alpha \\ &\leq L\delta\left(\sum_{l=k}^{m-1} \epsilon^{(l)}\right) + \alpha \leq L\delta\left(\sum_{l\geq n_\beta, l\in N_\alpha^c} \epsilon^{(l)}\right) + \alpha \leq \epsilon,\end{aligned}$$

hence the result. □

2.8.3 *Proof of Theorem 2.17*

Here we give the complete proof of the main convergence theorem.

Proof The proof of the first statement is based on classical theorems in the field of convex optimization (see Theorem A.8). The existence of a random variable $\boldsymbol{U}^{(k+1)}$ solution of Problem (2.24) is a consequence of the fact that the criterion to be minimized in (2.24) is a normal integrand, so that the arg min is closed-valued and measurable, and thus admits measurable selections (see [135, Theorem 14.37] for further details). The solution $\boldsymbol{U}^{(k+1)}$ is unique because K is strongly convex.

The proof of the last two statements involves four steps.

Select a Lyapunov function Λ. Let $u^\sharp \in U^\sharp$ be a solution of (2.22). We consider the so-called Lyapunov function $\Lambda : \mathbb{U} \to \mathbb{R}$, defined by

$$\Lambda(u) = K(u^\sharp) - K(u) - \left\langle \nabla K(u) \,, u^\sharp - u\right\rangle .$$

From the strong convexity of K, we have that

$$\frac{b}{2}\left\|u - u^\sharp\right\|^2 \leq \Lambda(u). \tag{2.40}$$

The Lyapunov function Λ is thus bounded from below and coercive.

Bound from above the variation of Λ. We consider the difference

$$\Delta^{(k)} = \Lambda(u^{(k+1)}) - \Lambda(u^{(k)}),$$

$\{u^{(k)}\}_{k\in\mathbb{N}}$ being the sequence of solutions generated by Algorithm 2.15 for a realization $(w^{(1)},\ldots,w^{(k)},\ldots)$ of the infinite-dimensional sample of $\boldsymbol{W}$:

$$\Delta^{(k)} = \underbrace{K(u^{(k)}) - K(u^{(k+1)}) - \left\langle \nabla K(u^{(k)})\,,\, u^{(k)} - u^{(k+1)}\right\rangle}_{T_1} + \underbrace{\left\langle \nabla K(u^{(k)}) - \nabla K(u^{(k+1)})\,,\, u^{\sharp} - u^{(k+1)}\right\rangle}_{T_2}.$$

- From the convexity of K, we have that

$$T_1 \leq 0.$$

- Let $r^{(k)} = \nabla_u j(u^{(k)}, w^{(k+1)})$. The optimality condition of Problem (2.23) writes

$$\left\langle \nabla K(u^{(k+1)}) + \epsilon^{(k)} r^{(k)} - \nabla K(u^{(k)})\,,\, u - u^{(k+1)}\right\rangle \geq 0,\ \ \forall u \in U^{\mathrm{ad}}. \tag{2.41}$$

  Evaluating (2.41) at $u = u^{\sharp}$ leads to

$$\begin{aligned} T_2 &\leq \epsilon^{(k)}\left\langle r^{(k)}\,,\, u^{\sharp} - u^{(k+1)}\right\rangle \\ &\leq \epsilon^{(k)}\underbrace{\left\langle r^{(k)}\,,\, u^{\sharp} - u^{(k)}\right\rangle}_{T_3} + \epsilon^{(k)}\underbrace{\left\langle r^{(k)}\,,\, u^{(k)} - u^{(k+1)}\right\rangle}_{T_4}. \end{aligned}$$

  - From the convexity of $j(\cdot, w^{(k+1)})$, we have that

$$T_3 \leq j(u^{\sharp}, w^{(k+1)}) - j(u^{(k)}, w^{(k+1)}).$$

  - The evaluation of (2.41) at $u = u^{(k)}$ and the strong monotonicity of ∇K imply that

$$b\left\|u^{(k+1)} - u^{(k)}\right\|^2 \leq \epsilon^{(k)}\left\langle r^{(k)}\,,\, u^{(k)} - u^{(k+1)}\right\rangle.$$

    Using the Schwartz inequality, we obtain

$$\left\|u^{(k+1)} - u^{(k)}\right\| \leq \frac{\epsilon^{(k)}}{b}\left\|r^{(k)}\right\|. \tag{2.42}$$

    Applying also the Schwartz inequality to the term T_4 and using (2.42) yield

$$T_4 \leq \frac{\epsilon^{(k)}}{b}\left\|r^{(k)}\right\|^2.$$

    An equivalent form for the LBG assumption is that there exist positive constants c_3 and c_4 such that $\left\|r^{(k)}\right\| \leq c_3\left\|u^{(k)} - u^{\sharp}\right\| + c_4$. Taking the square of the last inequality, using $(a+b)^2 \leq 2(a^2+b^2)$ as well as (2.40), we obtain that

$$\exists \alpha > 0,\ \exists \beta > 0,\ \forall k \in \mathbb{N},\ \left\| r^{(k)} \right\|^2 \leq \alpha \Lambda(u^{(k)}) + \beta,$$

and, consequently,

$$T_4 \leq \frac{\epsilon^{(k)}}{b} \left(\alpha \Lambda(u^{(k)}) + \beta \right).$$

Collecting the upper bounds obtained for T_1, T_3 and T_4, we deduce that

$$\Delta^{(k)} \leq \epsilon^{(k)} \left(j(u^\sharp, w^{(k+1)}) - j(u^{(k)}, w^{(k+1)}) \right) + \frac{\left(\epsilon^{(k)}\right)^2}{b} \left(\alpha \Lambda(u^{(k)}) + \beta \right).$$

Consider this inequality in terms of random variables. Taking the conditional expectation w.r.t. the σ-field $\mathcal{F}^{(k)}$ generated by $(\boldsymbol{W}^{(1)}, \ldots, \boldsymbol{W}^{(k)})$ on both sides, recalling that $\boldsymbol{W}^{(k+1)}$ is independent of the previous $\boldsymbol{W}^{(l)}$ and that $\boldsymbol{U}^{(k)}$ is $\mathcal{F}^{(k)}$-measurable, we obtain that

$$\begin{aligned} \mathbb{E}\big(\Lambda(\boldsymbol{U}^{(k+1)}) - \Lambda(\boldsymbol{U}^{(k)}) \mid \mathcal{F}^{(k)}\big) &\leq \alpha^{(k)} \mathbb{E}\big(\Lambda(\boldsymbol{U}^{(k)}) \mid \mathcal{F}^{(k)}\big) + \beta^{(k)} \\ &\quad + \epsilon^{(k)} \left(J(u^\sharp) - J(\boldsymbol{U}^{(k)}) \right), \end{aligned} \tag{2.43}$$

$\alpha^{(k)} = (\alpha/b)(\epsilon^{(k)})^2$ and $\beta^{(k)} = (\beta/b)(\epsilon^{(k)})^2$ being the terms of two convergent series. Thanks to the optimality of $u^\sharp$, we have that $J(u^\sharp) - J(\boldsymbol{U}^{(k)}) \leq 0$.

Convergence. A straightforward application of the Robbins-Siegmund Theorem 2.27 shows that the sequence $\left\{\Lambda(\boldsymbol{U}^{(k)})\right\}_{k\in\mathbb{N}}$ almost surely converges to a finite random variable $\boldsymbol{\Lambda}^\infty$, and that

$$\sum_{k=0}^{+\infty} \epsilon^{(k)} \big(J(\boldsymbol{U}^{(k)}) - J(u^\sharp)\big) < +\infty,\ \mathbb{P}\text{-a.s.}. \tag{2.44}$$

Sequence Limit. As proved in the previous step, the sequence $\left\{\Lambda(\boldsymbol{U}^{(k)})\right\}_{k\in\mathbb{N}}$ almost surely converges to a finite random variable, and hence is almost surely bounded. According to (2.40) and the LBG assumption, we deduce that both sequences $\{\boldsymbol{U}^{(k)}\}_{k\in\mathbb{N}}$ and $\left\{\nabla_u j(\boldsymbol{U}^{(k)}, \boldsymbol{W}^{(k+1)})\right\}_{k\in\mathbb{N}}$ are almost surely bounded. Thanks to (2.42), the same holds true for the sequence $\left\{\|\boldsymbol{U}^{(k+1)} - \boldsymbol{U}^{(k)}\|/\epsilon^{(k)}\right\}_{k\in\mathbb{N}}$. This last fact together with (2.44) make it possible to use Lemma 2.28 to claim that the sequence $\{J(\boldsymbol{U}^{(k)})\}_{k\in\mathbb{N}}$ almost surely converges to $J(u^\sharp)$.

Let Ω_0 denote the subset of Ω such that $\left\{\Lambda(\boldsymbol{U}^{(k)})\right\}_{k\in\mathbb{N}}$ is not bounded, and let Ω_1 denote the subset of Ω for which (2.44) does not hold: $\mathbb{P}(\Omega_0 \cup \Omega_1) = 0$. Pick some $\omega \notin \Omega_0 \cup \Omega_1$. The sequence of realizations $\{u^{(k)}\}_{k\in\mathbb{N}}$ of $\{\boldsymbol{U}^{(k)}\}_{k\in\mathbb{N}}$ associated with ω is bounded, and each $u^{(k)}$ belongs to the closed subset U^{ad}. By a compactness argument,[15] there exists a convergent subsequence $\{u^{(\Phi(k))}\}_{k\in\mathbb{N}}$ (note that the subsequence itself depends on ω); let $\bar{u}$ be the limit of this subsequence. By the lower semi-continuity of function J, we have that

$$J(\bar{u}) \leq \liminf_{k\to+\infty} J(u^{(\Phi(k))}) = J(u^{\sharp}).$$

We thus deduce that $\bar{u} \in U^{\sharp}$.

We ultimately consider the case when J is strongly convex with modulus a. Then Problem (2.22) has a unique solution $u^{\sharp}$. Thanks to the optimality condition (2.20), the strong convexity property of J writes

$$\begin{aligned} J(\boldsymbol{U}^{(k)}) - J(u^{\sharp}) &\geq \left\langle \nabla J(u^{\sharp}) \,, \boldsymbol{U}^{(k)} - u^{\sharp} \right\rangle + \frac{a}{2} \left\| \boldsymbol{U}^{(k)} - u^{\sharp} \right\|^2 \\ &\geq \frac{a}{2} \left\| \boldsymbol{U}^{(k)} - u^{\sharp} \right\|^2 . \end{aligned}$$

Since $J(\boldsymbol{U}^{(k)})$ converges almost surely to $J(u^{\sharp})$, we deduce that $\left\| \boldsymbol{U}^{(k)} - u^{\sharp} \right\|$ almost surely converges to zero. The proof is complete. □

[15] A subset of $\mathbb{U}$ is compact if it is closed and bounded, provided that $\mathbb{U}$ is a finite-dimensional space. If $\mathbb{U}$ is an infinite-dimensional Hilbert space, such a property remains true only in the weak topology, and the lower semi-continuity property of J is preserved in that topology because J is convex. See [64] for further details.

Part II
Decision Under Uncertainty and the Role of Information

Chapter 3
Tools for Information Handling

This chapter covers technical tools, that will be used throughout the book, on how to handle information in the mathematical formulations of stochastic optimization problems.

3.1 Introduction

A phenomenon is governed by an *issue* ω_0 belonging to a set Ω, representing *uncertainties*, or *states of Nature*, and called the *universe*. However, the issue ω_0 is unknown. In this chapter, we discuss how we can represent *information* about ω_0.

We review on binary relations and lattices in Sect. 3.2. Then, we shed light onto two complementary ways to grasp and tackle information, namely: "partitions and fields" and "mapping measurability".

In Sect. 3.3, we consider the "partitions and fields approach" to information handling. For this purpose, we recall the notions of *partitions*—and the associated *undistinguishability equivalence relations*. We introduce *partition fields*, or *π-fields*, as well as the more famous *σ-fields*. We highlight the property that the sets of partitions, of π-fields and of σ-fields, can be equipped with order relations and with lattice structures. This is how we can speak of more or less information, and how we can merge information from two sources (greatest lower bound) or extract common information (least upper bound). In the partitions and fields approach, information is handled by a collection $\mathcal{G}$ of subsets of the universe Ω satisfying specific axioms, and information about the uncertainty ω_0 in Ω is as follows. For any subset G in $\mathcal{G}$, one can answer "yes" or "no" to the question: "Does ω_0 belong to G?". The richer $\mathcal{G}$, the more information. Partition fields are closed under arbitrary union and intersection, countable or not, hence are special case of σ-fields, closed under countable union and intersection. Partition fields and σ-fields coincide when the underlying universe Ω is finite. Both notions have pros and cons. On the one hand, partition fields can raise difficulties in the stochastic case. As an illustration, the partition field generated by the singletons of the universe Ω is the complete field made of all subsets of Ω (being

P. Carpentier et al., *Stochastic Multi-Stage Optimization*, Probability Theory and Stochastic Modelling 75, DOI 10.1007/978-3-319-18138-7_3

made of all arbitrary union and intersection, countable or not, of singletons). Unfortunately, the complete field is too large to support a probability measure when Ω is not finite. On the other hand—in contrast to a widely accepted idea, and as noted by [58]—the use of σ-fields as the informational content of a signal is not without problems (a simple example is given in [58], and summarized in Remark 3.29, where the use of σ-fields as a model of information leads to a paradoxical conclusion: a decision maker prefers less information to more).

In Sect. 3.4, we turn the spotlight onto the "mapping measurability" approach to information handling. For this purpose, we recall the notion of *measurability* of mappings, in relation with either π-fields or σ-fields. In the mapping measurability approach, information is handled by mappings over the universe Ω (also called *signals*). We provide tools to pass from an *algebraic* aspect of measurability—where a mapping is measurable w.r.t. another one—to a *functional* one, where a mapping is a function of another one. The measurability framework developed will be widely used in the book.

In the stochastic case, when a probability $\mathbb{P}$ is given on Ω (equipped with a σ-field $\mathcal{A}$), information may be captured by conditional expectations, conditional probabilities, or stochastic kernels. In Sect. 3.5, we present the conditional expectation, leaving more elaborate definitions for Appendix B.

3.2 Basic Facts on Binary Relations and on Lattices

We review binary relations and lattices, both of which are useful to manipulate information structures.

3.2.1 Binary Relations

We follow here [85, 100]. A *binary relation* on the set A is a subset $\mathfrak{R}$ of $A^2 = A \times A$. As is traditional, we denote from now on

$$\forall(\alpha, \beta) \in A^2, \ \ \alpha \, \mathfrak{R} \, \beta \iff (\alpha, \beta) \in \mathfrak{R}. \tag{3.1}$$

When $\alpha \, \mathfrak{R} \, \beta$, we say that β is *related* to α. Well known binary relations are the *empty relation* $\emptyset$, the *universal relation* A^2, and the *equality* or *diagonal relation*

$$\Delta_A := \{(\alpha, \alpha) \mid \alpha \in A\}. \tag{3.2}$$

For each $\alpha \in A$, all the related elements define a subset $\alpha \, \mathfrak{R}$ of A by

$$\alpha \, \mathfrak{R} := \{\beta \in A \mid \alpha \, \mathfrak{R} \, \beta\}. \tag{3.3}$$

Thus, we have that

$$\forall(\alpha,\beta)\in A^2,\ \alpha\,\mathfrak{R}\,\beta \iff \beta\in\alpha\,\mathfrak{R}. \tag{3.4}$$

If B is a subset of A, we define the subset of all elements related to B by

$$B\,\mathfrak{R} := \bigcup_{\beta\in B}\beta\,\mathfrak{R}. \tag{3.5}$$

We also define symmetrically $\mathfrak{R}\,\beta$ and $\mathfrak{R}\,B$.

The set $\mathfrak{B}_A$ of all binary relations on A is equipped with the inclusion relation $\subset$:

$$\mathfrak{R}_\sharp\subset\mathfrak{R}_\flat \iff \Big(\forall(\alpha,\beta)\in A^2,\ \alpha\,\mathfrak{R}_\sharp\,\beta\Rightarrow\alpha\,\mathfrak{R}_\flat\,\beta\Big) \tag{3.6}$$

$$\iff \big(\forall\alpha\in A,\ \alpha\,\mathfrak{R}_\sharp\subset\alpha\,\mathfrak{R}_\flat\big)\,. \tag{3.7}$$

We define a relation $\preceq$ on the set $\mathfrak{B}_A$ of all binary relations by

$$\mathfrak{R}_\flat\preceq\mathfrak{R}_\sharp \iff \mathfrak{R}_\sharp\subset\mathfrak{R}_\flat, \tag{3.8}$$

and we say that the relation $\mathfrak{R}_\sharp$ is *finer* than $\mathfrak{R}_\flat$.

The *converse* $\mathfrak{R}^{-1}$ of a binary relation is

$$\forall(\alpha,\beta)\in A^2,\ \alpha\,\mathfrak{R}^{-1}\,\beta \iff \beta\,\mathfrak{R}\,\alpha. \tag{3.9}$$

We have that $\mathfrak{R}_\flat\subset\mathfrak{R}_\sharp\Rightarrow\mathfrak{R}_\flat^{-1}\subset\mathfrak{R}_\sharp^{-1}$.

The *directed graph* $G(\mathfrak{R})$ built from $\mathfrak{R}$ is $(A,\mathfrak{R})$, where elements of A are called *vertices* and those of $\mathfrak{R}$ *edges*. Thus, notions attached to graphs are easily transfered to relations.

A *chain* in a binary relation $\mathfrak{R}$ is a sequence $(\alpha_1,\ldots,\alpha_n)$ for some $n\geq 1$ such that $\alpha_i\,\mathfrak{R}\,\alpha_{i+1}$ for $i=1,\ldots,n-1$; this chain is said to be *from* α_1 to α_n, and its *length* is $n-1$. We also say that α_1 and α_n are *joined* by a chain of length $n-1$. A chain in a relation is the equivalent of a *path* in a graph. A chain is *simple* if the α_i's are all distinct.

The chain $(\alpha_1,\ldots,\alpha_n)$ is a *cycle* if $\alpha_n\,\mathfrak{R}\,\alpha_1$. A cycle is *trivial* if $n=1$ (length 0 and $\alpha_1\,\mathfrak{R}\,\alpha_1$), otherwise it is *nontrivial*. A binary relation $\mathfrak{R}$ is said to be *acyclic* if there is no cycle in $\mathfrak{R}$. This corresponds to acyclicity of the directed graph $G(\mathfrak{R})$ built from $\mathfrak{R}$.

The *composition* $\mathfrak{R}\circ\mathfrak{R}'$ of two binary relations is defined by

$$\forall(\alpha,\beta)\in A^2,\ \alpha(\mathfrak{R}\circ\mathfrak{R}')\beta \iff \exists\delta\in A,\ \alpha\,\mathfrak{R}\,\delta \text{ and } \delta\,\mathfrak{R}'\,\beta. \tag{3.10}$$

For the sake of simplicity, we abbreviate $\mathfrak{R}\circ\mathfrak{R}'$ into $\mathfrak{R}\mathfrak{R}'$. In the composition $\mathfrak{R}^2:=\mathfrak{R}\mathfrak{R}$, β is related to α if there is a chain of length 2 from α to β. We define as well $\mathfrak{R}^n := \underbrace{\mathfrak{R}\cdots\mathfrak{R}}_{n\text{ times}}$ for all $n\geq 1$ and the so-called *transitive closure* of the relation $\mathfrak{R}$ as [1]

$$\mathfrak{R}^{\infty} := \bigcup_{n \geq 1} \mathfrak{R}^{n}. \tag{3.11}$$

Two elements are related by $\mathfrak{R}^{\infty}$ if, and only if, they can be connected by a chain of any length:

$$\forall(\alpha, \beta) \in A^2, \ \alpha \mathfrak{R}^{\infty} \beta \iff \text{there exists a chain in } \mathfrak{R} \text{ from } \alpha \text{ to } \beta. \tag{3.12}$$

A binary relation $\mathfrak{R}$ is said to be

- *reflexive* if $\Delta_A \subset \mathfrak{R}$, that is,

$$\alpha \mathfrak{R} \alpha, \ \forall \alpha \in A, \tag{3.13}$$

- *symmetric* if $\mathfrak{R} \subset \mathfrak{R}^{-1}$, that is,

$$\forall(\alpha, \beta) \in A^2, \ (\alpha \mathfrak{R} \beta \Rightarrow \beta \mathfrak{R} \alpha), \tag{3.14}$$

  and we immediately have that $\mathfrak{R} = \mathfrak{R}^{-1}$,
- *antisymmetric* if $\mathfrak{R} \cap \mathfrak{R}^{-1} \subset \Delta_A$, that is,

$$\forall(\alpha, \beta) \in A^2, \ (\alpha \mathfrak{R} \beta \text{ and } \beta \mathfrak{R} \alpha \Rightarrow \alpha = \beta), \tag{3.15}$$

- *transitive* if $\mathfrak{R}^2 \subset \mathfrak{R}$, that is,

$$\forall(\alpha, \beta, \gamma) \in A^3, \ (\alpha \mathfrak{R} \beta \text{ and } \beta \mathfrak{R} \gamma \Rightarrow \alpha \mathfrak{R} \gamma). \tag{3.16}$$

The *transitive closure* of a binary relation is the smallest transitive binary relation which contains $\mathfrak{R}$. It is well defined because the intersection of transitive binary relations is transitive, and it coincides with $\mathfrak{R}^{\infty}$ in (3.11).

The *reflexive and transitive closure* $\mathfrak{R}^*$ of a binary relation $\mathfrak{R}$ is the smallest reflexive and transitive binary relation which contains $\mathfrak{R}$. It is well defined because the intersection of reflexive and transitive binary relations is reflexive and transitive, and we have that

$$\mathfrak{R}^* = \Delta_A \cup \mathfrak{R}^{\infty}. \tag{3.17}$$

An *equivalence relation* is a reflexive, symmetric and transitive binary relation. It is generally denoted by the symbol $\equiv$. For an equivalence relation $\mathfrak{R}$, the subsets $\alpha \mathfrak{R}$ for $\alpha \in A$ are called *equivalence classes*, and are denoted

$$A/\mathfrak{R} := \{\alpha \mathfrak{R} \mid \alpha \in A\}. \tag{3.18}$$

Anticipating Definition 3.2, the equivalence classes form a partition of A.

---

[1]Notice that $\mathfrak{R}^{\infty}$ is not the "limit" of $\mathfrak{R}^n$ when n goes to infinity.

A *pre-order* is a reflexive and transitive binary relation. An *order* is a reflexive, antisymmetric and transitive binary relation. A pre-order is generally denoted by the symbol $\preceq$. Any preorder $\preceq$ induces an order on the quotient set w.r.t. the equivalence relation $\alpha \equiv \beta \iff \alpha \preceq \beta$ or $\beta \preceq \alpha$. An order is said to be a *total order* when all pairs of elements can be compared, else it is a *partial order*. A set equipped with a total order relation consists of a single chain, whereas, equipped with a partial order relation, it may be represented by an acyclic graph.

The *complementary* relation $\mathfrak{R}^{c}$ of a binary relation $\mathfrak{R}$ is

$$\mathfrak{R}^{c} := A^{2} \backslash \mathfrak{R}, \tag{3.19}$$

or, equivalently,[2]

$$\alpha \, \mathfrak{R}^{c} \, \beta \iff \neg \, (\alpha \, \mathfrak{R} \, \beta) \iff (\alpha, \beta) \notin \mathfrak{R}. \tag{3.20}$$

An *ordering* of A is a bijection from $\{1, \ldots, n\}$ to A. The ordering $(\alpha_1, \ldots, \alpha_n)$ is said to be *strictly compatible* (resp. *compatible*) with a binary relation $\mathfrak{R}$ if $\alpha_i \, \mathfrak{R} \, \alpha_j \Rightarrow i < j$ (resp. $i \leq j$).

Proposition 3.1 *The following assertions are equivalent for a binary relation* $\mathfrak{R}$.

1. *There exists an ordering of* A *strictly compatible with* $\mathfrak{R}$.
2. *The complementary relation* $\mathfrak{R}^{c}$ *is reflexive, and the reflexive and transitive closure* $\mathfrak{R}^{*}$ *is an order.*
3. *The relation* $\mathfrak{R}$ *is acyclic.*
4. *The complementary relation* $(\mathfrak{R}^{\infty})^{c}$ *of the transitive closure* $\mathfrak{R}^{\infty}$ *is reflexive.*

Proof (1) $\Rightarrow$ (2). Consider an ordering $(\alpha_1, \ldots, \alpha_n)$ of A that is supposed to be strictly compatible with $\mathfrak{R}$, that is, $\alpha_i \, \mathfrak{R} \, \alpha_j \Rightarrow i < j$. Let $\mathfrak{O}$ be the order relation on A defined by $\alpha_i \, \mathfrak{O} \, \alpha_j \iff i \leq j$. By construction, the order relation $\mathfrak{O}$ is such that $\mathfrak{R} \subset \mathfrak{O} \backslash \Delta_A = \mathfrak{O} \cap \Delta_A^{c}$.
Thus, on the one hand, we have that $\mathfrak{R} \subset \mathfrak{O} \cap \Delta_A^{c} \Rightarrow \Delta_A \subset \mathfrak{R}^{c}$, and we conclude that $\mathfrak{R}^{c}$ is reflexive. On the other hand, $\mathfrak{R} \subset \mathfrak{O} \Rightarrow \mathfrak{R}^{*} \subset \mathfrak{O}^{*}$, where $\mathfrak{O}^{*} = \mathfrak{O}$ since $\mathfrak{O}$ is reflexive and transitive. We deduce that $\mathfrak{R}^{*} \cap (\mathfrak{R}^{*})^{-1} \subset \mathfrak{O} \cap \mathfrak{O}^{-1}$. Now, $\mathfrak{O} \cap \mathfrak{O}^{-1} \subset \Delta_A$ since $\mathfrak{O}$ is antisymmetric. We conclude that $\mathfrak{R}^{*} \cap (\mathfrak{R}^{*})^{-1} \subset \mathfrak{O} \cap \mathfrak{O}^{-1} \subset \Delta_A$. Thus, $\mathfrak{R}^{*} \cap (\mathfrak{R}^{*})^{-1} \subset \Delta_A$ and $\mathfrak{R}^{*}$ is antisymmetric.
(2) $\Rightarrow$ (3). Assume that $\mathfrak{R}$ is not acyclic, and let $(\alpha_1, \ldots, \alpha_k)$ denote a cycle. Notice that, for any $i = 1, \ldots, k$ and $j = 1, \ldots, k$, we have that $\alpha_i \, \mathfrak{R}^{\infty} \, \alpha_j$, and thus $\alpha_i \, \mathfrak{R}^{*} \, \alpha_j$ because $\mathfrak{R}^{\infty} \subset \mathfrak{R}^{*} = \Delta_A \cup \mathfrak{R}^{\infty}$ by (3.17). Since $\mathfrak{R}^{*}$ is an order by assumption, this implies that $\alpha_i = \alpha_j$. Thus, the cycle is necessarily reduced to a single element α_1 which satisfies $\alpha_1 \, \mathfrak{R} \, \alpha_1$: this contradicts the assumption that $\mathfrak{R}^{c}$ is reflexive, because we have that both $(\alpha_1, \alpha_1) \in \mathfrak{R}$ and $(\alpha_1, \alpha_1) \notin \mathfrak{R}$. Thus, the relation $\mathfrak{R}$ is acyclic.

[2] The symbol $\neg$ denotes the *negation operator*.

(3) $\Rightarrow$ (1). We know from graph theory that, when the directed graph $G(\mathfrak{R}) = (A, \mathfrak{R})$ is acyclic, it is possible to perform a topological sort of the nodes [42, p. 485], in other words, a strictly compatible ordering.

(3) $\iff$ (4). The relation $\mathfrak{R}$ is not acyclic if, and only if, there exists an $\alpha \in A$ and a chain from α to α, if and only if there exists an $\alpha \in A$ such that $\alpha\mathfrak{R}^{\infty}\alpha$, if, and only if, $(\mathfrak{R}^{\infty})^{c}$ is not reflexive.

This ends the proof. □

3.2.2 Lattices

Hereafter, we list a few basic notions from lattice theory [8, 144].

An *ordered set* is a set endowed with an order relation, denoted by $\preceq$.

Please note that the following elements do not necessarily exist. They are illustrated in Fig. 3.1, extracted from [8].

The *top element* (of an ordered set) is an element which is greater than any other element of the set; the top is denoted by $\top$. The *bottom element* $\bot$ (of an ordered set) has a similar definition.

A *maximum element* (of a subset) is an element of the subset which is greater than any other element of the subset. If it exists, it is unique, and it coincides with the top element if the subset is equal to the whole set. A *minimum element* (of a subset) has a similar definition.

A *maximal element* (of a subset) is an element of the subset which is not less than any other element of the subset.

A *majorant* (of a subset), also called *upper bound*, is an element not necessarily belonging to the subset, which is greater than any other element of the subset. If a majorant belongs to the subset, it is the maximum element. A *minorant* (of a subset), also called *lower bound*, has a similar definition.

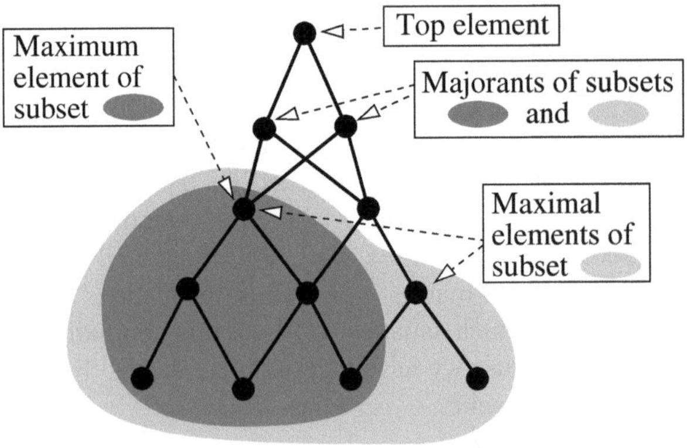

Fig. 3.1 Elements in an ordered set

A *least upper bound* (of a subset) is the least majorant, that is, the minimum element of the subset of majorants. A *greatest lower bound* has a similar definition.

A *sup-semilattice* is an ordered set such that there exists an upper bound for each pair of elements. An *inf-semilattice* has a similar definition. A *lattice* is an ordered set which is both a sup- and an inf-semilattice. A *complete sup-semilattice* is an ordered set such that there exists an upper bound for each finite or infinite subset. A *complete inf-semilattice* has a similar definition. A *complete lattice* has an obvious definition. A fundamental result asserts that a complete sup-semilattice which has a bottom element is a complete lattice.

If A is a lattice, and $(\alpha, \beta) \in A^2$, the *greatest lower bound* is denoted $\alpha \wedge \beta$ and the *least upper bound* $\alpha \vee \beta$.

3.3 Partitions and Fields Approach

In Sect. 3.3.1, we represent information by a *partition* or, equivalently, by an *undistinguishability equivalence relation*: two elements of the universe Ω are equivalent if they cannot be distinguished because they belong to the same subset of the partition. Then, we give a formal definition of information as a collection of subsets of the universe Ω satisfying the celebrated σ*-field* axioms in Sect. 3.3.3, or the less common *partition field* ones in Sect. 3.3.2. We highlight the property that the sets of partitions, π-fields and of σ-fields can be equipped with order relations and with lattice structures. This allows us to speak of more or less information and is how we can merge information from two sources (greatest lower bound) or extract common information (least upper bound). We show that partitions, equivalence relations and π-fields, are in one-to-one correspondence with mappings; these correspondences are both monotone (w.r.t. the orders) and preserve the lattice structures.

3.3.1 The Lattice of Partitions/Equivalence Relations

Partitions

We recall the definition of a partition.

Definition 3.2 A collection $\mathcal{P}$ of subsets of Ω is a *partition* if it consists of mutually disjoint nonempty subsets whose union is Ω:

$$\mathcal{P} = \{\Omega_i\}_{i \in I} \text{ with } \begin{cases} \Omega = \cup_{i \in I} \Omega_i , \\ \Omega_i \cap \Omega_j = \emptyset, \ \forall i \neq j, \\ \Omega_i \neq \emptyset, \ \forall i \in I. \end{cases} \tag{3.21}$$

Subsets of Ω whose union is Ω are also said to *cover* Ω, or to form a *covering* of Ω. Thus, a partition of Ω is a covering consisting of mutually disjoint nonempty subsets of Ω.

The *complete partition* (also said universal or discrete partition) $\{\{\omega\}\}_{\omega\in\Omega}$ consists of all singletons, while the *trivial partition* contains only Ω.

Undistinguishability Equivalence Relations

Any element in the universe Ω belongs to only one subset in a given partition $\mathcal{P}$.

Definition 3.3 Consider a partition $\mathcal{P} = \{\Omega_i\}_{i\in I}$. Two elements of the universe Ω are said to be *undistinguishable* w.r.t. $\mathcal{P}$ if they belong to the same subset of the partition $\mathcal{P}$. We define an equivalence relation on Ω, the *undistinguishability equivalence relation* $\mathfrak{R}_{\mathcal{P}}$, by

$$\omega\,\mathfrak{R}_{\mathcal{P}}\,\omega' \iff \exists i \in I,\ \omega \in \Omega_i \text{ and } \omega' \in \Omega_i. \tag{3.22}$$

It is easily seen that the equivalence classes (3.18) of the undistinguishability equivalence relation $\mathfrak{R}_{\mathcal{P}}$ are the elements of the partition $\mathcal{P}$:

$$\Omega/\mathfrak{R}_{\mathcal{P}} = \mathcal{P}. \tag{3.23}$$

On the other hand, the classes of any equivalence relation on Ω form a partition. Hence, partitions on Ω are in one-to-one correspondence with equivalence relations on Ω by the mappings

$$\mathcal{P} \mapsto \mathfrak{R}_{\mathcal{P}} \text{ and } \mathfrak{R} \mapsto \Omega/\mathfrak{R}. \tag{3.24}$$

A Partial Order and a Lattice Structure on Partitions

We define and characterize a partial order between partitions, and then we introduce a lattice structure.

Definition 3.4 We say that a partition $\mathcal{P}^\sharp$ is *finer* than a partition $\mathcal{P}^\flat$ if every element of $\mathcal{P}^\sharp$ is included in an element of $\mathcal{P}^\flat$. We denote this by $\mathcal{P}^\flat \preceq \mathcal{P}^\sharp$.

In the following proposition, we recall that a partition $\mathcal{P}^\sharp$ is finer than a partition $\mathcal{P}^\flat$ if, and only if, every element of $\mathcal{P}^\flat$ is the union of elements of $\mathcal{P}^\sharp$. The proof can be found in [144].

Proposition 3.5 *Consider two partitions* $\mathcal{P}^\flat = \{\Omega_i^\flat\}_{i\in I^\flat}$ *and* $\mathcal{P}^\sharp = \{\Omega_j^\sharp\}_{j\in I^\sharp}$. *The following assertions are equivalent:*

1. $\mathcal{P}^\flat \preceq \mathcal{P}^\sharp$;
2. $\forall j \in I^\sharp, \exists i \in I^\flat, \Omega_j^\sharp \subset \Omega_i^\flat$;
3. $\forall i \in I^\flat, \exists J \subset I^\sharp, \Omega_i^\flat = \bigcup_{j\in J} \Omega_j^\sharp$.

From the equivalence between item 1 and item 2, and from (3.22), we deduce that the order relations "finer" for the relations as defined in (3.8) and for the partitions in Definition 3.4 are compatible and coherent in the sense that

$$\mathcal{P}^{\flat} \preceq \mathcal{P}^{\sharp} \iff \mathfrak{R}_{\mathcal{P}^{\flat}} \preceq \mathfrak{R}_{\mathcal{P}^{\sharp}} . \tag{3.25}$$

The relation $\preceq$ on the set of all partitions on the universe Ω is a partial order. We leave the proof of the following proposition to the reader.

Proposition 3.6 *The relation $\preceq$ between partitions is a partial order on the set of all partitions on Ω.*

Equipped with the relation $\preceq$, the set of all partitions on the universe Ω is a lattice. We let the proof of the following proposition to the reader (see Fig. 3.2). A reference book is [144].

Proposition 3.7 *The set of all partitions on Ω is a lattice, with the operators $\wedge$ and $\vee$ as below. Let $\mathcal{P} = \{\Omega_i\}_{i \in I}$ and $\mathcal{P}' = \{\Omega'_j\}_{j \in I'}$ be two partitions on Ω.*

- *The* greatest lower bound *of the two partitions $\mathcal{P}$ and $\mathcal{P}'$ is the partition $\mathcal{P} \wedge \mathcal{P}'$ made of subsets of Ω which are both union of elements of $\mathcal{P}$ and union of elements of $\mathcal{P}'$: $G \in \mathcal{P} \wedge \mathcal{P}'$ if, and only if, $\exists J \subset I$ and $\exists J' \subset I'$ such that $G = \bigcup_{i \in J} \Omega_i = \bigcup_{j \in J'} \Omega'_j$.*
- *The* least upper bound *$\mathcal{P} \vee \mathcal{P}'$ of the two partitions $\mathcal{P}$ and $\mathcal{P}'$ is the partition made of all nonempty intersections between elements of $\mathcal{P}$ and elements of $\mathcal{P}'$: $G \in \mathcal{P} \vee \mathcal{P}'$ if, and only if, $\exists i \in I$ and $\exists j \in I'$ such that $G = \Omega_i \cap \Omega'_j$ and $G \neq \emptyset$.*

The bottom $\perp$ *of the lattice is the trivial partition Ω, whereas the complete partition $\{\{\omega\}\}_{\omega \in \Omega}$ is the* top $\top$.

Recall that partitions on Ω are in one-to-one correspondence with equivalence relations on Ω (see Definition 3.3). The following proposition details how the lattice operations upon partitions translate into operations upon equivalence relations. The proof can be found in [144].

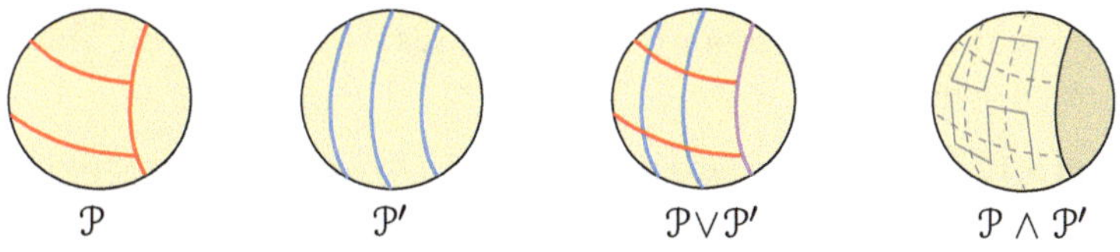

Fig. 3.2 Least upper and greatest lower bounds of partitions

Proposition 3.8 *Consider two partitions $\mathcal{P}$ and $\mathcal{P}'$ on Ω.*

- *The equivalence relation $\mathfrak{R}_{\mathcal{P}\wedge\mathcal{P}'}$ attached to the greatest lower bound $\mathcal{P}\wedge\mathcal{P}'$ is the transitive closure $(\mathfrak{R}_{\mathcal{P}}\cup\mathfrak{R}_{\mathcal{P}'})^{\infty}$ of the union of the equivalence relations $\mathfrak{R}_{\mathcal{P}}$ and $\mathfrak{R}_{\mathcal{P}'}$*
$$\mathfrak{R}_{\mathcal{P}\wedge\mathcal{P}'} = (\mathfrak{R}_{\mathcal{P}}\cup\mathfrak{R}_{\mathcal{P}'})^{\infty}. \tag{3.26}$$
- *The equivalence relation $\mathfrak{R}_{\mathcal{P}\vee\mathcal{P}'}$ attached to the least upper bound $\mathcal{P}\vee\mathcal{P}'$ is the intersection $\mathfrak{R}_{\mathcal{P}}\cap\mathfrak{R}_{\mathcal{P}'}$ of the equivalence relations $\mathfrak{R}_{\mathcal{P}}$ and $\mathfrak{R}_{\mathcal{P}'}$:*
$$\mathfrak{R}_{\mathcal{P}\vee\mathcal{P}'} = \mathfrak{R}_{\mathcal{P}}\cap\mathfrak{R}_{\mathcal{P}'}. \tag{3.27}$$

3.3.2 The Lattice of π-Fields (Partition Fields)

We define a partition field, or a π-field, as a collection of subsets of the universe Ω which is stable under arbitrary union and intersection (countable or not). In accordance with [58], we claim that partition fields are adequate to represent information. We show their specific connection to partitions. As for partitions, we refer to [148] for the proofs.

π-Fields or Partition Fields
The following definition is taken from [160].

Definition 3.9 A *partition field* (or π-field) on Ω is a nonempty collection $\mathcal{G}$ of subsets of Ω (identified with a subset $\mathcal{G}\subset 2^{\Omega}$) which is stable under complementation and unlimited union (hence, under unlimited intersection).

Notice that a partition field, being nonempty, necessarily contains Ω and $\emptyset$. The *trivial* partition field is $\{\emptyset, \Omega\}$, whereas the *complete* partition field (also called universal or discrete) is 2^{Ω}, the partition field consisting of *all* subsets of the universe Ω.

Definition 3.10 Let $\mathcal{G}'$ and $\mathcal{G}''$ be two collections of subsets of Ω: $\mathcal{G}'\subset 2^{\Omega}$ and $\mathcal{G}''\subset 2^{\Omega}$. Their union $\mathcal{G}'\cup\mathcal{G}''$ is the subset of 2^{Ω} made of subsets $G\subset\Omega$ such that either $G\in\mathcal{G}'$ or $G\in\mathcal{G}''$. Their intersection $\mathcal{G}'\cap\mathcal{G}''$ is the subset of 2^{Ω} made of subsets $G\subset\Omega$ such that both $G\in\mathcal{G}'$ and $G\in\mathcal{G}''$.

It is easily seen that the intersection of partition fields is a partition field. This justifies the following definition.

Definition 3.11 The partition field $\pi(\mathcal{C})$ *generated* by a collection $\mathcal{C}$ of subsets of Ω is the smallest partition field containing $\mathcal{C}$. It is also the intersection of all partition fields containing $\mathcal{C}$.

Remark 3.12 As an application, consider two sets Ω_1 and Ω_2, each equipped with a partition field $\mathcal{G}_i$, $i = 1, 2$. By definition, the *product partition field* $\mathcal{G}_1\otimes\mathcal{G}_2$ on $\Omega_1\times\Omega_2$ is the one generated by the rectangles, that is by the set
$\{G_1\times G_2 \mid G_i\in\mathcal{G}_i,\ i = 1, 2\}$. ◊

The partition field $\pi(\mathcal{C})$ may be a large collection of subsets, as illustrated by the following examples. This is the main reason why partition fields are not used in probability theory. Given "primitive events" in $\mathcal{C}$, the partition field that they generate is generally too large to support a probability measure.

Example 3.13 If the collection $\mathcal{C}$ contains all singletons, then $\pi(\mathcal{C}) = 2^{\Omega}$. Indeed, any subset of Ω may be written as the union (countable or uncountable) of its singletons. As an application, the partition field $\pi(\mathcal{T})$ generated by the Borelian topology $\mathcal{T}$ of $\mathbb{R}^d$ is the complete partition field. Indeed, $\pi(\mathcal{T})$ contains the closed sets (by complementing the open sets), hence the singletons. We deduce that $\pi(\mathcal{T}) = \pi(\mathcal{B}^{\text{o}}_{\mathbb{R}^d}) = 2^{\mathbb{R}^d}$. △

A Partial Order and a Lattice Structure on Partition Fields

We highlight the property that the set of π-fields can be equipped with order relation and with lattice structure. This allows us to speak of more or less information and is how we can merge information from two sources (greatest lower bound) or extract common information (least upper bound).

Definition 3.14 Consider two partition fields $\mathcal{G}^\flat$ and $\mathcal{G}^\sharp$ on Ω. When $\mathcal{G}^\flat \subset \mathcal{G}^\sharp$ (that is, $G \in \mathcal{G}^\flat \Rightarrow G \in \mathcal{G}^\sharp$), $\mathcal{G}^\sharp$ is said to be *finer* than $\mathcal{G}^\flat$, and this is denoted by $\mathcal{G}^\flat \preceq \mathcal{G}^\sharp$.

We let the proof of the following proposition to the reader.

Proposition 3.15 *The relation $\preceq$ between partition fields on Ω is a partial order relation.*

The set of all partition fields on Ω is a lattice, with the operators $\wedge$ and $\vee$ as below. Consider $\mathcal{G}$ and $\mathcal{G}''$ two partition fields on Ω.

- *The* greatest lower bound *of the partition fields $\mathcal{G}'$ and $\mathcal{G}''$ is $\mathcal{G}' \wedge \mathcal{G}'' = \mathcal{G}' \cap \mathcal{G}''$, made of subsets of Ω which belong both to $\mathcal{G}'$ and to $\mathcal{G}''$.*
- *The* least upper bound *of the partition fields $\mathcal{G}'$ and $\mathcal{G}''$ is $\mathcal{G}' \vee \mathcal{G}'' = \pi(\mathcal{G}' \cup \mathcal{G}'')$, the partition field generated by the subsets of Ω which belong either to $\mathcal{G}'$ or to $\mathcal{G}''$.*
- *The* bottom $\perp$ *of the lattice of all partitions fields on Ω is $\perp = \{\emptyset, \Omega\}$ the trivial partition field, whereas the complete partition field is the* top $\top = 2^{\Omega}$.

Partition of Atoms and Undistinguishability Equivalence Relation

We now shed light on the property that partition fields have non trivial atoms, which form a partition, and that partition fields are in one-to-one correspondence with partitions and with equivalence relations on Ω.

Definition 3.16 Consider a collection $\mathcal{G}$ of subsets of Ω. An *atom* of $\mathcal{G}$, or a $\mathcal{G}$-atom, is a nonempty subset $G \in \mathcal{G}$ such that $K \in \mathcal{G}$ and $K \subset G$ imply that $K = \emptyset$ or $K = G$.

Proposition 3.17 *Consider a collection $\mathcal{G}$ of subsets of Ω which is closed under intersection. Then, the atoms of $\mathcal{G}$ are mutually disjoint. If the union of the $\mathcal{G}$-atoms is equal to Ω, then the $\mathcal{G}$-atoms form a partition of Ω. As a consequence, if a collection of $\mathcal{G}$-atoms covers Ω, there are no $\mathcal{G}$-atoms apart from this collection.*

Proof Consider two distinct atoms $G' \in \mathcal{G}$ and $G'' \in \mathcal{G}$. Since, by assumption, $G = G' \cap G'' \in \mathcal{G}$ and $G \subset G'$, we deduce that $G = G'$ or $G = \emptyset$ since G' is a $\mathcal{G}$-atom. In the same way, we have that $G = G''$ or $G = \emptyset$. We conclude that $G = G' \cap G'' = \emptyset$, since G' and G'' are distinct. The last assertions follow easily. □

Proposition 3.18 *Consider $\mathcal{G}$ a partition field of Ω. The atoms of $\mathcal{G}$ form a partition of Ω, denoted by $\mathsf{part}(\mathcal{G})$. The partition field generated by the atoms of $\mathcal{G}$ is $\mathcal{G}$:*

$$\pi\big(\mathsf{part}(\mathcal{G})\big) = \mathcal{G}. \tag{3.28}$$

Proof For any $\omega \in \Omega$, let us denote by G_ω the intersection of all sets in $\mathcal{G}$ which contain ω (notice that G_ω is not empty since it contains ω). Since $\mathcal{G}$ is a partition field, this intersection G_ω belongs to $\mathcal{G}$: therefore, it is the smallest set in $\mathcal{G}$ which contains ω. We now prove that any G_ω is an atom.

Let $K \in \mathcal{G}$ be such that $K \subset G_\omega$. If $\omega \in K$, then $G_\omega \subset K$ by definition of G_ω since $K \in \mathcal{G}$, and thus $K = G_\omega$. If $\omega \notin K$, then $G_\omega \subset K^c$ by definition of G_ω, since $K^c \in \mathcal{G}$. Thus, we have that $K \subset G_\omega \subset K^c$. This implies that $K = \emptyset$, so that G_ω is an atom.

By construction, sets of the form G_ω cover Ω because their union contains Ω. We conclude with Proposition 3.17 that the atoms of $\mathcal{G}$ form a partition. □

Corollary 3.19 *The partition $\mathsf{part}(\mathcal{G})$ is the finest partition $\mathcal{P}$ such that $\mathcal{P} \subset \mathcal{G}$.*

Proof Denote by $\mathcal{P}_\mathcal{G}$ the least upper bound of the partitions $\mathcal{P}$ such that $\mathcal{P} \subset \mathcal{G}$. By construction, $\mathcal{P}_\mathcal{G}$ is the finest partition $\mathcal{P}$ such that $\mathcal{P} \subset \mathcal{G}$. We have that $\mathsf{part}(\mathcal{G}) \preceq \mathcal{P}_\mathcal{G}$. Take any $K \in \mathcal{P}_\mathcal{G}$. By Proposition 3.5, there exists $G \in \mathsf{part}(\mathcal{G})$ such that $K \subset G$. Now, we know that G is a $\mathcal{G}$-atom. Therefore, $K = G$ since $K \in \mathcal{G}$ and $K \neq \emptyset$ because $K \in \mathcal{P}_\mathcal{G}$, and we have proven that $\mathsf{part}(\mathcal{G}) = \mathcal{P}_\mathcal{G}$. □

As a consequence of Proposition 3.18, partition fields on Ω are in one-to-one correspondence with partitions of Ω (hence with equivalence relations on Ω as discussed in Sect. 3.3.1), by the mappings

$$\mathcal{G} \mapsto \mathsf{part}(\mathcal{G}) \text{ and } \mathcal{P} \mapsto \pi(\mathcal{P}). \tag{3.29}$$

Definition 3.20 A partition field $\mathcal{G}$ of Ω is said to be a *finite partition field* if the number of atoms in the partition $\mathsf{part}(\mathcal{G})$ is finite. It is said to be an *atomic partition field* if the number of atoms in the partition $\mathsf{part}(\mathcal{G})$ is countable (in bijection with a subset of $\mathbb{N}$, so that a finite partition field is an atomic partition field).

The following proposition expresses the property that the mapping $\mathcal{G} \mapsto \mathsf{part}(\mathcal{G})$ is monotone w.r.t. the orders and preserves the lattice structures.

Proposition 3.21 *Let $\mathcal{G}'$ and $\mathcal{G}''$ be π-fields.*

1. *The partition field $\mathcal{G}'$ is finer than $\mathcal{G}''$ if, and only if, every atom of $\mathcal{G}''$ is the union of $\mathcal{G}'$-atoms:*
$$\mathcal{G}' \preceq \mathcal{G}'' \iff \mathsf{part}(\mathcal{G}') \preceq \mathsf{part}(\mathcal{G}''). \tag{3.30}$$

2. *The partition $\mathsf{part}(\mathcal{G}' \wedge \mathcal{G}'')$ is made of subsets of Ω which are both union of $\mathcal{G}'$-atoms and union of $\mathcal{G}''$-atoms:*
$$\mathsf{part}(\mathcal{G}' \wedge \mathcal{G}'') = \mathsf{part}(\mathcal{G}') \wedge \mathsf{part}(\mathcal{G}''). \tag{3.31}$$

3. *The partition $\mathsf{part}(\mathcal{G}' \vee \mathcal{G}'')$ is made of all nonempty intersections between $\mathcal{G}'$-atoms and $\mathcal{G}''$-atoms:*
$$\mathsf{part}(\mathcal{G}' \vee \mathcal{G}'') = \mathsf{part}(\mathcal{G}') \vee \mathsf{part}(\mathcal{G}''). \tag{3.32}$$

Proof Notice that every element in a partition field can be written as a union of atoms of the partition field. Indeed, with the notations of the proof of Proposition 3.18, any $K \in \mathcal{G}$ can be written as $K = \cup_{\omega \in K} G_\omega$, where we recall that G_ω is the intersection of all sets in $\mathcal{G}$ which contain ω and is a $\mathcal{G}$-atom.

1. By Definition 3.14, $\mathcal{G}'$ is finer than $\mathcal{G}''$ if, and only if, every element of $\mathcal{G}''$ belongs to $\mathcal{G}'$. On the one hand, suppose that $\mathcal{G}'$ is finer than $\mathcal{G}''$. Therefore, every element of $\mathcal{G}''$ belongs to $\mathcal{G}'$ and, in particular, every atom of $\mathcal{G}''$ belongs to $\mathcal{G}'$, so that it can be expressed as a union of $\mathcal{G}'$-atoms. Since any element of $\mathcal{G}''$ is a union of $\mathcal{G}''$-atoms, we deduce that it is therefore a union of $\mathcal{G}'$-atoms. By Proposition 3.5, we conclude that $\mathsf{part}(\mathcal{G}') \preceq \mathsf{part}(\mathcal{G}'')$.
   On the other hand, if any atom of $\mathcal{G}''$ can be expressed as a union of $\mathcal{G}'$-atoms, then so can be any element of $\mathcal{G}''$ by Proposition 3.18: hence $\mathcal{G}'' \subset \mathcal{G}'$, that is, $\mathcal{G}'' \preceq \mathcal{G}'$.
2. By Proposition 3.15, the greatest lower bound of partition fields $\mathcal{G}'$ and $\mathcal{G}''$ is $\mathcal{G}' \wedge \mathcal{G}'' = \mathcal{G}' \cap \mathcal{G}''$. Therefore, it contains the subsets of Ω which are both in $\mathcal{G}'$ and in $\mathcal{G}''$, hence which are both union of $\mathcal{G}'$-atoms and union of $\mathcal{G}''$-atoms.
3. By Proposition 3.15, the least upper bound $\mathcal{G}' \vee \mathcal{G}'' = \pi(\mathcal{G}' \cup \mathcal{G}'')$.

   - First, we prove that any nonempty intersection $G = G' \cap G''$ between a $\mathcal{G}'$-atom G' and a $\mathcal{G}''$-atom G'' is an atom of $\mathcal{G}' \vee \mathcal{G}''$. Notice that the partition field $\mathcal{G}' \vee \mathcal{G}'' = \pi(\mathcal{G}' \cup \mathcal{G}'')$ is made of subsets of Ω which are arbitrary union and intersection of elements of $\mathcal{G}'$ and $\mathcal{G}''$, and therefore arbitrary union and intersection of $\mathcal{G}'$-atoms and $\mathcal{G}''$-atoms. With obvious notations, consider $K \in \pi(\mathcal{G}' \cup \mathcal{G}'')$ written as $K = \cup_{i \in I} \cap_{j \in J_i} G_{ij}$ where G_{ij} is either a $\mathcal{G}'$-atom or a $\mathcal{G}''$-atom. In fact, by Proposition 3.17, the set J_i may be reduced to two elements since two distinct $\mathcal{G}'$-atoms are necessarily disjoint, and the same for $\mathcal{G}''$-atoms. Therefore, we can write $K = \cup_{i \in I} G'_i \cap G''_i$ where G'_i is a $\mathcal{G}'$-atom and G''_i a $\mathcal{G}''$-atom. If $K \subset G = G' \cap G''$, then necessarily $G'_i \cap G''_i \subset G' \cap G''$ for all $i \in I$. As a consequence, $G'_i \cap G''_i \subset (G' \cap G'_i) \cap (G''_i \cap G'')$ where, again by Proposition 3.17, $G' \cap G'_i = \emptyset$ or $G' \cap G'_i = G'$, and $G'' \cap G''_i = \emptyset$ or $G'' \cap G''_i = G''$. Therefore, either $K = \emptyset$ or $K = \cup_{i \in I} G' \cap G'' = G$. We conclude that G is an atom of $\mathcal{G}' \vee \mathcal{G}''$.

- Second, we prove that all nonempty intersections between $\mathcal{G}'$-atoms and $\mathcal{G}''$-atoms form a partition. Indeed, on the one hand, two such distinct nonempty intersections are necessarily disjoint by Proposition 3.17. On the other hand, the union of all nonempty intersections between $\mathcal{G}'$-atoms and $\mathcal{G}''$-atoms is the union of all $\mathcal{G}'$-atoms (and also, of all $\mathcal{G}''$-atoms). Now, by Proposition 3.18, we know that the atoms of $\mathcal{G}'$ form a partition. Therefore, the union of all nonempty intersections between $\mathcal{G}'$-atoms and $\mathcal{G}''$-atoms is Ω. We conclude that all nonempty intersections between $\mathcal{G}'$-atoms and $\mathcal{G}''$-atoms form a partition. As a consequence, there can be no other atoms.

This ends the proof. □

3.3.3 The Lattice of σ-Fields

Less is required from σ-fields compared to π-fields, only stability by countable union and intersection. In contrast to a widely accepted idea, and as noted by [58], the use of σ-fields to handle information raises some problems. Indeed, in Remark 3.29, we discuss the fact that the inclusion order on σ-fields is not necessarily compatible with the order on partitions: this is why the use of σ-fields as the mathematical expression of information may be tricky. However, using π-fields can raise difficulties in the stochastic case. As an illustration, the π-field generated by the singletons of the universe Ω is the complete field made of all subsets of Ω (being made of all arbitrary union and intersection, countable or not, of singletons). Unfortunately, the complete field is too large to support a probability measure when Ω is not finite.

σ-Fields

Definition 3.22 A *σ-field* on Ω is a collection $\mathcal{G}$ of subsets of Ω (identified with a subset $\mathcal{G} \subset 2^{\Omega}$), containing Ω, and which is stable under complementation and countable union (hence under countable intersection).

Notice that a σ-field necessarily contains Ω and $\emptyset$. When Ω is finite, σ-fields and partition fields coincide. Every partition field is a σ-field, but the converse is generally false as illustrated by the examples below.

Example 3.23 Any σ-field which contains all singletons and which is not the complete σ-field 2^{Ω} cannot be a π-field (indeed, a π-field which contains all singletons is necessarily 2^{Ω} as seen in Example 3.13). Thus, the σ-field $\mathcal{B}^{o}_{\mathbb{R}}$ of Borel sets on $\mathbb{R}$ is not a π-field. The set made of all countable subsets of an uncountable set Ω and of all their complements is a σ-field, but is not a π-field. △

It is easy to see that the intersection of σ-fields (see Definition 3.10) is a σ-field. This justifies the following definition.

Definition 3.24 The σ-field $\sigma(\mathcal{C})$ *generated* by a collection $\mathcal{C}$ of subsets of Ω is the smallest σ-field containing $\mathcal{C}$. It is also the intersection of all σ-fields containing $\mathcal{C}$.

Remark 3.25 As an application (see also Sect. B.1.4), consider two sets Ω_1 and Ω_2, each equipped with a σ-field $\mathcal{G}_i$, $i = 1, 2$. By definition, the *product σ-field* $\mathcal{G}_1 \otimes \mathcal{G}_2$ on $\Omega_1 \times \Omega_2$ is the one generated by the rectangles $\{G_1 \times G_2 \mid G_i \in \mathcal{G}_i,\ i = 1, 2\}$. ♢

Definition 3.26 A σ-field $\mathcal{G}$ of Ω is said to be an *atomic σ-field* if it is generated by a countable partition $\{\Omega_n\}_{n \in N}$ where N is countable (in bijection with a subset of $\mathbb{N}$): $\mathcal{G} = \sigma(\{\Omega_n\}_{n \in N})$. In that case, the σ-field $\mathcal{G}$ is also a partition field (and an atomic one, as in Definition 3.20), and we have that $\mathcal{G} = \sigma(\{\Omega_n\}_{n \in N}) = \pi(\{\Omega_n\}_{n \in N})$ (see Definition 3.11). In the special case where the countable partition $\{\Omega_n\}_{n \in N}$ is finite, the σ-field $\mathcal{G}$ of Ω is said to be a *finite σ-field*.

π-Fields and σ-Fields

From the Definitions 3.9 of partition fields and 3.22 of σ-fields, we deduce the following inclusion.

Proposition 3.27 *For any collection $\mathcal{C}$ of subsets of Ω, we have that*

$$\sigma(\mathcal{C}) \subset \pi(\mathcal{C}). \tag{3.33}$$

Example 3.28 As an example, consider $\mathbb{R}$ with closed subsets $\mathcal{C}$ as "primitive events": $\sigma(\mathcal{C})$ is the set $\mathcal{B}^{o}_{\mathbb{R}}$ of Borelians, whereas, as seen in Example 3.13, $\pi(\mathcal{C}) = 2^{\mathbb{R}}$ since $\mathcal{C}$ contains all singletons. Therefore, any non Borelian subset belongs to $\pi(\mathcal{C})$ but not to $\sigma(\mathcal{C})$. △

Remark 3.29 In [58], the following example is analyzed to show that the use of σ-fields as information raises delicate issues. On $\Omega = [0, 1]$, consider the two partitions $\mathcal{C}^{\flat} = \{[0, 1/2],]1/2, 1]\}$ and the complete partition $\mathcal{C}^{\sharp} = \{\{\omega\}\}_{\omega \in [0,1]}$. By Definition 3.4, $\mathcal{C}^{\sharp}$ is finer than $\mathcal{C}^{\flat}$. The π-fields generated are $\pi(\mathcal{C}^{\flat}) = \{\emptyset, [0, 1], [0, 1/2],]1/2, 1]\}$ and $\pi(\mathcal{C}^{\sharp}) = 2^{[0,1]}$. Since $\mathcal{C}^{\flat}$ is a finite partition, the σ-field generated by $\mathcal{C}^{\flat}$ is $\sigma(\mathcal{C}^{\flat}) = \pi(\mathcal{C}^{\flat})$. It can be seen that the σ-field $\sigma(\mathcal{C}^{\sharp})$ generated by $\mathcal{C}^{\sharp}$ is made of subsets of $[0, 1]$ which are either countable or whose complement is countable. Hence, to summarize, we have that $\mathcal{C}^{\flat} \preceq \mathcal{C}^{\sharp}$ and $\pi(\mathcal{C}^{\flat}) \preceq \pi(\mathcal{C}^{\sharp})$, whereas $\sigma(\mathcal{C}^{\flat}) = \{\emptyset, [0, 1], [0, 1/2],]1/2, 1]\}$ and $\sigma(\mathcal{C}^{\sharp})$ are not comparable. Indeed, neither of the sets $\emptyset$, $[0, 1]$, $[0, 1/2]$, $]1/2, 1]$ is countable or has a countable complement. ♢

It can easily be proven that, in all generality,

$$\mathcal{C}^{\flat} \preceq \mathcal{C}^{\sharp} \Rightarrow \pi(\mathcal{C}^{\flat}) \preceq \pi(\mathcal{C}^{\sharp}). \tag{3.34}$$

However, this property is not true for σ-fields as the above example illustrates it. The inclusion order on σ-fields is not compatible with the order on partitions: this is why the use of σ-fields as the mathematical expression of information may be tricky.

A Partial Order and a Lattice Structure on σ-Fields
The set of all σ-fields on Ω is a *lattice*, with the operators $\wedge$ and $\vee$ defined as for partition fields in Proposition 3.15.

3.4 Mapping Measurability Approach

After having discussed the partitions and fields approach, we turn the spotlight onto the mapping measurability approach to information handling. For this purpose, we recall the notion of *measurability* of mappings, in relation with either partitions in Sect. 3.4.1, π-fields in Sect. 3.4.2 or σ-fields in Sect. 3.4.3. In the mapping measurability approach, information is handled by mappings over the universe Ω, also called *signals*. We provide tools to pass from an *algebraic* aspect of measurability—where a mapping is measurable w.r.t. another one—to a *functional* one, where a mapping is a function of another one. The measurability framework developed will be widely used in the book.

3.4.1 Measurability of Mappings w.r.t. Partitions

Any mapping defined over the universe Ω, whatever its image set, defines an equivalence relation on Ω as follows.

Definition 3.30 Le H be a mapping defined over the universe Ω. Two elements of Ω are equivalent or *undistinguishable* w.r.t. the mapping H when they share the same image. This defines the following equivalence relation $\mathfrak{R}_H$ on Ω:

$$\omega \, \mathfrak{R}_H \, \omega' \iff H(\omega) = H(\omega'). \tag{3.35}$$

This equivalence relation on Ω induces a partition of Ω into H-classes, in one-to-one correspondence with the image $\text{im}\, H$ of the mapping H. We say that this partition is the *partition generated by the mapping* H. We denote it by $\mathsf{part}(H)$, or by Ω/H when we want to emphasize the dependence upon the domain Ω:

$$\mathsf{part}(H) := \Omega/H := \left\{ H^{-1}(y) \mid y \in \text{im}\, H \right\} = H^{-1}(\{\{y\}\}_{y \in \text{im}\, H}). \tag{3.36}$$

Recall that $H^{-1}(y)$ is the subset $\{\omega \in \Omega \mid H(\omega) = y\}$ of Ω.

We say that the mapping H is *measurable w.r.t. a partition* $\mathcal{P}$, or is $\mathcal{P}$-measurable, if $\mathcal{P}$ is finer than the partition $\mathsf{part}(H)$ generated by the mapping H: $\mathsf{part}(H) \preceq \mathcal{P}$.

The following proposition is a straightforward consequence of the above definition.

Proposition 3.31 *A mapping defined over the domain Ω is measurable w.r.t. a partition if, and only if, it is constant on every element of this partition.*

Proof The mapping H is measurable w.r.t. the partition $\mathcal{P}$ if and only if $\mathsf{part}(H) \preceq \mathcal{P}$, that is, every element of $\mathsf{part}(H)$ is included in a element of $\mathcal{P}$ or, equivalently, is the union of elements of $\mathcal{P}$ by Proposition 3.5.

Since any element of $\mathsf{part}(H)$ is of the form $H^{-1}(y)$ for some $y \in \operatorname{im} H$ by (3.36), the mapping H is constant on every element of $\mathsf{part}(H)$, hence constant on every element of the partition $\mathcal{P}$.

On the other hand, if the mapping H is constant on every element of the partition $\mathcal{P}$, with any $G \in \mathcal{P}$ we associate $y \in \operatorname{im} H$ such that $G \subset H^{-1}(y)$. This means exactly that $\mathsf{part}(H) \preceq \mathcal{P}$. □

3.4.2 Measurability of Mappings w.r.t. π-Fields

In Definition 3.30, we introduced the notions of partition generated by a mapping, and of measurability of a mapping w.r.t. a partition. We now define the notions of partition field generated by a mapping, and of measurability of a mapping w.r.t. a partition field. Then, we relate both approaches.

Let $\mathbb{Y}$ be a set equipped with a partition field $\mathcal{Y}$. For any mapping $H : \Omega \to \mathbb{Y}$, the pre-image $H^{-1}(\mathcal{Y})$ of the partition field $\mathcal{Y}$ is easily seen to be a partition field of Ω. This property motivates the following definition.

Definition 3.32 Let Ω be equipped with a partition field $\mathcal{G}$, and let $\mathbb{Y}$ be another set equipped with a partition field $\mathcal{Y}$. For any mapping $H : \Omega \to \mathbb{Y}$, $H^{-1}(\mathcal{Y})$ is a partition field of Ω called the *partition field generated by H*, denoted by

$$\pi(H) := H^{-1}(\mathcal{Y}) = \left\{ H^{-1}(Y) \mid Y \in \mathcal{Y} \right\}. \tag{3.37}$$

The mapping H is said to be *measurable* w.r.t. the partition field $\mathcal{G}$, or $\mathcal{G}$-measurable, denoted by $H \preceq \mathcal{G}$, if $\mathcal{G}$ is finer than $\pi(H)$:

$$H \preceq \mathcal{G} \iff \pi(H) \preceq \mathcal{G} \iff \pi(H) \subset \mathcal{G}. \tag{3.38}$$

Remark 3.33 When not explicitely specified, the partition field $\mathcal{Y}$ on the image set $\mathbb{Y}$ is supposed to be the complete partition field $\{\{y\}\}_{y \in \mathbb{Y}}$, identified with $2^{\mathbb{Y}}$. ◊

In Definition 3.30, we introduced the partition $\mathsf{part}(H)$ generated by a mapping H. Partition and partition field generated by H are related by the following proposition.

Proposition 3.34 *Consider a mapping $H : \Omega \to \mathbb{Y}$. The atoms of the partition field $\pi(H)$ are the elements of the partition* $\mathsf{part}(H)$*:*

$$\mathsf{part}\big(\pi(H)\big) = \mathsf{part}(H). \tag{3.39}$$

The partition field $\pi(H)$ is generated by the partition $\mathsf{part}(H)$*:*

$$\pi(H) = \pi\big(\mathsf{part}(H)\big). \tag{3.40}$$

Proof As specified in Remark 3.33, the partition field $\mathcal{Y}$ on the image set $\mathbb{Y}$ is supposed to be the complete partition field $\{\{y\}\}_{y\in\mathbb{Y}}$.

By Definition 3.30, the elements of the partition $\mathsf{part}(H)$ are the subsets of Ω of the form $G = H^{-1}(y)$ for $y \in \operatorname{im} H$.

Now, a subset of Ω of the form $G = H^{-1}(y)$ for $y \in \operatorname{im} H$ is an atom of $\pi(H)$. Indeed, if $K \subset G$ and $K \in \pi(H)$, then we must have both $K = H^{-1}(J)$ for some $J \subset \operatorname{im} H$ by (3.37), and $K = H^{-1}(J) \subset H^{-1}(y) = G$. Therefore, $J \subset \{y\}$ and either $J = \{y\}$ and $K = G$, or $J = \emptyset$ and $K = \emptyset$.

By construction, sets of the form $G = H^{-1}(y)$ for $y \in \operatorname{im} H$ cover Ω, and we conclude with Proposition 3.17 that they form a partition of Ω, and that there are no other nonempty $\pi(H)$-atoms. Therefore, the atoms of the partition field $\pi(H)$ are the elements of the partition $\mathsf{part}(H)$. By Proposition 3.18, this is equivalent to (3.39).

Then, (3.40) is a consequence of Proposition 3.18 and of (3.28). Indeed, we obtain $\pi\Big(\mathsf{part}\big(\pi(H)\big)\Big) = \pi(H) = \pi\big(\mathsf{part}(H)\big)$. □

The counterpart of Proposition 3.31 is the following proposition.

Proposition 3.35 *Consider a mapping $H : \Omega \to \mathbb{Y}$ and a partition field $\mathcal{G}$ with associated partition* $\mathsf{part}(\mathcal{G})$*. The following assertions are equivalent.*

1. *The mapping H is measurable w.r.t. the partition field $\mathcal{G}$.*
2. *The mapping H is constant on every atom of the partition field $\mathcal{G}$.*
3. *The mapping H is constant on every element of the partition* $\mathsf{part}(\mathcal{G})$.
4. *The mapping H is measurable w.r.t. the partition* $\mathsf{part}(\mathcal{G})$.

Proof That item 2 is equivalent to item 3 results from Proposition 3.18. That item 3 is equivalent to item 4 results from Proposition 3.31.

Item 1 means that $\pi(H) \subset \mathcal{G}$. We deduce that $\mathsf{part}(H) \subset \pi(H) \subset \mathcal{G}$ and, by the last assertion of Proposition 3.18, we conclude that $\mathsf{part}(H) \preceq \mathsf{part}(\mathcal{G})$. Thus, we have proved item 4.

Item 4 means that $\mathsf{part}(H) \preceq \mathsf{part}(\mathcal{G})$. Therefore, there follows an inclusion between the partition fields generated: $\pi\big(\mathsf{part}(H)\big) \subset \pi\big(\mathsf{part}(\mathcal{G})\big)$. By Proposition 3.18, we conclude that $\pi(H) = \pi\big(\mathsf{part}(H)\big) \subset \mathcal{G} = \pi\big(\mathsf{part}(\mathcal{G})\big)$ by Proposition 3.34 and (3.40), and by Proposition 3.18. Thus, we have proved item 1. □

A Partial Order and a Lattice Structure on Mappings

The following definitions and propositions are taken from [33].

Definition 3.36 Let $H_i : \Omega \to \mathbb{Y}_i$, $i = 1, 2$. The mapping H_1 is said to be *measurable* w.r.t. the mapping H_2 if $\pi(H_1) \subset \pi(H_2)$. We denote this by $H_1 \preceq H_2$, or by $H_2 \succeq H_1$.

Remark 3.37 In Definition 3.44, we also find another notion of measurability of mappings w.r.t. a σ-field. In the book, the context will indicate which of both definitions is relevant. ◊

The following Proposition 3.38 is a tool to pass from an *algebraic* aspect of measurability—where a mapping is measurable w.r.t. another one—to a *functional* one, where a mapping is a function of another one (see item 4). An illustration is given in Fig. 3.3.

Proposition 3.38 *Consider two mappings $H_i : \Omega \to \mathbb{Y}_i$, $i = 1, 2$. The following conditions are equivalent characterizations of the fact that H_1 is measurable w.r.t. H_2:*

1. $H_1 \preceq H_2$*;*
2. $\forall (\omega, \omega') \in \Omega^2, \quad H_2(\omega) = H_2(\omega') \Rightarrow H_1(\omega) = H_1(\omega')$*;*
3. $\forall y_2 \in \operatorname{im} H_2, \ \exists y_1 \in \operatorname{im} H_1 : \ H_2^{-1}(y_2) \subset H_1^{-1}(y_1)$*;*
4. *there exists a unique mapping* $\mathfrak{f} : \operatorname{im} H_2 \to \operatorname{im} H_1$ *such that* $H_1 = \mathfrak{f} \circ H_2$*;*
5. $\pi(H_1) \subset \pi(H_2)$*;*
6. *every* $\pi(H_2)$*-atom is included in a* $\pi(H_1)$*-atom;*
7. *every* $\pi(H_1)$*-atom is the union of* $\pi(H_2)$*-atoms;*
8. $\Omega/H_1 \preceq \Omega/H_2$

Proof The equivalence of item 1 and item 5 is a consequence of Definition 3.36. Let us define the multi-application $\mathfrak{f}$ as:

$$\forall y_2 \in \operatorname{im} H_2, \ \mathfrak{f}(y_2) = \bigcup_{\omega \in H_2^{-1}(y_2)} \{H_1(\omega)\}.$$

From item 2, we deduce that $\mathfrak{f}$ is in fact a mapping ($\mathfrak{f}(y_2)$ contains a single element) satisfying item 4. The reverse implication is immediate.

The remaining items are equivalent by Definition 3.4 and by Proposition 3.18. □

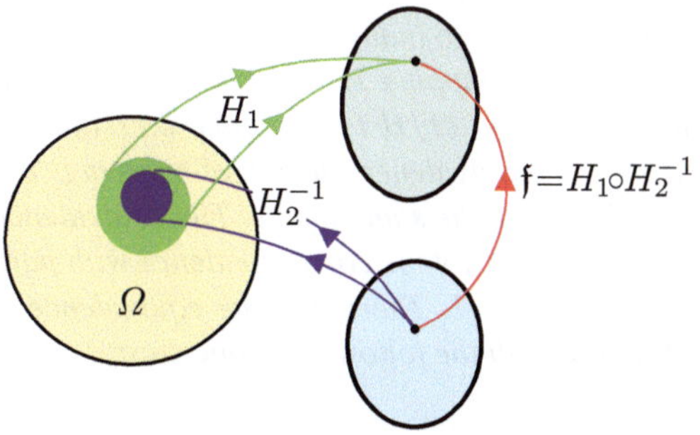

Fig. 3.3 Measurability relation between mappings

A consequence of Definition 3.36 is that the relation $\preceq$ between mappings is stable by right composition as follows.

Proposition 3.39 *Let* $H_i : \Omega \to \mathbb{Y}_i,\ i = 1, 2$. *Then,*

$$H_1 \preceq H_2 \Rightarrow \forall h : \Omega' \to \Omega,\ \ H_1 \circ h \preceq H_2 \circ h. \tag{3.41}$$

As stressed in Sect. 3.2.1, any preorder $\preceq$ induces an order on the quotient set w.r.t. the equivalence relation $H_1 \equiv H_2 \iff H_1 \preceq H_2$ or $H_2 \preceq H_1$. This motivates the following definition.

Definition 3.40 Let $H_i : \Omega \to \mathbb{Y}_i,\ i = 1, 2$. The mappings H_1 and H_2 are said to be *equivalent* if $\pi(H_1) = \pi(H_2)$. This is denoted by $H_1 \equiv H_2$.

Proposition 3.41 *Let* $H_i : \Omega \to \mathbb{Y}_i,\ i = 1, 2$. *The following conditions are equivalent characterizations of the fact that* H_1 *is equivalent to* H_2*:*

1. $H_1 \equiv H_2$;
2. $\forall(\omega, \omega') \in \Omega^2,\ \ H_2(\omega) = H_2(\omega') \iff H_1(\omega) = H_1(\omega')$;
3. $\Omega / H_1 = \Omega / H_2$;
4. *there exists an injection* $\mathfrak{f} : \mathrm{im}\, H_2 \to \mathbb{Y}_1$ *such that* $H_1 = \mathfrak{f} \circ H_2$;
5. *there exists a bijection* $\mathfrak{f} : \mathrm{im}\, H_2 \to \mathrm{im}\, H_1$ *such that* $H_1 = \mathfrak{f} \circ H_2$.

Proof We have that

$$H_1 \equiv H_2 \iff H_1 \preceq H_2 \text{ and } H_2 \preceq H_1 \iff \mathsf{part}(H_1) = \mathsf{part}(H_2)$$

if, and only if, by Proposition 3.38, there exist a mapping $\mathfrak{f} : \mathrm{im}\, H_2 \to \mathrm{im}\, H_1$ such that $H_1 = \mathfrak{f} \circ H_2$ and a mapping $\mathfrak{f}' : \mathrm{im}\, H_1 \to \mathrm{im}\, H_2$ such that $H_2 = \mathfrak{f}' \circ H_1$.

This ends the proof since $\mathfrak{f} : \mathrm{im}\, H_2 \to \mathrm{im}\, H_1$ is a bijection if, and only if, there exist $\mathfrak{f}' : \mathrm{im}\, H_1 \to \mathrm{im}\, H_2$ such that $\mathfrak{f} \circ \mathfrak{f}' = \mathrm{I}_{\mathrm{im}\, H_1}$ and $\mathfrak{f}' \circ \mathfrak{f} = \mathrm{I}_{\mathrm{im}\, H_2}$. □

As stressed above, the relation $\preceq$ defines a preorder for functions over Ω. Therefore, we deduce an order relation for the equivalence classes of functions over Ω. We leave the proof of the following proposition to the reader.

Proposition 3.42 *Over the equivalence classes of functions over* Ω*, the relation* $\preceq$ *of Definition 3.36 induces a partial order relation.*

With any mapping H *defined on* Ω*, we associate the partition* Ω / H *of* Ω*. The mapping* $H \mapsto \Omega / H$ *is monotone w.r.t. the relations* $\preceq$ *of Definitions 3.4 and 3.36. Moreover, equivalence classes of mappings over* Ω *w.r.t. the relation* $\equiv$ *of Definition 3.40 give the same image. The equivalence classes of mappings over* Ω *are in one-to-one monotone correspondence with partitions on* Ω *by the mapping induced by* $H \mapsto \Omega / H$*. Therefore, the equivalence classes of functions inherit a lattice structure with the following properties:*

- bottom element: *the class of* constant *functions on Ω;*
- top element: *the class of* injective *functions on Ω;*
- least upper bound: *the* least upper bound $H_1 \vee H_2$ *(class) can be represented by the couple mapping* (H_1, H_2).

There exists a *greatest lower bound* $H_1 \wedge H_2$, but there is no straightforward way of displaying a mapping which would represent the equivalence class of this lower bound.

In Fig. 3.2, one can visualize in the third drawing how the atoms of the partition associated with the least upper bound $H_1 \vee H_2$, represented by the couple of functions (H_1, H_2), are built as intersections of the atoms of the partitions associated with the functions H_1 and H_2. The fourth picture displays the partition associated with the greatest lower bound $H_1 \wedge H_2$. The associated undistinguishability equivalence relation (see Definition 3.3) is the transitive closure of the undistinguishability equivalence relations with the functions H_1 and H_2.

A Summary Table

We give in Table 3.1 correspondences between different ways of representing information.

Notice in Table 3.1 that the least upper bound of two partitions is easy to calculate by intersecting atoms, whereas computing the greatest lower bound is much harder

Table 3.1 Correspondences between partitions and fields and mapping measurability approaches to information

Description	Partitions and fields			Measurability
Information	Equivalence relation on Ω	Partition of Ω	Partition field on Ω	Mapping with domain Ω
Notation	$\mathfrak{R} \subset \Omega \times \Omega$	$\{\Omega_i\}_{i \in I} \subset 2^\Omega$	$\mathcal{G} \subset 2^\Omega$	$H : \Omega \to \mathbb{Y}$
Example	$\Omega \times \Omega$ complete	$\{\Omega\}$ trivial	$\{\emptyset, \Omega\}$ trivial	$H : \Omega \to \mathbb{Y}$ constant
Example	Δ_Ω diagonal	$\{\{\omega\}\}_{\omega \in \Omega}$ discrete	2^Ω discrete	$H = \mathrm{I}_\Omega$ identity
	Classes $\omega\,\mathfrak{R}$	Elements $\{\Omega_i\}_{i \in I}$	Atoms $G \in \mathcal{G}$	Pre-images $H^{-1}(\{y\})$
	Undistinguishable elements		Belong to	Same image
	$\omega' \,\mathfrak{R}\, \omega$	$\exists i, \{\omega, \omega'\} \subset \Omega_i$	same atom	$H(\omega) = H(\omega')$
Compare	$\mathfrak{R}^\flat \preceq \mathfrak{R}^\sharp$ classes inclusion	$\Omega_i^\flat \cap \Omega_j^\sharp \in \{\emptyset, \Omega_i^\flat\}$	$\mathcal{G}^\flat \subset \mathcal{G}^\sharp$	$\exists \mathfrak{f} : \mathbb{Y}^\sharp \to \mathbb{Y}^\flat$ $H^\flat = \mathfrak{f}(H^\sharp)$
Least upper bound	$\mathfrak{R}_1 \cap \mathfrak{R}_2$	$\Omega_i^1 \cap \Omega_j^2$	$\mathcal{G}_1 \vee \mathcal{G}_2$ generated by $\mathcal{G}_1 \cup \mathcal{G}_2$	$(H_1, H_2) : \Omega \to \mathbb{Y}_1 \times \mathbb{Y}_2$
	Intersection of classes			
Greatest lower bound	$(\mathfrak{R}_1 \cup \mathfrak{R}_2)^\infty$ transitive closure		$\mathcal{G}_1 \wedge \mathcal{G}_2 = \mathcal{G}_1 \cap \mathcal{G}_2$ intersection	

(it requires a transitive closure, once the identification with equivalence relations is made). The reverse holds true for partition fields: the greatest lower bound is easy to calculate by intersecting partition fields (taking the common elements), whereas this is not the case for the least upper bound.

The multiple equivalences seen above—between equivalence relations, partitions, partition fields, equivalence classes of mappings—reflect different ways of representing and handling information. However, when the universe Ω is not countable and is equipped with a probability distribution, the notion of partition field is not adapted. This is why we now turn to σ-fields, less easy to handle than partition fields, but tailored for probability theory.

3.4.3 Measurability of Mappings w.r.t. σ-Fields

Let Ω be equipped with a σ-field $\mathcal{G}$, and $\mathbb{Y}$ with a σ-field $\mathcal{Y}$.[3] The following notions of measurability for mappings from Ω to $\mathbb{Y}$ is classical.

Remark 3.43 Notice however that, in Definition 3.36, we use the same notation to define another notion of measurability of mappings (w.r.t. a π-field). In this book, the context will indicate which of both definitions is relevant. $\Diamond$

Definition 3.44 For any mapping $H : \Omega \to \mathbb{Y}$, $H^{-1}(\mathcal{Y})$ is a σ-field on Ω called the *σ-field generated by H*, denoted by $\sigma(H) := H^{-1}(\mathcal{Y})$.

The mapping H is said to be *measurable* w.r.t. the σ-field $\mathcal{G}$, or $\mathcal{G}$-measurable, denoted by $H \preceq \mathcal{G}$, if $\sigma(H) \subset \mathcal{G}$. In Probability Theory, a measurable mapping from $(\Omega, \mathcal{G})$ to $(\mathbb{Y}, \mathcal{Y})$ is called a *random variable*.

Definition 3.45 Let $H_i : \Omega \to \mathbb{Y}_i,\ i = 1, 2$. The mapping H_1 is said to be *measurable* w.r.t. the mapping H_2 if $\sigma(H_1) \subset \sigma(H_2)$. We denote this by $H_1 \preceq H_2$.

The following Proposition 3.46, due to Doob (see [53, Chap. 1, p. 18]), is a tool to pass from an *algebraic* aspect of measurability to a *functional* one.

Proposition 3.46 *Let $H_i : \Omega \to \mathbb{Y}_i,\ i = 1, 2$. Assume that $\mathbb{Y}_1$ is a separable complete metric space.*

The mapping H_1 is measurable w.r.t. H_2 if, and only if, there exists a measurable mapping $\mathfrak{f}$: $\text{im}\, H_2 \to \text{im}\, H_1$ such that $H_1 = \mathfrak{f} \circ H_2$.

Notice the difference between Propositions 3.38 and 3.46: the unique mapping $\mathfrak{f}$: $\text{im}\, H_2 \to \text{im}\, H_1$ such that $H_1 = \mathfrak{f} \circ H_2$ in Proposition 3.38 has to be measurable for Proposition 3.46 to hold true.

Remark 3.47 In the book, when we make use of Proposition 3.46, we implicitely assume that all spaces are separable complete and metric. $\Diamond$

---

[3]When $\mathbb{Y}$ is a Borel space (see Sect. B.6), it is implicitely assumed that $\mathcal{Y}$ is the σ-field $\mathcal{B}^{o}_{\mathbb{Y}}$ of Borel subsets.

We end up with a result that we will often use in the text.

Proposition 3.48 *Consider a random variable $H : \Omega \to \mathbb{Y}$, and a measurable one-to-one mapping $\varphi : \mathbb{Y} \to \mathbb{Z}$ with measurable inverse $\varphi^{-1} : \mathbb{Z} \to \mathbb{Y}$. Then, $\varphi \circ H$ is a random variable and $\sigma(H) = \sigma(\varphi \circ H)$.*

Proof It is implicit in the proposition that Ω is equipped with a σ-field $\mathcal{G}$, $\mathbb{Y}$ with a σ-field $\mathcal{Y}$, and $\mathbb{Z}$ with a σ-field $\mathcal{Z}$.

The random variable H is a measurable mapping $H : \Omega \to \mathbb{Y}$. Therefore, the composition $\varphi \circ H : \Omega \to \mathbb{Z}$ is a measurable mapping, hence is a random variable. We have that

$$\sigma(\varphi \circ H) = (\varphi \circ H)^{-1}(\mathcal{Z}) = H^{-1}\left(\varphi^{-1}(\mathcal{Z})\right) \subset H^{-1}(\mathcal{Y}) = \sigma(H),$$

where we have used the measurability of φ, namely $\varphi^{-1}(\mathcal{Z}) \subset \mathcal{Y}$. Now, since the mapping φ is one-to-one, we can write $H = \varphi^{-1} \circ \varphi \circ H$. Because the inverse φ^{-1} is measurable, we conclude from what we have just established that

$$\sigma(H) = \sigma(\varphi^{-1} \circ \varphi \circ H) \subset \sigma(\varphi \circ H).$$

This completes the proof. □

3.5 Conditional Expectation and Optimization

In what follows, our aim is to give an elementary and intuitive definition of the conditional expectation of a random variable w.r.t. information. When the universe Ω is finite, we have seen that the different ways to describe information—equivalence relations, partitions, partition fields, σ-fields, equivalence classes of mappings—are equivalent.

We assume in the next paragraphs that the set Ω is finite and that it carries a probability $\mathbb{P}$ (see Sect. B.1). We define the *conditional expectation* of a random variable w.r.t. a partition. Then, we provide a formula connecting optimization under measurability constraints and conditional expectation.

3.5.1 Conditional Expectation w.r.t. a Partition

Consider a function $\boldsymbol{X} : \Omega \to \mathbb{R}$. Because the set Ω is finite, the function $\boldsymbol{X}$ is measurable (where $\mathbb{R}$ is equipped with the Borelian σ-field $\mathcal{B}^{\mathrm{o}}_{\mathbb{R}}$), hence is a random variable. Taking a finite number of values, $\boldsymbol{X}$ is integrable, and its *mathematical expectation* is the scalar given by

$$\mathbb{E}(X) = \sum_{\omega \in \Omega} \mathbb{P}(\omega) X(\omega). \tag{3.42}$$

Consider a partition $\mathcal{P} = \{\Omega_i\}_{i \in I}$ of Ω, where we assume, for the sake of simplicity, that each Ω_i is such that $\mathbb{P}(\Omega_i) > 0$.

We define the *conditional probability* $\mathbb{P}_{|\Omega_i}$ knowing Ω_i by

$$\mathbb{P}_{|\Omega_i}(G) = \mathbb{P}(G \mid \Omega_i) = \frac{\mathbb{P}(G \cap \Omega_i)}{\mathbb{P}(\Omega_i)}, \quad \forall G \subset \Omega. \tag{3.43}$$

It can be proven that $\mathbb{P}_{|\Omega_i}$ satisfies the axioms of a probability (see Sect. B.1).

Let $X_{|\Omega_i}$ denote the restriction of the random variable X to Ω_i.

Definition 3.49 The *conditional expectation* of the random variable X w.r.t. the partition $\mathcal{P}$ is the $\mathcal{P}$-measurable random variable, denoted by $\mathbb{E}\left(X \mid \mathcal{P}\right)$ and having constant value $\mathbb{E}_{\mathbb{P}_{|\Omega_i}}(X_{|\Omega_i})$ on Ω_i, namely

$$\mathbb{E}\left(X \mid \mathcal{P}\right)(\omega) = \frac{\mathbb{E}\left(\mathbf{1}_{\Omega_i} X\right)}{\mathbb{P}(\Omega_i)}, \quad \forall \omega \in \Omega_i, \ \forall i \in I. \tag{3.44}$$

In case there would exist some Ω_i such that $\mathbb{P}(\Omega_i) = 0$, we can give any value to $\mathbb{E}\left(X \mid \mathcal{P}\right)$ on Ω_i, such as zero for instance. Hence, the random variable $\mathbb{E}\left(X \mid \mathcal{G}\right)$ is constant on every element Ω_i of the partition $\mathcal{P}$. Its value is the mean value of X restricted to Ω_i whenever $\mathbb{P}(\Omega_i) > 0$. Using indicator functions, the general formula is:

$$\mathbb{E}\left(X \mid \mathcal{P}\right) = \sum_{i \in I, \mathbb{P}(\Omega_i) > 0} \frac{\mathbb{E}\left(\mathbf{1}_{\Omega_i} X\right)}{\mathbb{P}(\Omega_i)} \mathbf{1}_{\Omega_i}. \tag{3.45}$$

By Proposition 3.35, the conditional expectation w.r.t. the partition $\mathcal{P}$ is measurable w.r.t. $\mathcal{P}$.

This elementary definition of conditional expectation w.r.t. a partition displays all the basic properties shared by the more general conditional expectations (see Sect. B.4). Let X and Y be two random variables on Ω, and λ be a real number. We have that

$$\mathbb{E}\left(\lambda X + Y \mid \mathcal{P}\right) = \lambda \mathbb{E}\left(X \mid \mathcal{P}\right) + \mathbb{E}\left(Y \mid \mathcal{P}\right), \tag{3.46a}$$

$$X \geq 0 \Rightarrow \mathbb{E}\left(X \mid \mathcal{P}\right) \geq 0, \tag{3.46b}$$

$$\mathbb{E}\left(\mathbb{E}\left(X \mid \mathcal{P}\right)\right) = \mathbb{E}\left(X\right). \tag{3.46c}$$

If Y is measurable w.r.t. $\mathcal{P}$, we have that

$$\mathbb{E}\left(YX \mid \mathcal{P}\right) = Y\mathbb{E}\left(X \mid \mathcal{P}\right). \tag{3.46d}$$

If $\mathcal{P}^{\flat}$, $\mathcal{P}^{\sharp}$ are two partitions of Ω, we have that

$$\mathcal{P}^{\flat} \preceq \mathcal{P}^{\sharp} \Rightarrow \mathbb{E}\left(\mathbb{E}\left(X \mid \mathcal{P}^{\sharp}\right) \mid \mathcal{P}^{\flat}\right) = \mathbb{E}\left(X \mid \mathcal{P}^{\flat}\right). \tag{3.46e}$$

Definition 3.50 When Y is a random variable, we define the conditional expectation of X w.r.t. Y by

$$\mathbb{E}\left(X \mid Y\right) := \mathbb{E}\left(X \mid \mathsf{part}(Y)\right). \tag{3.47}$$

By Proposition 3.38, the random variable $\mathbb{E}\left(X \mid Y\right)$ is measurable w.r.t. the random variable Y. Therefore, by Proposition 3.46, there exists a measurable mapping $\mathfrak{f} \colon \operatorname{im} Y \to \mathbb{R}$ such that

$$\mathbb{E}\left(X \mid Y\right) = \mathfrak{f} \circ Y. \tag{3.48}$$

Let us now give an explicit expression of the mapping $\mathfrak{f}$. Since $\mathsf{part}(Y) = \left\{Y^{-1}(y) \mid y \in \operatorname{im} Y\right\}$ by (3.36), the random variable $\mathbb{E}\left(X \mid Y\right)$ takes the value

$$\mathbb{E}\left(X \mid Y\right)(\omega) = \frac{\mathbb{E}\left(\mathbf{1}_{Y=y} X\right)}{\mathbb{P}\left(Y = y\right)}, \quad \forall \omega \text{ such that } Y(\omega) = y, \forall y \in \operatorname{im} Y. \tag{3.49}$$

One traditionally denotes

$$\mathbb{E}\left(X \mid Y = y\right) := \frac{\mathbb{E}\left(\mathbf{1}_{Y=y} X\right)}{\mathbb{P}\left(Y = y\right)}, \quad \forall y \in \operatorname{im} Y. \tag{3.50}$$

Putting $\mathfrak{f}(y) := \mathbb{E}\left(X \mid Y = y\right)$, we obtain (3.48). Notice however that the mapping $\mathfrak{f}$ depends on the couple of random variables (X, Y) by (3.50). Therefore, the equality (3.48) may be misleading, in letting one think that the dependence w.r.t. Y is only through the values $Y(\omega)$ in $\mathbb{E}\left(X \mid Y\right)(\omega) = \mathfrak{f}\big(Y(\omega)\big)$, whereas the dependence is also functional through the mapping $\mathfrak{f}$ in (3.50). We refer the reader back to the discussion at p. xv of the preamble on notations. To be very explicit, let us define

$$\mathfrak{f}_{[X,Y]}(y) := \mathbb{E}\left(X \mid Y = y\right), \quad \forall y \in \operatorname{im} Y. \tag{3.51}$$

Now, (3.48) writes $\mathbb{E}\left(X \mid Y\right) = \mathfrak{f}_{[X,Y]} \circ Y$, that is,

$$\mathbb{E}\left(X \mid Y\right)(\omega) = \mathfrak{f}_{[X,Y]}\big(Y(\omega)\big), \quad \forall \omega \in \Omega. \tag{3.52}$$

The issue raised here goes beyond notation problems, and is discussed in Sects. 4.2 and 9.6.

3.5.2 Interchanging Minimization and Conditional Expectation

Now, we provide a formula connecting optimization under measurability constraints and conditional expectation. To keep things simple, we suppose that the universe Ω and the control set $\mathbb{U}$ are both finite. Consider a criterion $j : \mathbb{U} \times \Omega \to \mathbb{R}$.

Let a partition $\mathcal{P} = \{\Omega_i\}_{i=1,\dots,n}$ of Ω be given. Since the universe Ω is finite, any mapping with domain Ω is a random variable. We define

$$\mathcal{U}^{\mathrm{ad}} := \{\boldsymbol{U} : \Omega \to \mathbb{U} \mid \boldsymbol{U} \preceq \mathcal{P}\}, \tag{3.53}$$

the set of random variables which are measurable w.r.t. the partition $\mathcal{P}$. By Proposition 3.35, any $\boldsymbol{U} \in \mathcal{U}^{\mathrm{ad}}$ is constant on every Ω_i, $i = 1, \dots, n$.

Now, we consider the following optimization problem under the measurability constraint $\boldsymbol{U} \preceq \mathcal{P}$:

$$\min_{\boldsymbol{U} \in \mathcal{U}^{\mathrm{ad}}} \mathbb{E}\big(j(\boldsymbol{U}(\cdot), \cdot)\big) = \min_{\boldsymbol{U} \preceq \mathcal{P}} \mathbb{E}\big(j(\boldsymbol{U}(\cdot), \cdot)\big). \tag{3.54}$$

To identify a solution, we introduce the following definition. Since $\mathbb{U}$ is finite, $\min_{u \in \mathbb{U}} \mathbb{E}(\mathbf{1}_{\Omega_i} j(u, \cdot))/\mathbb{P}(\Omega_i)$ is well defined for all $i \in= 1, \dots, n$. Let us denote by $\min_{u \in \mathbb{U}} \mathbb{E}(j(u, \cdot) \mid \mathcal{P})$ the random variable defined by

$$\min_{u \in \mathbb{U}} \mathbb{E}(j(u, \cdot) \mid \mathcal{P})(\omega) = \min_{u \in \mathbb{U}} \frac{\mathbb{E}(\mathbf{1}_{\Omega_i} j(u, \cdot))}{\mathbb{P}(\Omega_i)}, \quad \forall \omega \in \Omega_i, \ \forall i = 1, \dots, n. \tag{3.55}$$

Proposition 3.51 *Let $\boldsymbol{U}^\sharp$ be a $\mathcal{P}$-measurable random variable such that*

$$\boldsymbol{U}^\sharp(\omega) = u_i^\sharp \in \arg\min_{u \in \mathbb{U}} \frac{\mathbb{E}(\mathbf{1}_{\Omega_i} j(u, \cdot))}{\mathbb{P}(\Omega_i)}, \quad \forall \omega \in \Omega_i, \ \forall i = 1, \dots, n. \tag{3.56}$$

Then, we have that

$$\min_{\boldsymbol{U} \preceq \mathcal{P}} \mathbb{E}\big(j(\boldsymbol{U}(\cdot), \cdot)\big) = \mathbb{E}\big(\min_{u \in \mathbb{U}} \mathbb{E}(j(u, \cdot) \mid \mathcal{P})\big),$$

and the minimum is achieved for $\boldsymbol{U}^\sharp$.

Proof By Proposition 3.35, any random variable $\boldsymbol{U} \in \mathcal{U}^{\mathrm{ad}}$ is characterized by a n-uple $\{u_i\}_{i=1,\ldots,n}$, $u_i \in \mathbb{U}$ such that

$$\boldsymbol{U}_{|\Omega_i} \equiv u_i, \ \forall i = 1, \ldots, n. \tag{3.57}$$

We can assume that $\mathbb{P}(\Omega_i) > 0$ for all $i = 1, \ldots, n$.[4] By the definition (3.55) of $\min_{u\in\mathbb{U}} \mathbb{E}(j(u,\cdot) \mid \mathcal{P})$, we can write

$$\min_{u\in\mathbb{U}} \mathbb{E}(j(u,\cdot) \mid \mathcal{P})(\omega) = \sum_{i=1}^{n} \min_{u\in\mathbb{U}} \frac{\mathbb{E}(\mathbf{1}_{\Omega_i}(\cdot) j(u,\cdot))}{\mathbb{P}(\Omega_i)} \mathbf{1}_{\Omega_i}(\omega), \ \forall \omega \in \Omega.$$

Thus, for all $\{u_i\}_{i=1,\ldots,n}$, $u_i \in \mathbb{U}$, we have that

$$\min_{u\in\mathbb{U}} \mathbb{E}(j(u,\cdot) \mid \mathcal{P})(\omega) \leq \sum_{i=1}^{n} \mathbf{1}_{\Omega_i}(\omega) \frac{\mathbb{E}(\mathbf{1}_{\Omega_i}(\cdot) j(u_i,\cdot))}{\mathbb{P}(\Omega_i)}, \ \forall \omega \in \Omega.$$

With the random variable $\boldsymbol{U}$ such that $\boldsymbol{U}_{|\Omega_i} \equiv u_i, \ \forall i = 1, \ldots, n$, we get

$$\min_{u\in\mathbb{U}} \mathbb{E}(j(u,\cdot) \mid \mathcal{P}) \leq \sum_{i=1}^{n} \mathbf{1}_{\Omega_i} \frac{\mathbb{E}(\mathbf{1}_{\Omega_i} j(\boldsymbol{U}(\cdot),\cdot))}{\mathbb{P}(\Omega_i)},$$

and, using again (3.45), we obtain

$$\min_{u\in\mathbb{U}} \mathbb{E}(j(u,\cdot) \mid \mathcal{P}) \leq \mathbb{E}(j(\boldsymbol{U}(\cdot),\cdot) \mid \mathcal{P}). \tag{3.58}$$

By (3.46c), taking expectation on both sides of Inequality (3.58) we obtain

$$\mathbb{E}\big(\min_{u\in\mathbb{U}} \mathbb{E}(j(u,\cdot) \mid \mathcal{P})\big) \leq \mathbb{E}\big(j(\boldsymbol{U}(\cdot),\cdot)\big), \ \forall \boldsymbol{U} \preceq \mathcal{P}. \tag{3.59}$$

Now, if $\boldsymbol{U}$ is replaced by $\boldsymbol{U}^\sharp$ defined in (3.56), then inequalities are replaced by equalities in Eqs. (3.58) and (3.59). Thus, we conclude that

$$\mathbb{E}\big(\min_{u\in\mathbb{U}} \mathbb{E}(j(u,\cdot) \mid \mathcal{P})\big) = \min_{\boldsymbol{U}\preceq\mathcal{P}} \mathbb{E}\big(j(\boldsymbol{U}(\cdot),\cdot)\big) = \mathbb{E}\big(j(\boldsymbol{U}^\sharp(\cdot),\cdot)\big). \qquad \square$$

Let us consider the special case of the complete partition $\mathcal{P} = \{\{\omega\}\}_{\omega\in\Omega}$, which is a finite partition since Ω is supposed finite. Then, we have that $\mathbb{E}(j(u,\cdot) \mid \mathcal{P})(\omega) = j(u,\omega)$ and the previous result gives, as a special case, an interchange theorem for minimization and expectation:

---

[4]Indeed, if $\mathbb{P}(\Omega_j) = 0$, then the constant value u_j taken by the random variable $\boldsymbol{U}$ on $\mathbb{P}(\Omega_j)$ has no effect on the cost of the optimization problem (3.54).

$$\mathbb{E}\big(\min_{u\in\mathbb{U}} j(u,\cdot)\big) = \min_{\boldsymbol{U}} \mathbb{E}\big(j(\boldsymbol{U}(\cdot),\cdot)\big). \tag{3.60}$$

Here, the notation $\min_{\boldsymbol{U}}$ means that we consider any random variable $\boldsymbol{U}$, without measurability constraint.

3.5.3 Conditional Expectation as an Optimal Value of a Minimization Problem

The following basic example has already been examined in Sect. 1.3.1.

Consider a partition $\mathcal{P} = \{\Omega_i\}_{i=1,\dots,n}$ and a random variable $\boldsymbol{X}$. We use the interchange properties to characterize the conditional expectation w.r.t. the partition $\mathcal{P}$ as the minimizer of a minimization problem with measurability constraints. We consider Problem (3.54) with $j(u,\omega) = (u - \boldsymbol{X}(\omega))^2$, that is:

$$\min_{\boldsymbol{U}\preceq\mathcal{P}} \mathbb{E}\big((\boldsymbol{U}-\boldsymbol{X})^2\big). \tag{3.61}$$

Using Formula (3.51), we have to solve

$$\min_{u} \mathbb{E}((u-\boldsymbol{X})^2 \mid \mathcal{P}). \tag{3.62}$$

By (3.46a), we easily check that

$$\mathbb{E}\big((u-\boldsymbol{X})^2 \mid \mathcal{P}\big) = u^2 - 2u\mathbb{E}(\boldsymbol{X} \mid \mathcal{P}) + \mathbb{E}(\boldsymbol{X}^2 \mid \mathcal{P}).$$

For each Ω_i, the minimum of (3.62) is achieved at

$$u_i^\sharp = \frac{\mathbb{E}(\mathbf{1}_{\Omega_i}\boldsymbol{X})}{\mathbb{P}(\Omega_i)} = \arg\min_{u_i\in\mathbb{U}} \left\{u_i^2 - 2u_i\frac{\mathbb{E}(\mathbf{1}_{\Omega_i}\boldsymbol{X})}{\mathbb{P}(\Omega_i)} + \frac{\mathbb{E}(\mathbf{1}_{\Omega_i}\boldsymbol{X}^2)}{\mathbb{P}(\Omega_i)}\right\}.$$

By (3.44), we observe that $u_i^\sharp = \mathbb{E}\left(\boldsymbol{X} \mid \mathcal{P}\right)(\omega)$, $\forall\omega\in\Omega_i$. Therefore, the minimum of Problem (3.62) is equal to the *conditional variance*

$$\mathbb{E}\big((\mathbb{E}(\boldsymbol{X} \mid \mathcal{P}) - \boldsymbol{X})^2 \mid \mathcal{P}\big), \tag{3.63}$$

and is achieved by the $\mathcal{P}$-measurable random variable $\mathbb{E}(\boldsymbol{X} \mid \mathcal{P})$.

3.6 Conclusion

We have begun our exploration of tools for information handling by the *partitions and fields* approach. A *partition* splits the universe Ω in subsets of indistinguishable states of Nature. The set of all partitions is a *lattice*: the *least upper bound* of two partitions is made of all nonempty intersections between their respective elements; the definition of the *greatest lower bound* is more abstruse. Partitions are equivalent to partition fields, for which the greatest lower bound is easier to compute than the least upper bound. The axiomatics of partition fields is quite close to that of the well-known *σ-fields*. Partition fields are easier to handle than σ-fields, but σ-fields are tailored for Probability Theory.

In the *mapping measurability* approach, information is handled by mappings over the universe Ω. We have provided tools to pass from an *algebraic* aspect of measurability to a *functional* one. The measurability framework developed here will be widely used in the book. Table 3.1 sums up some correspondences between partitions and fields and mapping measurability approaches to information.

We have concluded with a non technical introduction of *conditional expectation* w.r.t. a partition. In this case, we easily obtain a formula connecting optimization under measurability constraints and conditional expectation. The possibility of *interchanging minimization and conditional expectation* will prove very useful.

In the rest of the book, we will consider either π-fields (partition fields) or σ-fields, depending on the context. In any case, once a π (or σ) field $\mathcal{A}$ is given on the universe Ω, we call *subfield* any π (or σ) field included in $\mathcal{A}$.

Chapter 4
Information and Stochastic Optimization Problems

4.1 Introduction

In Chap. 2, we presented static stochastic optimization problems with open-loop control solutions. In Chap. 3, we introduced various tools to handle information. Now, we examine *dynamic* stochastic decision issues characterized by the sequence: information $\rightarrow$ decision $\rightarrow$ information $\rightarrow$ decision $\rightarrow$ etc. This chapter focuses on the interplay between information and decision. First, we provide a "guided tour" of stochastic dynamic optimization issues by examining a simple one-dimensional, two-period linear dynamical system with a quadratic criterion. We examine the celebrated Witsenhausen counterexample, then describe how different information patterns deeply modify the optimal solutions. Second, we present the classical state control dynamical model. Within this formalism, when an optimal solution is searched for among functions of the state, optimization problems with time-additive criterion can be solved by Dynamic Programming (DP), by means of the well-known Bellman equation. This equation connects the value functions between two successive times by means of a static optimization problem over the control set and parameterized by the state. This provides an optimal feedback. We conclude this chapter with more advanced material. We present a more general form of optimal stochastic control problems relative to the state model. Following Witsenhausen, we recall that a Dynamic Programming equation also holds in such a context, due to sequentiality. This equation also connects the value functions between two successive times by means of a static optimization problem. However, the optimization is over a set of feedbacks, and it is parameterized by an information state, the dimension of which is much larger than that of the original state.

P. Carpentier et al., *Stochastic Multi-Stage Optimization*, Probability Theory and Stochastic Modelling 75, DOI 10.1007/978-3-319-18138-7_4

4.2 The Witsenhausen Counterexample

The Witsenhausen counterexample was already considered in Sect. 1.3.3. We examine it in more detail here.

4.2.1 A Simple Linear Quadratic Control Problem

Consider $(\Omega, \mathcal{A}, \mathbb{P})$ a probability space, and five (real) random variables related by the equations

$$X_1 = X_0 + U_0, \tag{4.1a}$$
$$X_2 = X_1 - U_1. \tag{4.1b}$$

Here U_0 and U_1 are called *control*, or *decision*, random variables, whereas X_0 is a *primitive* random variable (also called *noise*) called the *initial state*. The random variables X_1, X_2 are constructed by the dynamical system (4.1), and are called *states*.

Consider the optimization problem

$$\min_{U_0, U_1} \mathbb{E}\left(k^2 U_0^2 + X_2^2\right). \tag{4.2}$$

Since $k^2 U_0^2 + X_2^2$ is a nonnegative random variable, the expectation is well defined. As in Sect. 1.3.3, we interpret the optimization problem (4.1) and (4.2) as one where the first decision U_0 is costly, with quadratic cost $k^2 U_0^2$, whereas the penalty for the second decision U_1 to be far from X_1 is measured by the cost $X_2^2 = (X_1 - U_1)^2$. Of course, without restrictions on U_0 and U_1 (information pattern), the optimal solution is $U_0 = 0$ and $U_1 = X_1$, since this yields the best possible cost, namely 0.

Since we are tackling a stochastic optimization problem over a a probability space $(\Omega, \mathcal{A}, \mathbb{P})$, we represent information by means of σ-fields (see Sect. 3.3.3) or of random variables (see Sect. 3.4.3). In [155], Witsenhausen introduces the two *observation* random variables

$$Y_0 = X_0, \tag{4.3a}$$
$$Y_1 = X_1 + W_1, \tag{4.3b}$$

where W_1 is a *primitive* random variable (*noise*). We assume that[1] $\mathbb{E}\left(X_0^2\right) < +\infty$ and that

$$\sigma(X_0 + W_1) \neq \sigma(X_0, W_1). \tag{4.4}$$

[1] In [155], the random vector (X_0, W_1) is supposed to be Gaussian under the probability $\mathbb{P}$, but we do not need this assumption in what follows.

This measurability condition is satisfied if Ω is rich enough, for instance if $\Omega = \mathbb{R}^2$ and $\boldsymbol{X}_0$, $\boldsymbol{W}_1$ are the coordinates mappings.

Then, the so-called Witsenhausen counterexample is the optimization problem

$$\min_{\boldsymbol{U}_0 \preceq \boldsymbol{Y}_0, \boldsymbol{U}_1 \preceq \boldsymbol{Y}_1} \mathbb{E}\left(k^2 \boldsymbol{U}_0^2 + \boldsymbol{X}_2^2\right). \tag{4.5}$$

The *measurability constraints*

$$\boldsymbol{U}_0 \preceq \boldsymbol{Y}_0 \text{ and } \boldsymbol{U}_1 \preceq \boldsymbol{Y}_1 \tag{4.6}$$

have to be understood in the sense of Definition 3.44, that is, as $\sigma(\boldsymbol{U}_0) \subset \sigma(\boldsymbol{Y}_0)$ and $\sigma(\boldsymbol{U}_1) \subset \sigma(\boldsymbol{Y}_1)$. By Proposition 3.46, it means that the first decision $\boldsymbol{U}_0$ is a measurable function of $\boldsymbol{Y}_0$, and $\boldsymbol{U}_1$ of $\boldsymbol{Y}_1$:

$$\boldsymbol{U}_0 = \upsilon_0(\boldsymbol{Y}_0) = \upsilon_0(\boldsymbol{X}_0) \text{ and } \boldsymbol{U}_1 = \upsilon_1(\boldsymbol{Y}_1) = \upsilon_1(\boldsymbol{X}_1 + \boldsymbol{W}_1). \tag{4.7}$$

To alleviate notations, Witsenhausen proposes to replace the control $\boldsymbol{U}_0 = \upsilon_0(\boldsymbol{X}_0)$ by the control

$$\boldsymbol{U}_0' = \boldsymbol{X}_0 + \upsilon_0(\boldsymbol{X}_0), \tag{4.8}$$

so that the system becomes

$$\boldsymbol{Y}_0 = \boldsymbol{X}_0, \tag{4.9a}$$

$$\boldsymbol{Y}_1 = \boldsymbol{X}_1 + \boldsymbol{W}_1, \tag{4.9b}$$

$$\boldsymbol{X}_1 = \boldsymbol{U}_0', \;\; \boldsymbol{U}_0' \preceq \boldsymbol{Y}_0, \tag{4.9c}$$

$$\boldsymbol{X}_2 = \boldsymbol{X}_1 - \boldsymbol{U}_1, \;\; \boldsymbol{U}_1 \preceq \boldsymbol{Y}_1. \tag{4.9d}$$

These measurability constraints $\boldsymbol{U}_0' \preceq \boldsymbol{Y}_0$ and $\boldsymbol{U}_1 \preceq \boldsymbol{Y}_1$ form an *information pattern* where two decision makers jointly work as a team to minimize the expected cost in (4.5): the first one only observes the initial state $\boldsymbol{Y}_0 = \boldsymbol{X}_0$, whereas the second one observes a corrupted version $\boldsymbol{Y}_1 = \boldsymbol{X}_1 + \boldsymbol{W}_1$ of the state $\boldsymbol{X}_1$. Rewriting the expected cost in (4.5) as

$$\mathbb{E}\left(k^2 \boldsymbol{U}_0^2 + \boldsymbol{X}_2^2\right) = \mathbb{E}\left(k^2 (\boldsymbol{X}_0 - \boldsymbol{U}_0')^2 + (\boldsymbol{X}_1 - \boldsymbol{U}_1)^2\right), \tag{4.10}$$

we see that the first decision maker tries to cancel $\boldsymbol{X}_0$ with the control $\boldsymbol{U}_0'$, whereas the second one tries to cancel $\boldsymbol{X}_1$ with the control $\boldsymbol{U}_1$. The interaction comes from the dynamical and measurability constraints in (4.9).

4.2.2 Problem Transformation Exploiting Sequentiality

In Sect. 4.5, we present and study sequential systems. Anticipating on a formal definition, let us point out that the system (4.9) is sequential because the first observation $\boldsymbol{Y}_0 = \boldsymbol{X}_0$ does not depend on either of the two decisions $\boldsymbol{U}_0'$ and $\boldsymbol{U}_1$ (but only on the primitive random variable $\boldsymbol{X}_0$), whereas the second observation

$$\boldsymbol{Y}_1 = \boldsymbol{X}_1 + \boldsymbol{W}_1 = \boldsymbol{U}_0' + \boldsymbol{W}_1 \tag{4.11}$$

does not depend on $\boldsymbol{U}_1$, but does depend on $\boldsymbol{U}_0'$ (and on the primitive random variables $\boldsymbol{X}_0$ and $\boldsymbol{W}_1$).

This sequentiality property, and also time-additivity of the criterion in (4.5), make it possible to decompose the optimization problem (4.5)–(4.9) in two steps as follows.

Proposition 4.1 *We have that*

$$\begin{aligned} &\min_{\boldsymbol{U}_0 \preceq \boldsymbol{Y}_0, \boldsymbol{U}_1 \preceq \boldsymbol{Y}_1} \mathbb{E}\left(k^2 \boldsymbol{U}_0^2 + \boldsymbol{X}_2^2\right) \\ &\quad = \min_{\boldsymbol{U}_0' \preceq \boldsymbol{Y}_0} \mathbb{E}\big(k^2(\boldsymbol{X}_0 - \boldsymbol{U}_0')^2 + \mathrm{Var}\left(\boldsymbol{W}_1 \mid \boldsymbol{U}_0' + \boldsymbol{W}_1\right)\big), \end{aligned} \tag{4.12}$$

where the conditional variance *is given by*

$$\mathrm{Var}\left(\boldsymbol{W}_1 \mid \boldsymbol{U}_0' + \boldsymbol{W}_1\right) = \mathbb{E}\left(\boldsymbol{W}_1^2 \mid \boldsymbol{U}_0' + \boldsymbol{W}_1\right) - \left(\mathbb{E}\left(\boldsymbol{W}_1 \mid \boldsymbol{U}_0' + \boldsymbol{W}_1\right)\right)^2.$$

Proof Since $\boldsymbol{X}_2 = \boldsymbol{X}_1 - \boldsymbol{U}_1$, we have that

$$\begin{aligned} &\min_{\boldsymbol{U}_0 \preceq \boldsymbol{Y}_0, \boldsymbol{U}_1 \preceq \boldsymbol{Y}_1} \mathbb{E}\left(k^2 \boldsymbol{U}_0^2 + \boldsymbol{X}_2^2\right) \\ &\quad = \min_{\boldsymbol{U}_0' \preceq \boldsymbol{Y}_0, \boldsymbol{U}_1 \preceq \boldsymbol{Y}_1} \mathbb{E}\left(k^2(\boldsymbol{X}_0 - \boldsymbol{U}_0')^2 + (\boldsymbol{X}_1 - \boldsymbol{U}_1)^2\right), \end{aligned}$$

where, by sequentiality ($\boldsymbol{Y}_0$ does not depend upon $\boldsymbol{U}_1$),

$$= \min_{\boldsymbol{U}_0' \preceq \boldsymbol{Y}_0} \min_{\boldsymbol{U}_1 \preceq \boldsymbol{Y}_1} \mathbb{E}\left(k^2(\boldsymbol{X}_0 - \boldsymbol{U}_0')^2 + (\boldsymbol{X}_1 - \boldsymbol{U}_1)^2\right),$$

and, since $\boldsymbol{U}_0'$ does not depend on $\boldsymbol{U}_1$,

$$= \min_{\boldsymbol{U}_0' \preceq \boldsymbol{Y}_0} \left(k^2 \mathbb{E}\left((\boldsymbol{X}_0 - \boldsymbol{U}_0')^2\right) + \min_{\boldsymbol{U}_1 \preceq \boldsymbol{Y}_1} \mathbb{E}\left((\boldsymbol{X}_1 - \boldsymbol{U}_1)^2\right)\right).$$

Thus, by an interchange of minimization and expectation (see (3.51) in the finite case, and Sect. 8.3.5 in the general case), we are led to consider the problem

$$\min_{U_1 \preceq Y_1} \mathbb{E}\left((X_1 - U_1)^2\right) = \mathbb{E}\left(\min_{u_1} \mathbb{E}\left((X_1 - u_1)^2 \mid Y_1\right)\right). \tag{4.13}$$

The argmin of the inner optimization problem is $\mathbb{E}\left(X_1 \mid Y_1\right)$, and the associated optimal cost is, by definition (see (3.63)), the *conditional variance*

$$\text{Var}\left(X_1 \mid Y_1\right) := \mathbb{E}\left(X_1^2 \mid Y_1\right) - \left(\mathbb{E}\left(X_1 \mid Y_1\right)\right)^2. \tag{4.14}$$

Therefore, we have that

$$\begin{aligned}
&\min_{U_0' \preceq Y_0, U_1 \preceq Y_1} \mathbb{E}\left(k^2(X_0 - U_0')^2 + (X_1 - U_1)^2\right) \\
&= \min_{U_0' \preceq Y_0} \left(k^2 \mathbb{E}\left((X_0 - U_0')^2\right) + \mathbb{E}\left(\text{Var}\left(X_1 \mid Y_1\right)\right)\right) \\
&= \min_{U_0' \preceq Y_0} \mathbb{E}\left(k^2(X_0 - U_0')^2 + \text{Var}\left(X_1 \mid Y_1\right)\right) \\
&= \min_{U_0' \preceq Y_0} \mathbb{E}\left(k^2(X_0 - U_0')^2 + \text{Var}\left(X_1 \mid U_0' + W_1\right)\right) \quad \text{by (4.11)} \\
&= \min_{U_0' \preceq Y_0} \mathbb{E}\left(k^2(X_0 - U_0')^2 + \text{Var}\left(U_0' \mid U_0' + W_1\right)\right)
\end{aligned}$$

because $X_1 = U_0'$ in (4.9)

$$= \min_{U_0' \preceq Y_0} \mathbb{E}\left(k^2(X_0 - U_0')^2 + \text{Var}\left(U_0' - (U_0' + W_1) \mid U_0' + W_1\right)\right),$$

because $U_0' + W_1$ is a "constant", once conditioned on $U_0' + W_1$ (and the addition of a constant to a random variable does not change the variance)

$$\begin{aligned}
&= \min_{U_0' \preceq Y_0} \mathbb{E}\left(k^2(X_0 - U_0')^2 + \text{Var}\left(-W_1 \mid U_0' + W_1\right)\right), \\
&= \min_{U_0' \preceq Y_0} \mathbb{E}\left(k^2(X_0 - U_0')^2 + \text{Var}\left(W_1 \mid U_0' + W_1\right)\right),
\end{aligned} \tag{4.15}$$

because a random variable and its opposite have the same variance. □

4.2.3 The Dual Effect of the Initial Decision

As seen in (4.12), the control variable $\boldsymbol{U}_0'$ enters a conditioning term

$$\mathrm{Var}\left(\boldsymbol{X}_1 \mid \boldsymbol{Y}_1\right) = \mathrm{Var}\left(\boldsymbol{W}_1 \mid \boldsymbol{U}_0' + \boldsymbol{W}_1\right). \tag{4.16}$$

This is an example of the so called *dual effect* (see Chaps. 1 and 10) where the initial control $\boldsymbol{U}_0'$ has an impact on future decisions by providing more or less information inside the conditioning term in (4.16), in addition to contributing to the cost minimization directly through the term $k^2(\boldsymbol{X}_0 - \boldsymbol{U}_0')^2$ in (4.12). This dependence of a conditional expectation upon the control variable $\boldsymbol{U}_0'$ makes the resolution of the optimization problem (4.5) very delicate (see also the discussion on policy independence of conditional expectations in Sect. 9.6).

Remark 4.2 Let us materialize the dependence of (4.16) upon $\boldsymbol{U}_0'$ by considering the special case where

- the probability space is $\Omega = \mathbb{R}^2$, equipped with the Borel σ-field and the probability $\mathbb{P}$ Gaussian with zero mean and identity variance-covariance matrix,
- the random variables $(\boldsymbol{X}_0, \boldsymbol{W}_1)$ are the coordinates mappings on Ω.

In this setting, $(\boldsymbol{X}_0, \boldsymbol{W}_1)$ are independent Gaussian random variables (with zero mean and unit variance) under the probability $\mathbb{P}$. For $(y_0, y_1) \in \mathbb{R}^2$, let us set

$$\begin{aligned}
A_{[\boldsymbol{U_0}']}(y_0, y_1) &= \exp\Big(-(y_1^2 + \boldsymbol{U}_0'(y_0)^2 - 2y_1\boldsymbol{U}_0'(y_0))/2\Big), \\
B_{[\boldsymbol{U_0}']}(y_0, y_1) &= \frac{A_{[\boldsymbol{U_0}']}(y_0, y_1)}{\displaystyle\int_{-\infty}^{+\infty} A_{[\boldsymbol{U_0}']}(y_0', y_1)\,\mathrm{d}y_0'}, \\
C_{[\boldsymbol{U_0}']}(y_1) &= \int_{-\infty}^{+\infty} \boldsymbol{U}_0'(y_0')^2 B_{[\boldsymbol{U_0}']}(y_0', y_1)\,\mathrm{d}y_0' \\
&\quad -\left(\int_{-\infty}^{+\infty} \boldsymbol{U}_0'(y_0') B_{[\boldsymbol{U_0}']}(y_0', y_1)\,\mathrm{d}y_0'\right)^2,
\end{aligned}$$

where we refer the reader back to the discussion at p. xv of the preamble on notations, regarding the notation $\left[\boldsymbol{U_0}'\right]$. The calculus gives

$$\mathrm{Var}\left(\boldsymbol{W}_1 \mid \boldsymbol{U}_0' + \boldsymbol{W}_1\right) = C_{[\boldsymbol{U_0}']}(\boldsymbol{U}_0' + \boldsymbol{W}_1). \tag{4.17}$$

This equation has to be understood in the sense that

$$\mathrm{Var}\left(\boldsymbol{W}_1 \mid \boldsymbol{U}_0' + \boldsymbol{W}_1\right)(\omega) = C_{[\boldsymbol{U_0}']}\big(\boldsymbol{U}_0'(\omega) + \boldsymbol{W}_1(\omega)\big), \ \mathbb{P}\text{-a.s.} \tag{4.18}$$

as already stressed at the end of Sect. 3.5.1. ◊

4.3 Other Information Patterns

We take the basic one-dimensional two-period linear dynamical system with a quadratic criterion considered above as an opportunity to present how different information patterns deeply modify the optimal solutions.

In the Witsenhausen counterexample, we interpreted the information pattern as two decision makers working jointly as a team to minimize the expected cost. From now on, we interpret information patterns from the point of view of a single agent with different pieces of information at time $t = 0$ and $t = 1$.

Let $\mathcal{G}_0 = \sigma(\boldsymbol{Y}_0) = \sigma(\boldsymbol{X}_0)$ and $\mathcal{G}_1$ be two subfields of $\mathcal{A}$, and consider the following optimization problem, where $\boldsymbol{X}_1 = \boldsymbol{X}_0 + \boldsymbol{U}_0$ as in (4.1a) and $\boldsymbol{Y}_0 = \boldsymbol{X}_0$ as in (4.3a),

$$\min_{\boldsymbol{U}_0 \preceq \mathcal{G}_0, \boldsymbol{U}_1 \preceq \mathcal{G}_1} \mathbb{E}\left(k^2 \boldsymbol{U}_0^2 + (\boldsymbol{X}_1 - \boldsymbol{U}_1)^2\right). \tag{4.19}$$

In doing so, we restrict the class of decisions to those such that $\boldsymbol{U}_0$ is measurable w.r.t. the random variable $\boldsymbol{Y}_0$ ($\boldsymbol{U}_0 \preceq \boldsymbol{Y}_0$), and $\boldsymbol{U}_1$ is measurable w.r.t. the subfield $\mathcal{G}_1$ ($\boldsymbol{U}_1 \preceq \mathcal{G}_1$).

We now highlight that information patterns are important components of stochastic decision problems. Changing information pattern can turn an easy problem (Sect. 4.3.1) into an open problem (Sect. 4.3.5).

For all the information patterns (especially the subfield $\mathcal{G}_1$) below, we can show easily (mimicking the proof of Proposition 4.1) that

$$\min_{\boldsymbol{U}_0 \preceq \boldsymbol{Y}_0, \boldsymbol{U}_1 \preceq \mathcal{G}_1} \mathbb{E}\left(k^2 \boldsymbol{U}_0^2 + \boldsymbol{X}_2^2\right) = \min_{\boldsymbol{U}_0' \preceq \boldsymbol{Y}_0} \mathbb{E}\big(k^2(\boldsymbol{X}_0 - \boldsymbol{U}_0')^2 + \mathrm{Var}\left(\boldsymbol{X}_1 \mid \mathcal{G}_1\right)\big). \tag{4.20}$$

4.3.1 Full Noise Observation

Suppose that noises $\boldsymbol{X}_0$ and $\boldsymbol{W}_1$ are sequentially observed in the sense that

$$\mathcal{G}_0 = \sigma(\boldsymbol{X}_0) \text{ and } \mathcal{G}_1 = \sigma(\boldsymbol{X}_0, \boldsymbol{W}_1). \tag{4.21}$$

We claim that the couple

$$\boldsymbol{U}_0^\sharp = 0 \quad \text{and} \quad \boldsymbol{U}_1^\sharp = \boldsymbol{X}_0 = \boldsymbol{X}_0 + \boldsymbol{U}_0^\sharp = \boldsymbol{X}_1^\sharp \tag{4.22}$$

is optimal. Indeed, on the one hand, the cost in (4.19) is zero when $\boldsymbol{U}_0 = \boldsymbol{U}_0^\sharp = 0$ and $\boldsymbol{U}_1 = \boldsymbol{U}_1^\sharp = \boldsymbol{X}_1^\sharp$. On the other hand, the measurability constraints are satisfied since $\boldsymbol{U}_0^\sharp = 0 \preceq \mathcal{G}_0$ and $\boldsymbol{U}_1^\sharp = \boldsymbol{X}_0 \preceq \mathcal{G}_1 = \sigma(\boldsymbol{X}_0, \boldsymbol{W}_1)$.

Notice that the last term in (4.20) is now

$$\mathrm{Var}\left(\boldsymbol{X}_1 \mid \mathcal{G}_1\right) = \mathrm{Var}\left(\boldsymbol{X}_1 \mid \boldsymbol{X}_0, \boldsymbol{W}_1\right) = 0, \tag{4.23}$$

since $\boldsymbol{X}_1$—given by (4.1a) where $\boldsymbol{U}_0 \preceq \boldsymbol{X}_0$—is measurable w.r.t. $(\boldsymbol{X}_0, \boldsymbol{W}_1)$, hence is a "constant", once conditioned by $(\boldsymbol{X}_0, \boldsymbol{W}_1)$.

4.3.2 Classical Information Pattern

The classical pattern is defined as

$$\mathcal{G}_0 = \sigma(\boldsymbol{Y}_0) = \sigma(\boldsymbol{X}_0) \text{ and } \mathcal{G}_1 = \sigma(\boldsymbol{Y}_0, \boldsymbol{Y}_1), \tag{4.24}$$

where $\boldsymbol{Y}_1 = \boldsymbol{X}_1 + \boldsymbol{W}_1$ as in (4.3b). In the classical pattern, there is *perfect memory* in the sense that the observation $\boldsymbol{Y}_0$ at time $t = 0$ still is available at time $t = 1$. By explicitly writing

$$\mathcal{G}_1 = \sigma(\boldsymbol{Y}_0, \boldsymbol{Y}_1) = \sigma(\boldsymbol{X}_0, \boldsymbol{X}_0 + \boldsymbol{U}_0 + \boldsymbol{W}_1), \tag{4.25}$$

we see how the subfield $\mathcal{G}_1$ depends upon $\boldsymbol{U}_0$. However, we show now that $\mathcal{G}_1$ does not depend upon $\boldsymbol{U}_0$ when the control is restricted to $\boldsymbol{U}_0 \preceq \mathcal{G}_0$. Indeed, by Proposition 3.46, this last condition means that the first decision $\boldsymbol{U}_0 = \upsilon_0(\boldsymbol{Y}_0)$ is a measurable function of $\boldsymbol{Y}_0 = \boldsymbol{X}_0$. Therefore, when $\boldsymbol{U}_0 = \upsilon_0(\boldsymbol{X}_0)$, and using Proposition 3.48 with the mapping $\varphi(x_0, w_1) = (x_0, x_0 + \upsilon_0(w_0) + w_1)$, we obtain:

$$\mathcal{G}_1 = \sigma(\boldsymbol{X}_0, \boldsymbol{X}_0 + \upsilon_0(\boldsymbol{X}_0) + \boldsymbol{W}_1) = \sigma(\boldsymbol{X}_0, \boldsymbol{W}_1). \tag{4.26}$$

Therefore, the optimal solution is given by (4.22), being the same as in the full noise observation pattern case of Sect. 4.3.1. Notice that the subfield $\mathcal{G}_1$ does not depend on the control $\boldsymbol{U}_0$ (more generally, linear dynamical systems with linear observations and perfect memory also display observation fields that are independent of past controls [128]).

Notice that the last term in (4.20) is zero since the subfield $\mathcal{G}_1 = \sigma(\boldsymbol{X}_0, \boldsymbol{W}_1)$ is the same as in Sect. 4.3.1.

4.3.3 Markovian Information Pattern

The Markovian pattern is the case where the state is sequentially and perfectly observed (that is, without noise):

$$\mathcal{G}_0 = \sigma(\boldsymbol{X}_0) \text{ and } \mathcal{G}_1 = \sigma(\boldsymbol{X}_1)\,. \tag{4.27}$$

In contrast to the full noise observation pattern, the subfield

$$\mathcal{G}_1 = \sigma(\boldsymbol{X}_1) = \sigma(\boldsymbol{X}_0 + \boldsymbol{U}_0) \tag{4.28}$$

depends upon the initial decision $\boldsymbol{U}_0$, and the dual effect holds true. However, the optimal solution is given by (4.22). Indeed, the cost in (4.19) is zero and the measurability constraints are satisfied since $\boldsymbol{U}_0^\sharp = 0 \preceq \mathcal{G}_0 = \sigma(\boldsymbol{X}_0)$ and $\boldsymbol{U}_1^\sharp = \boldsymbol{X}_1^\sharp \preceq \mathcal{G}_1 = \sigma(\boldsymbol{X}_1^\sharp)$. The Markovian pattern is examined in more detail in Sect. 4.4.

Notice that the last term in (4.20) is now

$$\text{Var}\left(\boldsymbol{X}_1 \mid \mathcal{G}_1\right) = \text{Var}\left(\boldsymbol{X}_1 \mid \boldsymbol{X}_1\right) = 0, \tag{4.29}$$

since $\boldsymbol{X}_1$ is measurable w.r.t. $\sigma(\boldsymbol{X}_1)$. Therefore, although the dual effect holds true, and the conditioning is w.r.t. $\boldsymbol{X}_1 = \boldsymbol{X}_0 + \boldsymbol{U}_0$ which depends on the control $\boldsymbol{U}_0$, the resulting conditional expectation $\text{Var}\left(\boldsymbol{X}_1 \mid \boldsymbol{X}_1\right)$ does not depend on the control $\boldsymbol{U}_0$. The question of the policy independence of conditional expectations is discussed in more detail in Sect. 9.6.

4.3.4 Past Control Observation

In this nonclassical pattern defined by

$$\mathcal{G}_0 = \sigma(\boldsymbol{Y}_0) \text{ and } \mathcal{G}_1 = \sigma(\boldsymbol{U}_0, \boldsymbol{Y}_1), \tag{4.30}$$

the second decision maker observes a corrupted version $\boldsymbol{Y}_1 = \boldsymbol{X}_1 + \boldsymbol{W}_1$ of the state $\boldsymbol{X}_1$ and the decision $\boldsymbol{U}_0$ taken by the first decision maker. We have that

$$\mathcal{G}_1 = \sigma(\boldsymbol{U}_0, \boldsymbol{Y}_1) = \sigma(\boldsymbol{U}_0, \boldsymbol{X}_0 + \boldsymbol{U}_0 + \boldsymbol{W}_1) = \sigma(\boldsymbol{U}_0, \boldsymbol{X}_0 + \boldsymbol{W}_1). \tag{4.31}$$

By Proposition 3.46, we know that the first decision $\boldsymbol{U}_0 = \upsilon_0(\boldsymbol{Y}_0)$ is a measurable function of $\boldsymbol{Y}_0 = \boldsymbol{X}_0$. We have a dual effect, because

- if υ_0 is a one-to-one measurable mapping with measurable inverse υ_0^{-1}, then, using Proposition 3.48, we obtain

$$\mathcal{G}_1 = \sigma(\upsilon_0(\boldsymbol{X}_0), \boldsymbol{X}_0 + \boldsymbol{W}_1) = \sigma(\boldsymbol{X}_0, \boldsymbol{X}_0 + \boldsymbol{W}_1) = \sigma(\boldsymbol{X}_0, \boldsymbol{W}_1), \tag{4.32}$$

- if υ_0 is a constant mapping, we have that

$$\mathcal{G}_1 = \sigma(\upsilon_0(\boldsymbol{X}_0), \boldsymbol{X}_0 + \boldsymbol{W}_1) = \sigma(\boldsymbol{X}_0 + \boldsymbol{W}_1) \neq \sigma(\boldsymbol{X}_0, \boldsymbol{W}_1), \tag{4.33}$$

as we have assumed in (4.4).

In [24], it is shown that the optimization problem (4.19) with the information pattern (4.30) admits only ε-optimal decisions. Indeed, for $\varepsilon > 0$, take

$$\boldsymbol{U}_0 = \varepsilon \boldsymbol{Y}_0 = \varepsilon \boldsymbol{X}_0 \text{ and } \boldsymbol{U}_1 = \frac{\boldsymbol{U}_0}{\varepsilon} = \boldsymbol{X}_0,$$

which yields the cost

$$k^2 \boldsymbol{U}_0^2 + (\boldsymbol{X}_0 + \boldsymbol{U}_0 - \boldsymbol{U}_1)^2 = \varepsilon^2 (k^2 + 1) \boldsymbol{X}_0^2,$$

whose expectation is of order ε^2 (we have assumed that $\mathbb{E}\left(\boldsymbol{X}_0^2\right) < +\infty$). The only solution with zero cost is given by (4.22). However, the measurability constraints are not satisfied since $\boldsymbol{U}_0^\sharp = 0 \preceq \mathcal{G}_0 = \sigma(\boldsymbol{X}_0)$, but $\boldsymbol{U}_1^\sharp = \boldsymbol{X}_0 \npreceq \mathcal{G}_1 = \sigma(\boldsymbol{U}_0^\sharp, \boldsymbol{Y}_1{}^\sharp) = \sigma(\boldsymbol{X}_0 + \boldsymbol{W}_1)$.

Notice that the last term in (4.20) is now

$$\operatorname{Var}\left(\boldsymbol{X}_1 \mid \mathcal{G}_1\right) = \operatorname{Var}\left(\boldsymbol{X}_0 + \boldsymbol{U}_0 \mid \boldsymbol{U}_0, \boldsymbol{X}_0 + \boldsymbol{W}_1\right), \tag{4.34}$$

which cannot be computed easily.

4.3.5 The Witsenhausen Counterexample

The Witsenhausen counterexample, discussed in Sect. 4.2, is characterized by

$$\mathcal{G}_0 = \sigma(\boldsymbol{Y}_0) = \sigma(\boldsymbol{X}_0) \text{ and } \mathcal{G}_1 = \sigma(\boldsymbol{Y}_1) = \sigma(\boldsymbol{X}_0 + \boldsymbol{U}_0 + \boldsymbol{W}_1). \tag{4.35}$$

An optimal solution exists [155], but its expression is unknown.

4.4 State Model and Dynamic Programming (DP)

Here, we present the classical state control dynamical model. Within this formalism, optimization problems with time-additive criterion[2] can be solved by Dynamic Programming [15, 20, 127, 152, 153].

[2] In fact, although time-additive criteria are widely used, other criteria are also adapted to Dynamic Programming [20, 51, 152, 153].

In Sects. 4.2 and 4.3 above, the informational constraints were handled *algebraically* (some random variables must be *measurable* w.r.t. other random variables), whereas, in what follows, they are treated *functionally* (essentially, some random variables are represented as *functions* of other random variables, called *feedbacks*, *decision rules*, *policies*, or *strategies*).

4.4.1 State Model

We consider the *control state system* or *state model*:

$$x_{t+1} = f_t(x_t, u_t, w_{t+1}), \quad t = 0, \ldots, T-1. \tag{4.36}$$

We denote by

- the variable t, the *time index*, belonging to the subset $\{0, \ldots, T\}$ of the set $\mathbb{N}$ of integers; 0 is the *initial time* and $T > 0$ stands for the *time horizon*, that we suppose finite ($T < +\infty$);
- the variable u_t, the *control* or *decision*, chosen by the decision maker in the *decision space* denoted by $\mathbb{U}_t$; we consider a finite dimensional space, namely $\mathbb{U}_t = \mathbb{R}^{p_t}$;
- the variable w_t, the *uncertainty*, taking its values in a given set $\mathbb{W}_t = \mathbb{R}^{q_t}$;
- the variable x_t, the *"state"*,[3] embodying a set of variables which sum up the information needed together with the control and uncertainty to proceed from time t to the following time $t+1$; this state is an element of the *state space* denoted by $\mathbb{X}_t = \mathbb{R}^{n_t}$;
- the mappings $f_t : \mathbb{X}_t \times \mathbb{U}_t \times \mathbb{W}_{t+1} \to \mathbb{X}_{t+1}$, the *dynamics*, which represent the system evolution; in the sequel, we call *dynamics* the sequence $f := \{f_t\}_{t=0,\ldots,T-1}$; in many cases, the dynamics f_t does not depend on time t and is said to be *autonomous* or *stationary*;
- the variable x_0, the *initial state* or *initial condition*.

From now on, we suppose that all the spaces considered here—the decision spaces $\mathbb{U}_t = \mathbb{R}^{p_t}$, the state spaces $\mathbb{X}_t = \mathbb{R}^{n_t}$, and the uncertainty spaces $\mathbb{W}_t = \mathbb{R}^{q_t}$—are equipped with their corresponding Borel σ-fields (see Definition B.1), denoted, respectively, by $\mathcal{U}_t = \mathcal{B}^{\mathrm{o}}_{\mathbb{U}_t}$, $\mathcal{X}_t = \mathcal{B}^{\mathrm{o}}_{\mathbb{X}_t}$ and $\mathcal{W}_t = \mathcal{B}^{\mathrm{o}}_{\mathbb{W}_t}$. Proceeding along, we suppose that the dynamics f_t are measurable mappings. All the product sets below, $\prod_{t=0}^{T-1} \mathbb{U}_t$, $\prod_{t=0}^{T} \mathbb{X}_t$, and $\prod_{t=1}^{T} \mathbb{W}_t$ are equipped with their respective product σ-fields $\bigotimes_{t=0}^{T-1} \mathcal{U}_t$, $\bigotimes_{t=0}^{T} \mathcal{X}_t$, and $\bigotimes_{t=1}^{T} \mathcal{W}_t$.

A *state trajectory* is a sequence

$$x(\cdot) := (x_0, x_1, \ldots, x_{T-1}, x_T) \in \prod_{t=0}^{T} \mathbb{X}_t. \tag{4.37}$$

[3] See footnote 3 in Sect. 1.2.1.

A *control history* is a sequence

$$u(\cdot) := \big(u_0, u_1, \dots, u_{T-1}\big) \in \prod_{t=0}^{T-1} \mathbb{U}_t. \tag{4.38}$$

A *scenario* is a sequence

$$w(\cdot) := \big(w_1, w_2, \dots, w_T\big) \in \prod_{t=1}^{T} \mathbb{W}_t, \tag{4.39}$$

or it may also be a sequence

$$\big(x_0, w(\cdot)\big) := \big(x_0, w_1, w_2, \dots, w_T\big) \in \mathbb{X}_0 \times \prod_{t=1}^{T} \mathbb{W}_t, \tag{4.40}$$

when the initial state x_0 is supposed to be uncertain.

Remark 4.3 In practice, one often considers stationary spaces: $\mathbb{U}_t = \mathbb{U}$, $\mathbb{X}_t = \mathbb{X}$, $\mathbb{W}_t = \mathbb{W}$. In particular, the vocable "state" is usually related to a variable taking values in a space independent of t (see the discussion in Sect. 4.5.3). ◊

4.4.2 State Feedbacks, Decisions, State and Control Maps

When dealing with state models as in Sect. 4.4.1, it is customary to consider *state feedbacks*. In Sect. 4.5.4, we will see that, under the assumptions for which stochastic dynamic programming applies, there is no loss of optimality to restrict the search among state feedbacks instead of a larger class of strategies depending also on past noises.

Definition 4.4 A *state feedback* $\gamma = \{\gamma_t\}_{t=0,\dots,T-1}$ is a sequence of measurable mappings $\gamma_t : (\mathbb{X}_t, \mathcal{X}_t) \to (\mathbb{U}_t, \mathcal{U}_t)$. Accordingly, we set

$$\Gamma_t^{\mathcal{X}_t} := \Big\{\gamma_t : (\mathbb{X}_t, \mathcal{X}_t) \to (\mathbb{U}_t, \mathcal{U}_t) \;\Big|\; \gamma_t^{-1}(\mathcal{U}_t) \subset \mathcal{X}_t\Big\}, \tag{4.41}$$

for $t = 0, \dots, T-1$, and the set of all state feedbacks is denoted by

$$\Gamma_{\mathcal{X}} := \prod_{t=0}^{T-1} \Gamma_t^{\mathcal{X}_t}. \tag{4.42}$$

Remark 4.5 Selecting the class of state feedbacks reveals an implicit information structure: at time t, it is supposed that only the state x_t is available to produce a decision $u_t = \gamma_t(x_t)$. In Sect. 4.5.2, we consider another class of feedbacks, allowing for more or less information (than the state hereabove) to be handled at each time. For instance, we could have feedbacks $\gamma_t(x_0, \ldots, x_t)$ closed on the past states, or feedbacks $\gamma_t(x_0, w_1, \ldots, w_t)$ closed on the initial state and on past uncertainties. ◇

At this level of generality, no measurability assumptions are made. However, in the probabilistic setting that we consider in Sect. 4.4.4, we will introduce σ-fields with respect to which feedbacks will be supposed measurable.

For the sake of clarity, hereafter, we reserve the notation u_t for a control *variable* belonging to $\mathbb{U}_t$, $u_t \in \mathbb{U}_t$, whereas we denote by $\gamma = \{\gamma_t\}_{t=0,\ldots,T-1} \in \Gamma_{\mathcal{X}}$ a feedback *mapping*, with $\gamma_t(x) \in \mathbb{U}_t$.

The terminology *unconstrained case* covers the situation where the feedbacks γ_t can take any value in $\mathbb{U}_t$. The *control constraint case* restricts feedbacks to *admissible feedbacks* as follows

$$\Gamma_{\mathcal{X}}^{\mathrm{ad}} := \{\gamma \in \Gamma_{\mathcal{X}} \mid \gamma_t(x) \in C_t(x), \ \forall t = 0, \ldots, T-1, \ \forall x \in \mathbb{X}_t\}, \tag{4.43}$$

where $C_t(x)$ is a subset of the control set $\mathbb{U}_t$.

The following definitions are extensions of the notion of flow as solution of a differential equation.

Definition 4.6 For any given feedback $\gamma = \{\gamma_t\}_{t=0,\ldots,T-1} \in \Gamma_{\mathcal{X}}$, scenario $w(\cdot) \in \prod_{s=1}^{T} \mathbb{W}_s$, time $t \in \{0, \ldots, T\}$ and state $x \in \mathbb{X}_t$, we define the *state map* X_f by $X_f[t, x, \gamma, w(\cdot)] = (x_t, x_{t+1}, \ldots, x_T)$, the state trajectory solution of the dynamical equation

$$x_{s+1} = f_s\big(x_s, \gamma_s(x_s), w_{s+1}\big), \quad s = t, \ldots, T-1 \text{ with } x_t = x.$$

The *control map* U_f is defined in the same way by means of $U_f[t, x, \gamma, w(\cdot)]$, the associated decision path $(u_t, u_{t+1}, \ldots, u_{T-1})$, where $u_s = \gamma_s\big(x_s\big)$ for $s = t, \ldots, T-1$.

Remark 4.7 Consider two scenarios $w(\cdot)$ and $w'(\cdot)$ in $\prod_{s=1}^{T} \mathbb{W}_s$. If they coincide after time t, that is, if $(w_{t+1}, \ldots, w_T) = (w'_{t+1}, \ldots, w'_T)$, we can notice that

$$\begin{aligned} X_f[t, x, \gamma, w(\cdot)]_t &= X_f[t, x, \gamma, w'(\cdot)]_t = x, \\ X_f[t, x, \gamma, w(\cdot)]_s &= X_f[t, x, \gamma, w'(\cdot)]_s \text{ for } s = t+1, \ldots, T. \end{aligned}$$

This expresses a *causality* property: for $t > s$, the state $X_f[t, x, \gamma, w(\cdot)]_s$ depends upon the uncertainties $(w_{t+1}, \ldots, w_s)$ and not upon all $w(\cdot)$. The same property holds true for the control map U_f. ◇

4.4.3 Criterion

A *criterion* j is a measurable function

$$j : \prod_{t=0}^{T} \mathbb{X}_t \times \prod_{t=0}^{T-1} \mathbb{U}_t \times \prod_{t=1}^{T} \mathbb{W}_t \to \mathbb{R}, \tag{4.44}$$

which assigns a real number $j\big(x(\cdot), u(\cdot), w(\cdot)\big)$ to a state trajectory $x(\cdot) = \big(x_0, x_1, \dots, x_{T-1}, x_T\big)$, to a control history $u(\cdot) = \big(u_0, u_1, \dots, u_{T-1}\big)$, and to a scenario $w(\cdot) = \big(w_1, w_2, \dots, w_T\big)$.

For any feedback γ in (4.42) and scenario $w(\cdot)$ in (4.39), we evaluate the criterion (4.44) along the flow by

$$j_0^{\gamma}\big(x_0, w(\cdot)\big) := j\big(X_f[0, x_0, \gamma, w(\cdot)](\cdot), U_f[0, x_0, \gamma, w(\cdot)](\cdot), w(\cdot)\big), \tag{4.45}$$

where X_f, U_f are the state and control maps introduced in Definition 4.6. Thus, the *evaluation of the criterion along the flow* $j_0^{\gamma}\big(x_0, w(\cdot)\big)$ is the evaluation of the criterion j along the unique trajectory $x(\cdot) = X_f[0, x_0, \gamma, w(\cdot)](\cdot)$, $u(\cdot) = U_f[0, x_0, \gamma, w(\cdot)](\cdot)$, starting from the initial state x_0, and generated by the dynamic (4.36), driven by the feedback control $u_t = \gamma_t\big(x_t\big)$ and by the scenario $w(\cdot)$.

A traditional form of criterion is the *additive* and *time-separable* intertemporal criterion.

Definition 4.8 The criterion j in (4.44) is said to be *additive* and *time-separable* if it takes the form:

$$j\big(x(\cdot), u(\cdot), w(\cdot)\big) = \sum_{t=0}^{T-1} L_t\big(x_t, u_t, w_{t+1}\big) + K\big(x_T\big). \tag{4.46}$$

The function L_t is called the *integral cost* (or *instantaneous cost*), and K the *final cost*.

Remark 4.9 A criterion j in (4.44) is said to be *in Bolza form* if it is an additive and time-separable criterion with zero integral costs:

$$j\big(x(\cdot), u(\cdot), w(\cdot)\big) = K\big(x_T\big). \tag{4.47}$$

A state model as in Sect. 4.4.1 together with an additive and time-separable criterion as in (4.46) can be turned into an extended state model with new state

$$(x_t, c_t) \in \mathbb{X}_t \times \mathbb{R}, \tag{4.48}$$

new dynamics

$$\big((x_t, c_t), u_t, w_{t+1}\big) \mapsto \big(f_t(x_t, u_t, w_{t+1}), c_t + L_t(x_t, u_t, w_{t+1})\big), \tag{4.49}$$

and new criterion in Bolza form as follows

$$(x_T, c_T) \mapsto c_T + K(x_T). \tag{4.50}$$

This is why, sometimes, one only considers criterion in Bolza form. ◊

4.4.4 Stochastic Optimization Problem

Consider $(\Omega, \mathcal{A}, \mathbb{P})$ a probability space, and a random variable

$$\boldsymbol{W}(\cdot) : \Omega \to \prod_{t=1}^{T} \mathbb{W}_t. \tag{4.51}$$

Therefore, the evaluation $j_0^{\gamma}\big(x_0, \boldsymbol{W}(\cdot)\big)$ of the criterion along the flow in (4.45) is a random variable, since both the criterion j in (4.44) and the feedback γ in (4.42) are measurable. When this random variable is integrable, we define the *expected criterion* by:

$$\mathbb{E}\Big(j_0^{\gamma}\big(x_0, \boldsymbol{W}(\cdot)\big)\Big). \tag{4.52}$$

The *expected criterion* optimization problem is

$$\min_{\gamma \in \Gamma_{\mathbb{X}}^{\mathrm{ad}}} \mathbb{E}\Big(j_0^{\gamma}\big(x_0, \boldsymbol{W}(\cdot)\big)\Big), \tag{4.53}$$

and any $\gamma^{\sharp} \in \Gamma_{\mathbb{X}}^{\mathrm{ad}}$ which achieves the minimum is called an *optimal feedback*. Recall that feedbacks in (4.43) satisfy control constraints of the form $u_t \in C_t(x_t)$.

4.4.5 Stochastic Dynamic Programming

Let us suppose that the criterion j in (4.44) is additive and time-separable (see Definition 4.8, and also Footnote 2 in p. 104) and has the form (4.46). For any time $t = 0, \ldots, T$, we define

$$j_t\big(x_t, x_{t+1}, \ldots, x_T, u_t, u_{t+1}, \ldots, u_{T-1}, w_{t+1}, \ldots, w_T\big) := \sum_{s=t}^{T-1} L_s\big(x_s, u_s, w_{s+1}\big) + K\big(x_T\big), \tag{4.54}$$

and also, for any feedback γ in (4.42),

$$j_t^{\gamma}\big(x, w(\cdot)\big) := j_t\big(X_f[t, x, \gamma, w(\cdot)](\cdot), U_f[t, x, \gamma, w(\cdot)](\cdot), w(\cdot)\big), \tag{4.55}$$

where X_f, U_f are the state and control maps introduced in Definition 4.6. Notice that $j_t^{\gamma}\big(x, w(\cdot)\big)$ depends only upon the tail $\big(w_{t+1}, \ldots, w_T\big)$ of the scenario $w(\cdot)$ (see Remark 4.7).

In this Sect. 4.4.5, we make the following white noise assumption.

Assumption 4.10 (*White noise*) The random variables $\boldsymbol{W}_1, \ldots, \boldsymbol{W}_T$ are independent under the probability $\mathbb{P}$.

When the initial state $\boldsymbol{x}_0$ is supposed to be random, as in (4.40), we make the assumption that $\boldsymbol{x}_0, \boldsymbol{W}_1, \ldots, \boldsymbol{W}_T$ are independent under the probability $\mathbb{P}$.

Definition 4.11 The *value function* or *Bellman function* associated with the additive and time-separable criterion (4.46), the dynamics (4.36), and the control constraints (4.43), is defined by the *cost-to-go*

$$V_t(x) := \min_{\gamma \in \Gamma_{\mathbb{X}}^{\mathrm{ad}}} \mathbb{E}\Big(j_t^{\gamma}\big(x, \boldsymbol{W}(\cdot)\big)\Big), \quad \forall t = 0, \ldots, T-1, \ \forall x \in \mathbb{X}_t. \tag{4.56}$$

Notice that, by (4.54) and (4.55), the expectation $\mathbb{E}(\cdot)$ is taken w.r.t. the distribution of $\big(\boldsymbol{W}_{t+1}, \ldots, \boldsymbol{W}_T\big)$ under the probability $\mathbb{P}$.

We now present a discrete-time stochastic version of the famous *Bellman equation* [15]. This equation connects the value functions between time t and time $t+1$.

Proposition 4.12 *Under proper technical assumptions,*[4] *the value function introduced in Definition 4.11 satisfies the following backward induction, called* Dynamic Programming equation *or* Bellman equation

$$V_T(x) = K(x), \ \forall x \in \mathbb{X}_T, \tag{4.57a}$$

$$V_t(x) = \min_{u \in C_t(x)} \mathbb{E}\Big(L_t\big(x, u, \boldsymbol{W}_{t+1}\big) + V_{t+1}\big(f_t(x, u, \boldsymbol{W}_{t+1})\big)\Big), \ \forall x \in \mathbb{X}_t, \tag{4.57b}$$

where t runs from $T-1$ down to 0.

Notice that, in (4.57b), the expectation $\mathbb{E}(\cdot)$ is taken w.r.t. the distribution of $\boldsymbol{W}_{t+1}$ under the probability $\mathbb{P}$.

[4] To take the expectation in (4.57b), we need that the expression $V_{t+1}\big(f_t(x, u, \boldsymbol{W}_{t+1})\big)$ be measurable. For that, it suffices that the value function V_{t+1} be measurable. However, the proof that induction (4.57) preserves measurability is quite delicate, and requires proper technical assumptions. We refer the reader to [21, 63] for an in-depth analysis.

Proof In Sect. 4.5.4, we present results of Witsenhausen that establish a general Dynamic Programming equation for sequential systems. Then, the proof of this Proposition 4.12 is a consequence of Proposition 4.20, Eq. (4.105) and Proposition 4.22.

Else, more direct proofs may be found in many textbooks [15, 20, 51, 127, 152, 153]. We simply sketch the idea of the proof as follows. Suppose one starts at time t in state x, and picks up the decision $u \in C_t(x)$. For this, one pays the integral cost $L_t\big(x, u_t, \mathbf{W}_{t+1}\big)$, and this leads to the new state $f_t(x, u, \mathbf{W}_{t+1})$, from which the cost-to-go is, by definition, $V_{t+1}\big(f_t(x, u, \mathbf{W}_{t+1})\big)$. Indeed, the expectation w.r.t. the future noises $\mathbf{W}_{t+2}, \dots \mathbf{W}_T$ is legitimate in (4.56) because they are independent of the noise $\mathbf{W}_{t+1}$. Since the criterion (4.46) is additive, the total cost is $L_t\big(x, u_t, \mathbf{W}_{t+1}\big) + V_{t+1}\big(f_t(x, u, \mathbf{W}_{t+1})\big)$. Now, one has to consider the expectation w.r.t. the noise $\mathbf{W}_{t+1}$ to obtain the expected cost starting at time t in state x, and using the decision $u \in C_t(x)$. There remains to minimize w.r.t. $u \in C_t(x)$ to obtain the cost-to-go starting at time t in state x, namely the value function $V_t(x)$. □

Moreover, the Dynamic Programming induction (4.57b) reveals optimal feedbacks. Indeed, with the additional assumption that the infimum is achieved in (4.57) for at least one decision, and that a measurable selection (see Sect. 8.3.1) exists, we obtain an optimal feedback for the stochastic optimal control problem (4.53). Here again, we warn the reader that the issue of measurability in dynamic programming is quite delicate, and we refer her or him to [21, 63] for an in-depth analysis.

Proposition 4.13 *Suppose that, for any time $t \in \{0, \dots, T-1\}$ and state $x \in \mathbb{X}_t$, there exists a control $\gamma_t^\sharp(x) \in C_t(x)$ which achieves the minimum in (4.57b):*

$$\gamma_t^\sharp(x) \in \operatorname*{arg\,min}_{u \in C_t(x)} \mathbb{E}\Big(L_t\big(x, u, \mathbf{W}_{t+1}\big) + V_{t+1}\big(f_t(x, u, \mathbf{W}_{t+1})\big)\Big). \tag{4.58}$$

If, for each $t \in \{0, \dots, T-1\}$, the mapping $\gamma_t^\sharp : (\mathbb{X}_t, \mathfrak{X}_t) \to (\mathbb{U}_t, \mathfrak{U}_t)$ is measurable, then $\gamma^\sharp = \{\gamma_t^\sharp\}_{t=0,\dots,T-1}$ is an optimal feedback for the optimization problem (4.53), where the additive and time-separable criterion j is given by (4.46).

It should be noticed that $\gamma^\sharp$ is also an optimal feedback for all the optimization problems (4.54), that is, for every initial time $t \in \{0, \dots, T-1\}$ and initial state $x \in \mathbb{X}_t$.

4.5 Sequential Optimization Problems

We conclude this chapter with more advanced material. We present a more general form of optimal stochastic control problems than the state model considered in Sect. 4.4. We highlight the fact that a Dynamic Programming (DP) equation also holds true in such a context, due to sequentiality. This equation also connects the value functions between two successive times by means of a static optimization problem;

however, the optimization is over a set of feedbacks, and it is parameterized by an information state belonging to a set of measures, the dimension of which is much larger than that of the original finite dimensional state. We follow the exposition in [158].

4.5.1 Sequential Optimal Stochastic Control Problem

An optimal stochastic control problem is said to be *sequential* when it can be reduced, by elimination of intermediate variables, to the following form:

1. $(\Omega, \mathcal{A}, \mathbb{P})$ a probability space (randomness);
2. $(\mathbb{U}_0, \mathcal{U}_0), \ldots, (\mathbb{U}_{T-1}, \mathcal{U}_{T-1})$ measurable spaces (decision spaces);
3. $\mathcal{I}_t$ a subfield of $\mathcal{U}_0 \otimes \cdots \otimes \mathcal{U}_{t-1} \otimes \mathcal{A}$, for $t = 0, \ldots, T-1$ (information);
4. $\tilde{j} : \mathbb{U}_0 \times \cdots \times \mathbb{U}_{T-1} \times \Omega \to \mathbb{R}$ a measurable function (criterion).

In the above model, we have stochasticity with the triplet $(\Omega, \mathcal{A}, \mathbb{P})$, decisions with the control spaces $(\mathbb{U}_0, \mathcal{U}_0), \ldots, (\mathbb{U}_{T-1}, \mathcal{U}_{T-1})$, and information with the subfields $\mathcal{I}_0, \ldots, \mathcal{I}_{T-1}$. However, we have no explicit state, contrary to what was exposed in Sect. 4.4.1. Nevertheless, we have a kind of "state equation" as follows. Define a *decision rule*, or a *policy*, or a *strategy*, as a sequence

$$\lambda = \{\lambda_t\}_{t=0,\ldots,T-1} \tag{4.59}$$

of measurable mappings

$$\begin{aligned}
\lambda_0 &: (\Omega, \mathcal{A}) \to (\mathbb{U}_0, \mathcal{U}_0), \\
\lambda_1 &: (\mathbb{U}_0 \times \Omega, \mathcal{U}_0 \otimes \mathcal{A}) \to (\mathbb{U}_1, \mathcal{U}_1), \\
&\vdots \\
\lambda_{T-1} &: (\mathbb{U}_0 \times \cdots \times \mathbb{U}_{T-2} \times \Omega, \mathcal{U}_0 \otimes \cdots \otimes \mathcal{U}_{T-2} \otimes \mathcal{A}) \to (\mathbb{U}_{T-1}, \mathcal{U}_{T-1}).
\end{aligned}$$

With a policy λ, we associate the mapping $M_\lambda : \Omega \to \mathbb{U}_0 \times \cdots \times \mathbb{U}_{T-1}$ defined by

$$(u_0, \ldots, u_{T-1}) = M_\lambda(\omega) \iff \begin{cases} u_0 &= \lambda_0(\omega), \\ u_1 &= \lambda_1(u_0, \omega), \\ \vdots & \vdots \\ u_{T-1} &= \lambda_{T-1}(u_0, \cdots, u_{T-2}, \omega). \end{cases} \tag{4.60}$$

Notice that those recursive equations form a *strictly lower-triangular* system out of which a unique well-defined solution $(u_0, \ldots, u_{T-1})$ is entirely determined by the choice of ω. For nonsequential systems, the situation is more delicate, and is discussed in Sect. 9.3.

Let us introduce the mapping (*solution map*)

$$S_\lambda : \Omega \to \mathbb{U}_0 \times \cdots \times \mathbb{U}_{T-1} \times \Omega \tag{4.61}$$

defined by

$$S_\lambda(\omega) := \big(M_\lambda(\omega), \omega\big). \tag{4.62}$$

The element $S_\lambda(\omega) \in \mathbb{U}_0 \times \cdots \times \mathbb{U}_{T-1} \times \Omega$ may be seen as the "state trajectory" yielded by the policy λ (consider the analogy with the state map introduced in Definition 4.6, and with the solution map of Definition 9.11 to be seen later).

The optimization problem

$$\min_{\lambda_0 \preceq \mathfrak{I}_0, \dots, \lambda_{T-1} \preceq \mathfrak{I}_{T-1}} \mathbb{E}\Big(\tilde{j}\big(S_\lambda(\cdot)\big)\Big) \tag{4.63}$$

consists of finding $\lambda = \{\lambda_t\}_{t=0,\dots,T-1}$—with the informational restriction that λ_t is $\mathfrak{I}_t$-measurable for $t = 0, \dots, T-1$—to minimize $\mathbb{E}\big(\tilde{j}\big(S_\lambda(\cdot)\big)\big)$. The measurability of M_λ is established step by step, so that $\tilde{j} \circ S_\lambda$ is also proved to be measurable (see [158] for the details).

Example 4.14 Consider the following generic form of a 2-stage stochastic programming problem as exposed in [141, Chap. 3]:

$$\min_{u_0} L_0\big(u_0\big) + \mathbb{E}\Big(\min_{u_1} L_1\big(u_1, w_1\big)\Big), \tag{4.64}$$

where

$$u_0 \in C_0, \;\; u_1 \in C_1(u_0, w_1). \tag{4.65}$$

Notice that the formulation (4.64) and (4.65) does not make explicit reference to *information constraints*, but makes explicit reference to (set-membership) *decision constraints* in (4.65).[5] The form (4.64) corresponds to the decision spaces

$$(\mathbb{U}_0, \mathcal{U}_0) = (\mathbb{R}^{p_0}, \mathcal{B}^{\mathrm{o}}_{\mathbb{R}^{p_0}}) \text{ and } (\mathbb{U}_1, \mathcal{U}_1) = (\mathbb{R}^{p_1}, \mathcal{B}^{\mathrm{o}}_{\mathbb{R}^{p_1}}), \tag{4.66}$$

and to the probability space

$$\Omega = \mathbb{W}_1 = \mathbb{R}^{q_1}, \tag{4.67}$$

equipped with its Borel σ-field

$$\mathcal{A} = \mathcal{B}^{\mathrm{o}}_{\mathbb{W}_1} = \mathcal{B}^{\mathrm{o}}_{\mathbb{R}^{q_1}}, \tag{4.68}$$

[5] We do not treat the case with decision constraints in this section, but the extension is rather straightforward (for instance, decision constraints can be incorporated in the criterion with characteristic functions).

and with a probability $\mathbb{P}$. The following information fields

$$\mathcal{I}_0 = \{\emptyset, \Omega\} \text{ and } \mathcal{I}_1 = \mathcal{U}_0 \otimes \mathcal{A} \tag{4.69}$$

make explicit the implicit information pattern in 2-stage stochastic programming problems: at the first stage, there is no information whatsoever, whereas, at the second stage, the first decision and the state of Nature are available for decision-making. △

Example 4.15 We now consider the following generic form of a T-stage stochastic programming problem, with $T > 2$ (which displays more variety in possible information patterns than the 2-stage problem), written as in [141, Chap. 3]:

$$\min_{u_0} L_0(u_0) + \mathbb{E}\Big(\min_{u_1} L_1(u_1, w_1) + \mathbb{E}\Big(\cdots + \mathbb{E}\Big(\min_{u_{T-1}} L_{T-1}(u_{T-1}, w_{T-1}) \Big)\Big)\Big), \tag{4.70}$$

where

$$u_0 \in C_0, \; u_1 \in C_1(u_0, w_1), \; \ldots, \; u_{T-1} \in C_{T-1}(u_{T-2}, w_{T-1}). \tag{4.71}$$

The form (4.70) corresponds to the decision spaces

$$(\mathbb{U}_0, \mathcal{U}_0) = (\mathbb{R}^{p_0}, \mathcal{B}^{o}_{\mathbb{R}^{p_0}}), \; \ldots, \; (\mathbb{U}_{T-1}, \mathcal{U}_{T-1}) = (\mathbb{R}^{p_{T-1}}, \mathcal{B}^{o}_{\mathbb{R}^{p_{T-1}}}), \tag{4.72}$$

and to the probability space

$$\Omega = \prod_{t=1}^{T-1} \mathbb{W}_t \text{ with } \mathbb{W}_t = \mathbb{R}^{q_t}, \tag{4.73}$$

equipped with its Borel σ-field

$$\mathcal{A} = \bigotimes_{t=1}^{T-1} \mathcal{W}_t \text{ with } \mathcal{W}_t = \mathcal{B}^{o}_{\mathbb{W}_t}, \tag{4.74}$$

and with a probability $\mathbb{P}$.

For $t = 1, \ldots, T-1$, let us define the *past uncertainties σ-field*

$$\mathcal{A}_t = \mathcal{W}_1 \otimes \cdots \otimes \mathcal{W}_t \otimes \{\emptyset, \mathbb{W}_{t+1}\} \otimes \cdots \otimes \{\emptyset, \mathbb{W}_{T-1}\}, \tag{4.75}$$

and the *last uncertainty σ-field*

$$\overline{\mathcal{W}_t} = \{\emptyset, \mathbb{W}_1\} \otimes \cdots \{\emptyset, \mathbb{W}_{t-1}\} \otimes \mathcal{W}_t \otimes \{\emptyset, \mathbb{W}_{t+1}\} \otimes \cdots \otimes \{\emptyset, \mathbb{W}_{T-1}\}. \tag{4.76}$$

As for the 2-stage stochastic programming problem example, the formulation (4.70) and (4.71), taken from [141, Chap. 3], makes explicit reference to (set-membership) *decision constraints* (see Footnote 5 in p. 113), but does not make explicit reference to *information constraints*. We now propose three possible information patterns, compatible with the notation $u_t \in C_t(u_{t-1}, w_t)$ in (4.71), which express the fact that, at time t, *at least* the last control u_{t-1} and the last uncertainty w_t are available before making a decision.

- The last control and the last uncertainty are available for decision-making in the case where

$$\mathcal{I}_0 = \{\emptyset, \Omega\} \text{ and } \mathcal{I}_t = \{\emptyset, \mathbb{U}_0\} \otimes \cdots \otimes \{\emptyset, \mathbb{U}_{t-2}\} \otimes \mathcal{U}_{t-1} \otimes \overline{\mathcal{W}}_t. \tag{4.77}$$

- The last control and the past uncertainties are available for decision-making in the case where

$$\mathcal{I}_0 = \{\emptyset, \Omega\} \text{ and } \mathcal{I}_t = \{\emptyset, \mathbb{U}_0\} \otimes \cdots \otimes \{\emptyset, \mathbb{U}_{t-2}\} \otimes \mathcal{U}_{t-1} \otimes \mathcal{A}_t. \tag{4.78}$$

- The past controls and the past uncertainties are available for decision-making in the case where

$$\mathcal{I}_0 = \{\emptyset, \Omega\} \text{ and } \mathcal{I}_t = \mathcal{U}_0 \otimes \cdots \otimes \mathcal{U}_{t-1} \otimes \mathcal{A}_t. \tag{4.79}$$

The authors of [141] call *implementable policy* what we call *decision rule*, or *policy*, or *strategy* in (4.59). △

4.5.2 Optimal Stochastic Control Problem in Standard Form

Now, we consider a formulation where state and dynamics are explicit.

An optimal stochastic control problem is said to be in *standard form* when it can be reduced, by a change of variables, to the following form:

1. $(\mathbb{X}_0, \mathcal{X}_0), \ldots, (\mathbb{X}_T, \mathcal{X}_T)$ are measurable spaces ("state" spaces)[6];
2. $(\mathbb{U}_0, \mathcal{U}_0), \ldots, (\mathbb{U}_{T-1}, \mathcal{U}_{T-1})$ are measurable spaces (decision spaces);
3. $\mathcal{I}_t$ is a subfield of $\mathcal{X}_t$, for $t = 0, \ldots, T-1$ (information);
4. $f_t : (\mathbb{X}_t \times \mathbb{U}_t, \mathcal{X}_t \otimes \mathcal{U}_t) \to (\mathbb{X}_{t+1}, \mathcal{X}_{t+1})$ is measurable, for $t = 0, \ldots, T-1$ (dynamics);
5. π_0 is a probability on $(\mathbb{X}_0, \mathcal{X}_0)$ (randomness);
6. $j : (\mathbb{X}_T, \mathcal{X}_T) \to \mathbb{R}$ is a measurable function (criterion).

[6] See footnote 3 in Sect. 1.2.1.

A *feedback* $\gamma = \{\gamma_t\}_{t=0,\dots,T-1}$ is a sequence of measurable mappings $\gamma_t : (\mathbb{X}_t, \mathfrak{X}_t) \to (\mathbb{U}_t, \mathfrak{U}_t)$. For $t = 0, \dots, T-1$, we set

$$\Gamma_t^{\mathfrak{I}_t} := \left\{ \gamma_t : (\mathbb{X}_t, \mathfrak{X}_t) \to (\mathbb{U}_t, \mathfrak{U}_t) \;\middle|\; \gamma_t^{-1}(\mathfrak{U}_t) \subset \mathfrak{I}_t \right\}, \tag{4.80}$$

the set of feedbacks measurable w.r.t. the information $\mathfrak{I}_t$. The set of *admissible feedbacks* is

$$\Gamma_{\mathfrak{I}} := \prod_{t=0}^{T-1} \Gamma_t^{\mathfrak{I}_t}. \tag{4.81}$$

Admissible feedbacks are the feedbacks measurable w.r.t. the information for all times.

For any feedback $\gamma = \{\gamma_t\}_{t=0,\dots,T-1} \in \Gamma_{\mathfrak{I}}$ and $t = 0, \dots, T-1$, the *closed-loop dynamics* $f_t^{\gamma_t} : (\mathbb{X}_t, \mathfrak{X}_t) \to (\mathbb{X}_{t+1}, \mathfrak{X}_{t+1})$ is defined by

$$f_t^{\gamma_t}(x) := f_t\big(x, \gamma_t(x)\big), \quad \forall x \in \mathbb{X}_t. \tag{4.82}$$

The optimization problem

$$\min_{\gamma \in \Gamma_{\mathfrak{I}}} \int_{\mathbb{X}_0} j(X_T^{\gamma}(x_0))\, \mathrm{d}\pi_0(x_0) \tag{4.83}$$

consists in finding a feedback $\gamma \in \Gamma_{\mathfrak{I}}$ to minimize $\int_{\mathbb{X}_0} j(X_T^{\gamma}(x_0))\, \mathrm{d}\pi_0(x_0)$, where the final state $x_T^{\gamma} : (\mathbb{X}_0, \mathfrak{X}_0) \to (\mathbb{X}_T, \mathfrak{X}_T)$ is defined by induction by

$$X_0^{\gamma}(x_0) = x_0 \text{ and } X_{t+1}^{\gamma}(x_0) = f_t\Big(X_t^{\gamma}(x_0), \gamma_t\big(X_t^{\gamma}(x_0)\big)\Big). \tag{4.84}$$

Example 4.16 Though rather abstract, the above formulation includes, after transformation and change of notations, the state model exposed in Sects. 4.4.1–4.4.4. However, the same notations do not denote the same objects. To avoid confusion, we denote, in this example, by $\overline{x}_t$, $\overline{\mathbb{X}}_t$, $\overline{\mathfrak{X}}_t$, the generic "state", the "state" space and the "state" field of the optimal stochastic control problem in standard form,[7] as well as information fields $\overline{\mathfrak{I}}_t$ and dynamics $\overline{f}_t$.

For simplicity, and without loosing generality (see Remark 4.9), we suppose that the additive and time-separable criterion in (4.44) is in Bolza form, to be consistent with item 6 in the standard form Sect. 4.5.2. A possible standard form (among many) for the optimal stochastic control problem exposed in Sects. 4.4.1–4.4.4 is given by the following elements.

[7] Concerning those quotes around the word *state*, we refer the reader to Footnote 3 of Chap. 1 and to Sect. 4.5.3.

1. The "state" spaces and fields

$$\overline{x}_t = (x_t, w_{t+1}, \ldots, w_T), \tag{4.85a}$$

$$\overline{\mathbb{X}}_t = \mathbb{X}_t \times \prod_{s=t+1}^{T} \mathbb{W}_s, \tag{4.85b}$$

$$\overline{\mathfrak{X}}_t = \mathfrak{X}_t \otimes \bigotimes_{s=t+1}^{T} \mathcal{W}_s, \tag{4.85c}$$

for $t = 0, \ldots, T$, built from the spaces introduced in Sect. 4.4.1. It may seem weird to have future uncertainties $w_{t+1}, \ldots, w_T$ in the "state" $\overline{x}_t$ at time t. However, recall that, for an optimal stochastic control problem in standard form, the "state" is not necessarily perfectly observed. Only the part x_t of $\overline{x}_t$ may be available for feedback. We will come back to this point in item 4.
2. The same decision spaces and fields $(\mathbb{U}_0, \mathcal{U}_0), \ldots, (\mathbb{U}_{T-1}, \mathcal{U}_{T-1})$ as those introduced in Sect. 4.4.1.
3. The information fields

$$\overline{\mathfrak{I}}_t = \mathfrak{X}_t \otimes \bigotimes_{s=t+1}^{T} \{\emptyset, \mathbb{W}_s\} \subset \overline{\mathfrak{X}}_t = \mathfrak{X}_t \otimes \bigotimes_{s=t+1}^{T} \mathcal{W}_s, \tag{4.86}$$

for $t = 0, \ldots, T-1$, corresponding to the information structure revealed by the choice of feedbacks in Sect. 4.4.2. Indeed, we thus express that feedbacks can only depend on the (original) state x_t. More formally, for $t = 0, \ldots, T-1$, we set, following (4.80),

$$\Gamma_t^{\overline{\mathfrak{I}}_t} := \left\{ \overline{\gamma}_t : (\overline{\mathbb{X}}_t, \overline{\mathfrak{X}}_t) \to (\mathbb{U}_t, \mathcal{U}_t) \;\middle|\; \overline{\gamma}_t^{-1}(\mathcal{U}_t) \subset \overline{\mathfrak{I}}_t \right\}, \tag{4.87}$$

the set of feedbacks measurable w.r.t. the information $\overline{\mathfrak{I}}_t$. Following (4.81), the set of admissible feedbacks is

$$\Gamma_{\overline{\mathfrak{I}}} := \prod_{t=0}^{T-1} \Gamma_t^{\overline{\mathfrak{I}}_t}. \tag{4.88}$$

4. The measurable dynamics $\overline{f}_t : (\overline{\mathbb{X}}_t \times \mathbb{U}_t, \overline{\mathfrak{X}}_t \otimes \mathcal{U}_t) \to (\overline{\mathbb{X}}_{t+1}, \overline{\mathfrak{X}}_{t+1})$ given by

$$\begin{aligned} \overline{f}_t(\overline{x}_t, u_t) &= \overline{f}_t(x_t, w_{t+1}, \ldots, w_T, u_t) \\ &= (\underbrace{f_t(x_t, u_t, w_{t+1})}_{x_{t+1}}, w_{t+2}, \ldots, w_T), \end{aligned} \tag{4.89}$$

for $t = 0, \dots, T-1$, built from the original dynamics introduced in Sect. 4.4.1. The fact that the original dynamics $f_t(x_t, u_t, w_{t+1})$ explicitly depends on w_{t+1} makes that w_{t+1} must be part of the new state $\overline{x}_t$ in (4.85). We deduce that w_{t+2} must be part of the new state $\overline{x}_{t+1}$ But then, since the new dynamics $\overline{f}_t(\overline{x}_t, u_t)$ arrives in the state space $\overline{\mathbb{X}}_{t+1}$ with state $\overline{x}_{t+1}$ which includes w_{t+2}, then w_{t+2} has also to be part of the new state $\overline{x}_t$ in (4.85). This is how we arrive at the new state $(x_t, w_{t+1}, \dots, w_T)$.
5. The probability $\overline{\pi}_0 = \delta_{x_0} \otimes \mathbb{P}_{\boldsymbol{W}(\cdot)}$ on $(\overline{\mathbb{X}}_0, \overline{\mathcal{X}}_0) = (\mathbb{X}_0 \times \prod_{s=1}^T \mathbb{W}_s, \mathcal{X}_0 \otimes \bigotimes_{s=1}^T \mathcal{W}_s)$, product of a Dirac measure on $\mathbb{X}_0$ with the probability image (see (B.8)) of $\mathbb{P}$ on $\prod_{t=1}^T \mathbb{W}_t$ by the random variable $\boldsymbol{W}(\cdot) : \Omega \to \prod_{t=1}^T \mathbb{W}_t$ in (4.51), as exposed in Sect. 4.4.4. We can also take $\overline{\pi}_0 = \pi_0 \otimes \mathbb{P}_{\boldsymbol{W}(\cdot)}$ when the initial state x_0 is supposed to be uncertain, as in (4.40).
6. The same criterion $j = K : (\mathbb{X}_T, \mathcal{X}_T) \to \mathbb{R}$ as the one introduced in Sect. 4.4.3 (notice indeed that $\overline{\mathbb{X}}_T = \mathbb{X}_T$, and recall that we supposed the criterion in (4.44) to be in Bolza form (see Remark 4.9)).

Notice that the formulation discussed here is more general than the state model formulation in Sect. 4.4 regarding information patterns. Indeed, in Sect. 4.4.2, the feedbacks in (4.42) are state feedbacks, which corresponds here to the information

$$\overline{\mathfrak{I}}_t = \mathcal{X}_t \otimes \bigotimes_{s=t+1}^{T} \{\emptyset, \mathbb{W}_s\}, \tag{4.90}$$

whereas item 3 in p. 115 only assumes that

$$\overline{\mathfrak{I}}_t \subset \mathcal{X}_t \otimes \bigotimes_{s=t+1}^{T} \mathcal{W}_s, \tag{4.91}$$

in the definition of an optimal stochastic control problem in standard form at the beginning of 4.5.2. △

As Witsenhausen points it out in [158], "standard problems appear to form a very special class since the only source of randomness is the initial state, the controller has no memory and the cost depends only upon the last state". However, he shows the generality of standard problems.

Proposition 4.17 ([158]) *Any sequential control problem may be reduced to standard form.*

With the notations of Sect. 4.5.1, the proof consists in taking

- $(\mathbb{X}_0, \mathcal{X}_0) = (\Omega, \mathcal{A})$, $\pi_0 = \mathbb{P}$,
- $(\mathbb{X}_t, \mathcal{X}_t) = (\mathbb{U}_0 \times \cdots \times \mathbb{U}_{t-1} \times \Omega, \mathcal{U}_0 \otimes \cdots \otimes \mathcal{U}_{t-1} \otimes \mathcal{A})$, with generic element $x_t = (u_0, \dots, u_{t-1}, \omega)$,
- $f_t(x_t, u_t) = (u_0, \dots, u_{t-1}, u_t, \omega)$,
- $j(x_T) = \tilde{j}(u_0, \dots, u_{T-1}, \omega)$.

The proof introduces the "state" $x_t = (u_0, \ldots, u_{t-1}, \omega)$. We now discuss what is generally covered by the term "state".

4.5.3 What Is a State?

An informal definition of a state may be found in [152] as a sufficient statistic for the uncertainty ω and past controls. Define a *history* as a sequence $(u_0, \ldots, u_{t-1}, \omega)$. Quoting Whittle, suppose there is a variable x_t which summarizes the past history in that, given time t and the value of x_t, one can calculate the optimal u_t and also x_{t+1} without knowledge of the whole history $(u_0, \ldots, u_{t-1}, \omega)$, for all t. Such a variable is termed *sufficient*. While history takes value in an increasing space as t increases, a sufficient variable taking values in a space independent of t is called a *state variable*.

In Sect. 9.3.3, we briefly discuss the abstract notion of state for nonsequential systems, as suggested by Witsenhausen. We also refer the reader to Sect. 9.6.2.

4.5.4 Dynamic Programming Equations

Here, we follow the exposition in [158]. Witsenhausen points out how a very general Dynamic Programming equation holds true for sequential systems. Specific forms of this Dynamic Programming equation, like the Bellman equation (4.57), appear as special cases under specific information patterns. We follow the notations of Sect. 4.5.2.

Equivalent Deterministic Optimization Problem from the Standard Form
Witsenhausen asserts that, from a problem in the standard form exposed in Sect. 4.5.2, "one can obtain an equivalent deterministic problem by considering the (unconditional) distribution π_t of the state x_t as the new state".

Let us define a pairing $\langle \cdot \,, \cdot \rangle$ between the Banach space Π_t of signed measures on $(\mathbb{X}_t, \mathcal{X}_t)$ (with the total variation norm) and the Banach space Φ_t of bounded measurable functions on $(\mathbb{X}_t, \mathcal{X}_t)$ (with the supremum norm) as follows. For any signed measure $\pi \in \Pi_t$ and any bounded measurable function $\varphi \in \Phi_t$, we set

$$\langle \pi \,, \varphi \rangle := \int_{\mathbb{X}_t} \varphi(x) \, \mathrm{d}\pi(x). \tag{4.92}$$

For a given feedback $\gamma = \{\gamma_t\}_{t=0,\ldots,T-1}$ in (4.81), the (nonlinear) closed-loop dynamics (4.82) induces two linear operators as follows. The linear operator $\left(f_t^{\gamma_t}\right)_\star$ from Π_t to Π_{t+1}, defined by

$$\left(f_t^{\gamma_t}\right)_\star \pi := \pi \circ (f_t^{\gamma_t})^{-1}, \tag{4.93}$$

maps any signed measure $\pi \in \Pi_t$ onto its image measure by the closed-loop dynamics $f_t^{\gamma_t}$ defined in (4.82) (the notation (4.93) is coherent with the notation (B.9) for the probability image). The dual operator $\left(f_t^{\gamma_t}\right)^\star$ from Φ_{t+1} to Φ_t, given by

$$\left(f_t^{\gamma_t}\right)^\star \varphi := \varphi \circ f_t^{\gamma_t}, \tag{4.94}$$

maps functions φ over $\mathbb{X}_{t+1}$ onto functions over $\mathbb{X}_t$.

From (B.10), we have the duality relation

$$\left\langle \left(f_t^{\gamma_t}\right)_\star \pi \,, \varphi \right\rangle = \left\langle \pi \,, \left(f_t^{\gamma_t}\right)^\star \varphi \right\rangle, \tag{4.95}$$

for any signed measure $\pi \in \Pi_t$ and function $\varphi \in \Phi_{t+1}$.

Proposition 4.18 ([158]) *Consider an optimal stochastic control problem in standard form as in Sect. 4.5.2. If the feedback $\gamma^\sharp = \left\{\gamma_t^\sharp\right\}_{t=0,\dots,T-1}$ is optimal for (4.83), then one has that (with $X_T^\gamma(x_0)$ given by (4.84))*

$$\min_{\gamma \in \Gamma_{\mathfrak{I}}} \int_{\mathbb{X}_0} j\left(X_T^\gamma(x_0)\right) \mathrm{d}\pi_0(x_0) = \left\langle \pi_t^\sharp \,, \varphi_t^\sharp \right\rangle \tag{4.96a}$$

$$= \min_{\gamma_t \in \Gamma_t^{\mathfrak{I}_t}} \left\langle \pi_t^\sharp \,, \left(f_t^{\gamma_t}\right)^\star \varphi_{t+1}^\sharp \right\rangle \tag{4.96b}$$

$$= \min_{\gamma_t \in \Gamma_t^{\mathfrak{I}_t}} \left\langle \left(f_t^{\gamma_t}\right)_\star \pi_t^\sharp \,, \varphi_{t+1}^\sharp \right\rangle, \tag{4.96c}$$

for all $t = 0, \dots, T-1$, and where the (new) state sequence $\left\{\pi_t^\sharp\right\}_{t=0,\dots,T-1}$ solves the forward state linear equation *(Fokker-Planck equation)*

$$\pi_{t+1}^\sharp = \left(f_t^{\gamma_t^\sharp}\right)_\star \pi_t^\sharp, \; t = 0, \dots, T-1, \; \pi_0^\sharp = \pi_0, \tag{4.97}$$

and the co-state sequence $\left\{\varphi_t^\sharp\right\}_{t=0,\dots,T-1}$ solves the backward co-state linear equation

$$\varphi_t^\sharp = \left(f_t^{\gamma_t^\sharp}\right)^\star \varphi_{t+1}^\sharp, \; t = T-1, \dots, 0, \; \varphi_T^\sharp = j. \tag{4.98}$$

The structure of these equations evokes the two-point boundary equations in deterministic optimal control "à la Pontryagin" (Hamilton equations). The analogy of structure between the Fokker-Planck equation in stochastic process and the Dynamic Programming equation in optimal control has been raised in [52]. The (new) state $\pi_t \in \Pi_t$ is also called the *information state*.

Dynamic Programming Equation in Standard Form

We conclude by exposing how Witsenhausen provides a general dynamic programming equation in the framework of Sect. 4.5.2. For all $t = 0, \dots, T-1$, let $r_t(\pi_0) \subset \Pi_t$ denote the set of all states π_t reachable at time t from any initial state $\pi_0 \in \Pi_0$:

$$r_t(\pi_0) := \left\{ \left(f_t^{\gamma_t}\right)_\star \cdots \left(f_0^{\gamma_0}\right)_\star \pi_0 \;\middle|\; \gamma_0 \in \Gamma_0^{\mathfrak{J}_0}, \dots, \gamma_t \in \Gamma_t^{\mathfrak{J}_t} \right\}. \tag{4.99}$$

Dually, for any final co-state $\varphi_T \in \Phi_T$, let us define the co-reachable set $\rho_t(\varphi_T)$ in Φ_t as

$$\rho_t(\varphi_T) := \left\{ \left(f_t^{\gamma_t}\right)^\star \cdots \left(f_{T-1}^{\gamma_{T-1}}\right)^\star \varphi_T \;\middle|\; \gamma_{T-1} \in \Gamma_{T-1}^{\mathfrak{J}_{T-1}}, \dots, \gamma_t \in \Gamma_t^{\mathfrak{J}_t} \right\}. \tag{4.100}$$

Definition 4.19 The "support function"[8] of the co-reachable set $\rho_t(\varphi_T)$ yields the *value function*

$$\widetilde{V}_t(\pi_t; \varphi_T) := \inf_{\varphi_t \in \rho_t(\varphi_T)} \langle \pi_t \,, \varphi_t \rangle \,, \quad \forall \pi_t \in \Pi_t. \tag{4.101}$$

Proposition 4.20 ([158]) *The value functions (4.101) are related by the backward Dynamic Programming equation*

$$\widetilde{V}_T(\pi_T; \varphi_T) = \langle \pi_T \,, \varphi_T \rangle \,, \quad \forall \pi_T \in \Pi_T, \tag{4.102a}$$

$$\widetilde{V}_t(\pi_t; \varphi_T) = \min_{\gamma_t \in \Gamma_t^{\mathfrak{J}_t}} \widetilde{V}_{t+1}\big(\left(f_t^{\gamma_t}\right)_\star \pi_t; \varphi_T\big), \quad \forall \pi_t \in \Pi_t. \tag{4.102b}$$

Dually, Witsenhausen introduces the support function $\widetilde{W}_t$ of the reachable set $r_t(\pi_0) \subset \Pi_t$

$$\widetilde{W}_t(\varphi_t; \pi_0) := \inf_{\pi_t \in r_t(\pi_0)} \langle \pi_t \,, \varphi_t \rangle \,, \quad \forall \varphi_t \in \Phi_t, \tag{4.103}$$

and shows a *forward Dynamic Programming equation*

$$\widetilde{W}_0(\varphi_0; \pi_0) = \langle \pi_0 \,, \varphi_0 \rangle \,, \tag{4.104a}$$

$$\widetilde{W}_{t+1}(\varphi_{t+1}; \pi_0) = \min_{\gamma_t \in \Gamma_t^{\mathfrak{J}_t}} \widetilde{W}_t(\left(f_t^{\gamma_t}\right)^\star \varphi_{t+1}; \pi_0), \quad \forall \varphi_{t+1} \in \Phi_{t+1}. \tag{4.104b}$$

[8] We use the terminology "support function" despite, traditionally, the support function of a subset is a *supremum* of linear functions over this subset, whereas we consider here an *infimum*. Indeed, our optimization problem (4.96a) is one of minimization, and not of maximization.

Witsenhausen claims that the optimal cost (4.83) for an optimal stochastic control problem in standard form as in Sect. 4.5.2 is given by

$$\min_{\gamma \in \Gamma_{\mathfrak{I}}} \int_{\mathbb{X}_0} j\big(X_T^{\gamma}(x_0)\big) \, \mathrm{d}\pi_0(x_0) = \widetilde{W}_T(j; \pi_0) = \widetilde{V}_0(\pi_0; j). \tag{4.105}$$

The Classical Bellman Dynamic Programming Equation

In Example 4.16, we realized a conversion of the optimal stochastic control problem in *state form* discussed in Sects. 4.4.1–4.4.4 into an optimal stochastic control problem in *standard form* as in Sect. 4.5.2.

For a problem in standard form, we just saw how Witsenhausen obtains a general Dynamic Programming equation (4.102). We now discuss how the (classical) Bellman Dynamic Programming equation (4.57) matches with the Witsenhausen Dynamic Programming equation (4.102).

We make the white noise Assumption 4.10 (crucial in establishing the Bellman equation (4.57)). Since the random variables $\boldsymbol{W}_1, \dots, \boldsymbol{W}_T$ are independent under the probability $\mathbb{P}$, then the probability image of $\mathbb{P}$ on $\prod_{t=1}^{T} \mathbb{W}_t$ by the random variable $\boldsymbol{W}(\cdot) : \Omega \to \prod_{t=1}^{T} \mathbb{W}_t$ in (4.51) can be written as (see Sect. B.1.4 for product probabilities)

$$\mathbb{P}_{\boldsymbol{W}(\cdot)} = \bigotimes_{t=1}^{T} \mathbb{P}_{\boldsymbol{W}_t} . \tag{4.106}$$

Definition 4.21 We define, by backward induction, the *classical Bellman value function*

$$V_T(x) = K(x) = \varphi_T(x), \quad \forall x \in \mathbb{X}_T \tag{4.107a}$$

$$V_t(x) = \min_{u \in \mathbb{U}_t} \int_{\mathbb{W}_{t+1}} \mathbb{P}_{\boldsymbol{W}_{t+1}}(\mathrm{d}w_{t+1}) V_{t+1}\big(f_t(x, u, w_{t+1})\big), \quad \forall x \in \mathbb{X}_t. \tag{4.107b}$$

Notice that the above equation is the classical Bellman Dynamic Programming equation (4.57), with zero integral costs since the criterion in (4.44) is in Bolza form (see Remark 4.9), and with final cost $\varphi_T = K$.

We adopt the notations of Example 4.16: to avoid confusion, recall that we denote by $\overline{x}_t, \overline{\mathbb{X}}_t, \overline{\mathfrak{X}}_t$, the generic "state", the "state" space and the "state" field of the optimal stochastic control problem in standard form, as well as dynamics $\overline{f}_t$, information fields $\overline{\mathfrak{I}}_t$, feedbacks $\overline{\gamma}_t$, measures $\overline{\pi}_t$ and reachable sets $\overline{r}_t$.

Proposition 4.22 *Assume that, for all $t \in \{0, \dots, T-1\}$, there exists a control $\gamma_t^\sharp(x) \in \mathbb{U}_t$ which achieves the minimum in (4.107b) and is such that the mapping $\gamma_t^\sharp : (\mathbb{X}_t, \mathfrak{X}_t) \to (\mathbb{U}_t, \mathfrak{U}_t)$ is measurable. Then, for all probability $\overline{\pi}_t$ of the form*

$$\overline{\pi}_t(\mathrm{d}\overline{x}_t) = \pi_t(\mathrm{d}x_t) \otimes \bigotimes_{s=t+1}^{T} \mathbb{P}_{\mathbf{W}_s}, \tag{4.108}$$

we have that

$$\widetilde{V}_t(\overline{\pi}_t; \varphi_T) = \langle \pi_t \, , V_t \rangle \, , \tag{4.109}$$

where the Witsenhausen value function $\widetilde{V}_t$ is defined in (4.101) and the Bellman value function V_t is given by (4.107).

Proof The proof is by induction. But we need preliminary results. This is why we break the proof in several steps.

1. First, we show that the feedbacks sets $\Gamma_t^{\mathfrak{X}_t}$ in (4.41) and $\Gamma_t^{\overline{\mathfrak{I}}_t}$ in (4.87) are in one-to-one correspondence, for any $t = 0, \ldots, T-1$.
   Any measurable $\gamma_t : (\mathbb{X}_t, \mathfrak{X}_t) \to (\mathbb{U}_t, \mathfrak{U}_t)$ in $\Gamma_t^{\mathfrak{X}_t}$ can be extended into a feedback $\widehat{\gamma}_t : (\overline{\mathbb{X}}_t, \overline{\mathfrak{X}}_t) \to (\mathbb{U}_t, \mathfrak{U}_t)$ by setting

$$\widehat{\gamma}_t(\overline{x}_t) = \widehat{\gamma}_t(x_t, w_{t+1}, \ldots, w_T) = \gamma_t(x_t), \tag{4.110}$$

   where we recall that $\overline{x}_t = (x_t, w_{t+1}, \ldots, w_T)$, by (4.85).
   Since, by (4.41), we have that $\gamma_t^{-1}(\mathfrak{U}_t) \subset \mathfrak{X}_t$, we deduce that $\widehat{\gamma}_t^{-1}(\mathfrak{U}_t) \subset \overline{\mathfrak{I}}_t = \mathfrak{X}_t \otimes \bigotimes_{s=t+1}^{T} \{\emptyset, \mathbb{W}_s\}$ by (4.86). In other words, $\widehat{\gamma}_t$ is an admissible feedback, measurable w.r.t. the information $\overline{\mathfrak{I}}_t$ given by (4.86).
   Now, take an admissible feedback $\overline{\eta}_t \in \Gamma_t^{\overline{\mathfrak{I}}_t}$.[9] Since $\overline{\eta}_t^{-1}(\mathfrak{U}_t) \subset \overline{\mathfrak{I}}_t = \mathfrak{X}_t \otimes \bigotimes_{s=t+1}^{T} \{\emptyset, \mathbb{W}_s\}$, we deduce (by Proposition 3.46 with projection mapping) that there exists a measurable $\gamma_t : (\mathbb{X}_t, \mathfrak{X}_t) \to (\mathbb{U}_t, \mathfrak{U}_t)$ such that

$$\overline{\eta}_t(\overline{x}_t) = \overline{\eta}_t(x_t, w_{t+1}, \ldots, w_T) = \gamma_t(x_t), \tag{4.111}$$

   and therefore $\overline{\eta}_t = \widehat{\gamma}_t$.
2. Second, we show that, when $\widehat{\gamma}_t$ is given by (4.110), the closed-loop dynamics (4.82) is given by, for any $t = 0, \ldots, T-1$,

$$\overline{f}_t^{\widehat{\gamma}_t}(\overline{x}_t) = (f_t^{\gamma_t}(x_t, w_{t+1}), w_{t+2}, \ldots, w_T). \tag{4.112}$$

   Indeed, we have that

$$\overline{f}_t^{\widehat{\gamma}_t}(\overline{x}_t) = \overline{f}_t\big(\overline{x}_t, \widehat{\gamma}_t(\overline{x}_t)\big),$$

[9] To avoid confusion, we temporarily adopt the notation $\overline{\eta}_t$ for a generic feedback in $\Gamma_t^{\overline{\mathfrak{I}}_t}$, instead of $\overline{\gamma}_t$.

by definition (4.82) of $\overline{f}_t^{\widehat{\gamma}_t}$,

$$=\overline{f}_t\big((x_t, w_{t+1}, \dots, w_T), \gamma_t(x_t)\big),$$

by definition (4.85) of $\overline{x}_t$ and by definition (4.110) of $\widehat{\gamma}_t$,

$$=\Big(f_t\big(x_t, \gamma_t(x_t), w_{t+1}\big), w_{t+1}, \dots, w_T\Big),$$

by definition (4.89) of $\overline{f}_t$,

$$=\big(f_t^{\gamma_t}(x_t, w_{t+1}), w_{t+2}, \dots, w_T\big),$$

by definition (4.82) of $f_t^{\gamma_t}$.

3. Third, we deduce from the closed-loop dynamics expression (4.112) that, for any $t = 0, \dots, T-1$,

$$\Big(\overline{f}_t^{\widehat{\gamma}_t}\Big)_\star \Big(\pi_t(\mathrm{d}x_t) \otimes \bigotimes_{s=t+1}^{T} \mathbb{P}_{\mathbf{W}_s}\Big) = \underbrace{\big(f_t^{\gamma_t}\big)_\star \big(\pi_t(\mathrm{d}x_t) \otimes \mathbb{P}_{\mathbf{W}_{t+1}}\big)}_{\pi_{t+1}(\mathrm{d}x_{t+1})} \otimes \bigotimes_{s=t+2}^{T} \mathbb{P}_{\mathbf{W}_s}. \tag{4.113}$$

Notice that this formula is made possible because of the white noise Assumption 4.10.

4. Fourth, we show that, for any $t = 0, \dots, T-1$,

$$\min_{\gamma_t \in \Gamma_t^{\mathbb{X}_t}} \int_{\mathbb{X}_t} \pi_t(\mathrm{d}x) \int_{\mathbb{W}_{t+1}} \mathbb{P}_{\mathbf{W}_{t+1}}(\mathrm{d}w_{t+1}) V_{t+1}\Big(f_t\big(x, \gamma_t(x), w_{t+1}\big)\Big) = \int_{\mathbb{X}_t} \pi_t(\mathrm{d}x) \min_{u \in \mathbb{U}_t} \int_{\mathbb{W}_{t+1}} \mathbb{P}_{\mathbf{W}_{t+1}}(\mathrm{d}w_{t+1}) V_{t+1}\big(f_t(x, u, w_{t+1})\big). \tag{4.114}$$

Recall that, by assumption, the control $\gamma_t^\sharp(x) \in \mathbb{U}_t$ achieves the minimum in (4.107b), that is,

$$\min_{u \in \mathbb{U}_t} \int_{\mathbb{W}_{t+1}} \mathbb{P}_{\mathbf{W}_{t+1}}(\mathrm{d}w_{t+1}) V_{t+1}\big(f_t(x, u, w_{t+1})\big) = \int_{\mathbb{W}_{t+1}} \mathbb{P}_{\mathbf{W}_{t+1}}(\mathrm{d}w_{t+1}) V_{t+1}\Big(f_t\big(x, \gamma_t^\sharp(x), w_{t+1}\big)\Big).$$

Therefore, for any $\gamma_t \in \Gamma_t^{\mathbb{X}_t}$, we have that

$$\begin{aligned}
\int_{\mathbb{W}_{t+1}} \mathbb{P}_{\mathbf{W}_{t+1}}(\mathrm{d}w_{t+1}) V_{t+1}\Big(f_t\big(x, \gamma_t(x), w_{t+1}\big)\Big) \\
&\geq \min_{u \in \mathbb{U}_t} \int_{\mathbb{W}_{t+1}} \mathbb{P}_{\mathbf{W}_{t+1}}(\mathrm{d}w_{t+1}) V_{t+1}\big(f_t(x, u, w_{t+1})\big) \\
&= \int_{\mathbb{W}_{t+1}} \mathbb{P}_{\mathbf{W}_{t+1}}(\mathrm{d}w_{t+1}) V_{t+1}\Big(f_t\big(x, \gamma_t^\sharp(x), w_{t+1}\big)\Big).
\end{aligned}$$

Integrating these inequalities with respect to $\pi_t(\mathrm{d}x)$ and then taking the minimum over $\gamma_t \in \Gamma_t^{\mathbb{X}_t}$, we deduce that

$$\begin{aligned}
\min_{\gamma_t \in \Gamma_t^{\mathbb{X}_t}} \int_{\mathbb{X}_t} \pi_t(\mathrm{d}x) \int_{\mathbb{W}_{t+1}} \mathbb{P}_{\mathbf{W}_{t+1}}(\mathrm{d}w_{t+1}) V_{t+1}\Big(f_t\big(x, \gamma_t(x), w_{t+1}\big)\Big) \\
&\geq \int_{\mathbb{X}_t} \pi_t(\mathrm{d}x) \min_{u \in \mathbb{U}_t} \int_{\mathbb{W}_{t+1}} \mathbb{P}_{\mathbf{W}_{t+1}}(\mathrm{d}w_{t+1}) V_{t+1}\big(f_t(x, u, w_{t+1})\big) \\
&= \int_{\mathbb{X}_t} \pi_t(\mathrm{d}x) \int_{\mathbb{W}_{t+1}} \mathbb{P}_{\mathbf{W}_{t+1}}(\mathrm{d}w_{t+1}) V_{t+1}\Big(f_t\big(x, \gamma_t^\sharp(x), w_{t+1}\big)\Big).
\end{aligned}$$

Since $\gamma_t^\sharp \in \Gamma_t^{\mathbb{X}_t}$ by assumption, we obtain (4.114).
Notice that, to interchange minimization and expectation, we could also have used (3.51) in the finite case, and Sect. 8.3.5 in the general case, under proper assumptions.

5. Fifth, we prove (4.109) by backward induction. We have that $\widetilde{V}_t(\overline{\pi}_T; \varphi_T) = \langle \pi_T \, , V_T \rangle$, by (4.107a). Then, assume that (4.109) holds true for $t+1$. We have

$$\widetilde{V}_t(\overline{\pi}_t; \varphi_T) = \min_{\overline{\gamma}_t \in \Gamma_t^{\mathbb{J}_t}} \widetilde{V}_{t+1}\left(\left(\overline{f}_t^{\overline{\gamma}_t}\right)_\star \overline{\pi}_t; \varphi_T\right),$$

by the general DP equation (4.102b),

$$= \min_{\gamma_t \in \Gamma_t^{\mathbb{X}_t}} \widetilde{V}_{t+1}\left((f_t^{\gamma_t})_\star (\pi_t(\mathrm{d}x_t) \otimes \mathbb{P}_{\mathbf{W}_{t+1}}) \otimes \bigotimes_{s=t+2}^{T} \mathbb{P}_{\mathbf{W}_s}; \varphi_T \right),$$

by (4.113), because $\overline{\pi}_t = \pi_t(\mathrm{d}x_t) \otimes \bigotimes_{s=t+1}^{T} \mathbb{P}_{\mathbf{W}_s}$, and by the correspondence between $\Gamma_t^{\mathbb{J}_t}$ and $\Gamma_t^{\mathbb{X}_t}$ induced by (4.110),

$$= \min_{\gamma_t \in \Gamma_t^{\mathbb{X}_t}} \left\langle \underbrace{(f_t^{\gamma_t})_\star (\pi_t(\mathrm{d}x_t) \otimes \mathbb{P}_{\mathbf{W}_{t+1}})}_{\pi_{t+1}(\mathrm{d}x_{t+1})} \, , V_{t+1} \right\rangle,$$

by the induction assumption (4.109), because of the property (4.113),

$$= \min_{\gamma_t \in \Gamma_t^{\mathfrak{X}_t}} \left\langle \pi_t(\mathrm{d}x_t) \otimes \mathbb{P}_{\mathbf{W}_{t+1}} , V_{t+1} \circ f_t^{\gamma_t} \right\rangle ,$$

by the duality relation (4.95),

$$= \min_{\gamma_t \in \Gamma_t^{\mathfrak{X}_t}} \int_{\mathbb{X}_t} \pi_t(\mathrm{d}x) \int_{\mathbb{W}_{t+1}} \mathbb{P}_{\mathbf{W}_{t+1}}(\mathrm{d}w_{t+1}) V_{t+1}\Big(f_t\big(x, \gamma_t(x), w_{t+1}\big)\Big),$$

$$= \int_{\mathbb{X}_t} \pi_t(\mathrm{d}x) \min_{u \in \mathbb{U}_t} \int_{\mathbb{W}_{t+1}} \mathbb{P}_{\mathbf{W}_{t+1}}(\mathrm{d}w_{t+1}) V_{t+1}\big(f_t(x, u, w_{t+1})\big),$$

by (4.114),

$$= \int_{\mathbb{X}_t} \pi_t(\mathrm{d}x) V_t(x),$$

by the classical Bellman Dynamic Programming equation (4.107b),

$$= \langle \pi_t , V_t \rangle .$$

Therefore, we have finally proven (4.109) by backward induction. □

Notice that the classical Bellman value function V_t in (4.107) and the Witsenhausen value function $\widetilde{V}_t$ in (4.102) are related by

$$V_t(x) = \widetilde{V}_t\left(\delta_x \otimes \bigotimes_{s=t+1}^{T} \mathbb{P}_{\mathbf{W}_s}\right). \tag{4.115}$$

Another form of DP equation (9.68) is discussed in Sect. 9.6.2

Why Can We Restrict the Search to Optimal State Feedbacks?
In Sect. 4.4.2, we only considered state feedbacks as information pattern. In other words, we restricted our available information to the current state, neglecting the richer information provided by past noises. Supposing again that we make the white noise Assumption 4.10, we will now sketch why there is no loss of optimality in restricting the search to state feedbacks.

We define

$$h_t = \big(x_0, w_1, w_2, \ldots, w_t, x_t\big), \tag{4.116}$$

the collection of initial state, past noises and current state, which belongs to the set

$$\mathbb{H}_t = \mathbb{X}_0 \times \prod_{s=1}^{t} \mathbb{W}_s \times \mathbb{X}_t, \tag{4.117}$$

equipped with the product σ-field

$$\mathcal{H}_t = \mathcal{X}_0 \otimes \bigotimes_{t=1}^{t} \mathcal{W}_s \otimes \mathcal{X}_t. \tag{4.118}$$

Definition 4.23 A *closed-loop strategy* $\upsilon = \{\upsilon_t\}_{t=0,\dots,T-1}$ is a sequence of measurable mappings $\upsilon_t : (\mathbb{H}_t, \mathcal{H}_t) \to (\mathbb{U}_t, \mathcal{U}_t)$. Accordingly, we set

$$\Upsilon_t^{\mathcal{H}_t} := \left\{ \upsilon_t : (\mathbb{H}_t, \mathcal{H}_t) \to (\mathbb{U}_t, \mathcal{U}_t) \;\middle|\; \upsilon_t^{-1}(\mathcal{U}_t) \subset \mathcal{H}_t \right\}, \tag{4.119}$$

for $t = 0, \dots, T-1$, and the set of all closed-loop strategies is denoted by

$$\Upsilon_{\mathcal{H}} := \prod_{t=0}^{T-1} \Upsilon_t^{\mathcal{H}_t}. \tag{4.120}$$

Obviously, the class of closed-loop strategies (4.120) is larger than that of state feedbacks (4.42):

$$\Gamma_{\mathcal{X}} \subset \Upsilon_{\mathcal{H}}. \tag{4.121}$$

In Proposition 4.22, we will prove that optimizing over closed-loop strategies provides the same value as optimizing over state feedbacks.

For simplicity, and without loosing generality (see Remark 4.9), we suppose that the additive and time-separable criterion in (4.44) is in Bolza form, to be consistent with item 6 in the standard form p. 115.

To account for closed-loop strategies in the definition of an optimal stochastic control problem in standard form (see the beginning of 4.5.2), we introduce—as in Example 4.16—new "state" $\overline{x}_t$, new "state" space $\overline{\mathbb{X}}_t$, new "state" field $\overline{\mathcal{X}}_t$, as well as information fields $\overline{\mathcal{I}}_t$ and dynamics $\overline{f}_t$. We propose the following standard form, only stressing the differences with Example 4.16.

1. The "state" spaces and fields

$$\overline{x}_t = (x_0, w_1, \dots, w_T, x_t), \tag{4.122a}$$

$$\overline{\mathbb{X}}_t = \mathbb{X}_0 \times \prod_{s=1}^{T} \mathbb{W}_s \times \mathbb{X}_t, \tag{4.122b}$$

$$\overline{\mathcal{X}}_t = \mathcal{X}_0 \otimes \bigotimes_{s=1}^{T} \mathcal{W}_s \otimes \mathcal{X}_t , \tag{4.122c}$$

for $t = 0, \ldots, T$, built from the spaces introduced in Sect. 4.4.1.

2. As in Example 4.16.
3. The information fields

$$\overline{\mathfrak{I}}_t = \mathcal{X}_0 \otimes \bigotimes_{s=1}^{t} \mathcal{W}_s \otimes \bigotimes_{s=t+1}^{T} \{\emptyset, \mathbb{W}_s\} \otimes \mathcal{X}_t$$
$$\subset \overline{\mathcal{X}}_t = \mathcal{X}_0 \otimes \bigotimes_{s=1}^{t} \mathcal{W}_s \otimes \bigotimes_{s=t+1}^{T} \mathcal{W}_s \otimes \mathcal{X}_t , \tag{4.123}$$

   for $t = 0, \ldots, T-1$, corresponding to the information structure revealed by the choice of feedbacks in Definition 4.23. Indeed, we thus express that feedbacks can only depend on $h_t = \big(x_0, w_1, w_2, \ldots, w_t, x_t\big)$ and not on future noises $w_{t+1}, \ldots, w_T$.
4. The measurable dynamics $\overline{f}_t : (\overline{\mathbb{X}}_t \times \mathbb{U}_t, \overline{\mathcal{X}}_t \otimes \mathcal{U}_t) \to (\overline{\mathbb{X}}_{t+1}, \overline{\mathcal{X}}_{t+1})$ given by

$$\begin{aligned} \overline{f}_t(\overline{x}_t, u_t) &= \overline{f}_t(x_0, w_1, \ldots, w_T, x_t, u_t) \\ &= \big(x_0, w_1, \ldots, w_T, \underbrace{f_t(x_t, u_t, w_{t+1})}_{x_{t+1}}\big) , \end{aligned} \tag{4.124}$$

   for $t = 0, \ldots, T-1$, built from the original dynamics introduced in Sect. 4.4.1.
5. As in Example 4.16.
6. As in Example 4.16.

To Proposition 4.22 now corresponds the following Proposition 4.24. We only sketch the proof, stressing the differences between the two Propositions.

Proposition 4.24 *Assume that, for all $t \in \{0, \ldots, T-1\}$, there exists a control $\gamma_t^\sharp(x) \in \mathbb{U}_t$ which achieves the minimum in (4.107b) and is such that the mapping $\gamma_t^\sharp : (\mathbb{X}_t, \mathcal{X}_t) \to (\mathbb{U}_t, \mathcal{U}_t)$ is measurable. Then, for all probability $\overline{\pi}_t$ of the form*

$$\overline{\pi}_t(\mathrm{d}\overline{x}_t) = \pi_0(\mathrm{d}x_0) \otimes \bigotimes_{s=1}^{T} \mathbb{P}_{\mathbf{W}_s} \otimes \pi_t(\mathrm{d}x_t) \tag{4.125}$$

we have that

$$\widetilde{V}_t(\overline{\pi}_t; \varphi_T) = \langle \pi_t \, , V_t \rangle , \tag{4.126}$$

where the Witsenhausen value function $\widetilde{V}_t$ is defined in (4.101) and the Bellman value function V_t is given by (4.107).

As a consequence, optimizing over closed-loop strategies provides the same value as optimizing over state feedbacks. In the course of the proof, we see that the state feedback $\gamma^\sharp$ achieves the optimum among closed-loop strategies.

Proof The proof is by induction. But we need preliminary results. This is why we break the proof in several steps.

1. First, we show that the closed-loop strategies set $\Upsilon_t^{\mathcal{H}_t}$ in (4.119) and the feedbacks set $\Gamma_t^{\overline{\mathfrak{I}}_t}$—in (4.41) with information field (4.123)—are in one-to-one correspondence.
   For this, we map any $\upsilon_t : (\mathbb{H}_t, \mathcal{H}_t) \to (\mathbb{U}_t, \mathcal{U}_t)$ into a feedback $\widehat{\upsilon}_t : (\overline{\mathbb{X}}_t, \overline{\mathcal{X}}_t) \to (\mathbb{U}_t, \mathcal{U}_t)$ by setting
$$\widehat{\upsilon}_t(\overline{x}_t) = \widehat{\upsilon}_t(x_0, w_1, \ldots, w_T, x_t) = \upsilon_t(x_0, w_1, \ldots, w_t, x_t). \tag{4.127}$$

2. Second, for any $\upsilon_t : (\mathbb{H}_t, \mathcal{H}_t) \to (\mathbb{U}_t, \mathcal{U}_t)$, we define
$$f_t^{\upsilon_t}(x_0, w_1, \ldots, w_t, w_{t+1}, x_t)) = f_t\big(x_t, \underbrace{\upsilon_t(x_0, w_1, \ldots, w_t, x_t)}_{u_t}, w_{t+1}\big). \tag{4.128}$$
   Then we show that, when $\widehat{\upsilon}_t$ is given by (4.127), the closed-loop dynamics (4.82) is expressed by
$$\overline{f}_t^{\widehat{\upsilon}_t}(\overline{x}_t) = \big(x_0, w_1, \ldots, w_T, \underbrace{f_t^{\upsilon_t}(x_0, w_1, \ldots, w_t, w_{t+1}, x_t)}_{x_{t+1}}\big). \tag{4.129}$$

3. Third, we deduce from the closed-loop dynamics expression (4.129) that
$$\begin{aligned}&\Big(\overline{f}_t^{\widehat{\upsilon}_t}\Big)_\star \Big(\pi_0(\mathrm{d}x_0) \otimes \bigotimes_{s=1}^{T} \mathbb{P}_{\mathbf{W}_s} \otimes \pi_t(\mathrm{d}x_t)\Big) \\ &= \pi_0(\mathrm{d}x_0) \otimes \bigotimes_{s=1}^{T} \mathbb{P}_{\mathbf{W}_s} \otimes \underbrace{\big(f_t^{\upsilon_t}\big)_\star \Big(\pi_0(\mathrm{d}x_0) \otimes \bigotimes_{s=1}^{t+1} \mathbb{P}_{\mathbf{W}_s} \otimes \pi_t(\mathrm{d}x_t)\Big)}_{\pi_{t+1}(\mathrm{d}x_{t+1})}.\end{aligned} \tag{4.130}$$

4. Fourth, we show that, for any $t = 0, \ldots, T-1$,
$$\min_{\upsilon_t \in \Upsilon_t^{\mathcal{H}_t}} \int_{\mathbb{X}_t} \pi_t(\mathrm{d}x) \int_{\prod_{s=1}^{t+1} \mathbb{W}_s} \pi_0(\mathrm{d}x_0) \otimes \bigotimes_{s=1}^{t+1} \mathbb{P}_{\mathbf{W}_s}(\mathrm{d}w_s) \\ V_{t+1}\Big(f_t\big(x, \upsilon_t(x_0, w_1, \ldots, w_t, x), w_{t+1}\big)\Big)$$

$$= \int_{\mathbb{X}_t} \pi_t(\mathrm{d}x) \min_{u \in \mathbb{U}_t} \int_{\mathbb{W}_{t+1}} \mathbb{P}_{\mathbf{W}_{t+1}}(\mathrm{d}w_{t+1}) V_{t+1}\big(f_t(x, u, w_{t+1})\big). \tag{4.131}$$

Recall that, by assumption, the control $\gamma_t^\sharp(x) \in \mathbb{U}_t$ achieves the minimum in (4.107b), that is,

$$\min_{u \in \mathbb{U}_t} \int_{\mathbb{W}_{t+1}} \mathbb{P}_{\mathbf{W}_{t+1}}(\mathrm{d}w_{t+1}) V_{t+1}\big(f_t(x, u, w_{t+1})\big) = \int_{\mathbb{W}_{t+1}} \mathbb{P}_{\mathbf{W}_{t+1}}(\mathrm{d}w_{t+1}) V_{t+1}\Big(f_t\big(x, \gamma_t^\sharp(x), w_{t+1}\big)\Big).$$

Therefore, for any $v_t \in \Upsilon_t^{\mathcal{H}_t}$, we have that

$$\begin{aligned}
&\int_{\mathbb{W}_{t+1}} \mathbb{P}_{\mathbf{W}_{t+1}}(\mathrm{d}w_{t+1}) V_{t+1}\Big(f_t\big(x, v_t(x_0, w_1, \ldots, w_t, x), w_{t+1}\big)\Big) \\
&\qquad \geq \min_{u \in \mathbb{U}_t} \int_{\mathbb{W}_{t+1}} \mathbb{P}_{\mathbf{W}_{t+1}}(\mathrm{d}w_{t+1}) V_{t+1}\big(f_t(x, u, w_{t+1})\big) \\
&\qquad = \int_{\mathbb{W}_{t+1}} \mathbb{P}_{\mathbf{W}_{t+1}}(\mathrm{d}w_{t+1}) V_{t+1}\Big(f_t\big(x, \gamma_t^\sharp(x), w_{t+1}\big)\Big).
\end{aligned}$$

Integrating these inequalities with respect to $\pi_t(\mathrm{d}x) \otimes \pi_0(\mathrm{d}x_0) \otimes \bigotimes_{s=1}^{t} \mathbb{P}_{\mathbf{W}_s}(\mathrm{d}w_s)$ and then taking the minimum over $v_t \in \Upsilon_t^{\mathcal{H}_t}$, we deduce that

$$\begin{aligned}
&\min_{v_t \in \Upsilon_t^{\mathcal{H}_t}} \int_{\mathbb{X}_t} \pi_t(\mathrm{d}x) \int_{\prod_{s=1}^{t+1} \mathbb{W}_s} \pi_0(\mathrm{d}x_0) \otimes \bigotimes_{s=1}^{t+1} \mathbb{P}_{\mathbf{W}_s}(\mathrm{d}w_s) \\
&\qquad\qquad V_{t+1}\Big(f_t\big(x, v_t(x_0, w_1, \ldots, w_t, x), w_{t+1}\big)\Big) \\
&\quad \geq \int_{\mathbb{X}_t} \pi_t(\mathrm{d}x) \min_{u \in \mathbb{U}_t} \int_{\mathbb{W}_{t+1}} \mathbb{P}_{\mathbf{W}_{t+1}}(\mathrm{d}w_{t+1}) V_{t+1}\big(f_t(x, u, w_{t+1})\big) \\
&\quad = \int_{\mathbb{X}_t} \pi_t(\mathrm{d}x) \int_{\mathbb{W}_{t+1}} \mathbb{P}_{\mathbf{W}_{t+1}}(\mathrm{d}w_{t+1}) V_{t+1}\Big(f_t\big(x, \gamma_t^\sharp(x), w_{t+1}\big)\Big).
\end{aligned}$$

Since $\gamma_t^\sharp \in \Gamma_t^{\mathcal{X}_t}$ by assumption, and since $\Gamma_t^{\mathcal{X}_t} \subset \Upsilon_t^{\mathcal{H}_t}$ by (4.121), we obtain (4.131). In other words, the state feedback $\gamma^\sharp$ achieves the optimum among closed-loop strategies.

5. Fifth, we prove (4.109) by backward induction.
We have that $\widetilde{V}_t(\overline{\pi}_T; \varphi_T) = \langle \pi_T \,, V_T \rangle$, by (4.107a). Then, assume that (4.109) holds true for $t+1$. We have

$$\widetilde{V}_t(\overline{\pi}_t; \varphi_T) = \min_{\overline{\gamma}_t \in \Gamma_t^{\mathcal{I}_t}} \widetilde{V}_{t+1}\Big(\big(\overline{f}_t^{\overline{v}_t}\big)_\star \overline{\pi}_t; \varphi_T\Big),$$

by the general DP equation (4.102b),

$$= \min_{v_t \in \Upsilon_t^{\mathcal{H}_t}} \widetilde{V}_{t+1}\left(\pi_0(\mathrm{d}x_0) \otimes \bigotimes_{s=1}^{T} \mathbb{P}_{\mathbf{W}_s} \otimes \right. \left. (f_t^{v_t})_\star \left(\pi_0(\mathrm{d}x_0) \otimes \bigotimes_{s=1}^{t+1} \mathbb{P}_{\mathbf{W}_s} \otimes \pi_t(\mathrm{d}x_t)\right); \varphi_T\right),$$

by (4.130), because $\overline{\pi}_t(\mathrm{d}\overline{x}_t) = \pi_0(\mathrm{d}x_0) \otimes \bigotimes_{s=1}^{T} \mathbb{P}_{\mathbf{W}_s} \otimes \pi_t(\mathrm{d}x_t)$, and by the correspondence induced by (4.127),

$$= \min_{v_t \in \Upsilon_t^{\mathcal{H}_t}} \left\langle \underbrace{(f_t^{v_t})_\star \left(\pi_0(\mathrm{d}x_0) \otimes \bigotimes_{s=1}^{t+1} \mathbb{P}_{\mathbf{W}_s} \otimes \pi_t(\mathrm{d}x_t)\right)}_{\pi_{t+1}(\mathrm{d}x_{t+1})}, V_{t+1} \right\rangle,$$

by the induction assumption (4.109), because of the property (4.130),

$$= \min_{v_t \in \Upsilon_t^{\mathcal{H}_t}} \left\langle \pi_0(\mathrm{d}x_0) \otimes \bigotimes_{s=1}^{t+1} \mathbb{P}_{\mathbf{W}_s} \otimes \pi_t(\mathrm{d}x_t), V_{t+1} \circ f_t^{v_t} \right\rangle,$$

by the duality relation (4.95),

$$= \min_{v_t \in \Upsilon_t^{\mathcal{H}_t}} \int_{\mathbb{X}_t} \pi_t(\mathrm{d}x) \int_{\prod_{s=1}^{t+1} \mathbb{W}_s} \pi_0(\mathrm{d}x_0) \otimes \bigotimes_{s=1}^{t+1} \mathbb{P}_{\mathbf{W}_s}(\mathrm{d}w_s) \, V_{t+1}\Big(f_t\big(x, v_t(x_0, w_1, \ldots, w_t, x), w_{t+1}\big)\Big)$$

by (4.128)

$$= \int_{\mathbb{X}_t} \pi_t(\mathrm{d}x) \min_{u \in \mathbb{U}_t} \int_{\mathbb{W}_{t+1}} \mathbb{P}_{\mathbf{W}_{t+1}}(\mathrm{d}w_{t+1}) V_{t+1}\big(f_t(x, u, w_{t+1})\big)$$

by (4.131),

$$= \int_{\mathbb{X}_t} \pi_t(\mathrm{d}x) V_t(x)$$

by the classical Bellman Dynamic Programming equation (4.107b)

$$= \langle \pi_t , V_t \rangle .$$

Therefore, we have finally proven (4.126) by backward induction. □

4.6 Conclusion

This chapter introduced dynamical stochastic optimization issues, with an emphasis upon the interplay between information and decision. We have highlighted how information patterns are important components of stochastic decision problems: changing information patterns can turn an easy problem (like the classical linear-quadratic optimization one) into an open problem (like the Witsenhausen Counterexample). We have seen how stochastic optimization problems formulated within the framework of state models can be solved by Dynamic Programming when an optimal solution is searched among functions of the state. The celebrated Bellman equation for time-additive criterion and independent noises has been given. At last, following Witsenhausen, we have shown that a Dynamic Programming equation is a general feature of sequential systems. However, the value function is defined over an infinite dimensional space of signed measures, and the DP equation involves a minimum over feedbacks. Specific forms of this Dynamic Programming equation, like the Bellman equation, appear as special cases under specific information patterns. Interestingly, the Bellman value function is defined over a finite dimensional space, and the Bellman equation involves a minimum over controls and not over feedbacks. A guiding thread throughout this chapter is the notion of state. A state summarizes the past history, being a sufficient statistic for the uncertainty and past controls. Whereas history takes value in an increasing space as time t increases, a state variable remains in a space independent of t. A state is a way of compressing information for calculating optimal controls.

Chapter 5
Optimality Conditions for Stochastic Optimal Control (SOC) Problems

5.1 Introduction

In this chapter, we consider Stochastic Optimal Control (SOC) problems anew, as introduced in Sect. 1.2.1 (see (1.1)) and discussed in Chap. 4. In that chapter, optimality conditions were derived by using the Dynamic Programming (DP) approach. This approach is based on a *functional* point of view upon informational constraints, as opposed to the *algebraic* point of view: see the discussions at the end of Sect. 1.1.3 and in Chap. 3 which led to Table 3.1. Essentially, as long as informational constraints are initially formulated as *measurability* conditions between decision variables and available observations (this is basically the *algebraic* viewpoint), Proposition 3.46 showed that there exists a translation of those constraints in terms of looking for the decision variables as (measurable) *functions* of the observations.

Another important ingredient of the DP approach is the notion of *state* that has been discussed in Sects. 4.4 and 4.5. When available, this notion eliminates the need to use functions of the whole set of past observations because, then, the state variables represent a sufficient summary.

In the following, we come back to the algebraic viewpoint because it is well suited to a *variational* approach of the optimization problem. By "variational approach", we refer to the attempt to characterize the optimal solution by a set of "stationarity conditions à la Kuhn-Tucker", that is, essentially, to express the orthogonality of the gradient of the cost function to the admissible set at the optimal point. When applied to optimal control problems, this approach leads to some kind of Pontryagin's optimality conditions [124]. Indeed, in *deterministic* and *continuous time* optimal control problems, Pontryagin's Maximum (or Minimum) Principle is a powerful necessary condition. Here, we consider *stochastic* and *discrete time* control problems, and we aim at obtaining a set of stationarity conditions that are used in Chap. 7 to derive numerical resolution methods.

To that purpose, we must express the gradient of the cost function w.r.t. control variables, but since the cost function also involves intermediate "state" variables through the dynamic equations, we must use a well known technique to differentiate

P. Carpentier et al., *Stochastic Multi-Stage Optimization*, Probability Theory and Stochastic Modelling 75, DOI 10.1007/978-3-319-18138-7_5

this composition of functions (cost and dynamics): this goes through the use of "adjoint state (or co-state)" variables. However, all those variables (controls, states, co-states) are indeed discrete time stochastic processes (that, mathematically, are considered as members of L^2 Hilbert spaces) and there are measurability constraints between them to be taken into account when deriving those optimality conditions. Fundamentally, measurability constraints define linear subspaces, and projection operators onto those linear subspaces are *conditional expectations* (see Sects. 1.3.1 and 3.5). Therefore, one may expect to finally come up with an adequate combination of state and co-state equations, and orthogonality conditions for the gradient of the cost functions w.r.t. controls, all involving conditional expectations.

As we shall see, conditional expectations can be embedded more or less deeply into those equations. This yields several forms of the optimality conditions (which in turn provide different numerical schemes). The derivation of those conditions can, however, only be achieved in the situation of Static Information Structure (SIS): refer to Sect. 1.2.2 for this notion. The main reason is that, with SIS, conditional expectations are w.r.t. a "static conditioning", that is, w.r.t. exogenous random variables or processes (noises) which do not depend on the decision variables. Indeed, if that were not the case, we would have to differentiate w.r.t. the conditioning since this conditioning would change with the decisions (see Sect. 4.2.3). But we do not know of a technique to perform this operation, nor of a theory to provide a background to this notion of differentiation.

Nevertheless, despite this difficulty of differentiating expressions involving conditional expectations when the conditioning is w.r.t. *variables* instead of exogenous *data*, we present, under appropriate assumptions, optimality conditions which do involve conditional expectations w.r.t. state variables (the assumptions are the same as those required by the DP approach in order to allow for finite dimensional state variables, that is, the so-called "Markovian case"). Indeed, this is not contradictory since those optimality conditions are *first* obtained by using only "static" conditioning, and then, *afterwards*, by using the appropriate assumptions, it is shown that some proved measurability properties make their transformation into the final form with conditioning w.r.t. state variables possible.

Consequently, some connection with the functional point of view, and in particular with the DP approach, becomes possible. But, as we will discuss in Chap. 7, the differences may provide some numerical advantages to the variational approach over the DP technique in certain circumstances.

5.2 SOC Problems, Formulation and Assumptions

We consider again problems as (1.1) over a fixed time horizon T with, here, some additional constraints upon the control variables. All random variables are defined over a probability space $(\Omega, \mathcal{A}, \mathbb{P})$ and they are assumed to be square integrable.

5.2.1 Dynamics

The system dynamics follows the equations (which must be understood $\mathbb{P}$-a.s.):

$$\boldsymbol{X}_0 = \boldsymbol{W}_0, \tag{5.1a}$$

$$\boldsymbol{X}_{t+1} = f_t(\boldsymbol{X}_t, \boldsymbol{U}_t, \boldsymbol{W}_{t+1}), \ \ \forall t = 0, \ldots, T-1. \tag{5.1b}$$

In these equations,

- the variable $\boldsymbol{W}_t \in \mathcal{W}_t$ is the noise variable at time t, a random variable with values in $\mathbb{W}_t := \mathbb{R}^{d_{w_t}}$; $\mathcal{W}_t = L^2(\Omega, \mathcal{A}, \mathbb{P}; \mathbb{W}_t)$ and we set $\mathcal{W} := \mathcal{W}_0 \times \cdots \times \mathcal{W}_T$; although this is not of a major importance, for simplicity we assume that all dimensions d_{w_t} are equal to the same d_w, except perhaps for d_{w_0} which must match the dimension of $\boldsymbol{X}_0$ (because of (5.1a)); we make the same simplifying assumption for the other variables hereafter;
- the variable $\boldsymbol{U}_t \in \mathcal{U}_t$ is the control variable at time t, a random variable with values in $\mathbb{U}_t := \mathbb{R}^{d_u}$; $\mathcal{U}_t = L^2(\Omega, \mathcal{A}, \mathbb{P}; \mathbb{U}_t)$ and we set $\mathcal{U} := \mathcal{U}_0 \times \cdots \times \mathcal{U}_{T-1}$;
- the variable $\boldsymbol{X}_t \in \mathcal{X}_t$ is the "state" variable at time t, a random variable with values in $\mathbb{X}_t := \mathbb{R}^{d_x}$; $\mathcal{X}_t = L^2(\Omega, \mathcal{A}, \mathbb{P}; \mathbb{X}_t)$ and we set $\mathcal{X} := \mathcal{X}_0 \times \cdots \times \mathcal{X}_T$.

Of course, due to (5.1a), $\mathbb{X}_0$ is the same as $\mathbb{W}_0$.

Whenever we refer to $\boldsymbol{W}$ without subscript t (and likewise for $\boldsymbol{U}$ and $\boldsymbol{X}$), we refer to an element of $\mathcal{W}$, that is, the stochastic process $(\boldsymbol{W}_0, \ldots, \boldsymbol{W}_T)$.

Functions f_t are assumed to be continuously differentiable with respect to their first two arguments (state and control) and to be normal integrands (see Definition 8.22).

Remark 5.1 As in footnote 3 of Chap. 1, here we use the term "state" in a rather loose sense. The definition of this term will be sharpened (as is the case in Sect. 4.4) when moving to the Markovian case later on in Sect. 5.5. ◊

5.2.2 Cost Function

The performance of the system is measured by the expectation of a cost function involving an integral and a final costs:

$$\widetilde{j}(u, x, w) := \sum_{t=0}^{T-1} L_t(x_t, u_t, w_{t+1}) + K(x_T), \tag{5.2a}$$

$$\widetilde{J}(\boldsymbol{U}, \boldsymbol{X}) := \mathbb{E}\big(\widetilde{j}(\boldsymbol{U}, \boldsymbol{X}, \boldsymbol{W})\big). \tag{5.2b}$$

We delay to Remark 5.4 comments that this latter equation deserves in connection with the discussion in p. xv of the preamble on notations.

Functions L_t are assumed to be continuously differentiable with respect to their first two arguments (state and control) and to be normal integrands. Function K is also assumed to be continuously differentiable.

5.2.3 Constraints

The control variables $\boldsymbol{U}_t$, in the hand of the decision maker whose purpose is to minimize the cost function (5.2), are subject to two kinds of constraints.[1] On the one hand, each $\boldsymbol{U}_t$ is subject to almost-sure constraints described hereafter. On the other hand, decisions are based on information progressively available on line, and these constraints are expressed by measurability constraints upon the control process.

More precisely, for every t, consider a set-valued mapping $\boldsymbol{C}_t : \Omega \rightrightarrows \mathbb{U}_t$. Then, $\boldsymbol{U}_t$ is subject to the almost sure constraint:

$$\boldsymbol{U}_t(\omega) \in \boldsymbol{C}_t(\omega) \ \mathbb{P}\text{-a.s.}. \tag{5.3}$$

The subsets $\boldsymbol{C}_t(\omega)$ are assumed to be $\mathbb{P}$-a.s. closed and convex subsets. Then, the subset $\mathcal{U}_t^{\mathrm{a}} \subset \mathcal{U}_t$ of controls satisfying (5.3) is also closed and convex. We set $\mathcal{U}^{\mathrm{a}} := \mathcal{U}_0^{\mathrm{a}} \times \cdots \times \mathcal{U}_{T-1}^{\mathrm{a}}$.

As discussed in Sect. 1.2.1, the control process must at least be *causal* (or *nonanticipative*). In order to express this property mathematically, we introduce the σ-field $\mathcal{F}_t$ generated by the $\boldsymbol{W}_s$ from $s = 0$ to t (see (1.1c)—as mentioned there, the collection $\mathcal{F}$ of subfields $\mathcal{F}_t$ of $\mathcal{A}$ which is such that $\mathcal{F}_t \subset \mathcal{F}_{t+1}$ is a *filtration*). Then, $\boldsymbol{U}_t$ must at least be measurable w.r.t. $\mathcal{F}_t$, which we write $\boldsymbol{U}_t \preceq \mathcal{F}_t$ (see Definition 3.44). This is the case when the decision maker has a complete observation and a perfect memory of past noises.

However, it may be that the information available to the decision maker at time t is smaller than $\mathcal{F}_t$. We thus introduce another subfield $\mathcal{G}_t \subset \mathcal{F}_t$ and we require that

$$\boldsymbol{U}_t \preceq \mathcal{G}_t. \tag{5.4}$$

This constraint defines a linear subspace of $\mathcal{U}_t$ denoted $\mathcal{U}_t^{\mathrm{b}}$ and we set $\mathcal{U}^{\mathrm{b}} = \mathcal{U}_0^{\mathrm{b}} \times \cdots \times \mathcal{U}_{T-1}^{\mathrm{b}}$. We set $\mathcal{G} := \{\mathcal{G}_t\}_{t=0,\dots,T-1}$.

Remark 5.2 Stochastic processes belonging to $\mathcal{U}^{\mathrm{b}}$ are called "$\mathcal{G}$-adapted" even though we do not necessarily assume that $\mathcal{G}$ is a filtration, that is, that memory is perfect (some information may be "forgotten" when time passes). ◊

As mentioned in Sect. 1.2.1, the subfield $\mathcal{G}_t$ may be generated by an *observation* variable $\boldsymbol{Y}_t$, representing the collection of *all* observations available to the decision

[1] Other types of constraints advocated in Sect 1.1.2 might also be considered but we limit ourselves to those introduced hereafter.

maker at time t. What is important to stress here is that $\boldsymbol{Y}_t$ or $\mathcal{G}_t$ should *not* depend upon past decisions. This is a restriction, but an essential one to stay in the realm of SIS as opposed to DIS (see Chap. 1).

Finally, the control process is constrained to belong to $\mathcal{U}^{\mathrm{a}} \cap \mathcal{U}^{\mathrm{b}}$ defined as the combination of (5.3) and (5.4) for $t = 0, \ldots, T-1$. However, an additional assumption has to be made in order to prevent this feasible set from being empty. From the intuitive point of view, one cannot reasonably impose constraints like (5.3), in which the right-hand side is a random set, if this random set is not fully determined by the information available to the decision maker at time t. Otherwise stated, it is wise to require the following.

Assumption 5.3 The set-valued random mapping $\boldsymbol{C}_t$ assumes nonempty, closed, convex values $\mathbb{P}$-a.s. and is $\mathcal{G}_t$-measurable for all t.

About such multivalued mappings and their measurability, refer to Sect. 8.3.1.

From the mathematical point of view, assuming for example that $\mathcal{G}_t$ is generated by an observation $\boldsymbol{Y}_t$, for any pair (ω, ω') such that $\boldsymbol{Y}_t(\omega) = \boldsymbol{Y}_t(\omega')$, then $\boldsymbol{U}_t(\omega)$ must be equal to $\boldsymbol{U}_t(\omega')$ according to (5.4). But at the same time, if $\boldsymbol{C}_t(\omega) \cap \boldsymbol{C}_t(\omega') = \emptyset$, and if this situation happens with a positive probability, then constraints (5.3) and (5.4) are incompatible. This is prevented by the above assumption which implies that $\boldsymbol{C}_t(\omega) = \boldsymbol{C}_t(\omega')$ for any such pair (ω, ω').

We refer the reader back to Remark 1.1 regarding the decision-hazard scheme adopted here.

5.2.4 The Stochastic Programming (SP) Version

The problem of minimizing the cost function (5.2) under (5.1), (5.3) and (5.4) is amenable to an SP formulation (see Sect. 1.2.2 and Problem (1.3)) by considering that $\boldsymbol{X}$ is just an intermediate stochastic process which is completely determined by $\boldsymbol{U}$ and $\boldsymbol{W}$ through (5.1) (this is related to the notion of "state map"—see Definition 4.6—except that, here, the state trajectory is generated by a sequence of controls $\boldsymbol{U}_t$ rather than by a sequence of feedbacks γ_t). Following this path, we define $j(\boldsymbol{U}, \boldsymbol{W})$ as the value of $\tilde{j}(\boldsymbol{U}, \boldsymbol{X}, \boldsymbol{W})$ (see (5.2a)) once $\boldsymbol{X}$ has been expressed as a function of $\boldsymbol{U}$ and $\boldsymbol{W}$, and

$$J(\boldsymbol{U}) := \mathbb{E}\big(j(\boldsymbol{U}, \boldsymbol{W})\big). \tag{5.5}$$

Then, the SOC problem boils down to

$$\min_{\boldsymbol{U} \in \mathcal{U}} J(\boldsymbol{U}) \quad \text{s.t.} \quad \boldsymbol{U} \in \mathcal{U}^{\mathrm{a}} \cap \mathcal{U}^{\mathrm{b}}. \tag{5.6}$$

Remark 5.4 We refer the reader back to the discussion in p. xv of the preamble on notations: according to that discussion, instead of (5.6), we should write

$$\min_{[U]\in\mathcal{U}} J\big([U]\big) \quad \text{s.t.} \quad [U] \in \mathcal{U}^{\mathrm{a}} \cap \mathcal{U}^{\mathrm{b}},$$

and $J\big([U]\big)$ in (5.5) too, in order to stress the difference between the meaning the symbol $\boldsymbol{U}$ has there (which refers to the stochastic process *as a whole*) and the meaning it has in $j(\boldsymbol{U}, \boldsymbol{W})$ in which $\boldsymbol{U}$ and $\boldsymbol{W}$ are shortcuts for the values $\boldsymbol{U}(\omega)$ and $\boldsymbol{W}(\omega)$, ω being the variable w.r.t. which the mathematical expectation (that is, an integral) is computed. Without this subtle difference in mind, an equation such as (5.5) is hardly intelligible. As stated earlier, we nevertheless keep notations simple as in (5.5) and (5.6). ◇

Necessary optimality conditions for this problem essentially express that the gradient of J at a solution $\boldsymbol{U}^\sharp$, denoted $\nabla J(\boldsymbol{U}^\sharp)$, must belong to the cone orthogonal to the constraints at the point $\boldsymbol{U}^\sharp$, which can also be expressed with help of the orthogonal projector onto the feasible set at this point. Therefore, we need two kinds of results:

- how to compute the gradient of J which has been defined rather implicitly;
- how to operate a projection onto the intersection of subsets

$$\mathcal{U}^{\mathrm{ad}} := \mathcal{U}^{\mathrm{a}} \cap \mathcal{U}^{\mathrm{b}}.$$

The next section is devoted to the latter issue, and the former is addressed in Sect. 5.4.1.

5.3 Optimality Conditions for the SP Formulation

We first examine a specific property of the projection operation onto the feasible set involved in Problem (5.6), and then use it to express optimality conditions for this problem.

5.3.1 Projection on the Feasible Set

Projection on Constraints (5.4)
Recall that $\mathcal{U}$ is considered to be the Hilbert space which is the Cartesian product of $\mathcal{U}_t = L^2(\Omega, \mathcal{A}, \mathbb{P}; \mathbb{U}_t)$ for $t = 0, \ldots, T-1$. Constraints (5.4) define a closed linear subspace $\mathcal{U}^{\mathrm{b}}$ of $\mathcal{U}$, namely the stochastic processes which are $\mathcal{G}$-adapted. This subspace $\mathcal{U}^{\mathrm{b}}$ may also be considered as the Cartesian product of subspaces $\mathcal{U}_t^{\mathrm{b}}$. The projection on $\mathcal{U}_t^{\mathrm{b}}$, that is, on the subspace of $\mathcal{G}_t$-measurable random variables, is well-known: it is the conditional expectation $\mathbb{E}(\cdot \mid \mathcal{G}_t)$ (see Definition B.5).

Projection on Constraints (5.3)

Constraints (5.3) define another closed convex subset $\mathcal{U}^{\mathrm{a}}$ of $\mathcal{U}$ which can also be considered as the Cartesian product of subsets $\mathcal{U}_t^{\mathrm{a}}$, namely the subsets of random variables taking their values in $\boldsymbol{C}_t$, $\mathbb{P}$-a.s., for which the projection also operates separately for each value of t. This projection is denoted $\mathrm{proj}_{\mathcal{U}_t^{\mathrm{a}}}$.

The next lemma shows that $\mathrm{proj}_{\mathcal{U}_t^{\mathrm{a}}}$ operates "ω per ω" (pointwise). In this lemma, we drop the subscript t to alleviate notation.

Lemma 5.5 *For any random variable* $\boldsymbol{U}$, $\mathrm{proj}_{\mathcal{U}^{\mathrm{a}}}\big(\boldsymbol{U}\big)$ *is the random variable* $\omega \mapsto \mathrm{proj}_{\boldsymbol{C}(\omega)}\big(\boldsymbol{U}(\omega)\big)$.

Proof By definition, $\mathrm{proj}_{\mathcal{U}^{\mathrm{a}}}\big(\boldsymbol{U}\big)$ is the random variable $\boldsymbol{V}$ which minimizes the distance to $\boldsymbol{U}$ and such that $\boldsymbol{V}(\omega) \in \boldsymbol{C}(\omega)$, $\mathbb{P}$-a.s.. For any subset A, we introduce the characteristic function $\chi_A(x)$ (see (A.2)). Then $\mathrm{proj}_{\mathcal{U}^{\mathrm{a}}}\big(\boldsymbol{U}\big)$ is the arg min of the minimization problem:

$$\min_{\boldsymbol{V}\in\mathcal{U}} \frac{1}{2}\left\|\boldsymbol{V}-\boldsymbol{U}\right\|_{\mathcal{U}}^2 + \chi_{\mathcal{U}^{\mathrm{a}}}(\boldsymbol{V}).$$

According to Remark 8.28, this problem can also be formulated as:

$$\min_{\boldsymbol{V}\in\mathcal{U}} \int \left(\frac{1}{2}\left\|\boldsymbol{V}(\omega)-\boldsymbol{U}(\omega)\right\|_{\mathbb{U}}^2 + \chi_{\boldsymbol{C}(\omega)}\big(\boldsymbol{V}(\omega)\big)\right) \mathrm{d}\mathbb{P}(\omega).$$

Using the interchange of minimization and integration (see Sect. 8.3.5), the minimization can be done in a pointwise manner under the integral sign, which completes the proof. □

Projection on the Intersection $\mathcal{U}^{\mathrm{a}} \cap \mathcal{U}^{\mathrm{b}}$

The projection on $\mathcal{U}^{\mathrm{a}} \cap \mathcal{U}^{\mathrm{b}}$ is composed of the projections on $\mathcal{U}_t^{\mathrm{a}} \cap \mathcal{U}_t^{\mathrm{b}}$ for $t = 0, \dots T-1$. We are going to show that these projections on intersections of two subsets can be expressed in a simple way using the projectors on the two subsets separately, which is not generally true. This particular property arises from a special feature established by the next lemma.

Suppose that, in Lemma 5.5, $\boldsymbol{U} \in \mathcal{U}^{\mathrm{b}}$, that is, $\boldsymbol{U}_t$ is $\mathcal{G}_t$-measurable for all t. Then the following lemma shows that $\mathrm{proj}_{\mathcal{U}_t^{\mathrm{a}}}\big(\boldsymbol{U}_t\big)$ also belongs to $\mathcal{U}_t^{\mathrm{b}}$. Again, in this lemma, we drop the subscript t.

Lemma 5.6 *Under Assumption 5.3,* $\mathrm{proj}_{\mathcal{U}^{\mathrm{a}}}\big(\mathcal{U}^{\mathrm{b}}\big) \subset \mathcal{U}^{\mathrm{b}}$.

Proof Consider any $\boldsymbol{U} \in \mathcal{U}^{\mathrm{b}}$. From Lemma 5.5, we know that

$$\big(\mathrm{proj}_{\mathcal{U}^{\mathrm{a}}}\big(\boldsymbol{U}\big)\big)(\omega) = \arg\min_{v\in\boldsymbol{C}(\omega)} \frac{1}{2}\left\|v-\boldsymbol{U}(\omega)\right\|_{\mathbb{U}}^2, \quad \mathbb{P}\text{-a.s..}$$

In the right-hand side, all random elements, namely $\boldsymbol{U}$ and $\boldsymbol{C}$, are $\mathcal{G}$-measurable. From [7, Theorem 8.2.11] (measurability of marginal functions), we conclude that the arg min $\text{proj}_{\mathcal{U}^{\text{a}}}\big(\boldsymbol{U}\big)$ is also a $\mathcal{G}$-measurable function, which means that $\text{proj}_{\mathcal{U}^{\text{a}}}\big(\boldsymbol{U}\big) \in \mathcal{U}^{\text{b}}$. □

We can now state the main proposition about the structure of the projection on $\mathcal{U}^{\text{a}} \cap \mathcal{U}^{\text{b}}$ (subscript t is once again dropped).

Proposition 5.7 *Consider a closed linear subspace $\mathcal{U}^{\text{b}}$ of a Hilbert space $\mathcal{U}$ and a closed convex subset $\mathcal{U}^{\text{a}}$ of $\mathcal{U}$. Assume that* $\text{proj}_{\mathcal{U}^{\text{a}}}\big(\mathcal{U}^{\text{b}}\big) \subset \mathcal{U}^{\text{b}}$. *Then*

$$\text{proj}_{\mathcal{U}^{\text{a}} \cap \mathcal{U}^{\text{b}}} = \text{proj}_{\mathcal{U}^{\text{a}}} \circ \text{proj}_{\mathcal{U}^{\text{b}}}. \tag{5.7}$$

Proof Let $x \in \mathcal{U}$, $y \in \mathcal{U}^{\text{a}} \cap \mathcal{U}^{\text{b}}$, $z = \text{proj}_{\mathcal{U}^{\text{b}}}(x)$ and $v = \text{proj}_{\mathcal{U}^{\text{a}}}(z)$. The following variational inequality characterizes the projection v of z (see Example A.11):

$$\langle z - v, s - v \rangle \leq 0, \quad \forall s \in \mathcal{U}^{\text{a}}. \tag{5.8}$$

Therefore,

$$\begin{aligned} \langle x - v, y - v \rangle &= \langle x - z, y - v \rangle + \langle z - v, y - v \rangle \\ &\leq \langle x - z, y - v \rangle \end{aligned}$$

since the second scalar product in the right-hand side of the former line above is non positive (see (5.8) with y, which does belong to $\mathcal{U}^{\text{a}}$, instead of s). Now, since $z \in \mathcal{U}^{\text{b}}$ and thanks to the assumption that $\text{proj}_{\mathcal{U}^{\text{a}}}\big(\mathcal{U}^{\text{b}}\big) \subset \mathcal{U}^{\text{b}}$, then $v \in \mathcal{U}^{\text{b}}$; moreover, $y - v \in \mathcal{U}^{\text{b}}$ too, since $y \in \mathcal{U}^{\text{b}}$ and this is a linear subspace. On the other hand, since $z = \text{proj}_{\mathcal{U}^{\text{b}}}(x)$, then $x - z$ is orthogonal to the subspace $\mathcal{U}^{\text{b}}$, hence $\langle x - z, y - v \rangle = 0$. Finally,

$$\langle x - v, y - v \rangle \leq 0, \quad \forall x \in \mathcal{U}, \quad \forall y \in \mathcal{U}^{\text{a}} \cap \mathcal{U}^{\text{b}},$$

an inequality which characterizes $v = \text{proj}_{\mathcal{U}^{\text{a}}} \circ \text{proj}_{\mathcal{U}^{\text{b}}}(x)$ as the projection of x on $\mathcal{U}^{\text{a}} \cap \mathcal{U}^{\text{b}}$. □

5.3.2 Stationary Conditions

Consider the minimization of a convex differentiable function $G : \mathcal{U} \to \mathbb{R}$ over a closed convex subset $\mathcal{U}^{\text{ad}}$ of a Hilbert space $\mathcal{U}$. We consider the characteristic function $\chi_{\mathcal{U}^{\text{ad}}}$ as defined by (A.2). Its subdifferential $\partial\chi_{\mathcal{U}^{\text{ad}}}(u)$ is the normal cone to $\mathcal{U}^{\text{ad}}$ at u (see Example A.6).

The following statements are three equivalent, necessary, and sufficient conditions for $u^\sharp \in \mathcal{U}^{\text{ad}}$ to be optimal (see Eq. (A.7) for those optimality conditions and Sect. A.1.4 for the notation ∇ and the gradient notion):

$$\forall u \in \mathcal{U}^{\text{ad}}, \ \left\langle \nabla G(u^\sharp) \, , u - u^\sharp \right\rangle \geq 0, \tag{5.9a}$$

$$\nabla G(u^\sharp) \in -\partial \chi_{\mathcal{U}^{\text{ad}}}(u^\sharp), \tag{5.9b}$$

$$\forall \varepsilon \geq 0, \ u^\sharp = \text{proj}_{\mathcal{U}^{\text{ad}}} \left(u^\sharp - \varepsilon \nabla G(u^\sharp) \right). \tag{5.9c}$$

Coming back to (5.6) and using (5.7) in (5.9c), we get

$$\begin{aligned} \forall \varepsilon \geq 0, \ \boldsymbol{U}^\sharp &= \text{proj}_{\mathcal{U}^{\text{a}} \cap \mathcal{U}^{\text{b}}} \left(\boldsymbol{U}^\sharp - \varepsilon \nabla J(\boldsymbol{U}^\sharp) \right) \\ &= \text{proj}_{\mathcal{U}^{\text{a}}} \circ \text{proj}_{\mathcal{U}^{\text{b}}} \left(\boldsymbol{U}^\sharp - \varepsilon \nabla J(\boldsymbol{U}^\sharp) \right) \\ &= \text{proj}_{\mathcal{U}^{\text{a}}} \left(\boldsymbol{U}^\sharp - \varepsilon \, \text{proj}_{\mathcal{U}^{\text{b}}} \left(\nabla J(\boldsymbol{U}^\sharp) \right) \right) \end{aligned} \tag{5.10a}$$

using the linearity of $\text{proj}_{\mathcal{U}^{\text{b}}}$. By (5.9), the last equality is equivalent to

$$\text{proj}_{\mathcal{U}^{\text{b}}} \left(\nabla J(\boldsymbol{U}^\sharp) \right) \in -\partial \chi_{\mathcal{U}^{\text{a}}}(\boldsymbol{U}^\sharp). \tag{5.10b}$$

We now reintroduce the subscript t. As mentioned earlier, $\text{proj}_{\mathcal{U}_t^{\text{b}}}(\cdot) = \mathbb{E}(\cdot \mid \mathcal{G}_t)$. Hence (5.10) can more explicitly be written: for $t = 0, \ldots, T-1$,

$$\boldsymbol{U}_t^\sharp = \text{proj}_{\mathcal{U}_t^{\text{a}}} \left(\boldsymbol{U}_t^\sharp - \varepsilon \mathbb{E} \left(\nabla_{u_t} J(\boldsymbol{U}^\sharp) \mid \mathcal{G}_t \right) \right), \tag{5.11a}$$

$$\mathbb{E} \left(\nabla_{u_t} J(\boldsymbol{U}^\sharp) \mid \mathcal{G}_t \right) \in -\partial \chi_{\mathcal{U}_t^{\text{a}}}(\boldsymbol{U}_t^\sharp). \tag{5.11b}$$

These formulas can indeed be applied not only for each value of t but also for almost all ω (see Lemma 5.5).

5.4 Optimality Conditions for the SOC Formulation

In this section, we come back to SOC problems described in Sect. 5.2 and we use the results obtained so far for the SP formulation in order to derive optimality conditions for those problems. To that purpose, we must first examine how cost gradients can be computed in SOC problems. Then, optimality conditions in two different forms are derived.

5.4.1 Computation of the Cost Gradient

The SP formulation (5.6) has been obtained by conceptually substituting $\boldsymbol{X}$ into the cost function (5.2) as a function of $\boldsymbol{U}$ and $\boldsymbol{W}$ defined by the recursive dynamic Eq. (5.1), and by taking the expectation to obtain the cost $J(\boldsymbol{U})$. This implicit definition of the cost function requires a special technique in order to compute the gradient w.r.t. the decision variable.

Indeed, the random context can be momentarily forgotten in that gradients are first calculated for each sample path ω. In addition, the "noise" $\boldsymbol{W}$ plays no particular role in this matter and can also be provisionally omitted. Therefore, let u denote a whole control history $\{\boldsymbol{U}_t(\omega)\}_{t=0,\dots,T-1}$ for a sample ω and x denote the corresponding state trajectory. The dynamics (5.1) defines a pseudo-implicit vector equation:

$$f(u, x) = 0. \tag{5.12}$$

We say "pseudo-implicit" in that those recursive equations form a *strictly lower-triangular* system out of which a unique well-defined solution x is entirely determined by the choice of u (and indeed also by w that we omitted—see the notion of state map in Sect. 4.4.2). Let now $g(u, x)$ be the corresponding cost (more precisely, the value of $\widetilde{j}(u, x, w)$ (see (5.2a)) for a given sample ω and the corresponding noise history). Since x is uniquely, but implicitly, defined by (5.12) as a function of u only, then $g(u, x)$ can also be considered as a function of u, say $G(u)$. The following lemma shows how one can compute the gradient $\nabla G(u)$ (assuming, of course, all differentiability properties required for this gradient to exist).[2]

Lemma 5.8 *We assume that the implicit Eq. (5.12) admits a unique solution in x for any given u. Let $G(u)$ be the value of $g(u, x)$ when x is so determined by u. Assuming the existence of all derivatives involved, the gradient $\nabla G(u)$ is given by the following rule:*

$$\nabla G(u) = \nabla_u g(u, x) + \nabla_u f(u, x)\lambda \tag{5.13a}$$

in which x is drawn from (5.12) and λ is in the same space as x and such that

$$\nabla_x g(u, x) + \nabla_x f(u, x)\lambda = 0. \tag{5.13b}$$

Proof Since x follows from u out of (5.12), an infinitesimal variation δu causes a variation δx such that

$$\big(\nabla_u f(u, x)\big)^\top \delta u + \big(\nabla_x f(u, x)\big)^\top \delta x = 0, \tag{5.14}$$

and a variation of the cost

$$\big(\nabla G(u)\big)^\top \delta u = \big(\nabla_u g(u, x)\big)^\top \delta u + \big(\nabla_x g(u, x)\big)^\top \delta x. \tag{5.15}$$

Consider any vector λ of the same dimension as $f(u, x)$ or x. We may combine the right-hand side of (5.15) with the scalar product of λ by the left-hand side of (5.14)

---

[2]The reader may refer to Sect. A.1.4 for notations such as $\nabla_u g$, $\nabla_x g$, $\nabla_u f$ and $\nabla_x f$, and especially to Remark A.4 regarding the last two ones.

since the latter is identically zero. This yields:

$$\big(\nabla G(u)\big)^\top \delta u = \big(\nabla_u g(u,x) + \nabla_u f(u,x)\lambda\big)^\top \delta u \\ + \big(\nabla_x g(u,x) + \nabla_x f(u,x)\lambda\big)^\top \delta x. \quad (5.16)$$

If λ is chosen in such a way that (5.13b) is satisfied, then (5.13a) is derived by identifying both sides of (5.16) which holds true for any infinitesimal δu.[3] This completes the proof. □

The result of Lemma 5.8 can be summarized as the following "recipe":

- form the "pseudo-Lagrangian" $\ell(u,x;\lambda) = g(u,x) + \lambda^\top f(u,x)$;
- draw x from $\nabla_\lambda \ell(u,x;\lambda) = 0$ (which does not depend on λ);
- draw λ from $\nabla_x \ell(u,x;\lambda) = 0$;
- with those values, compute the gradient $\nabla_u \ell(u,x;\lambda)$ to obtain $\nabla G(u)$.

This is reminiscent of some sort of Kuhn-Tucker stationary conditions but the gradient is valid here for *any* u, not only for the arg min of some optimization problem.

We now apply this recipe to the expression of the cost $\widetilde{j}(u,x,w)$ (see (5.2a)) (which plays the role of $g(u,x)$ in the lemma) when x is derived from (5.1) playing the role of (5.12). Thus, we consider:

$$\ell(u,x;\lambda) = K(x_T) \\ + \sum_{t=0}^{T-1} \Big(L_t(x_t,u_t,w_{t+1}) + \lambda_{t+1}^\top \big(f_t(x_t,u_t,w_{t+1}) - x_{t+1}\big)\Big) \\ + \lambda_0(w_0 - x_0). \quad (5.17)$$

Of course, $\nabla_\lambda \ell = 0$ yields (5.1) again whereas $\nabla_x \ell = 0$ yields the well-known co-state backward dynamic equations:

$$\lambda_T = \nabla K(x_T), \quad (5.18a)$$

and for $t = 1, \dots, T-1$,

$$\lambda_t = \nabla_x f_t(x_t,u_t,w_{t+1})\lambda_{t+1} + \nabla_x L_t(x_t,u_t,w_{t+1}). \quad (5.18b)$$

The equation for λ_0 has been skipped since λ_0 is not involved in the following cost gradient expression:

$$\nabla_{u_t} j(u,w) = \nabla_u L_t(x_t,u_t,w_{t+1}) \\ + \nabla_u f_t(x_t,u_t,w_{t+1})\lambda_{t+1}, \quad t = 0,\dots,T-1. \quad (5.19)$$

---

[3] Observe that this holds true even if λ out of (5.13b) is non unique, that is, even if $\nabla_x f(u,x)$ is singular.

5.4.2 Optimality Conditions with Non-adapted Co-States

Now that we have an explicit expression of the cost gradient for the SOC formulation, we can specialize the optimality conditions (5.11b). Recall that those conditions are necessary and also sufficient in the convex case (see Theorem A.10). We drop the superscript $\sharp$ but of course all variables are supposed here to assume their optimal values:

$$\boldsymbol{X}_0 = \boldsymbol{W}_0, \tag{5.20a}$$

$$\boldsymbol{X}_{t+1} = f_t(\boldsymbol{X}_t, \boldsymbol{U}_t, \boldsymbol{W}_{t+1}), \ t = 0, \dots, T-1, \tag{5.20b}$$

$$\boldsymbol{\lambda}_T = \nabla K(\boldsymbol{X}_T), \tag{5.20c}$$

$$\boldsymbol{\lambda}_t = \nabla_x f_t(\boldsymbol{X}_t, \boldsymbol{U}_t, \boldsymbol{W}_{t+1})\boldsymbol{\lambda}_{t+1} + \nabla_x L_t(\boldsymbol{X}_t, \boldsymbol{U}_t, \boldsymbol{W}_{t+1}), \quad t = 1, \dots, T-1, \tag{5.20d}$$

$$\mathbb{E}\big(\nabla_u L_t(\boldsymbol{X}_t, \boldsymbol{U}_t, \boldsymbol{W}_{t+1}) + \nabla_u f_t(\boldsymbol{X}_t, \boldsymbol{U}_t, \boldsymbol{W}_{t+1})\boldsymbol{\lambda}_{t+1} \mid \mathcal{G}_t\big) \in -\partial\chi_{\mathcal{U}_t^{\mathrm{a}}}(\boldsymbol{U}_t), \quad t = 0, \dots, T-1. \tag{5.20e}$$

These conditions involve a co-state stochastic process $\boldsymbol{\lambda}$ which is *not* $\mathcal{F}$-adapted since the dynamics (5.20c) and (5.20d) propagate backwards and therefore $\boldsymbol{\lambda}_t$ is not in general $\mathcal{F}_t$-measurable. The next section shows that it is possible to obtain optimality conditions involving another $\mathcal{F}$-adapted co-state process. Let us explain why this is natural.

Indeed, the conditions (5.20) may be viewed as Kuhn-Tucker stationary conditions in which the dynamics (5.1) have been dualized using the $\boldsymbol{\lambda}_t$ as multipliers to form a Lagrangian similar to (5.17) but with random variables as arguments. Let us focus on the particular duality term

$$\big\langle f_t(\boldsymbol{X}_t, \boldsymbol{U}_t, \boldsymbol{W}_{t+1}) - \boldsymbol{X}_{t+1} \,, \boldsymbol{\lambda}_{t+1}\big\rangle_{\mathcal{X}_{t+1}} = \int \Big\langle f_t\big(\boldsymbol{X}_t(\omega), \boldsymbol{U}_t(\omega), \boldsymbol{W}_{t+1}(\omega)\big) - \boldsymbol{X}_{t+1}(\omega) \,, \boldsymbol{\lambda}_{t+1}(\omega)\Big\rangle_{\mathbb{R}^{d_x}} \mathrm{d}\mathbb{P}(\omega).$$

The left-hand side factor of scalar products is $\mathcal{F}_{t+1}$-measurable; therefore, considering a decomposition of $\boldsymbol{\lambda}_{t+1}$ into its $\mathcal{F}_{t+1}$-measurable component, namely

$$\boldsymbol{\Lambda}_{t+1} := \mathbb{E}(\boldsymbol{\lambda}_{t+1} \mid \mathcal{F}_{t+1}), \tag{5.21}$$

on the one hand, and its orthogonal complement $\boldsymbol{\lambda}_{t+1} - \boldsymbol{\Lambda}_{t+1}$on the other hand, only the former component contributes to the scalar product. Hence it should be possible to get optimality conditions involving only that $\boldsymbol{\Lambda}_{t+1}$. This is what we show now.

5.4.3 Optimality Conditions with Adapted Co-States

Recall that $\mathcal{G}_t \subset \mathcal{F}_t \subset \mathcal{F}_{t+1}$. Therefore $\mathbb{E}(\cdot \mid \mathcal{G}_t) = \mathbb{E}\big(\mathbb{E}(\cdot \mid \mathcal{F}_{t+1}) \bigm| \mathcal{G}_t\big)$. We make use of this property in the left-hand side member of (5.20e)

$$\mathbb{E}\Big(\mathbb{E}\big(\nabla_u L_t(\boldsymbol{X}_t, \boldsymbol{U}_t, \boldsymbol{W}_{t+1}) + \nabla_u f_t(\boldsymbol{X}_t, \boldsymbol{U}_t, \boldsymbol{W}_{t+1})\boldsymbol{\lambda}_{t+1} \mid \mathcal{F}_{t+1}\big) \Bigm| \mathcal{G}_t\Big)$$

and we notice that, except for $\boldsymbol{\lambda}_{t+1}$, the other terms involved in the conditional expectation are $\mathcal{F}_{t+1}$-measurable and thus, they "get out" of $\mathbb{E}(\cdot \mid \mathcal{F}_{t+1})$ (see Proposition B.22) which yields

$$\mathbb{E}\Big(\nabla_u L_t(\boldsymbol{X}_t, \boldsymbol{U}_t, \boldsymbol{W}_{t+1}) + \nabla_u f_t(\boldsymbol{X}_t, \boldsymbol{U}_t, \boldsymbol{W}_{t+1})\mathbb{E}(\boldsymbol{\lambda}_{t+1} \mid \mathcal{F}_{t+1}) \Bigm| \mathcal{G}_t\Big),$$

that is, only $\boldsymbol{\lambda}_{t+1}$ is affected by this operator and this makes precisely $\boldsymbol{\Lambda}_{t+1}$, defined by (5.21), pop up in this expression. Hence, the orthogonality condition (5.20e) can equivalently be written:

$$\mathbb{E}\big(\nabla_u L_t(\boldsymbol{X}_t, \boldsymbol{U}_t, \boldsymbol{W}_{t+1}) + \nabla_u f_t(\boldsymbol{X}_t, \boldsymbol{U}_t, \boldsymbol{W}_{t+1})\boldsymbol{\Lambda}_{t+1} \mid \mathcal{G}_t\big) \in -\partial\chi_{\mathcal{U}_t^{\mathrm{a}}}(\boldsymbol{U}_t), \quad t = 0, \ldots, T-1. \tag{5.22a}$$

It remains to show that $\boldsymbol{\Lambda}$ can equally replace $\boldsymbol{\lambda}$ wherever the latter appears in the remaining of the optimality conditions (5.20). There is nothing to say about (5.20a) and (5.20b), nor really about (5.20c) since the right-hand side is obviously $\mathcal{F}_T$-measurable, hence $\boldsymbol{\lambda}_T = \boldsymbol{\Lambda}_T$, and (5.20c) may as well be written:

$$\boldsymbol{\Lambda}_T = \nabla K(\boldsymbol{X}_T). \tag{5.22b}$$

Therefore, we consider applying the operator $\mathbb{E}(\cdot \mid \mathcal{F}_t)$ to both sides of (5.20d) in order to see how $\boldsymbol{\Lambda}_t$ can be expressed. Again, since $\mathcal{F}_t \subset \mathcal{F}_{t+1}$, we use the rule $\mathbb{E}(\cdot \mid \mathcal{F}_t) = \mathbb{E}\big(\mathbb{E}(\cdot \mid \mathcal{F}_{t+1}) \bigm| \mathcal{F}_t\big)$. Except for $\boldsymbol{\lambda}_{t+1}$, all other terms in this right-hand side are $\mathcal{F}_{t+1}$-measurable, hence those terms get out of the inner conditional expectation and $\boldsymbol{\Lambda}_{t+1}$ appears, which yields:

$$\boldsymbol{\Lambda}_t = \mathbb{E}\big(\nabla_x f_t(\boldsymbol{X}_t, \boldsymbol{U}_t, \boldsymbol{W}_{t+1})\boldsymbol{\Lambda}_{t+1} + \nabla_x L_t(\boldsymbol{X}_t, \boldsymbol{U}_t, \boldsymbol{W}_{t+1}) \bigm| \mathcal{F}_t\big), \quad t = 1, \ldots, T-1. \tag{5.22c}$$

To summarize, Eqs. (5.20a), (5.20b) and (5.22) are an alternative set of optimality conditions which involve a co-state process $\boldsymbol{\Lambda}$ which is $\mathcal{F}$-adapted. Note that (5.22a) involves a conditional expectation w.r.t. $\mathcal{G}_t$ whereas, in (5.22c), the conditioning is w.r.t. $\mathcal{F}_t$.

5.5 The Markovian Case

In this section, we specialize the optimality conditions obtained so far to the so-called Markovian problems which are described in the following subsection.

5.5.1 Markovian Setting and Assumptions

In this section, additional assumptions are introduced in order to fit in the situation considered in Sect. 4.4 in which the terminology "state" gets its full meaning. So far, we have seen that stochastic processes such as $\boldsymbol{X}$ and $\boldsymbol{\Lambda}$ are $\mathcal{F}$-adapted (whereas $\boldsymbol{U}$ was constrained to be $\mathcal{G}$-adapted). By Proposition 3.46, this implies that there exists, for example, a measurable function Ξ_t such that $\boldsymbol{X}_t = \Xi_t(\boldsymbol{W}_0, \dots, \boldsymbol{W}_t)$ (and likewise for $\boldsymbol{\Lambda}_t$). An important drawback from the practical point of view of such a situation is that the number of arguments of those functions keeps on increasing with t.

The proper notion of "state" (see Sect. 4.5.3) aims precisely at introducing a stochastic process, generally denoted $\boldsymbol{X}$, such that, at every time instant t, knowing $\boldsymbol{X}_t$ is "enough", that is, this information summarizes the whole past information that is *useful* for the decision problem at hand. Of course, for this notion to be non trivial, this "state variable" $\boldsymbol{X}_t$ should be of a dimension which is *stable* (contrary to $(\boldsymbol{W}_0, \dots, \boldsymbol{W}_t)$ the dimension of which steadily increases) and also as small as possible.

As announced in Remark 5.1 and illustrated by the previous observations, the process $\boldsymbol{X}$ manipulated so far in this chapter did not deserve the terminology of "state" in this more precise meaning. We now introduce the assumptions which endow $\boldsymbol{X}$ with its full role of state variable. Those assumptions are of two kinds:

- on the one hand, the exogenous noise process $\boldsymbol{W}$ should, in some sense, "forget the past", which translates into an independence assumption between successive realizations, known as a "white noise" assumption;
- on the other hand, constraints on the decision process $\boldsymbol{U}$ should not reintroduce past observations.

Assumption 5.9 (*White noise*) The random variables $\boldsymbol{W}_0, \dots, \boldsymbol{W}_T$ are independent.

This assumption may be alleviated but at the price of increasing the state vector dimension. For example, $\boldsymbol{W}$ may be an ARMA[4] process but the additional memory introduced by this process must be incorporated into the state vector [20].

[4] Auto-Regressive Moving Average.

Assumption 5.10 (*Decision constraints*) For $t = 0, \ldots, T-1$,

1. $\mathcal{G}_t = \mathcal{F}_t$.
2. the set-valued mappings $\boldsymbol{C}_t$ involved in constraints (5.3) are constant, that is, deterministic, and are thus denoted C_t.

The former part of this assumption refers to the information available to the decision maker at t: she has full observation of past noise realizations and complete memory of those observations. Under this assumption and the "white noise" Assumption 5.9, it is possible to prove that all past observations can potentially be summarized by $\boldsymbol{X}_t$ only. The additional instantaneous constraints (5.3) on decision values should not reintroduce the past noises, hence the Assumption 5.10-2.

Assumption 5.10-2 may have been less restrictive by assuming that $\boldsymbol{C}_t$ is $\boldsymbol{X}_t$-measurable (as it is allowable in the DP approach [20]). However, such a more general assumption would have reintroduced some dynamics through the constraints as long as the feasible set for $\boldsymbol{U}_t$ would have been dependent on past decisions $\boldsymbol{U}_s$, $s < t$, through $\boldsymbol{X}_t$. In order to avoid this additional source of complexity, we limit ourselves to the present Assumption 5.10-2 (this may be considered a restriction brought by the variational approach followed here).

With the previous assumptions at hand, we revisit the optimality conditions obtained in Sect. 5.4 and show additional measurability properties of the stochastic processes involved.

5.5.2 Optimality Conditions with Non-adapted Co-States

We consider here the optimality conditions (5.20) obtained in Sect. 5.4.2.

Theorem 5.11 *Under the foregoing Assumptions 5.9 and 5.10, considering the stochastic processes involved in (5.20),*

1. *for $t = 1, \ldots, T$, $\boldsymbol{\lambda}_t$ is measurable w.r.t. the subfield generated by $(\boldsymbol{X}_t, \boldsymbol{W}_{t+1}, \ldots, \boldsymbol{W}_T)$;*
2. *for $t = 1, \ldots, T-1$, $\boldsymbol{U}_t$ is measurable w.r.t. the subfield generated by $\boldsymbol{X}_t$;*
3. *as a consequence, Condition (5.20e) (in which $\mathcal{G}_t = \mathcal{F}_t$ according to Assumption 5.10-1) can be replaced by:*

$$\mathbb{E}\big(\nabla_u L_t(\boldsymbol{X}_t, \boldsymbol{U}_t, \boldsymbol{W}_{t+1}) + \nabla_u f_t(\boldsymbol{X}_t, \boldsymbol{U}_t, \boldsymbol{W}_{t+1})\boldsymbol{\lambda}_{t+1} \mid \boldsymbol{X}_t\big) \\ \in -\partial\chi_{\mathcal{U}_t^{\mathrm{a}}}(\boldsymbol{U}_t), \quad t = 0, \ldots, T-1. \tag{5.23}$$

Proof The proof is by induction. Observe first that the first statement is true for $t = T$ since (5.20c) shows that $\boldsymbol{\lambda}_T$ is a deterministic function of $\boldsymbol{X}_T$. We now carry the induction backwards by assuming that, for some t, $\boldsymbol{\lambda}_{t+1}$ is $(\boldsymbol{X}_{t+1}, \boldsymbol{W}_{t+2}, \ldots, \boldsymbol{W}_T)$-measurable, and we prove that the same property holds true one step backward. We also prove that $\boldsymbol{U}_t$ is $\boldsymbol{X}_t$-measurable.

The induction assumption is equivalent to saying that there exist measurable functions μ_{t+1} and ν_{t+1} such that

$$\begin{aligned}\boldsymbol{\lambda}_{t+1} &= \mu_{t+1}(\boldsymbol{X}_{t+1}, \boldsymbol{W}_{t+2}, \ldots, \boldsymbol{W}_T)\\ &= \mu_{t+1}\big(f_t(\boldsymbol{X}_t, \boldsymbol{U}_t, \boldsymbol{W}_{t+1}), \boldsymbol{W}_{t+2}, \ldots, \boldsymbol{W}_T\big)\\ &= \nu_{t+1}(\boldsymbol{X}_t, \boldsymbol{U}_t, \boldsymbol{W}_{t+1}, \ldots, \boldsymbol{W}_T).\end{aligned}$$

We use this equation in (5.20e) (with $\mathcal{G}_t = \mathcal{F}_t$). This yields an equation of the form:

$$\mathbb{E}\big(h_t(\boldsymbol{X}_t, \boldsymbol{U}_t, \boldsymbol{W}_{t+1}, \ldots, \boldsymbol{W}_T) \bigm| \mathcal{F}_t\big) \in -\partial\chi_{\mathcal{U}_t^{\mathrm{a}}}(\boldsymbol{U}_t). \tag{5.24}$$

At the left-hand side of this equation, $\boldsymbol{X}_t$ and $\boldsymbol{U}_t$ are $\mathcal{F}_t$-measurable, so that they are "conditionally deterministic" whereas the random variables $\boldsymbol{W}_s, s = t+1, \ldots, T$, are *independent* of $\mathcal{F}_t$ according to Assumption 5.9; therefore, the conditional expectation amounts to an expectation (or integral) operating on those variables $\boldsymbol{W}_s$ while $\boldsymbol{X}_t$ and $\boldsymbol{U}_t$ are provisionally "frozen" at parametric values x and u (see Proposition B.22).

On the other hand, the multi-valued function present at the right-hand side of (5.24) is entirely determined by the parametric value u according to Assumption 5.10-2. Therefore, this equation may be considered as an implicit equation in u parametrized by x. Hence there should exist a measurable selection of the solution $\boldsymbol{U}_t$ w.r.t. $\boldsymbol{X}_t$.[5]

In order to complete the proof by induction, we refer back to (5.20d) to see that, in the right-hand side of this equation, the whole expression is a function of $\boldsymbol{X}_t, \boldsymbol{W}_{t+1}, \ldots, \boldsymbol{W}_T$.

Finally, going back to (5.24) which was a sketchy representation of (5.20e), one realizes that the expression subject to the conditional expectation $\mathbb{E}(\cdot \mid \mathcal{F}_t)$ involves variables which are either $\boldsymbol{X}_t$-measurable or independent of $\mathcal{F}_t$ (and thus of $\boldsymbol{X}_t$ too): hence, by Proposition B.22, $\mathbb{E}(\cdot \mid \mathcal{F}_t)$ can be replaced by $\mathbb{E}(\cdot \mid \boldsymbol{X}_t)$, which leads to (5.23). □

Remark 5.12 Conditions (5.20) were established by starting from the general optimality conditions (5.9) and by using

- on the one hand, the special property of the projection on the feasible set defined by constraints (5.3) and (5.4) (see (5.7)), which led us to (5.11);
- and, on the other hand, the more explicit definition of the cost function (5.5) in the SOC formulation and of its gradient computation as described in Sect. 5.4.1.

Condition (5.23) has been further derived from (5.20e) in Theorem 5.11 thanks to the measurability properties established in this theorem for $\boldsymbol{\lambda}$ and $\boldsymbol{U}$ by using Assumptions 5.9 and 5.10. In particular, the weaker property $\boldsymbol{U}_t \preceq \boldsymbol{X}_t$ instead of the initial constraint $\boldsymbol{U}_t \preceq \mathcal{F}_t$ is a valuable result.

---

[5]When Ω is finite, this claim is straightforward; otherwise it is rather technical to prove [149].

However, we may have started with the constraint $\boldsymbol{U}_t \preceq \boldsymbol{X}_t$ replacing (5.4) in the definition of $\mathcal{U}^{\mathrm{b}}$, and with Assumption 5.10-2 in constraint (5.3) for the definition of $\mathcal{U}^{\mathrm{a}}$, and we may have attempted to establish the corresponding optimality conditions for this problem directly. This would have been a difficult challenge because, through $\boldsymbol{X}_t$, the feasible set $\mathcal{U}_t^{\mathrm{a}} \cap \mathcal{U}_t^{\mathrm{b}}$ now itself depends on the solution $\boldsymbol{U}$. The indirect path we have followed to derive those optimality conditions has been a way to circumvent this difficulty. The fact that the seemingly implicit definition of the feasible set is only "pseudo-implicit" was probably helpful: indeed, the feasible set for $\boldsymbol{U}_t$ depends upon past decision variables $\boldsymbol{U}_s$, $s < t$, through $\boldsymbol{X}_t$, but not on $\boldsymbol{U}_t$ itself. ◊

5.5.3 Optimality Conditions with Adapted Co-States

The co-state $\boldsymbol{\lambda}$ involved in Theorem 5.11 is again a *noncausal* stochastic process since $\boldsymbol{\lambda}_t$ depends on future noises $\boldsymbol{W}_{t+1}, \dots, \boldsymbol{W}_T$. However, for the same reason already given at the end of Sect. 5.4.2, and as shown in Sect. 5.4.3, it is possible to derive optimality conditions in which this process is replaced by a causal one, namely $\boldsymbol{\Lambda}$, such that $\boldsymbol{\Lambda}_t = \mathbb{E}(\lambda_t \mid \boldsymbol{X}_t)$. Then, of course $\boldsymbol{\Lambda}_t$ is $\boldsymbol{X}_t$-measurable, as was already $\boldsymbol{U}_t$ in Theorem 5.11.

We start again from (5.22) with $\mathcal{G}_t = \mathcal{F}_t$ in (5.22a). Those conditions, which are of course still valid in the Markovian case, involve a co-state process $\boldsymbol{\Lambda}$ defined by $\boldsymbol{\Lambda}_t = \mathbb{E}(\boldsymbol{\lambda}_t \mid \mathcal{F}_t)$. However, in the Markovian case, we know that $\boldsymbol{\lambda}_t \preceq (\boldsymbol{X}_t, \boldsymbol{W}_{t+1}, \dots, \boldsymbol{W}_T)$ and that $(\boldsymbol{W}_{t+1}, \dots, \boldsymbol{W}_T)$ are independent of $\mathcal{F}_t$ (and a fortiori of $\boldsymbol{X}_t$). Hence $\mathbb{E}(\boldsymbol{\lambda}_t \mid \mathcal{F}_t)$ boils down to $\mathbb{E}(\boldsymbol{\lambda}_t \mid \boldsymbol{X}_t)$ and is of course $\boldsymbol{X}_t$-measurable.

Similar arguments as those used in the proof of Theorem 5.11 can also be used to replace $\mathbb{E}(\cdot \mid \mathcal{F}_t)$ by $\mathbb{E}(\cdot \mid \boldsymbol{X}_t)$ in the right-hand side of (5.22a)–(5.22c), which finally yields the following conditions:

$$\mathbb{E}\big(\nabla_u L_t(\boldsymbol{X}_t, \boldsymbol{U}_t, \boldsymbol{W}_{t+1}) + \nabla_u f_t(\boldsymbol{X}_t, \boldsymbol{U}_t, \boldsymbol{W}_{t+1})\boldsymbol{\Lambda}_{t+1} \mid \boldsymbol{X}_t\big) \in -\partial\chi_{\mathcal{U}_t^{\mathrm{a}}}(\boldsymbol{U}_t), \quad t = 0, \dots, T-1, \tag{5.25a}$$

$$\boldsymbol{\Lambda}_T = \nabla K(\boldsymbol{X}_T), \tag{5.25b}$$

$$\boldsymbol{\Lambda}_t = \mathbb{E}\big(\nabla_x f_t(\boldsymbol{X}_t, \boldsymbol{U}_t, \boldsymbol{W}_{t+1})\boldsymbol{\Lambda}_{t+1} + \nabla_x L_t(\boldsymbol{X}_t, \boldsymbol{U}_t, \boldsymbol{W}_{t+1}) \mid \boldsymbol{X}_t\big), \quad t = 1, \dots, T-1. \tag{5.25c}$$

As we shall see later on, the fact that all processes involved in the optimality conditions (5.25) are $\boldsymbol{X}_t$-measurable provides a connection with Dynamic Programming. Still, those optimality conditions involve *conditional* expectations. When one comes to the stage of numerical resolution, a topic that is addressed in Chap. 7, the presence of those conditional expectations sounds as a drawback since the literature on conditional expectation approximation only offers *biased* estimators with asymptotic convergence rates that do depend on the dimension of the conditioning term, namely $\boldsymbol{X}$ [86]. On the contrary, the approximation of expectations through Monte-Carlo techniques involves *unbiased* estimates, the variance of which converges to zero at a rate which is better and does not involve a consideration of dimension.

In the next section, in the Markovian case, we move to another approach of the SOC problem which ultimately gets rid of conditional expectations and which involves only expectations. This approach may be termed a *"functional point of view"*.

5.5.4 Optimality Conditions from a Functional Point of View

Since $\boldsymbol{U}_t$ and $\boldsymbol{\Lambda}_t$ involved in (5.25) are $\boldsymbol{X}_t$-measurable, under proper assumptions (see Proposition 3.46), there exist measurable mappings $\phi_t : \mathbb{X}_t \to \mathbb{U}_t$ and $\Lambda_t : \mathbb{X}_t \to \mathbb{X}_t$ such that $\boldsymbol{U}_t = \phi_t(\boldsymbol{X}_t)$ and $\boldsymbol{\Lambda}_t = \Lambda_t(\boldsymbol{X}_t)$. In particular, $\Lambda_T = \nabla K$ according to (5.25b).

Using those expressions of $\boldsymbol{U}_t$ and $\boldsymbol{\Lambda}_t$ in (5.25a) and (5.25c), we get:

$$\mathbb{E}\Big(\nabla_u L_t\big(\boldsymbol{X}_t, \phi_t(\boldsymbol{X}_t), \boldsymbol{W}_{t+1}\big) + \nabla_u f_t\big(\boldsymbol{X}_t, \phi_t(\boldsymbol{X}_t), \boldsymbol{W}_{t+1}\big) \Lambda_{t+1}\Big(f_t\big(\boldsymbol{X}_t, \phi_t(\boldsymbol{X}_t), \boldsymbol{W}_{t+1}\big)\Big) \;\Big|\; \boldsymbol{X}_t\Big) \in -\partial\chi_{\mathcal{U}_t^{\mathrm{a}}}\big(\phi_t(\boldsymbol{X}_t)\big),$$

$$\Lambda_t(\boldsymbol{X}_t) = \mathbb{E}\Big(\nabla_x f_t\big(\boldsymbol{X}_t, \phi_t(\boldsymbol{X}_t), \boldsymbol{W}_{t+1}\big)\Lambda_{t+1}\Big(f_t\big(\boldsymbol{X}_t, \phi_t(\boldsymbol{X}_t), \boldsymbol{W}_{t+1}\big)\Big) + \nabla_x L_t\big(\boldsymbol{X}_t, \phi_t(\boldsymbol{X}_t), \boldsymbol{W}_{t+1}\big) \;\Big|\; \boldsymbol{X}_t\Big).$$

Observe that, in both equations, the expressions subject to the conditional expectation $\mathbb{E}(\cdot \mid \boldsymbol{X}_t)$ involve random variables which are either entirely determined by the values x assumed by $\boldsymbol{X}_t$ or which are independent from this variable. Therefore, the conditional expectation reduces to a simple expectation (over the distribution of $\boldsymbol{W}_{t+1}$) for every given x. We may thus rewrite (5.20a), (5.20b) and (5.25) as functional equations as follows:

$$\boldsymbol{X}_0 = \boldsymbol{W}_0, \tag{5.26a}$$

$$\boldsymbol{X}_{t+1} = f_t\big(\boldsymbol{X}_t, \phi_t(\boldsymbol{X}_t), \boldsymbol{W}_{t+1}\big), \quad t = 0, \dots, T-1, \tag{5.26b}$$

$$\mathbb{E}\Big(\nabla_u L_t\big(\cdot, \phi_t(\cdot), \boldsymbol{W}_{t+1}\big) + \nabla_u f_t\big(\cdot, \phi_t(\cdot), \boldsymbol{W}_{t+1}\big)\Lambda_{t+1}\Big(f_t\big(\cdot, \phi_t(\cdot), \boldsymbol{W}_{t+1}\big)\Big)\Big) \in -\partial\chi_{\mathcal{U}_t^{\mathrm{a}}}\big(\phi_t(\cdot)\big), \quad t = 0, \dots, T-1, \tag{5.26c}$$

$$\Lambda_T(\cdot) = \nabla K(\cdot), \tag{5.26d}$$

$$\Lambda_t(\cdot) = \mathbb{E}\Big(\nabla_x f_t\big(\cdot, \phi_t(\cdot), \boldsymbol{W}_{t+1}\big)\Lambda_{t+1}\big(f_t\big(\cdot, \phi_t(\cdot), \boldsymbol{W}_{t+1}\big)\big) + \nabla_x L_t\big(\cdot, \phi_t(\cdot), \boldsymbol{W}_{t+1}\big)\Big), \quad t = 1, \ldots, T-1. \tag{5.26e}$$

In the present Markovian case, the Dynamic Programming equation (4.57) is another characterization of the optimal feedback solution. We repeat it hereafter for convenience:

$$V_T(\cdot) = K(\cdot), \tag{5.27a}$$

$$V_t(\cdot) = \min_{u_t \in \boldsymbol{C}_t} \mathbb{E}\Big(L_t\big(\cdot, u_t, \boldsymbol{W}_{t+1}\big) + V_{t+1}\big(f_t\big(\cdot, u_t, \boldsymbol{W}_{t+1}\big)\big)\Big), \quad t = 0, \ldots, T-1. \tag{5.27b}$$

Observe that, according to (5.27a) and (5.26d), Λ_T is nothing but ∇V_T, that is, the gradient of the cost-to-go or Bellman function. This interpretation which is valid at time T carries over at other time stages as long as V_t remains differentiable: differentiability is assumed here for simplicity although more sophisticated notions of derivatives (subgradient in the convex case, Clarke's derivative, etc.) could be used to cover more general cases.

The argument is by induction backwards in time. The statement is true at T. We assume it is true from T to $t+1$. At t, we consider (5.27b) as a so-called "marginal function" of the form:

$$V_t(x) = \min_{u_t \in \boldsymbol{C}_t} F_{t+1}(x, u_t)$$

with an obvious definition of F_{t+1}. Since Danskin [49] and his seminal result on sensitivity of marginal functions, a result further extended and generalized by many subsequent authors, we know that, if the marginal function above is differentiable, its gradient is given by the following formula:

$$\nabla V_t(x) = \nabla_x F_{t+1}(x, u_t^\sharp),$$

where $u_t^\sharp$ is any element of the arg min. Then, comparing this formula with (5.26e), given the definition of F_{t+1} resulting from (5.27b), the argument is complete.

5.6 Conclusions

In this chapter, several sets of optimality conditions with a Lagrange-Kuhn-Tucker (or variational) flavor have been presented for SOC problems in discrete time. The first group of conditions presented in Sects. 5.4.2 and 5.4.3 does not require the white noise Assumption 5.9 and allows for more general information patterns than those imposed by Assumption 5.10. Conditions (5.20) involve an anticipative co-state process whereas (5.22) provides alternative conditions with an adapted co-state. But both involve conditional expectation operations with respect to σ-fields generated by random variables with dimensions increasing over time, which makes their numerical implementation, considered in Chap. 7, a difficult problem.

With the more restrictive assumptions of the so-called Markovian case discussed in Sect. 5.5.1, the conditional expectations involved in either (5.23) or (5.25) are now with respect to a σ-field generated by the state variable which remains of constant dimension over time. This alleviates the problem of their numerical implementation, but still *conditional* expectation approximation remains a more difficult problem than simple expectation numerical implementation.

By exploiting their measurability property, we thus introduced an additional step which consists in replacing the stochastic processes of control and co-state by sequences of deterministic functions operating upon the state process. This enabled us to obtain conditions (5.26) which involved only expectations. At the same time, this point of view brings us closer to the Dynamic Programming (DP) approach with its potential drawback of the "curse of dimensionality" [15]. We discuss the differences in numerical implementation between the resolution of (5.26) versus DP in Chap. 7.

Part III
Discretization and Numerical Methods

Chapter 6
Discretization Methodology for Problems with Static Information Structure (SIS)

In this chapter, we consider problems formulated as in (1.3), which we repeat here for convenience

$$\min_{\boldsymbol{U} \preceq \mathcal{G}} \mathbb{E}\big(j(\boldsymbol{U}, \boldsymbol{W})\big) \tag{6.1a}$$

where $\mathcal{G}$ is a σ-field, or

$$\min_{\boldsymbol{U} \preceq \boldsymbol{Y}} \mathbb{E}\big(j(\boldsymbol{U}, \boldsymbol{W})\big), \tag{6.1b}$$

where $\boldsymbol{Y}$ is a random variable (called observation). Both $\mathcal{G}$ and $\boldsymbol{Y}$ are *static*, that is, they do not depend on the control $\boldsymbol{U}$ (in Sect. 1.2.2, we used the acronym SIS for this situation). Recall that problems with DIS (see again Sect. 1.2.2), but no dual effect, are also amenable to this formulation (such situations are considered in Sect. 10.3).

We are mainly interested in devising systematic approaches to the discretization of such problems in order to solve them numerically with the help of a computer. Essentially, in the discretized problem, any random variable, be it part of the data as $\boldsymbol{W}$, or of the unknowns as $\boldsymbol{U}$, is represented by a finite set of values (e.g. $\{w^i\}_{i=1,\dots,N}$), and its associated probability law is represented by a sum of atomic measures (Dirac measures δ_{w^i} located at w^i) with positive weights p^i summing up to 1. Consequently, in the discrete problem, expectations reduce to finite sums, and optimization is w.r.t. a finite set $\{u^i\}$ of variables.

Before we can address this main topic, the next section briefly discusses the theory of *quantization* which is essentially a tool to derive approximate, but finite, representations of random variables, and which provides a framework in which to discuss the quality of those approximations.

P. Carpentier et al., *Stochastic Multi-Stage Optimization*, Probability Theory and Stochastic Modelling 75, DOI 10.1007/978-3-319-18138-7_6

6.1 Quantization

When trying to solve stochastic optimization problems numerically, one may have to manipulate approximate, but *finite*, representations of random variables. The quantization technique reduces the amount of information necessary to represent a random variable while trying to preserve as much as possible of the original random variable. It has its origin in Communication Theory [68], in which random signals must be sent through a channel with limited bandwidth. By reducing the amount of information necessary to describe, and thus transmit, the signal, one hopes to increase the flow of signals sent through the channel. At the same time, the signals should be distorted as little as possible. There is clearly a trade-off here.

In this text, we do not address this trade-off directly; we rather assume that the amount of information retained to represent a random variable is given,[1] and we try to minimize the distortion in the representation of the random variable under this constraint.

Indeed, we first start with set-theoretic notions that are limited to algebraic aspects of quantization. Then, we move to the more quantitative notion of *optimal* quantization, where the set over which quantization is considered must be a *normed vector space*.

6.1.1 Set-Theoretic Quantization

A random variable $\boldsymbol{W}$ is a measurable mapping from a probability space $(\Omega, \mathcal{A}, \mathbb{P})$ to a measurable space $(\mathbb{W}, \mathcal{W})$. In this subsection, the probability law plays no role, but it is used in the next subsection.

Consider a *projection* $Q : \mathbb{W} \to \mathbb{W}$, that is, a measurable mapping such that $Q \circ Q = Q$. Assume, moreover, that its image $\operatorname{im} Q$ has a finite cardinality. That is, it contains a finite number N of distinct values. We call Q a *quantization* and $Q \circ \boldsymbol{W}$ a *quantized* approximation of $\boldsymbol{W}$.

We may consider Q as factorized into two mappings,

$$Q = d \circ e,$$

where

1. $e : \mathbb{W} \to \{1, \dots, N\}$ is called the *encoding*;
2. $d : \{1, \dots, N\} \to \operatorname{im} Q \subset \mathbb{W}$ is a *bijection*, which is called the *decoding*.

In a communication context, instead of sending values $w \in \mathbb{W}$ over the communication channel, only the *code* $i = e(w) \in \{1, \dots, N\}$ is sent; at the other end of the

[1] This is likely to determine the complexity of the discretized optimization problem we seek to formulate.

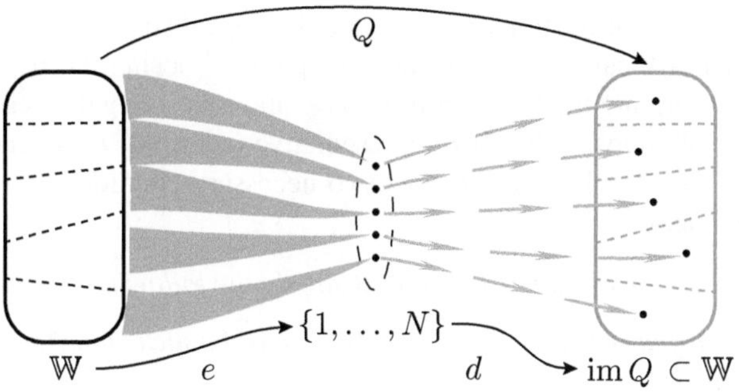

Fig. 6.1 Quantization Q, encoding e, decoding d

channel, the message i is *decoded* by using $d(i) \in \text{im } Q$; for this reason, im $Q \subset \mathbb{W}$ is called the *codebook*: this is a collection of N values in $\mathbb{W}$.

Of course, $\Omega/(Q \circ \boldsymbol{W})$ is a partition[2] of Ω with N elements. Observe that this partition which defines the information carried by the quantized random variable is independent of the decoding d (as long as this is a bijection), it only depends on the encoding e. Otherwise stated, $\Omega/(Q \circ \boldsymbol{W}) = \Omega/(e \circ \boldsymbol{W})$.

Similarly,

$$\mathbb{W}/Q = \mathbb{W}/e, \tag{6.2}$$

and this defines a partition of $\mathbb{W}$, the N elements of which are called *quantization cells*. All elements in cell i are represented by the same *representative* $d(i)$ in the codebook. In summary, while the encoding e defines the cells (and thus the information carried by the quantized variable), the decoding d defines the representative in each cell (called the *centroid* of the cell), which has an importance as long as the values w are physical quantities (for example, consumption of energy, prices, etc.). The whole situation is illustrated by Fig. 6.1.

6.1.2 Optimal Quantization in Normed Vector Spaces

We assume now that $\mathbb{W}$ is a normed vector space (the norm is denoted $\|\cdot\|$). Given a value of N, the idea is to choose the least distorted quantized variable. Distortion may be defined by the L^2-distance between the original and the quantized random variable, which is equal to the square root of the *Mean Quadratic Error* (MQE):

$$\text{MQE} := \mathbb{E}\big(\|\boldsymbol{W} - Q \circ \boldsymbol{W}\|^2\big). \tag{6.3}$$

[2]The notation $\Omega/(Q \circ \boldsymbol{W})$ was introduced in Definition 3.30.

Optimizing the quantization amounts to reducing this distortion measure to the minimum. This task can be split up into two parts: choosing the best encoding e, or equivalently defining the best partition $\mathbb{W}/e$; and specifying the best decoding d, which consists of choosing the best representative (centroid) w^i in each cell C^i of this partition. The next lemma provides two necessary conditions related to those two optimal choices.

Lemma 6.1 *An optimal quantization must satisfy the following two conditions:*

1. *given the centroids* $\{w^i\}_{i=1,\dots,N}$*, the cells* C^i *must be such that,* $\mathbb{P}_{\mathbf{W}}$*-almost surely, if* $w \in C^i$*, then* $\|w - w^i\| \leq \|w - w^j\|, \forall j \neq i$*;*
2. *given the cells* C^i*, the centroid* w^i *is equal to* $\mathbb{E}\big(\mathbf{W} \mid \mathbf{W}^{-1}(C^i)\big)$.

Proof The MQE can be written as follows

$$\mathbb{E}\big(\|\mathbf{W} - Q \circ \mathbf{W}\|^2\big) = \mathbb{E}\Big(\sum_{i=1}^{N} \mathbb{E}\Big(\|\mathbf{W} - w^i\|^2 \;\Big|\; \mathbf{W}^{-1}(C^i)\Big)\Big).$$

Suppose the first condition of the lemma is not satisfied. Then, since N is finite, it means that there exist a subset of $\mathbb{W}$ with *positive* probability for $\mathbb{P}_{\mathbf{W}}$ and two indices i and j such that every w in that subset belongs to C^i whereas $\|w - w^j\| < \|w - w^i\|$. Then, by changing the definition of cells so that this whole subset is moved from C^i to C^j, the above performance index is improved, which contradicts optimality.

Now, considering the ith conditional expectation in the right-hand side above, it is well known that $w^i = \mathbb{E}\big(\mathbf{W} \mid \mathbf{W}^{-1}(C^i)\big)$ is the value which minimizes that expression (see Definition B.5). □

The first condition in this lemma defines what is known as a *Voronoi diagram* or *tessellation* (see Fig. 6.2). This condition may be called "*the nearest neighbor*" condition in that any w should be represented by its nearest neighbor in the codebook.

Using (6.2), observe that

$$\mathbb{E}\big(Q(\mathbf{W})\big) = \sum_{i=1}^{N} \mathbb{P}_{\mathbf{W}}(C^i) w^i = \mathbb{E}\Big(\mathbb{E}\big(\mathbf{W} \mid \mathbf{W}^{-1}(\mathbb{W}/e)\big)\Big) = \mathbb{E}(\mathbf{W}), \tag{6.4}$$

that is, an optimal quantized variable is respectful of the first moment of the original random variable. However, the next lemma shows that the second order moment or the variance is underestimated when replacing the original variable by its optimal quantized version.

Lemma 6.2 *For an optimal quantization* Q*, with* MQE *defined by* (6.3)*, one has that:*

$$\mathbb{E}(\mathbf{W}) = \mathbb{E}\big(Q(\mathbf{W})\big), \tag{6.5a}$$

Fig. 6.2 Voronoi tessellation

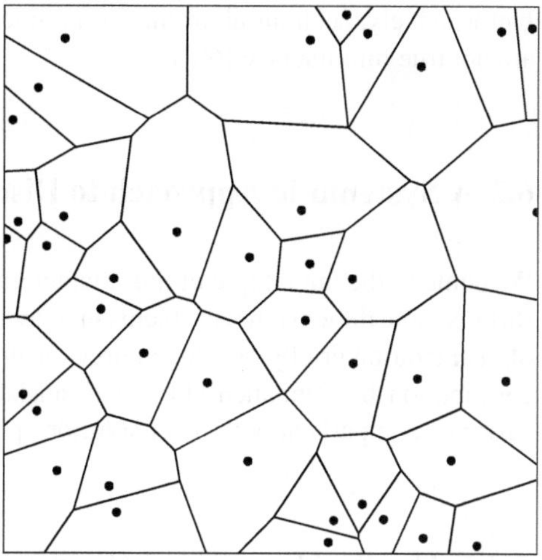

$$\text{Var}\left(\boldsymbol{W}\right) = \text{Var}\left(Q(\boldsymbol{W})\right) + \text{MQE}. \tag{6.5b}$$

Proof The former claim is a repetition of (6.4). We concentrate on the latter.

$$\begin{aligned}\text{Var}\left(\boldsymbol{W}\right) &= \mathbb{E}\left(\left\|\boldsymbol{W} - \mathbb{E}(\boldsymbol{W})\right\|^2\right)\\ &= \mathbb{E}\left(\left\|\boldsymbol{W} - Q(\boldsymbol{W}) + Q(\boldsymbol{W}) - \mathbb{E}(\boldsymbol{W})\right\|^2\right)\\ &= \mathbb{E}\left(\left\|\boldsymbol{W} - Q(\boldsymbol{W})\right\|^2\right) + \mathbb{E}\left(\left\|Q(\boldsymbol{W}) - \mathbb{E}(\boldsymbol{W})\right\|^2\right)\\ &\quad + 2\mathbb{E}\left(\sum_{i=1}^{N} \mathbb{E}(\langle \boldsymbol{W} - w^i\,, w^i - \mathbb{E}(\boldsymbol{W})\rangle \mid \boldsymbol{W}^{-1}(C^i))\right).\end{aligned}$$

The first term is precisely the MQE; the second term is the variance of $Q(\boldsymbol{W})$ thanks to (6.5a); the third term is zero since, in the scalar product, the second factor is constant over C^i whereas, the first factor has a zero conditional expectation given the value of w^i (see Lemma 6.1). □

The two necessary optimality conditions in Lemma 6.1 may be used to build up an iterative (Lloyd's) algorithm to find an optimal quantization of a given random variable. This algorithm proceeds by alternating the following two stages:

1. given a collection of N centroids w^i, the Voronoi cells are (re)drawn using the first conditon in Lemma 6.1; this amounts to defining all the half spaces delimited by the medians (hyperplanes) of all segments (w^i, w^j) (it involves manipulating affine inequalities);
2. given the cells, the centroids are (re)defined by using the second optimality condition in the lemma (it involves computing integrals over the cells).

Unfortunately, such an algorithm sometimes tends to stop on local minima which are not true minima (see [68]).

6.2 A Systematic Approach to Discretization

We come to the main topic of this chapter, namely the reformulation of problems as (6.1) as finite dimensional problems in order to solve them numerically. The language of quantization briefly developed previously is used throughout this section, and the contribution by Pennanen [110] is "translated" in this language in order to make easier its comparison with other solutions proposed hereafter.

6.2.1 The Problematics of Discretization

As we have seen in Sect. 1.4, the resolution of the discrete problem does not directly provide an acceptable answer in that what is expected as a "solution" to (6.1) is a random variable over the *original* probability space, satisfying, in addition, the measurability conditions imposed in the formulation (6.1). Therefore, some "reconstruction" is needed after the discrete problem has been solved, and examples of this reconstruction were given in Sect. 1.4 (see (1.15) and Fig. 1.3).

Notice that, even if the formulation of the discrete problem seemingly only requires the consideration of random variables assuming finitely many distinct values, these random variables must be embedded into the original space by defining "cells" around the atomic support of the discrete probability distribution defined by the weights $\{p^i\}_{i=1,\dots,N}$. Then, a reasonable requirement for convergence of the reconstructed solution towards the optimal original problem solution is that each p^i gets arbitrarily close to the true probability mass of the cell around atom i as the number N of cells goes to infinity.

However, as shown by the discussion around examples in Sect. 1.4, it is not enough to handle the approximation of mathematical expectations in a sound manner: some caution is also required to properly represent the essential contraints of *information structure* in the discrete problem. Here again, cells around atoms, or as otherwise stated, partitions of the original spaces, play a part. This section concentrates on this particular issue.

Pennanen [110] was probably the first author to envisage this topic in a systematic way. Regarding the approximation techniques for expectations, he considered not only usual Monte Carlo sampling but also Quasi-Monte Carlo and other sophisticated quadrature techniques. In order to be able to give an asymptotic epi-convergence theorem as the number of samples goes to infinity, he imposed a rather strong condition which tightly coordinates the samples used to approximate expectations with the way informational constraints[3] are translated in the discretized problem. This condition naturally leads to the construction of scenario trees that were introduced in the end of Chap. 1 (see Fig. 1.5) and that are very popular in the SP community. However, as we shall see by the end of this chapter, the technique of scenario trees is hindered by the poor convergence rate of the underlying Monte Carlo approximation technique (see Sect. 6.3.2), and therefore it is important to show that other more flexible conditions can also be used that still enable a convergence proof to be given.

In this chapter, the focus is on the links between the sampling technique and the translation of informational constraints in the discrete problem; convergence issues are discussed in Chap. 8. We also consider more general informational constraints beyond non anticipativity constraints. Finally, we use the language of lattice of partition fields introduced in Chap. 3, which provides a formalism making clearer the comparison of Pennanen's work with alternative methods we propose in Sect. 6.2.3.

Remark 6.3 Although the issue of convergence is deferred to Chap. 8, the following observation may be kept in mind as a safeguard against unreasonable discretization schemes. Remember that, according to Sect. 3.5.2, Problem (6.1) can be reformulated as

$$\mathbb{E}\Big(\min_u \mathbb{E}\big(j(u, \boldsymbol{W}) \mid \mathcal{G}\big)\Big) \quad \text{or} \quad \mathbb{E}\Big(\min_u \mathbb{E}\big(j(u, \boldsymbol{W}) \mid \boldsymbol{Y}\big)\Big). \tag{6.6}$$

Therefore, in any discretized version derived for the original problem, one should try to identify an expression which serves as an approximation of the conditional expectation of the cost and check that this approximation is sound enough. An application of this observation is encountered, for example, in Remark 6.13. ◊

6.2.2 The Approach Inspired by Pennanen's Work

In [110], sequential stochastic optimization problems are considered with non anticipativity constraints: in the framework (6.1b), this amounts to considering that $\boldsymbol{Y} = h(\boldsymbol{W})$ with $\boldsymbol{Y} = \{\boldsymbol{Y}_t\}_{t=1,2,\ldots}$, $\boldsymbol{W} = \{\boldsymbol{W}_t\}_{t=1,2,\ldots}$ and $\boldsymbol{Y}_t = h_t(\boldsymbol{W}) = (\boldsymbol{W}_1, \ldots, \boldsymbol{W}_t)$.

In what follows, we forget about the time index t which plays no particular part as long as $\boldsymbol{Y}$ does not depend on $\boldsymbol{U}$ (the SIS assumption) and we just retain that $\boldsymbol{Y} = h(\boldsymbol{W})$ where h is a function (generally non injective) from $\mathbb{Y}$ to $\mathbb{W}$.

Remark 6.4 In (6.1b), the cost function j depends on the two random variables $\boldsymbol{U}$ and $\boldsymbol{W}$. The observation $\boldsymbol{Y}$ is another random variable which may, or may not, be in relation with $\boldsymbol{W}$. In the present framework, since we assume that $\boldsymbol{Y} = h(\boldsymbol{W})$, we may consider all involved random variables (including $\boldsymbol{U}$) as measurable mappings from $\mathbb{W}$ to another space, that is, we may consider that $\Omega = \mathbb{W}$. Then $\boldsymbol{W}$ is the identity mapping from $\mathbb{W}$ to itself and $\boldsymbol{Y}$ is identical to the mapping h. This is what is assumed hereafter unless otherwise explicitly stated. ◊

[3] Actually, Pennanen considers *non anticipativity* constraints only.

First Stage

The formulation of a discretized problem involves the definition of finite sets to approximate the spaces $\mathbb{Y}$ and $\mathbb{W}$. This is formalized by defining *quantizations* as defined in Sect. 6.1. We first consider $\boldsymbol{W}_N$, a quantized version of $\boldsymbol{W}$. More precisely, we consider the following applications and properties:

- $q_{\mathbb{W}} : \mathbb{W} \to \mathbb{W}_N$ where $\mathbb{W}_N$ is a finite subset of cardinality N of $\mathbb{W}$; we require that $q_{\mathbb{W}}(w) = w$ whenever $w \in \mathbb{W}_N$;[4]
- $\iota_{\mathbb{W}}$, the canonical injection of $\mathbb{W}_N$ into $\mathbb{W}$; then,
  - $q_{\mathbb{W}} \circ \iota_{\mathbb{W}}$ is the identity $\mathrm{I}_{\mathbb{W}_N}$ in $\mathbb{W}_N$;
  - $Q_{\mathbb{W}} = \iota_{\mathbb{W}} \circ q_{\mathbb{W}}$ is then a projection in $\mathbb{W}$: this is indeed a quantization as defined in Sect. 6.1, and $\boldsymbol{W}_N = Q_{\mathbb{W}} \circ \boldsymbol{W} = Q_{\mathbb{W}}$ (according to Remark 6.4) is the quantized random variable we were looking for;
  - $\mathbb{W}/q_{\mathbb{W}} = \mathbb{W}/Q_{\mathbb{W}}$ is a partition of $\mathbb{W}$ into N cells.

Next, in order to obtain a quantized version of $\boldsymbol{Y}$, Pennanen proceeds in the following way. He first considers the discrete random variable $\boldsymbol{Y}_N$ defined as $h \circ \boldsymbol{W}_N$. Notice that since h is not injective in general, it may happen that the set of values $\mathbb{Y}_N = h \circ Q_{\mathbb{W}}(\mathbb{W})$ has a cardinality smaller than N (despite the subscript N in this notation). Then, consider:

- $\iota_{\mathbb{Y}}$, the canonical injection of $\mathbb{Y}_N$ into $\mathbb{Y}$;
- $h_N : \mathbb{W}_N \to \mathbb{Y}_N$ such that $\iota_{\mathbb{Y}} \circ h_N = h \circ \iota_{\mathbb{W}}$.

Remark 6.5 This h_N can be obtained as

$$h_N = \iota_{\mathbb{Y}}^{-1} \circ h \circ \iota_{\mathbb{W}}, \tag{6.7}$$

where $\iota_{\mathbb{Y}}^{-1}$ is any mapping such that $\iota_{\mathbb{Y}} \circ \iota_{\mathbb{Y}}^{-1} = \mathrm{I}_{\mathbb{Y}_N}$, the identity over $\mathbb{Y}_N$. This $\iota_{\mathbb{Y}}^{-1}$ is not uniquely defined: any mapping from $\mathbb{Y}$ to $\mathbb{Y}_N$ can play the role of $\iota_{\mathbb{Y}}^{-1}$, as long as its restriction to $\mathbb{Y}_N$ behaves as the identity. Nevertheless, h_N is well defined since precisely, in (6.7), only the restriction of $\iota_{\mathbb{Y}}^{-1}$ to $\mathbb{Y}_N$ is involved. ◊

Therefore, $\boldsymbol{Y}_N$ is well defined:

$$\boldsymbol{Y}_N = h(\boldsymbol{W}_N) = h \circ Q_{\mathbb{W}} = \iota_{\mathbb{Y}} \circ h_N \circ q_{\mathbb{W}}. \tag{6.8}$$

The expression $h \circ Q_{\mathbb{W}}$ is also well defined once $Q_{\mathbb{W}}$ has been chosen (while h is given). The situation is summarized in Fig. 6.3.

Obviously, $\boldsymbol{Y}_N \preceq \boldsymbol{W}_N$. However, $\boldsymbol{Y}_N$ is not necessarily a quantized version of $\boldsymbol{Y}$ and, in particular, one cannot claim that $\boldsymbol{Y}_N \preceq \boldsymbol{Y}$. Here is a counterexample.

[4]The finite set $\mathbb{W}_N$ generally results from some sampling of the noise $\boldsymbol{W}$ or alternative quadrature methods. But, as recognized by Pennanen himself, in order to define a consistent discretization scheme, it is not enough to introduce the finite set $\mathbb{W}_N$, but it is also necessary to define how the whole original set $\mathbb{W}$ is mapped onto that finite set: this is why $q_{\mathbb{W}}$ must be defined.

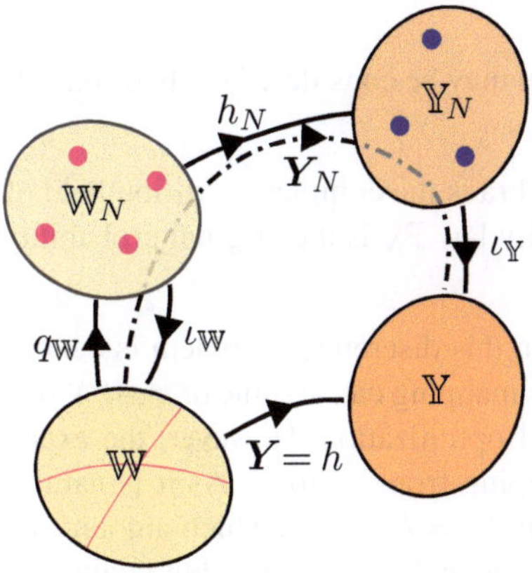

Fig. 6.3 Discretization according to Pennanen, stage 1

Example 6.6 Consider

$$\mathbb{W} = [-2, 2] \times [-2, 2], \quad q_{\mathbb{W}}(w_1, w_2) = \begin{pmatrix} \mathrm{sign}(w_1) \\ \mathrm{sign}(w_2) \end{pmatrix}, \quad \mathbb{W}_N = \{(\pm 1, \pm 1)\},$$
$$h(w_1, w_2) = w_1 + w_2, \quad \mathbb{Y}_N = \{-2, 0, 2\}.$$

Observe that $h(-1, 2) = h(1/2, 1/2) = 1$ whereas $\boldsymbol{Y}_N(-1, 2) = \mathrm{sign}(-1) + \mathrm{sign}(2) = 0$ is different from $\boldsymbol{Y}_N(1/2, 1/2) = \mathrm{sign}(1/2) + \mathrm{sign}(1/2) = 2$. According to Proposition 3.38 (item 2), $\boldsymbol{Y}_N$ is not measurable with respect to $\boldsymbol{Y}$. The two dots with coordinates $(-1, 2)$ and $(1/2, 1/2)$ are represented in Fig. 6.4. This figure also displays three partitions of $\mathbb{W} = [-2, 2] \times [-2, 2]$. The first partition corresponds to $\mathbb{W}/h$, and it has infinitely many elements (only a few are represented in the figure). The second partition corresponds to $\mathbb{W}/Q_{\mathbb{W}}$ and it has four elements (the values of $Q_{\mathbb{W}}$ belonging to $\mathbb{W}_N$ are indicated). The third partition, namely $\mathbb{W}/(h \circ Q_{\mathbb{W}})$, has only three elements corresponding to the three elements of $\mathbb{Y}_N$ indicated in the figure. △

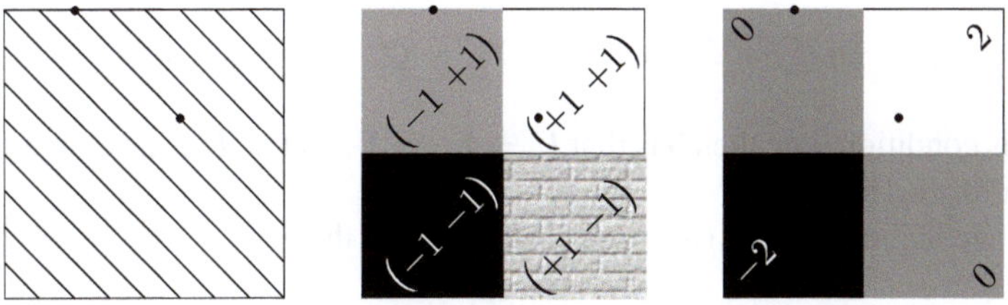

Fig. 6.4 Representations of the partitions corresponding to $\mathbb{W}/h$, $\mathbb{W}/Q_{\mathbb{W}}$ and $\mathbb{W}/(h \circ Q_{\mathbb{W}})$ in Example 6.6

Second Stage

The discretized problem may be considered to "live" on a discrete probability space:

- Ω_N is $\mathbb{W}_N$;
- the associated σ-algebra is the complete partition field of this set;
- the discrete probability law $\mathbb{P}_N$ is the original probability law $\mathbb{P}$ transported from $\mathbb{W}$ to $\mathbb{W}_N$ by $q_{\mathbb{W}}$.

The decision variable in this discretized problem is denoted $\boldsymbol{U}_N$: this is a mapping from $\mathbb{W}_N$ to $\mathbb{U}$, and this mapping can assume *at most* N distinct values; these values result from a numerical optimization. However, the expected answer is a random variable, that is, a mapping from $\mathbb{W}$ to $\mathbb{U}$. As it is natural, $\boldsymbol{U}_N$ is extended to the whole $\mathbb{W}$ by considering $\boldsymbol{U} = \boldsymbol{U}_N \circ q_{\mathbb{W}}$, which amounts to building up a piecewise constant function using the cells of $\mathbb{W}/q_{\mathbb{W}}$ (but again, although there are N cells, there might be less than N distinct values of the control).

The searched solution $\boldsymbol{U}$ should also satisfy some measurability requirements in order to reflect the information structure of the original problem. As a straightforward translation of the informational constraint in (6.1b), in the discrete problem T. Pennanen requires that

$$\boldsymbol{U}_N \preceq h_N. \tag{6.9}$$

Lemma 6.7 *The condition* (6.9) *implies that* $\boldsymbol{U} \preceq \boldsymbol{Y}_N$.

Proof Indeed, $\boldsymbol{U}_N \preceq h_N$ implies that $\boldsymbol{U}_N \circ q_{\mathbb{W}} \preceq h_N \circ q_{\mathbb{W}}$ (see (3.41)), and $h_N \circ q_{\mathbb{W}} \equiv \iota_{\mathbb{Y}} \circ h_N \circ q_{\mathbb{W}}$ since $\iota_{\mathbb{Y}}$ is injective (see Proposition 3.41). The latter is just $\boldsymbol{Y}_N$ (see (6.8)). □

Condition (6.9) implies that there exists a mapping $\gamma_N : \mathbb{Y}_N \to \mathbb{U}$ such that $\boldsymbol{U}_N = \gamma_N \circ h_N$. However, since $\boldsymbol{Y}_N$ is not necessarily measurable w.r.t. $\boldsymbol{Y}$ (as shown by Example 6.6), the proposed $\boldsymbol{U}$ is not necessarily measurable w.r.t. $\boldsymbol{Y}$ either, and, in this case, it would not be an admissible solution for problem (6.1b). Therefore, Pennanen finally requires the following additional condition:

$$\boldsymbol{Y}_N \preceq \boldsymbol{Y} \tag{6.10a}$$

which is equivalent (see Proposition 3.46) to

$$\exists Q_{\mathbb{Y}} : \operatorname{im} \mathbb{Y} \to \mathbb{Y} \quad \text{such that} \quad \boldsymbol{Y}_N = Q_{\mathbb{Y}} \circ \boldsymbol{Y} = Q_{\mathbb{Y}} \circ h. \tag{6.10b}$$

With this condition, one then has that $\boldsymbol{U} \preceq \boldsymbol{Y}_N \preceq \boldsymbol{Y}$, that is, $\boldsymbol{U}$ is now an admissible solution for (6.1b).

Put together, conditions (6.8) and (6.10b) imply that

$$h \circ Q_{\mathbb{W}} = Q_{\mathbb{Y}} \circ h. \tag{6.11}$$

Lemma 6.8 *Equation* (6.11) *implies that* $Q_{\mathbb{Y}}$ *(with domain* $\operatorname{im} h$*) is a projection; hence,* $\boldsymbol{Y}_N = Q_{\mathbb{Y}} \circ \boldsymbol{Y}$ *is a quantization of* $\boldsymbol{Y}$*. Therefore,* $Q_{\mathbb{Y}}$ *can be factorized as* $\iota_{\mathbb{Y}} \circ q_{\mathbb{Y}}$ *where* $q_{\mathbb{Y}} : \mathbb{Y} \to \mathbb{Y}_N$ *and* $\iota_{\mathbb{Y}}$ *is again the canonical injection of* $\mathbb{Y}_N$ *into* $\mathbb{Y}$*. Then,* (6.11) *is equivalent to*

$$h_N = q_{\mathbb{Y}} \circ h \circ \iota_{\mathbb{W}}. \tag{6.12}$$

Proof We must show that $Q_{\mathbb{Y}} \circ Q_{\mathbb{Y}} \circ h = Q_{\mathbb{Y}} \circ h$. Indeed, with (6.11) used repeatedly, and the fact that $Q_{\mathbb{W}}$ itself is a projection, one has that

$$Q_{\mathbb{Y}} \circ Q_{\mathbb{Y}} \circ h = Q_{\mathbb{Y}} \circ h \circ Q_{\mathbb{W}} = h \circ Q_{\mathbb{W}} \circ Q_{\mathbb{W}} = h \circ Q_{\mathbb{W}} = Q_{\mathbb{Y}} \circ h.$$

Now, with reference to the right-hand side of (6.8), (6.11) can be written as

$$\iota_{\mathbb{Y}} \circ h_N \circ q_{\mathbb{W}} = \iota_{\mathbb{Y}} \circ q_{\mathbb{Y}} \circ h, \tag{6.13}$$

which is equivalent to

$$h_N \circ q_{\mathbb{W}} = q_{\mathbb{Y}} \circ h,$$

since $\iota_{\mathbb{Y}}$ is injective, and this is again equivalent to (6.12) when composing both sides of the above equation with $\iota_{\mathbb{W}}$ (which is injective) to the right hand and remembering that $q_{\mathbb{W}} \circ \iota_{\mathbb{W}}$ is nothing but the identity in $\mathbb{W}_N$. □

The situation is summarized as follows, and it is illustrated by Fig. 6.5. Given the noise $\boldsymbol{W}$ and the observation $\boldsymbol{Y} = h(\boldsymbol{W})$,

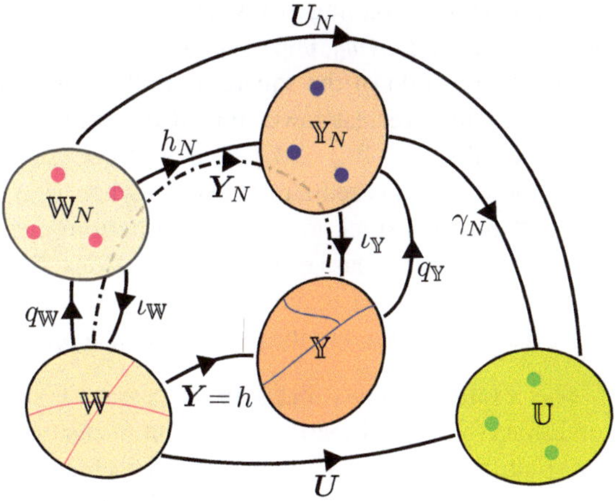

Fig. 6.5 Discretization according to Pennanen, stage 2

- a quantized noise $\boldsymbol{W}_N = Q_{\mathbb{W}} \circ \boldsymbol{W}$ is first defined;
- a discrete random variable $\boldsymbol{Y}_N = h \circ \boldsymbol{W}_N$ is next introduced;
- this discrete random variable is given the status of a quantized observation by imposing Condition (6.10);
- the last two steps finally result in Condition (6.11).

Discussion

Equation (6.11) says that the quantized observation must be the observation of the quantized noise. This condition is intuitively appealing. However, it is unclear how one can ensure it in a systematic construction of a discretization scheme in this general setting. In the particular case of *non anticipativity* constraints, Pennanen [111] proposed a procedure that we briefly discuss in Sect. 6.3. What makes things rather locked in general is the initial requirement (6.8) that the observation function h_N in the discrete model should be intimately related to the original observation function h. This is precisely the condition we relax later on.

Equation (6.12) should be compared with Eq. (6.7): this shows that the choice of the mapping $\iota_{\mathbb{Y}}^{-1}$ in the latter equation is nothing but the choice of the quantization map $q_{\mathbb{Y}}$ which defines the cells in $\mathbb{Y}$ (as long as the set of centroids $\mathbb{Y}_N$ is already defined).

Observe that (6.8) implies that $\boldsymbol{Y}_N \preceq \boldsymbol{W}_N$ whereas, by (6.10a), $\boldsymbol{Y}_N \preceq \boldsymbol{Y}$, hence $\boldsymbol{Y}_N \preceq \boldsymbol{W}_N \wedge \boldsymbol{Y}$ (see Chap. 3). Then, consider the following example.

Example 6.9 To stay close to the sequential situation considered by Pennanen and still maintain simplicity, we consider (1.8) again. All sample trajectories (w_0^i, w_1^i) of the noise $\boldsymbol{W} = (\boldsymbol{W}_0, \boldsymbol{W}_1)$ are represented as dots in the square $\mathbb{W} = [-1, 1] \times [-1, 1]$ with coordinates (w_0^i, w_1^i).

A quantization based on such a sampling may be obtained by drawing the Voronoi tessellation corresponding to this set of dots (Fig. 6.2 illustrates the partition $\Omega / \boldsymbol{W}_N$). On the other hand, $h(w_0, w_1) = w_0 \in \mathbb{Y} = [-1, 1]$. The partition $\Omega / \boldsymbol{Y}$ corresponds to a decomposition of the square into all vertical segments it contains. According to the way the greatest lower bound of partitions is obtained (see Sect. 3.3.1), it is realized that $\Omega / (\boldsymbol{W}_N \wedge \boldsymbol{Y})$ is likely to consist of the whole square as the single element, and this remains true even when N goes to infinity. Otherwise stated, $\boldsymbol{W}_N \wedge \boldsymbol{Y}$ remains stuck to the class of bottom elements in the lattice of functions over Ω, namely the class of *constant* functions. Since the "solution" $\boldsymbol{U}$ produced is constrained to be measurable w.r.t. $\boldsymbol{Y}_N \preceq \boldsymbol{W}_N \wedge \boldsymbol{Y}$ (see Lemma 6.7), it cannot be better than the solution in the class of *open-loop* controls. △

This example shows that some necessary conditions derived from Pennanen's conditions (6.8)–(6.10) are not sufficient to ensure convergence of the discrete problem solution towards that of the original problem as N goes to infinity (convergence that Pennanen could prove however). Therefore, Pennanen's conditions are strong enough to avoid the pitfall described in the previous example. The main practical difficulty is that the quantized observation $\boldsymbol{Y}_N$ is not mastered directly (that is, a priori and directly derived from $\boldsymbol{Y}$ by quantization, which would make things a lot easier to design).

Let us explain why condition (6.8), in fact, reflects a stochastic tree structure as depicted by Fig. 1.5. To show this, one must imagine that in our previous model, $\boldsymbol{W}$ represents a stochastic process $\{\boldsymbol{W}_s\}_{s=1,2}$, whereas $\boldsymbol{Y}$ represents the same stochastic process truncated at the first stage, that is, $\boldsymbol{Y} = \boldsymbol{W}_1$.[5] The finite set $\mathbb{W}_N$ is represented by N nodes: each such node carries a *pair* of values (w_1^i, w_2^i) for $i = 1, \ldots, N$. The finite set $\mathbb{Y}_N$ corresponds to the discrete representation of the truncated process $\boldsymbol{W}_1$. It is also represented by a finite set of nodes corresponding to the distinct values found in $\mathbb{Y}_N$: the cardinality of this set is M, which is less than or equal to N; each node carries a value y^j for $j = 1, \ldots, M$. Now, condition (6.8) says two things:

1. the set of N nodes at the second stage is partitioned into M disjoint subsets, each subset being in relation with a node at the first stage: this is the translation of $\boldsymbol{Y}_N \preceq \boldsymbol{W}_N$; otherwise stated, there exists a mapping $\mathfrak{f}$ from $\{1, \ldots, N\}$ to $\{1, \ldots, M\}$ which defines the *preceding* node of each leaf in the tree;
2. moreover, $y^{\mathfrak{f}(i)} = w_1^i$ for $i = 1, \ldots, N$ according to (6.8); as a consequence, it is not necessary to attach a pair of values (w_1^i, w_2^i) to leaf i but attaching w_2^i only is enough since w_1^i can already be read on the preceding node $\mathfrak{f}(i)$ of leaf i as the value $y^{\mathfrak{f}(i)}$.

It should be noticed that while the former item above involves only the encoding parts of the quantizations $Q_{\mathbb{W}}$ and $Q_{\mathbb{Y}}$ (see Sect. 6.1.1) which determine the *topology* of the tree, the latter also involves the decoding parts of those quantizations, that is the *numerical values of samples* attached to nodes.

Equation (6.11) claims that $h \circ Q_{\mathbb{W}} \preceq h$. In a representation such as Figs. 1.2 or 1.3, in which realizations of $\boldsymbol{W}$ are represented as dots in a square, whereas $\boldsymbol{Y} = h(\boldsymbol{W})$ is the corresponding abscissas of those dots, the previous measurability condition says that if two dots are aligned vertically ($h(w) = h(w')$), then their quantized representations are also aligned vertically ($h \circ Q_{\mathbb{W}}(w) = h \circ Q_{\mathbb{W}}(w')$). This only leaves room for quantizations of $\boldsymbol{W}$ which look like that of the left-hand side of Fig. 6.6, with groups of samples aligned vertically and corresponding cells also lined up vertically. The right-hand side of the figure depicts the corresponding stochastic tree. Figure 6.7 shows the Voronoi tessellation that would correspond to the same sample set $\mathbb{W}_N$, but this is not permitted in Pennanen's approach.

[5] The following explanation can then be easily extended by considering $\boldsymbol{W} = \{\boldsymbol{W}_s\}_{s=1,\ldots,T}$ and $\boldsymbol{Y} = \{\boldsymbol{W}_s\}_{s=1,\ldots,t}$ for any intermediate $t < T$. The truncation operator (which retains only the *prefix* of the process up to t) stands for the observation function h of the general theory.

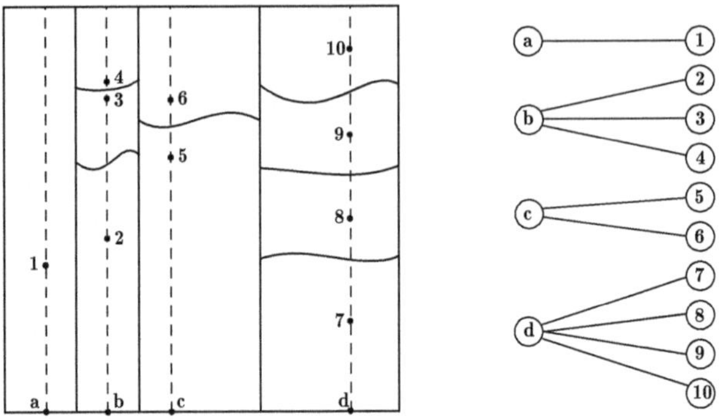

Fig. 6.6 Noise quantization that leads to a stochastic tree

Fig. 6.7 Voronoi tessellation corresponding to the samples of Fig. 6.6

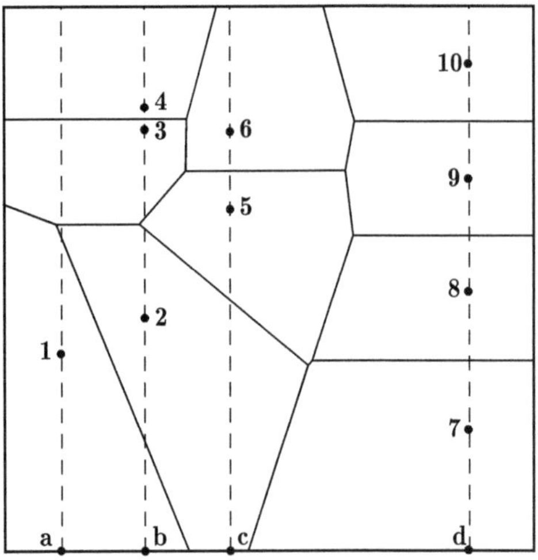

6.2.3 A Constructive Proposal

In this subsection, the formulation (6.1b) is considered anew, but $\boldsymbol{Y}$ may or may not be a function h of $\boldsymbol{W}$. In the latter case, a possible choice of Ω is $\mathbb{W} \times \mathbb{Y}$; in the former case, one can again choose $\Omega = \mathbb{W}$.

Remark 6.10 The following observation was already mentioned in Remark 1.4. There is no fundamental difference between the situation when $\boldsymbol{Y}$ is a function of $\boldsymbol{W}$ and the situation when it is not. Indeed, in the latter case, one can redefine the exogeneous noise as the pair $(\boldsymbol{Y}, \boldsymbol{W})$ (this is the new $\boldsymbol{W}$) and then, h is just the linear operator which extracts the first component of this vector (whereas the cost function depends only on the second component of this new $\boldsymbol{W}$).

It is more fundamental to realize that an *optimal* quantization of a pair of random variables $(\boldsymbol{Y}, \boldsymbol{W})$ is generally *not* the Cartesian product of the two optimal quantizations of $\boldsymbol{W}$ and $\boldsymbol{Y}$ obtained separately (using their marginal probability laws), *even if they are independent random variables.*[6] $\Diamond$

First Version

The main departure from Pennanen's approach is that, now, irrespective of the fact that $\boldsymbol{Y}$ is, or is not, a function h of $\boldsymbol{W}$, these two random variables are quantized independently. Therefore, we introduce:

- $q_{\mathbb{W}} : \mathbb{W} \to \mathbb{W}_N$ where $\mathbb{W}_N$ is a finite subset of cardinality N of $\mathbb{W}$; we require that $q_{\mathbb{W}}(w) = w$ whenever $w \in \mathbb{W}_N$;
- $\iota_{\mathbb{W}}$, the canonical injection of $\mathbb{W}_N$ into $\mathbb{W}$; then,
  - $q_{\mathbb{W}} \circ \iota_{\mathbb{W}}$ is the identity $\mathrm{I}_{\mathbb{W}_N}$ in $\mathbb{W}_N$;
  - $Q_{\mathbb{W}} = \iota_{\mathbb{W}} \circ q_{\mathbb{W}}$ is a quantization and $Q_{\mathbb{W}} \circ \boldsymbol{W}$ is the quantized noise $\boldsymbol{W}_N$;
  - $\mathbb{W}/q_{\mathbb{W}} = \mathbb{W}/Q_{\mathbb{W}}$ is a partition of $\mathbb{W}$ into N cells;
- $q_{\mathbb{Y}} : \mathbb{Y} \to \mathbb{Y}_M$ where $\mathbb{Y}_M$ is a finite subset of cardinality M of $\mathbb{Y}$; we require that $q_{\mathbb{Y}}(y) = y$ whenever $y \in \mathbb{Y}_M$;
- $\iota_{\mathbb{Y}}$, the canonical injection of $\mathbb{Y}_M$ into $\mathbb{Y}$; then,
  - $q_{\mathbb{Y}} \circ \iota_{\mathbb{Y}}$ is the identity $\mathrm{I}_{\mathbb{Y}_M}$ in $\mathbb{Y}_M$;
  - $Q_{\mathbb{Y}} = \iota_{\mathbb{Y}} \circ q_{\mathbb{Y}}$ is a quantization and $Q_{\mathbb{Y}} \circ \boldsymbol{Y}$ is the quantized observation $\boldsymbol{Y}_M$;
  - $\mathbb{Y}/q_{\mathbb{Y}} = \mathbb{Y}/Q_{\mathbb{Y}}$ is a partition of $\mathbb{Y}$ into M cells.

Then, in the discretized problem, the decision variable $\boldsymbol{U}$ is subject to the constraint $\boldsymbol{U} \preceq \boldsymbol{Y}_M$, that is, there exists a feedback $\gamma_M : \mathbb{Y}_M \to \mathbb{U}$ such that $\boldsymbol{U} = \gamma_M(\boldsymbol{Y}_M) = \gamma_M \circ q_{\mathbb{Y}} \circ \boldsymbol{Y}$. Of course, this constraint automatically produces an admissible solution for the original problem. The situation is illustrated by Fig. 6.8 (compare with Fig. 6.5).

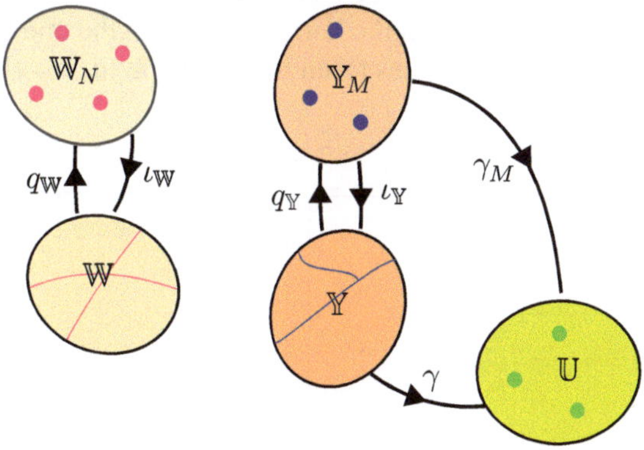

Fig. 6.8 Independent quantization of $\boldsymbol{W}$ and $\boldsymbol{Y}$

[6]Considering two *scalar* random variables with *uniform* distributions over bounded intervals, it can be checked that for the same number of cells, a pavement of a large surface in the plane with hexagons is more efficient in terms of the criterion (6.3) than a pavement with squares. The former cannot obviously be obtained as the Cartesian product of two one-dimensional quantizations.

At this stage, since there is no connection between $\boldsymbol{W}_N$ and $\boldsymbol{Y}_M$ (even if there is one between $\boldsymbol{W}$ and $\boldsymbol{Y}$), the appropriate Ω to consider in the discretized problem is $\mathbb{Y}_M \times \mathbb{W}_N$—this finite set has a maximum of MN elements—with the probability law transported from the original Ω to $\mathbb{Y} \times \mathbb{W}$ by the mapping $(q_{\mathbb{Y}}, q_{\mathbb{W}})$.

To make things more concrete, we consider an example.

Example 6.11 We remain in the context of Example 6.9, with a two-dimensional $\boldsymbol{W}$ and with $\boldsymbol{Y}$ being the first coordinate $\boldsymbol{W}_1$; we use the same representation as in Fig. 6.6. The left-hand part of Fig. 6.9 represents the quantization of $\boldsymbol{W}$ (with $N = 8$). The middle part of that figure represents the quantization of $\boldsymbol{Y}$ on the x-axis (with $M = 5$). Since $\boldsymbol{W}$ and $\boldsymbol{Y}$ are not independent variables here, all the MN combinations of w^i with y^k are not possible, that is, the probability law transported from $\mathbb{W} \times \mathbb{Y}$ to $\mathbb{W}_N \times \mathbb{Y}_M$ by $(q_{\mathbb{W}}, q_{\mathbb{Y}})$ has only 21 non-zero atoms (out of 40): these are the probability masses of the cells depicted in the right-hand side of Fig. 6.9. An alternative representation is that of Fig. 6.10 which depicts all possible pairs of realizations of $(\boldsymbol{Y}_M, \boldsymbol{W}_N)$ in the discrete model: contrary to the situation of Fig. 6.6, there is no longer a tree structure involved now. The formulation of the discretized problem is as follows:

$$\min_{\{u^k\}} \sum_{k \in \{\mathbf{a}, \dots, \mathbf{e}\}} \sum_{i=1}^{8} p^{ik} \, j(u^k, w^i) \tag{6.14}$$

in which p^{ik} is the probability weight of the cell ik ($i = 1, \dots, 8$ and $k \in \{\mathbf{a}, \dots, \mathbf{e}\}$) in the right-hand side of Fig. 6.9. Again, only 21 of those p^{ik} are not zero, but any approximation of the probability masses of the cells that would converge asymptotically to the true values as N and M go to infinity would also be acceptable. △

Remark 6.12 Contrary to the scheme inspired by Pennanen's work described in Sect. 6.2.2, it should be clear that only the *encoding* part of $Q_{\mathbb{Y}}$ matters here. That is, only the cells on the horizontal axis in the middle part of Fig. 6.9 are important, not the precise values taken by y^k for $k \in \{\mathbf{a}, \dots, \mathbf{e}\}$. ◇

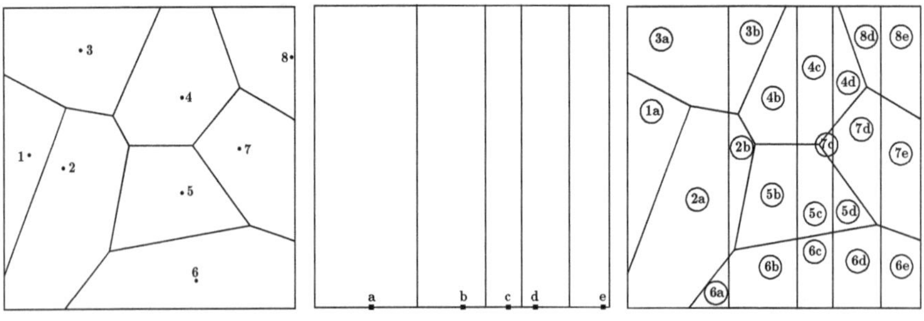

Fig. 6.9 Independent quantizations of $\boldsymbol{W}$ and $\boldsymbol{Y}$: an example

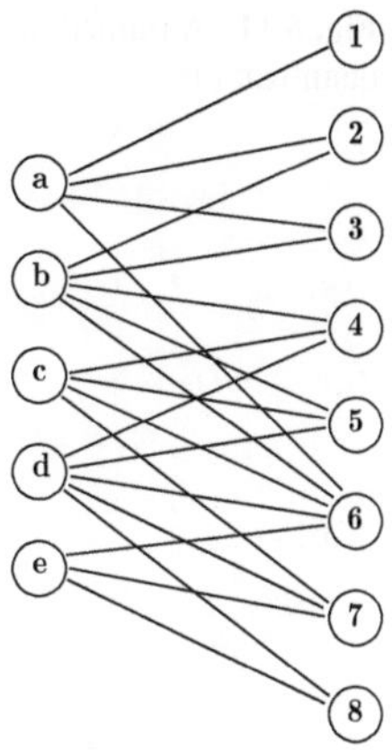

Fig. 6.10 Possible pairs of observations and noises

Remark 6.13 In application of Remark 6.3 to (6.14), for $k \in \{\mathbf{a}, \ldots, \mathbf{e}\}$, the approximation of $\mathbb{E}(j(\boldsymbol{U}, \boldsymbol{W}) \mid \boldsymbol{Y} = y^k)$ in the discrete problem is given by the expression

$$\frac{1}{\sum_{i \in I(k)} p^{ik}} \Big(\sum_{i \in I(k)} p^{ik} \, j(u^k, w^i) \Big), \tag{6.15}$$

where $I(k)$ is the subset of $\{1, \ldots, 8\}$ such that $p^{ik} \neq 0$. Therefore, not only each subset $I(k)$ must be non empty, but its cardinality should asymptotically go to infinity. That is, each vertical strip in Fig. 6.9 should intersect asymptotically an infinite number of Voronoi cells: generically, this should be the case when the y^k's and w^i's are sampled independently and when their numbers go to infinity. But this is not the case in the situation illustrated by Fig. 1.3 where the cardinality of each $I(k)$ remains equal to 1 asymptotically, even when the number of samples went to infinity (see also Sect. 8.5.4 for a related discussion). ◇

Second Version

In Example 6.11, there exists a mapping h such that $\boldsymbol{Y} = h(\boldsymbol{W})$ (namely $\boldsymbol{Y} = \boldsymbol{W}_1$), but there is none which relates the quantized observation $\boldsymbol{Y}_M$ to the quantized noise. Whenever $\boldsymbol{Y} = h(\boldsymbol{W})$, there is a way to recover such a mapping at the price of redefining the quantized noise.

As shown in Example 6.11 (see also the right-hand side of Fig. 6.9), as long as $\boldsymbol{U} \preceq \boldsymbol{Y}_M$, the minimal partition of Ω generating a partition field with respect to which the random variable $(\boldsymbol{U}, \boldsymbol{W}_N)$, hence also $j(\boldsymbol{U}, \boldsymbol{W}_N)$, becomes measurable is that associated with $\boldsymbol{W}_N \vee \boldsymbol{Y}_M = Q_{\mathbb{W}} \circ \boldsymbol{W} \vee Q_{\mathbb{Y}} \circ \boldsymbol{Y}$. Under the assumption that $\boldsymbol{Y} = h(\boldsymbol{W})$, this may be considered a partition of $\mathbb{W}$ (that is, $\boldsymbol{W}_N \preceq \boldsymbol{W}$ and $\boldsymbol{Y}_M \preceq \boldsymbol{Y} \preceq \boldsymbol{W}$; according to Proposition 3.7 and Fig. 3.2, the partition on $\mathbb{W}$ is obtained by superposing the previous partition defined by $Q_{\mathbb{W}}$ and that brought back

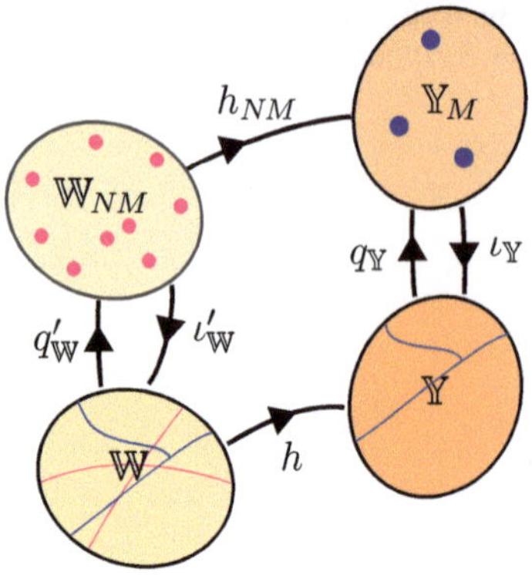

Fig. 6.11 A refined noise quantization

from $\mathbb{Y}$ to $\mathbb{W}$ by the generally multi-valued mapping h^{-1}—see the space $\mathbb{W}$ in the lower left-hand side corner of Fig. 6.11).

So, we introduce a new quantized noise $Q'_{\mathbb{W}}(\boldsymbol{W})$, such that $Q'_{\mathbb{W}}(\boldsymbol{W}) \equiv Q_{\mathbb{W}} \circ \boldsymbol{W} \vee Q_{\mathbb{Y}} \circ \boldsymbol{Y}$ (the symbol $\equiv$ is to be taken in the sense of Proposition 3.41). Since the encoding part is already defined, in order to complete the definition of $Q'_{\mathbb{W}}$, it remains to define the decoding part, which amounts to choosing centroids in the cells depicted in the right-hand side of Fig. 6.9. The new quantized noise $Q'_{\mathbb{W}}(\boldsymbol{W})$ is denoted $\boldsymbol{W}_{NM}$, but $N \times M$ is just an upper bound of the cardinality of the new discrete noise set. Notice that

- $\boldsymbol{W}_N \preceq \boldsymbol{W}_{NM} \equiv \boldsymbol{W}_N \vee \boldsymbol{Y}_M$ (the new quantized noise is "finer" that the previous one);
- $\boldsymbol{Y}_M = Q_{\mathbb{Y}} \circ \boldsymbol{Y} \preceq Q_{\mathbb{W}} \circ \boldsymbol{W} \vee Q_{\mathbb{Y}} \circ \boldsymbol{Y} \equiv Q'_{\mathbb{W}} \circ \boldsymbol{W}$, that is, $\boldsymbol{Y}_M \preceq \boldsymbol{W}_{NM}$, hence, by Proposition 3.46, there exists $h_{NM} : \mathbb{Y}_M \to \mathbb{W}_{NM}$ such that $Y_M = h_{NM}(\boldsymbol{W}_{NM})$.

Therefore, we are able to express the quantized observation $\boldsymbol{Y}_M$ as a function h_{NM} of this finer quantized noise $\boldsymbol{W}_{NM}$. In Fig. 6.11, $Q'_{\mathbb{W}}$ is the composition $\iota'_{\mathbb{W}} \circ q'_{\mathbb{W}}$, where $q'_{\mathbb{W}} : \mathbb{W} \to \mathbb{W}_{NM}$ is such that $q'_{\mathbb{W}} \equiv Q'_{\mathbb{W}}$ and $\iota'_{\mathbb{W}}$ is the canonical injection of $\mathbb{W}_{NM}$ into $\mathbb{W}$. By definition of h_{NM}, we have that

$$Q_{\mathbb{Y}} \circ h = \iota_{\mathbb{Y}} \circ h_{NM} \circ q'_{\mathbb{W}}. \tag{6.16}$$

This is similar to (6.13) and, as in Lemma 6.8, it can be proved that (6.16) is equivalent to

$$h_{NM} = q_{\mathbb{Y}} \circ h \circ \iota'_{\mathbb{W}}. \tag{6.17}$$

Discussion

The connection between the original observation function h and that of the discrete model h_{NM} is weaker than it was in Pennanen's approach: essentially, we have nothing similar to (6.8) here.

Returning to Example 6.11, in order to completely define $Q'_{\mathbb{W}}$, we must draw a dot in each cell of the right-hand side part of Fig. 6.9. Those dots represent 2-dimensional vectors w^{ik} (namely, the dots with coordinates (w_1^{ik}, w_2^{ik})) with $i \in \{1, \dots, 8\}$ and $k \in \{\mathbf{a}, \dots, \mathbf{e}\}$, but not all pairs out of this cartesian product are present (only 21 out of 40). This collection of vectors $\{w^{ik}\}$ is the set $\mathbb{W}_{NM}$. According to (6.17), $h_{NM} : \mathbb{W}_{NM} \to \mathbb{Y}_M$ is such that $h_{NM}(w^{ik}) = y^k$ with $k \in \{\mathbf{a}, \dots, \mathbf{e}\}$. But obviously, the precise values of those y^k play no particular role: what matters is the partition of $\mathbb{Y}$ generated by $\boldsymbol{Y}_M$ (the encoding part), the codebook $\{y^k\}_{k \in \{\mathbf{a},\dots,\mathbf{e}\}}$ is not relevant (the only constraint being that each dot belongs to its corresponding cell).

Therefore, since the codebooks of quantized noises and observations are somewhat flexible, we may use this flexibility to try to get closer to Pennanen's scheme. Graphically, we may try to move the dots y^k within their cells in $\mathbb{Y}$, and simultaneously choose dots $w^1, \dots, w^{21}$ within the cells 1**a**, ..., 8**e** in the right-hand side of Fig. 6.9 so that each dot representing a quantized observation be vertically aligned with a subset of the dots representing quantized noises as shown in Fig. 6.6. Observe that this is not necessarily possible: if we restrict our attention to the vertical strip labelled **a**, there is *no* vertical line in this strip that crosses simultaneously the cells 1**a** and 6**a**.

Mathematically, the issue is that of choosing $Q_{\mathbb{Y}}$ and $Q_{\mathbb{W}}$ (in which only the encoding parts are important), so that it becomes possible, a posteriori, to choose the decoding parts of $Q_{\mathbb{Y}}$ and of $Q'_{\mathbb{W}}$ (with the constraint that $Q'_{\mathbb{W}} \equiv Q_{\mathbb{W}} \vee Q_{\mathbb{Y}} \circ h$) in such a way that (compare to (6.8))

$$h \circ Q'_{\mathbb{W}} = \iota_{\mathbb{Y}} \circ h_{NM} \circ q'_{\mathbb{W}} \tag{6.18a}$$

$$= Q_{\mathbb{Y}} \circ h, \tag{6.18b}$$

the latter equation using (6.16).

At this moment, we do have that

$$Q_{\mathbb{Y}} \circ h \circ Q'_{\mathbb{W}} = Q_{\mathbb{Y}} \circ h, \tag{6.19}$$

which is a weaker property than (6.18): indeed, by composing (6.16) with $Q'_{\mathbb{W}} = \iota'_{\mathbb{W}} \circ q'_{\mathbb{W}}$ to the right, (6.19) is derived. But we do not know of a constructive method to ensure (6.18) itself.

In the next section, we briefly sketch the procedure proposed by Pennanen [111] in the particular case when informational constraints reduce to non anticipativity constraints. This procedure can be related to a special Monte Carlo technique for approximating the expectation of a function of several *independent* random variables. However, we are going to show that this special scheme has a rather bad rate of convergence when compared with the usual Monte Carlo scheme. This is why it is important to be able to get rid of the scenario tree structure as a way to translate informational constraints in the discrete problem.

6.3 A Handicap of the Scenario Tree Approach

In this section, we explain Pennanen's technique [111] to sample noise processes (in the simplest case of *two-stage white noise* processes—see Assumption 5.9) in order to obtain scenario trees,[7] and we make the connection of this technique to a particular way of numerically estimating the expectation of a function of two independent random variables. We then show that this particular (unbiased) estimation technique is not efficient, in terms of the variance of its error, w.r.t. the classical Monte Carlo estimation technique.

6.3.1 How to Sample Noises to Get Scenario Trees

As illustrated by Fig. 6.6, scenario trees for two-stage stochastic processes are related to the fact that dots in a two-dimensional space, which represent sample trajectories, are grouped in vertical clusters. That is, there are several trajectories which share common first-stage values. As discussed earlier, the probability that this occurs naturally for continuous-valued stochastic processes is zero when trajectories are generated by pseudo-random Monte Carlo sampling using the probability law of the process. Therefore, the scenario tree structure can be obtained

- either by manipulating a bunch of Monte Carlo samples (or scenarios recorded in the real life) in order to force the tree structure,[8]—but, then, the original sample set must be altered in a way which is not necessarily respectful of the underlying probability law;
- or by making use of special sampling procedures (assuming the underlying probability distribution is known). The latter option is proposed by Pennanen in [111]. We now give a sketch of this idea.

If the stochastic process $\boldsymbol{W}$ is a "white noise", that is, the random variables $\boldsymbol{W}_t$, $\boldsymbol{W}_{t+1}$, ..., are all independent, then the procedure amounts to

- drawing N_0 sample values w_0^i, $i = 1, \ldots, N_0$, of $\boldsymbol{W}_0$ according to the probability distribution of this random variable;
- for each such w_0^i, obtaining N_1 sample values w_1^{ij}, $j = 1, \ldots, N_1$, by Monte Carlo sampling according to the distribution of $\boldsymbol{W}_1$, and associating them with that value w_0^i to form two-stage sample trajectories (w_0^i, w_1^{ij}) (thus, there are $N_0 \times N_1$ such trajectories);
- repeating this process over the whole time horizon.

[7]Other references dealing with scenario tree generation will be mentioned at Sect. 7.4.1.

[8]Optimal quantization (see Sect. 6.1.2) may be used to that purpose and many authors proposed various techniques to build up such trees—see e.g. [12, 61, 71, 114].

Observe that the clusters of N_1 values w_1^{ij} associated with the w_0^i's may be all identical (in which case, the notation w_1^{ij} can be reduced to w_1^j) or different. We examine later on what is the impact of either choice.

If the stochastic process $\boldsymbol{W}$ is not a white noise, Pennanen assumes that it can be modelled by a recurrent dynamic equation driven by a white noise. Then, the above procedure is used for the driving white noise, and the sample noise trajectories are then obtained by propagating these trajectories through the dynamic equation.

In the rest of this section, to keep things simple, we limit ourselves to the discussion of white noise processes.

6.3.2 Variance Analysis

As discussed throughout this chapter, the discretization of stochastic optimization problems with SIS involves some sort of noise sampling as well as the sound translation of informational constraints in the discrete problem formulation. So far in this chapter, we have given attention to the latter aspect. But it should be clear that the quality of approximation of the mathematical expectations (and *conditional* mathematical expectations, as underlined in Remark 6.3) involved in the problem is also important. If the cost function is badly approximated, one cannot expect that a good approximation of the optimal solution can be derived from the discrete problem solution, whatever care is exercised about the other aspects (in particular, informational constraints) of the problem.

In this subsection, we concentrate on this aspect of the approximation: more precisely, we consider any real-valued function f of two scalar variables, such that the mathematical expectation $\mathbb{E}\big(f(\boldsymbol{X}, \boldsymbol{Y})\big)$, where $\boldsymbol{X}$ and $\boldsymbol{Y}$ are *independent* random variables, makes sense. In relation with the previous discussion, f should be thought of as the cost function. Respectively, $\boldsymbol{X}$ and $\boldsymbol{Y}$, should be interpreted as $\boldsymbol{W}_0$ and $\boldsymbol{W}_1$, the two first stages of a white noise stochastic process. No decision variable appears here since we forget about optimization to pay attention to the quality of approximation of the expectation. The discussion is limited to two-time stages only, but the generalization to several time stages should be straightforward.

In a standard Monte Carlo procedure, N sample values (x^i, y^i) of the pair of random variables $(\boldsymbol{X}, \boldsymbol{Y})$ are generated according to their joint probability distribution, or they have been recorded from real data. An unbiased estimate of $\mathbb{E}\big(f(\boldsymbol{X}, \boldsymbol{Y})\big)$ (which is denoted simply $\mathbb{E}f$ for short) is provided by the arithmetic mean

$$\frac{1}{N}\sum_{i=1}^{N} f(x^i, y^i). \tag{6.20}$$

It is well known that the variance of this estimate is an $\mathrm{O}(1/N)$ when N is the number of samples.[9]

[9]For the notation O, see footnote 3 in Chap. 2.

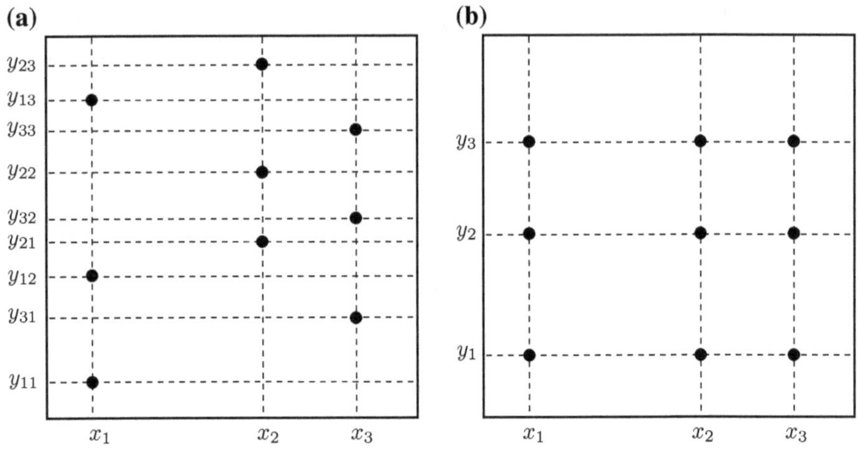

Fig. 6.12 Two ways of sampling to get scenario trees (option **a** *left-hand side* and option **b** *right-hand side*)

In Pennanen's procedure described in Sect. 6.3.1, N_x samples are generated for $\boldsymbol{X}$. With each such sample value x^i, a group of samples y^j with $j \in J(i)$, is associated: these samples are also generated by Monte Carlo sampling according to the probability distribution of $\boldsymbol{Y}$. To make things simpler (but this is not essential), we assume that all such index sets $J(i)$ have the same cardinality N_y. Moreover, as suggested earlier, there are two options to consider:

option (a): N_x sample groups of cardinality N_y are generated independently;
option (b): the same group $\{y^j\}_{j \in J}$ is associated with all samples x^i.

Pictorially, those options are illustrated by Fig. 6.12 (with $N_x = N_y = 3$).

In both cases, overall, $N_x \times N_y$ samples are used to produce the following estimate of $\mathbb{E}f$:

$$\frac{1}{N_x \times N_y} \sum_{i=1}^{N_x} \sum_{j \in J(i)} f(x^i, y^j), \tag{6.21}$$

where $J(i)$ is indeed independent of i in option (b) (that is, $J(i) = \{1, \ldots, N_y\}$ for all i), whereas, in option (a), the $J(i)$'s should be viewed as disjoint subsets of indices to translate the fact that the N_x groups $\{y^j\}_{j \in J(i)}$ have been sampled independently.

Clearly, the estimate (6.21) of $\mathbb{E}f$ is also unbiased, and if we want to compare it with (6.20) from the point of view of its variance, we must assume that $N = N_x \times N_y$. We first study option (b). Recall that, in this option, all subsets $J(i)$ involved in (6.21) coincide with $\{1, \ldots, N_y\}$.

Proposition 6.14 *Given the value of* $N = N_x \times N_y$, *with option (b), the variance of estimate* (6.21) *is minimal when* $N_y = N_x$ *and this variance is of order* $\mathrm{O}(1/N_x)$; *therefore the variance is of order* $\mathrm{O}(1/\sqrt{N})$.[10]

[10]The authors are indebted to Prof. Benjamin Jourdain for preliminary results in this direction.

Proof Let $J(i) = \{1, \ldots, N_y\}$ for all i. The variance of the estimate (6.21) is

$$\sigma^2 = \mathbb{E}\Bigg(\bigg(\frac{1}{N_x N_y}\sum_{i=1}^{N_x}\sum_{j=1}^{N_y}\big(f(\boldsymbol{X}^i, \boldsymbol{Y}^j) - \mathbb{E}f\big)\bigg)\bigg)$$
$$\times\bigg(\frac{1}{N_x N_y}\sum_{k=1}^{N_x}\sum_{l=1}^{N_y}\big(f(\boldsymbol{X}^k, \boldsymbol{Y}^l) - \mathbb{E}f\big)\bigg). \tag{6.22}$$

In this expression, the outer expectation is with respect to the probability distributions of independent random variables $\{\boldsymbol{X}^i\}_{i=1,\ldots,N_x}$, $\{\boldsymbol{X}^k\}_{k=1,\ldots,N_x}$, $\{\boldsymbol{Y}^j\}_{j=1,\ldots,N_y}$ and $\{\boldsymbol{Y}^l\}_{l=1,\ldots,N_y}$ (replicating $\boldsymbol{X}$ and $\boldsymbol{Y}$), the realizations of which are the samples x^i, x^k, y^j and y^l used in the estimate.

If the expression (6.21) is expanded, this yields $N_x^2 N_y^2$ products of random variables (with zero mean) of the type

$$\frac{1}{N_x^2 N_y^2}\mathbb{E}\Big(\big(f(\boldsymbol{X}^i, \boldsymbol{Y}^j) - \mathbb{E}f\big)\big(f(\boldsymbol{X}^k, \boldsymbol{Y}^l) - \mathbb{E}f\big)\Big). \tag{6.23}$$

We split up this set of products into four subsets:

1. the subset for which $i = k$ and $j = l$, of cardinality $N_x N_y$; for this subset, all products of the type (6.23) are squares and their sum contributes to σ^2 (in (6.22)) for a term of order $O(1/N_x N_y)$;
2. the subset for which $i \neq k$ and $j \neq l$: the cardinality of this subset is $N_x(N_x - 1)N_y(N_y - 1)$; since this subset contains only products of random variables of the type (6.23) which have zero mean and are mutually independent, it contributes for 0 to (6.22);
3. the subset for which $i = k$ but $j \neq l$, of cardinality $N_x N_y(N_y - 1)$ that we study later on;
4. symmetrically, the subset for which $i \neq k$ but $j = l$ of cardinality $N_x(N_x - 1)N_y$.

It can be checked that the sum of cardinalities of the four subsets is equal to $N_x^2 N_y^2$, as it must be.

It remains to study the contribution of products in the third and fourth subsets. Those products involve either the same $\boldsymbol{X}^i$ but different $\boldsymbol{Y}^j$ and $\boldsymbol{Y}^l$, or symmetrically, the same $\boldsymbol{Y}^j$, but different $\boldsymbol{X}^i$ and $\boldsymbol{X}^k$. We only study the former subset since conclusions also apply to the latter by symmetry. We first prove that all terms of the type (6.23) such that $\boldsymbol{X}^i$ is the same but $\boldsymbol{Y}^j$ is independent of $\boldsymbol{Y}^l$ have nonnegative expectations. Indeed, with the short-hand notation

$$\mathbb{E}^{\boldsymbol{X}} f := \mathbb{E}\big(f(\boldsymbol{X}, \boldsymbol{Y}) \mid \boldsymbol{X}\big),$$

one has that

$$\mathbb{E}\Big(\big(f(\boldsymbol{X}^i,\boldsymbol{Y}^j)-\mathbb{E}f\big)\big(f(\boldsymbol{X}^i,\boldsymbol{Y}^l)-\mathbb{E}f\big)\Big)=$$

$$\mathbb{E}\Bigg(\mathbb{E}^{\boldsymbol{X}^i}\Big(\big(\underbrace{f(\boldsymbol{X}^i,\boldsymbol{Y}^j)-\mathbb{E}^{\boldsymbol{X}^i}f}_{\boldsymbol{B}^j}+\underbrace{\mathbb{E}^{\boldsymbol{X}^i}f-\mathbb{E}f}_{\boldsymbol{C}}\big)$$

$$\times\big(\underbrace{f(\boldsymbol{X}^i,\boldsymbol{Y}^l)-\mathbb{E}^{\boldsymbol{X}^i}f}_{\boldsymbol{B}^l}+\underbrace{\mathbb{E}^{\boldsymbol{X}^i}f-\mathbb{E}f}_{\boldsymbol{C}}\big)\Big)\Bigg).$$

The independence of $\boldsymbol{Y}^j$ and $\boldsymbol{Y}^l$, and therefore of $\boldsymbol{B}^j$ and $\boldsymbol{B}^l$, and the fact that the latter variables have zero conditional means, imply that $\mathbb{E}^{\boldsymbol{X}^i}\big(\boldsymbol{B}^j\boldsymbol{B}^l\big)=0$. Moreover, since $\boldsymbol{C}$ is $\boldsymbol{X}^i$-measurable,

$$\mathbb{E}^{\boldsymbol{X}^i}(\boldsymbol{B}^j\boldsymbol{C})=\mathbb{E}^{\boldsymbol{X}^i}(\boldsymbol{B}^j)\times\boldsymbol{C}=0.$$

The same applies to $\mathbb{E}^{\boldsymbol{X}^i}(\boldsymbol{C}\,\boldsymbol{B}^l)$. The only nonzero term is thus the nonnegative term $\mathbb{E}(\boldsymbol{C}^2)$, which is the variance of $\mathbb{E}^{\boldsymbol{X}^i}f$ (generically of order $\mathrm{O}(1)$).

Finally, the terms in the third and fourth subsets above contribute all together for a nonnegative term of order $\mathrm{O}\Big(\big((N_x-1)+(N_y-1)\big)/N_xN_y\Big)\sim\mathrm{O}(1/N_x+1/N_y)$. This contribution is added to that of the first subset which was $\mathrm{O}(1/N_xN_y)$. For the comparison of (6.20) and (6.21), we assume that $N=N_xN_y$: for N given, the variance of the estimate (6.21) is minimal when $N_x=N_y=\sqrt{N}$, and this variance is of order $\mathrm{O}(1/\sqrt{N})$, to be compared with $\mathrm{O}(1/N)$ of the standard Monte Carlo estimate (6.20). □

It is easy to figure out how this result extends to the case of T stages instead of 2: the variance of the "tree" estimate (6.21) is of order $\mathrm{O}(1/\sqrt[T]{N})$ instead of $\mathrm{O}(1/N)$ for (6.20). Needless to say, this quickly becomes a dramatic loss of quality of the tree estimate as T keeps growing.

Consider now option (a). The calculations in the proof of Proposition 6.14 must be adapted in the following way. First, (6.21) is now valid with subsets $J(i)$ which must be considered as disjoint subsets (that is, the corresponding subsets of random variables $\{\boldsymbol{Y}^j\}_{j\in J(i)}$ are independent). Consequently, (6.22) must be replaced by

$$\sigma^2=\mathbb{E}\Bigg(\Bigg(\frac{1}{N_xN_y}\sum_{i=1}^{N_x}\sum_{j\in J(i)}\big(f(\boldsymbol{X}^i,\boldsymbol{Y}^j)-\mathbb{E}f\big)\Bigg)\Bigg)$$

$$\times\Bigg(\frac{1}{N_xN_y}\sum_{k=1}^{N_x}\sum_{l\in J(k)}\big(f(\boldsymbol{X}^k,\boldsymbol{Y}^l)-\mathbb{E}f\big)\Bigg). \tag{6.24}$$

Among the four subsets considered in the previous proof, the first and the third ones have the same cardinalities as earlier and contribute to σ^2 for the same amounts as in that proof, namely, $O(1/N_x N_y)$ and $O\big((N_y - 1)/N_x N_y\big)$ respectively. Since $i \neq k$ implies that $j \neq l$ (because $J(i) \cap J(k)$ should be considered as empty), then the fourth subset (which earlier contained $N_x(N_x - 1)N_y$ elements) is now empty whereas the cardinality of the second subset increases to $N_x(N_x - 1)N_y^2$ (that is, the elements of the fourth subset which previously contributed for nonnegative terms are transferred to the second subset whose terms contribute for 0). Finally, with option (a), the variance of (6.21) is of order $O(1/N_x N_y) + O\big((N_y - 1)/N_x N_y\big)$.

Under the constraint that $N = N_x N_y$ (to make the comparison with (6.20) possible), we reach the conclusion that the best (minimal) variance is obtained when $N_y = 1$ and $N_x = N$. Notice that this is no longer a tree, but indeed N independent scenarios, that is (6.20) and (6.21) actually coincide. This just says that, from only the point of view of minimizing the variance of the estimate of the cost function over the whole time horizon, the structure of N independent scenarios is far better than a tree structure. But of course, one should again recall Remark (6.3): if $N_y = 1$, the conditional expectation of the cost (knowing $\boldsymbol{W}_0$) which serves as the objective function in the minimization problem when choosing $\boldsymbol{U}_0$ is approximated with help of a *single* sample of $\boldsymbol{W}_1$, which is very bad. To avoid this, we should have put a lower bound on N_y in order to bound the variance of this conditional expectation, and in this simple case, it is clear that, with the constraint $N = N_x N_y$, the best trade-off between the variance of the estimate of the expected cost over the two-stage horizon and the variance of the estimate of the conditional expectation restrained to the second stage only is again to take $N_x = N_y = \sqrt{N}$.

The conclusion of this rough variance analysis is that the tree structure, which is one way to represent informational (or simply, non anticipativity) constraints, is not very efficient from the point of view of the variance of estimates it provides. Therefore, we should avoid the tree structure and find another way to translate the informational constraints in the discrete problem. The methodology presented in Sect. 6.2.3 suggested that this is indeed possible. A more concrete technique in the context of stochastic optimal control problems is presented in Chap. 7.

6.4 Conclusion

In this chapter, we have proposed a methodology to derive approximate finite-dimensional versions of the generic stochastic optimization problem (6.1) which involves *static* informational constraints (the so-called SIS—see Sect. 1.4). This discretization stage should not only limit the computations to finite-dimensional objects (probability measures, decision variables, etc.) but it should also translate the original informational constraints in a way which makes it possible to build up a *feasible* solution for the original problem after the discrete finite-dimensional problem has been solved. In addition, one expects that the performance of this feasible solution approaches the true optimal performance when the dimension of the approximate optimization problem goes to infinity. This convergence issue has not been explicitly considered in this chapter and is deferred to Chap. 8.

A popular technique to formulate discrete optimization problems taking care at least of non anticipativity constraints (the minimal form of informational constraints in multi-stage problems) is the so-called scenario tree technique. Pennanen [110] proposed a complete study of this technique, including the reconstruction of a feasible solution for the original problem and its asymptotic convergence to the optimal solution. We have given a description of his approach using the language of quantization presented in the first part of this chapter. We then have proposed other approaches which attempt to relax some of the constraints imposed by Pennanen's approach, and more generally by the scenario tree technique. Indeed, as explained in the end of the chapter, the convergence speed of this technique is seriously handicapped by the variance of the expectation estimates it produces, which is typically in $O(\sqrt[T]{N})$ when using N sample trajectories for a problem with T time stages. This result is also mentioned by Shapiro [139] who uses large deviation techniques to establish it.

The argument above provides motivation to eliminate the scenario tree structure and to find alternative ways to account for informational constraints. The approach described in this chapter is a first proposal in this new direction, and it must be confirmed by the convergence study presented in Chap. 8. Chapter 7 presents another approach, however, that is not directly comparable with the one introduced here because they proceed along two different paths. In this chapter, we have formulated a discretized optimization problem which attempts to mimic the original problem and we have derived the proposed solution from the resolution of this problem. In the next chapter, we consider the optimality conditions of the infinite-dimensional original problem (established in Chap. 5), and we propose discretization schemes in order to approximately solve those conditions.

Chapter 7
Numerical Algorithms

7.1 Introduction

In this chapter, we consider the issue of numerical approximation of the solution of stochastic optimal control (SOC) problems for which optimality conditions were studied in Chap. 5. In a more abstract setting of stochastic optimization problems, this issue was already addressed in Chap. 6. Already in that chapter, we faced the question of combining two types of concerns:

- on the one hand, expectations or conditional expectations must be evaluated using finite computational procedures, which appeals to some sort of Monte Carlo or quasi-Monte Carlo sampling;
- on the other hand, informational constraints must be correctly translated into the discrete problem.

We have seen that those two concerns may interact with each other. For example, with the goal of properly handling these two aspects of the problem, we described in Chap. 6 the systematic approach proposed by Pennanen [110, 111], and we showed that this approach leads to the "scenario tree technology". This technology has been very popular in the Stochastic Programming community as a way to translate non anticipativity constraints (see e.g. [61, 71, 114]). Unfortunately, regarding the other goal of approximating (conditional) expectations, as explained in Sect. 6.3, this approach turns out to be rather inefficient. We come back to this observation once again in this chapter.

In Sect. 6.2.3, we outlined discretization procedures which do not necessarily result in the use of scenario trees. However, those descriptions may have remained rather conceptual and abstract. By starting from the optimality conditions obtained in Chap. 5, we end up here with more concrete numerical schemes. Indeed, there have always been two main paths towards numerical resolution of infinite dimensional optimization problems:

- either one starts with a discretization of the problem itself, that is, by giving finite dimensional approximate expressions of variables, cost function, constraints, etc.,

P. Carpentier et al., *Stochastic Multi-Stage Optimization*, Probability Theory and Stochastic Modelling 75, DOI 10.1007/978-3-319-18138-7_7

and then by solving the optimality conditions of this finite dimensional optimization problem;

- or, one first starts by obtaining the optimality conditions in the original infinite dimensional setting, and then devises a numerical scheme to solve them by necessarily using finite dimensional calculations.

Obviously, the scenario tree technology proceeds from the former point of view, whereas what is discussed in this chapter stems from the latter. The advantage of the former is that a clear optimization problem is solved, namely the discretized one. Hence, the concern is on how well this discretized problem is representative of the original one. In the latter approach, the goal remains to solve the optimality conditions of the "true" problem but this is done only approximately. Ideally, one would like to be in the situation of following either of the two paths of a *commutative* diagram, namely "discretize then optimize" or "optimize then discretize" and of reaching the *same* final point, but this is rarely the case. That is, the optimality conditions of the discretized problem can hardly be interpreted as the discretization of the optimality conditions of an infinite dimensional optimization problem.

Indeed, what is important to check is the performance of the *final* "solution" obtained, once this solution is implemented upon the "real" system (as illustrated in Sect. 1.4.2). But this statement presumes that the so-called solution is *implementable* on the real system, which requires that it is at least feasible, that is respectful of the constraints (in particular the informational constraints). We insisted in Chap. 6 that it is not enough to obtain optimal decisions at the nodes of a scenario tree because an implementable and feasible solution of a SOC problem, for example, consists of a feedback strategy that delivers decisions as a function of *any* observation one *can* and *may* obtain in a real-life situation. Hence, the numerical resolution of the optimization problem formulated on the scenario tree should be followed by a post-processing of the "results", a phase that we call "feedback synthesis". We illustrate why this stage is ill-posed in the scenario tree approach.

As previously discussed, taking care of the informational constraints may be done using two points of view:

- in the *algebraic* point of view, those constraints are translated by *measurability* constraints between random variables, and the problem is thus to find an adequate transposition of these kind of constraints in the discrete problem;
- in the *functional* point of view, as long as measurability of a variable $\boldsymbol{Y}$ w.r.t. another variable $\boldsymbol{X}$ is equivalent to the existence of a measurable function f such that $\boldsymbol{Y} = f(\boldsymbol{X})$ (according to Proposition 3.46), the solution is searched for directly as a function of observations.

In the scenario tree approach, the tree structure is supposed to approximately translate the former algebraic point of view by its discrete structure (each node, corresponding to a single path upstream, carries a decision which is thus indexed by this past history but which may face several possible paths in the future, in fact those conditionally possible given the past history). The Dynamic Programming (DP) approach is clearly using the latter functional point of view by manipulating a cost-to-go (Bellman) function and a feedback function, both depending on a state variable (in the context of the Markovian framework—see Sects. 4.4 and 5.5.1—in which this state variable is an appropriate summary of the useful information to be used to make decisions).

In the present chapter, we start from the optimality conditions obtained in Chap. 5 and derive a numerical resolution method. In so doing, the situation is less clear-cut regarding the previous two algebraic versus functional options. In the initial formulation of the problem, informational constraints are dealt with algebraically as measurability restrictions. Then, in the process of deriving the corresponding optimality conditions, those constraints are expressed with the help of *projection* operators, that is, here, *conditional expectations*. Finally, approximating those conditional expectations amounts to obtaining discrete representations of *functions* of the conditioning variables. Therefore, following this path, we end up with a mix of Monte Carlo sampling techniques and functional approximations.

As we shall see, the main departure from the DP-based approach, which proceeds from the discretization of the state space directly, is that the proposed technique builds up an adaptive mesh in the state space which takes care of the actual probability distribution at the optimum. Such a feature may be quite beneficial when it happens that this optimal distribution concentrates in a narrow region of the space. This feature is similar to that proposed by Broadie and Glasserman [31], although we consider here more general SOC problems in which the state probability distribution is solution dependent. In American option problems, the optimal decision, namely the optimal stopping time, does not affect this distribution prior to this terminal time.

In the rest of this chapter, we briefly recall how SOC problems are discretized and solved by DP and by the scenario tree approach before presenting numerical methods based on the optimality conditions obtained in Chap. 5. In order to illustrate and compare those various approaches, we start with a simple benchmark problem, namely a SOC problem (reservoir management) with a one-dimensional state variable.

7.2 A Simple Benchmark Problem

We consider the production management of a hydro-electric dam formulated as a SOC problem over a probability space $(\Omega, \mathcal{A}, \mathbb{P})$ as follows.

7.2.1 Formulation

The horizon is 24 h long with time steps of one hour each. Therefore, the time index t ranges from 0 to T with $T = 24$.

The water volume stored in the dam at t is denoted $\boldsymbol{X}_t$. This and all other random variables introduced hereafter, are considered to be elements of $L^2(\Omega, \mathcal{A}, \mathbb{P}; \mathbb{R})$ (according to the framework already adopted in Sect. 5.2.1). This storage $\boldsymbol{X}_t$ is physically constrained to remain between a lower bound $\underline{x}$ (a minimal volume—possibly zero—kept in the dam) and an upper bound $\overline{x}$ (maximal water volume the dam can contain; beyond that volume, any additional inflow causes an overflow and a loss of water).

Let $\boldsymbol{U}_t$ denote the *desired* volume of water one would like to turbinate at stage t in order to produce electricity. Due to the physical bounds on the water storage, the *actual* volume which can effectively be turbinated, denoted $\boldsymbol{E}_{t+1}$, may be different from the desired one: when the water storage hits its lower bound, $\boldsymbol{E}_{t+1}$ may be less than $\boldsymbol{U}_t$ in order to satisfy this lower bound. Let $\boldsymbol{A}_{t+1}$ denote the random water inflow into the dam during stage t brought by a river flow and/or rainfall (the shift in index t reflects the *decision-hazard* framework already adopted in Sect. 1.2.1—see also Remark 1.1). The sequence $\{\boldsymbol{A}_t\}_{t=1,\dots,T}$ is a stochastic process with known distribution. Then, setting

$$\boldsymbol{E}_{t+1} = \min(\boldsymbol{U}_t, \boldsymbol{X}_t + \boldsymbol{A}_{t+1} - \underline{x}), \tag{7.1}$$

prevents $\boldsymbol{X}_{t+1}$ from falling below $\underline{x}$. Indeed, the dynamics (5.1) of the dam read

$$\boldsymbol{X}_{t+1} = \min(\boldsymbol{X}_t - \boldsymbol{E}_{t+1} + \boldsymbol{A}_{t+1}, \overline{x}),$$

since $\boldsymbol{X}_{t+1}$ cannot exceed the upper bound $\overline{x}$ due to the possible overflow. Finally, the intermediate variable $\boldsymbol{E}_{t+1}$ can be eliminated using (7.1) and the dynamics can be written as

$$\boldsymbol{X}_{t+1} = \min\left(\max(\boldsymbol{X}_t - \boldsymbol{U}_t + \boldsymbol{A}_{t+1}, \underline{x}), \overline{x}\right). \tag{7.2}$$

Observe that this equation prevents the storage to take values outside the interval $[\underline{x}, \overline{x}]$.

The control variables $\boldsymbol{U}_t$ are subject to the following bound constraints: for all $t = 0, \dots, T-1$,

$$\underline{u} \leq \boldsymbol{U}_t \leq \overline{u}. \tag{7.3}$$

The effectively turbinated water volume $\boldsymbol{E}_{t+1}$ (see (7.1)) produces a certain amount of electric power denoted $\boldsymbol{P}_{t+1}$ which also depends on the water storage (indeed, on the water level in the dam, due to the fall height effect):

$$\boldsymbol{P}_{t+1} = g(\boldsymbol{X}_t, \boldsymbol{E}_{t+1}). \tag{7.4}$$

Let $\{\boldsymbol{D}_t\}_{t=1,\dots,T}$ denote the electricity demand, another stochastic process with known distribution. In our decision-hazard framework, the production $\boldsymbol{P}_{t+1}$ has to meet demand $\boldsymbol{D}_{t+1}$: either $\boldsymbol{P}_{t+1} \geq \boldsymbol{D}_{t+1}$ and the production excess is sold on the electricity market, or $\boldsymbol{P}_{t+1} \leq \boldsymbol{D}_{t+1}$ and the gap must be compensated for, either by buying power on the market or by paying a penalty. The associated cost is given by the function c_t applied to the gap $\boldsymbol{D}_{t+1} - \boldsymbol{P}_{t+1}$:

$$c_t(\boldsymbol{D}_{t+1} - \boldsymbol{P}_{t+1}). \tag{7.5}$$

The final stock $\boldsymbol{X}_T$ is given a value $-K(\boldsymbol{X}_T)$, which is subtracted from the cost function.

The initial condition $\boldsymbol{X}_0$ is a random variable with known distribution, and we define the noise random process $\{\boldsymbol{W}_t\}_{t=0,\dots,T}$ by

$$\begin{aligned} \boldsymbol{W}_0 &= \boldsymbol{X}_0, \\ \boldsymbol{W}_t &= (\boldsymbol{A}_t, \boldsymbol{D}_t), \ t = 1, \dots, T. \end{aligned} \tag{7.6}$$

We assume that this noise process is fully observed and that the control variables are measurable with respect to the past noises. The dam management problem is finally formulated as follows:

$$\min \mathbb{E}\left(\sum_{t=0}^{T-1} c_t\Big(\boldsymbol{D}_{t+1} - g\big(\boldsymbol{X}_t, \min(\boldsymbol{U}_t, \boldsymbol{X}_t + \boldsymbol{A}_{t+1} - \underline{x})\big)\Big) + K(\boldsymbol{X}_T)\right), \tag{7.7}$$

subject to the constraints (7.2) and (7.3) and to the measurability constraints

$$\boldsymbol{U}_t \preceq (\boldsymbol{W}_0, \dots, \boldsymbol{W}_t), \ t = 0, \dots, T-1. \tag{7.8}$$

Remark 7.1 Several functional expressions above involve the max and/or the min operator(s), which are not everywhere differentiable. We approximate the nonsmooth operator min by the following operator depending on a smoothing parameter s:

$$\min(x, y) \approx \begin{cases} y & \text{if } y \le x - s, \\ (x+y)/2 - (x-y)^2/(4s) - s/4 & \text{if } x - s \le y \le x + s, \\ x & \text{if } y \ge x + s. \end{cases}$$

(and likewise for max since $\max(a, b) = -\min(-a, -b)$) in order to recover a differentiable problem. Figure 7.1 shows plots of the function $y \mapsto \min(1-y, y)$ for $y \in [0, 1]$ and of the smoothed min with $s = 0.2$. $\lozenge$

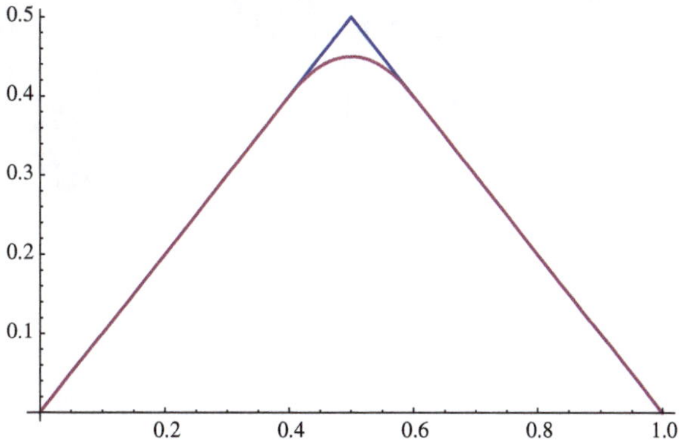

Fig. 7.1 $\min(1-y, y)$ for $y \in [0, 1]$ and its smoothed counterpart with $s = 0.2$

7.2.2 Numerical and Functional Data

Both electricity demand and water inflows are supposed to be white noise processes (see Assumption 4.10). A discrete valued stationary disturbance is added around a deterministic mean trajectory which evolves throughout the 24 h. Two hundred ($N = 200$) sample trajectories, denoted $\{\boldsymbol{A}_t^k\}_{t=1,\dots,T}^{k=1,\dots,N}$ and $\{\boldsymbol{D}_t^k\}_{t=1,\dots,T}^{k=1,\dots,N}$, are randomly generated for those inflow and demand processes: they are depicted in Fig. 7.2. They are referred to as "scenarios" in that they are the basis for a discrete representation of probability distributions.

The initial state $\boldsymbol{X}_0$ follows a uniform distribution over $[\underline{x}, \overline{x}] = [0, 2]$. We also generate N "particles" $\{\boldsymbol{X}_0^k\}^{k=1,\dots,N}$ for the initial state, each one being associated with one of the previous noise trajectories with the same index k to form a scenario $\{\boldsymbol{W}_t^k\}_{t=0,\dots,T}$. The control variables $\boldsymbol{U}_t$ are subject to the bounds $[\underline{u}, \overline{u}] = [0, 1]$ for $t = 0, \dots, T-1$.

Function g involved in (7.4) is given by:

$$g(x, y) = y \frac{x + \overline{x} - 2\underline{x}}{2(\overline{x} - \underline{x})} .$$

It is proportional to y and affine in x, and it ranges from $y/2$ to y when x varies from $\underline{x}$ to $\overline{x}$.

The cost function c_t involved in (7.5) is given by:

$$c_t(y) = \tau_t (e^y - 1),$$

where τ_t is the market price of electricity at stage t. The variation of this price is depicted in Fig. 7.3.

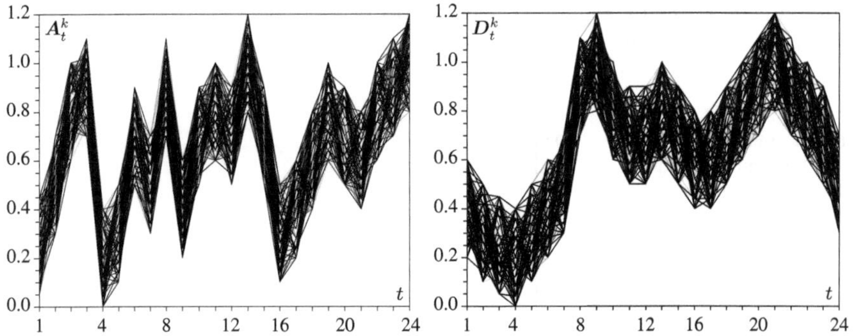

Fig. 7.2 Water inflow and electricity demand trajectories

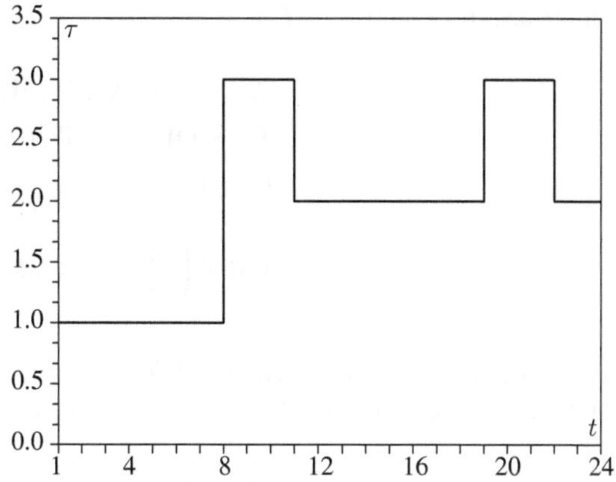

Fig. 7.3 Electricity price τ_t

The final cost K in (7.7) reflects an incentive to fill up the dam at the end of the day:

$$K(x) = 12(x - \overline{x})^2.$$

7.3 Manipulating Functions with a Computer and Implementation in Dynamic Programming (DP)

In this section, we discuss how functions can be represented and manipulated with a computer, and we illustrate these options by considering the numerical resolution of the Dynamic Programming equation. Several other techniques to attempt to compute approximate solutions to this equation (Stochastic Dual Dynamic Programming (SDDP) or Approximate Dynamic Programming (ADP) to name but a few) are proposed in the literature. We will briefly come back on them in the conclusion of this chapter.

7.3.1 The DP Equation

The problem described in Sect. 7.2 is a simple instance of a generic SOC problem in discrete time, with fixed and final horizon, as introduced earlier in this book (see Sect. 4.4 and Chap. 5, in particular Eqs. (5.1) and (5.2)). We recall this formulation once again here for convenience:

$$\boldsymbol{X}_0 = \boldsymbol{W}_0, \tag{7.9a}$$

and $\mathbb{P}$-a.s., for $t = 0, \ldots, T-1$,

$$X_{t+1} = f_t(X_t, U_t, W_{t+1}), \tag{7.9b}$$
$$U_t \preceq (W_0, \ldots, W_t), \tag{7.9c}$$
$$U_t \in C_t, \tag{7.9d}$$
$$\min \mathbb{E}\left(\sum_{t=0}^{T-1} L_t(X_t, U_t, W_{t+1}) + K(X_T) \right). \tag{7.9e}$$

In the Markovian setting (see Sect. 5.5), W is supposed to be a white noise process, and the DP equation provides the solution (see (4.57)):

$$V_T(x) = K(x), \quad \forall x \in \mathbb{X}, \tag{7.10a}$$

for $t = 0, \ldots, T-1$,

$$V_t(x) = \min_{u \in C_t} \mathbb{E}\Big(L_t(x, u, W_{t+1}) + V_{t+1}\big(f_t(x, u, W_{t+1})\big)\Big), \quad \forall x \in \mathbb{X}. \tag{7.10b}$$

The optimal feedback law $\gamma_t(\cdot)$ is obtained as the arg min in the latter equation and is a function of x which ranges in $\mathbb{X}$ for all t. Hence, solving the DP equation amounts to computing two sequences of functions, the optimal feedback laws $\{\gamma_t\}_{t=0,\ldots,T-1}$ and the Bellman functions $\{V_t\}_{t=0,\ldots,T}$, both over their whole domain $\mathbb{X}$.

For example, in the case of the benchmark problem in Sect. 7.2, this domain is one-dimensional and can be reduced to the simple segment [0, 2]. Nevertheless, such functions are *infinite dimensional* objects, and the next subsection briefly describes how they can be manipulated with a computer.

7.3.2 Discrete Representation of a Function

The problem of approximately describing functions of continuous-valued arguments by a finite amount of information may receive several answers. Of course, functions having a closed-form mathematical expression, as polynomial or trigonometric functions for example, do not raise a difficulty, at least from the user point of view, since computers have built-in high precision representations of most basic mathematical functions. So, one way to extend the catalog of functions that can be manipulated "exactly", at least seemingly form the user point of view, is to consider various combinations of functions already present in the basic catalog. Let us limit ourselves to the simplest kind of combinations, namely linear combinations. Then, apart from knowing the catalog itself, one has to store and manipulate the weights or coefficients of those linear combinations. Numerical algorithms should be described in terms of operations to be performed on those coefficients.

There is however a major drawback in such an approach: given a necessarily finite and limited catalog of basic functions, with linear combinations, one cannot get out of the linear subspace generated by this subset of functions. Hence, if the solution one is looking for lies in this subspace, or at least can be closely approximated by an element of this subspace, the approach is going to be successful. But this generally requires a sharp a priori knowledge of the shape of this solution, or one must manipulate a very large catalog of functions, which means a large dimensional problem in terms of unknown coefficients, in order to reduce the risk of grossly missing the target. This risk is entirely taken at the initial stage of choosing the catalog of basic functions with no subsequent recourse.

The previous approach is often referred to as the "parametric" point of view in that the unknowns of the problem are parameters or coefficients. At the opposite side, the non parametric approach does not presume a shape or subspace to which the solution should belong. In order to represent a function on some domain, it proceeds rather by collecting pointwise values of the function (necessarily a finite number of such values) and by interpolating the values elsewhere when needed. In fact, there is no clear cut between this point of view and the previous one in that one may consider that functions are now approximated by combinations of "local" functions, that is, functions with a narrow support around each particular point. The supports may, or may not, overlap (at least they should make it possible to represent functions which can take nonzero values anywhere in the appropriate domain).

Grid, Particle

We refer the set of points at which solutions are explicitly evaluated as the *grid*. This grid is denoted $\boldsymbol{x} = \{x^i\}_{i=1,\dots,N} \in \mathbb{X}^N$ and any point x^i is called a *particle*. Obviously, the denser the grid, the more accurate is the approximation of the solution, but also the higher is the dimension of the numerical problem to solve (the larger N is). The relation between how dense the grid is and how large N is depends on two factors: the dimension of the embedding space $\mathbb{X}$ and also the region of this space it is a priori useful to cover for a specific problem. We come back to the latter point in Sect. 7.5.2.

Trace, Trace Operator

Given a grid $\{x^i\}_{i=1,\dots,N}$ and a function $\gamma : \mathbb{X} \to \mathbb{U}$, we define the *trace* of γ on the grid as the trace $\boldsymbol{u} = \{\gamma(x^i)\}_{i=1,\dots,N} \in \mathbb{U}^N$. The *trace operator* is the operator which associates with a function $\gamma : \mathbb{X} \to \mathbb{U}$ its finite dimensional representation $(\boldsymbol{x}, \boldsymbol{u}) = \{x^i, \gamma(x^i)\}_{i=1,\dots,N}$. We denote this trace operator by $\mathfrak{T}_{\mathbb{U}} : \mathbb{U}^{\mathbb{X}} \to \mathbb{X}^N \times \mathbb{U}^N$.

Interpolation-Regression

In addition to the grid, one must define an *interpolation* or *regression* method in order to evaluate the function at points which do not belong to the grid. This is generally needed even to compute an approximation of the solution at a particle. This is a reciprocal operator of the trace operator, called *interpolation-regression* operator and denoted $\mathfrak{R}_{\mathbb{U}} : \mathbb{X}^N \times \mathbb{U}^N \to \mathbb{U}^{\mathbb{X}}$. But this is not in general the inverse operator of the trace operator, that is, $\mathfrak{R}_{\mathbb{U}} \circ \mathfrak{T}_{\mathbb{U}}$ is not in general the identity over $\mathbb{U}^{\mathbb{X}}$. Even $\mathfrak{T}_{\mathbb{U}} \circ \mathfrak{R}_{\mathbb{U}}$ may not be the identity over $\mathbb{X}^N \times \mathbb{U}^N$. That is, starting with the trace

$(\boldsymbol{x}, \boldsymbol{u}) = \{x^i, \gamma(x^i)\}_{i=1,\dots,N}$ and obtaining an approximation $\widetilde{\gamma}(\cdot) = \mathfrak{R}_{\mathbb{U}}(\boldsymbol{x}, \boldsymbol{u})(\cdot)$ of $\gamma(\cdot)$, it may be that

$$\widetilde{\gamma}(x^i) \neq \gamma(x^i). \tag{7.11}$$

There are various ways of defining the interpolation-regression operator:

polynomial and spline interpolation which consists of approximating the function in the vicinity of some particle by finding a polynomial of a certain degree which minimizes some distance taking into account the values of the trace at a certain number of neighboring particles; this is the topic of an abundant literature (see e.g. [125]);

kernel regression which amounts to representing the function as a weighted sum of its trace values; the relative weights are provided by a kernel function which plays the role of a distance from the point where the function must be evaluated to the various particles; seminal references are Nadaraya [107] and Watson [151] going back to 1964, followed here again by an abundant literature (see for example Wand and Jones [150]);

nearest neighbor which yields a piecewise constant approximation of a function based on the values collected at the particles; each particle is the centroid of a Voronoi cell (see Sect. 6.1.2) and the interpolated function is considered constant over each cell, taking the value the trace assumes at the corresponding particle; indeed, one may consider that this technique falls into both previous categories, that is, it uses polynomials of degree 0 or kernels which are piecewise constant with 0–1 values (indeed membership functions of cells); more general smoother kernels may thus be viewed as "fuzzy set" membership functions.

7.3.3 The Discrete DP Equation

With the trace and interpolation-regression operators at hand, we can now revisit the DP equation (7.10) and examine how it can be approximately solved with a computer. At each time stage t, two functions must essentially be obtained, namely the Bellman function V_t and the feedback function γ_t, both with a domain lying in $\mathbb{X}$.

Hence, a grid $\boldsymbol{x}_t$ must be first defined at each time stage to cover this domain. With no a priori knowledge of the solution, a regular grid with particles evenly distributed over that domain is a natural choice. However the grid must have a finite number of particles, but this is generally possible only if the domain can be a priori bounded. This is, for example, the case with the problem of Sect. 7.2 since the dynamics (7.2) cannot produce a state out of the segment $[\underline{x}, \overline{x}] \subset \mathbb{R}$. But otherwise, this may be difficult since no subsets of the domain the state vector may visit can be ignored or left unexplored.

If $\mathbb{X} = \mathbb{R}^d$, the regular grid (covering a bounded subset as we assume it from now on) is generally obtained as the Cartesian product of d one-dimensional grids, say, with M_t particles in each dimension, which results in $N_t = M_t^d$ particles overall at each stage. This is the source of the "curse of dimensionality", which refers to the exponential growth rate of the number of particles with the state vector dimension d for a given and fixed precision.

Once the grids $\{\boldsymbol{x}_t\}_{t=0,\dots,T}$ are defined, the basis of the approximation of the Bellman functions V_t, and of the feedback functions γ_t, are the traces $\{\boldsymbol{v}_t\}_{t=0,\dots,T}$ and $\{\boldsymbol{u}_t\}_{t=0,\dots,T-1}$ with v_t^i as an estimation of $V_t(x_t^i)$ and u_t^i as an estimation of $\gamma_t(x_t^i)$ for $i = 1, \dots, N_t$. But the DP equation (7.10) also involves an expectation operation which cannot generally be performed analytically. We assume that this expectation is approximated by the Monte Carlo method. We thus consider N scenarios $\{w_t^j\}_{t=0,\dots,T}^{j=1,\dots,N}$ of the random noise process which are supposed to have been drawn independently according to the probability distribution of this process.

Finally, the discrete DP equation reads:
for $i = 1 \dots, N_T$,

$$v_T^i = K(x_T^i), \tag{7.12a}$$

for $t = 0, \dots, T-1$, and $i = 1, \dots, N_t$,

$$\widetilde{V}_{t+1} = \mathfrak{R}_{\mathbb{R}}(\boldsymbol{x}_{t+1}, \boldsymbol{v}_{t+1}), \tag{7.12b}$$

$$v_t^i = \min_{u \in C_t} \frac{1}{N} \sum_{j=1}^{N} \Big(L_t\big(x_t^i, u, w_{t+1}^j\big) + \widetilde{V}_{t+1}\big(f_t(x_t^i, u, w_{t+1}^j)\big)\Big), \tag{7.12c}$$

$$u_t^i = \arg\min_{u \in C_t} \frac{1}{N} \sum_{j=1}^{N} \Big(L_t\big(x_t^i, u, w_{t+1}^j\big) + \widetilde{V}_{t+1}\big(f_t(x_t^i, u, w_{t+1}^j)\big)\Big), \tag{7.12d}$$

$$\widetilde{\gamma}_t = \mathfrak{R}_{\mathbb{U}_t}(\boldsymbol{x}_t, \boldsymbol{u}_t). \tag{7.12e}$$

Remark 7.2 Observe that the interpolated feedback functions provided by (7.12e) are not used in the backward recursion in time, but they are required as the expected practical solution of the problem. On the contrary, the interpolated Bellman functions provided by (7.12b) are needed to proceed with (7.12c) since the arguments $f_t(x_t^i, u, w_{t+1}^j)$ at which those functions are evaluated do not, in general, coincide with grid points x_{t+1}^i. The only exception is about V_T which is identical to the final cost function K supposed to be known analytically as part of the problem data. The interpolated function $\widetilde{V}_0$ is only needed if one is willing to estimate the optimal cost value, namely $\mathbb{E}\big(\widetilde{V}_0(\boldsymbol{X}_0)\big) = \mathbb{E}\big(\widetilde{V}_0(\boldsymbol{W}_0)\big)$. ◊

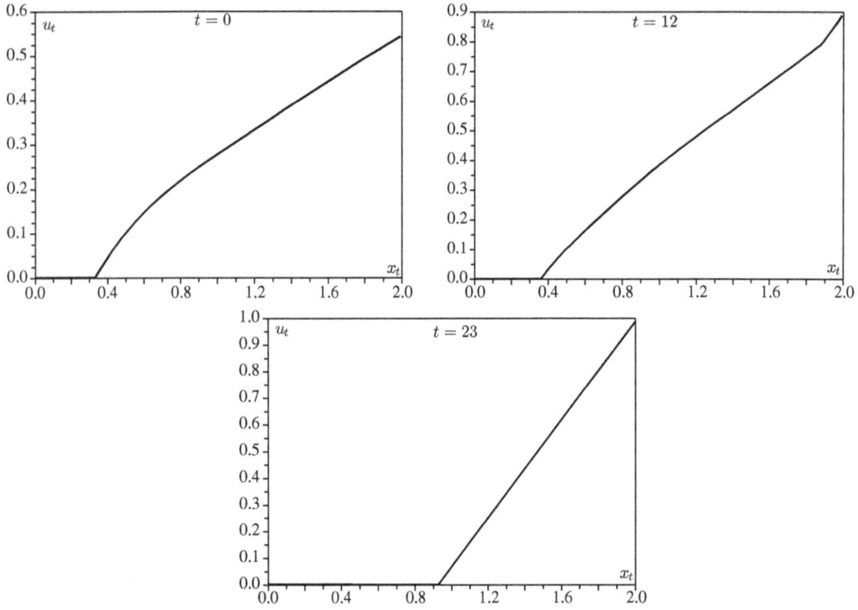

Fig. 7.4 Optimal feedback evaluated by DP for three time instants

7.3.4 Application to the Benchmark Problem

In the problem of Sect. 7.2, the state vector is of dimension 1, and, in fact, it is confined to the segment [0, 2]: this makes it quite easy to solve numerically using Eq. (7.12), whatever interpolation-regression is used, since one can afford a grid with a rather high number of particles (200 points are used and the interpolation-regression operators are piecewise linear: the value of a function outside the grid is obtained as the weighted mean of the two surrounding grid points).

For later reference, Fig. 7.4 shows the shape of the optimal feedback $\widetilde{\gamma}_t$ obtained for three particular values of t, namely $t \in \{0, 12, 23\}$.

7.4 Resolution by the Scenario Tree Technique

In this section, we consider the resolution of SOC problems formulated as finite-dimensional optimization problems over a scenario tree. Actually, we skip the post-processing of the results derived for this finite-dimensional optimization problem, post-processing which should lead to a feedback law that would be implementable in the real system. Instead, for the benchmark problem of Sect. 7.2, we evaluate the quality of the results so obtained by means of a more visual graphical representation, and we draw some interesting conclusions about the drawbacks of the scenario tree approach.

7.4.1 General Considerations

Unlike other methods presented in this chapter, which proceed by first obtaining optimality conditions and then solving those conditions numerically, the scenario tree approach first derives a finite dimensional optimization problem which is supposed to represent a sound approximation of Problem (7.9) and then obtains a solution of this finite dimensional problem by standard mathematical programming methods. Recall, however, that a final stage is required to provide feedback functions that can be implemented on the original system.

The tree architecture is characterized by the fact that each node of the tree corresponds to a unique past noise history but is generally followed by several possible future histories. Therefore, the decision attached to each node may be considered as indexed by this unique past history but still faces some uncertainties in future outcomes, these possible outcomes being "conditioned" by the past observed so far. This feature is supposed to translate the information constraint (7.9c) in the finite dimensional problem. As discussed in Sect. 6.2.2, Pennanen [110] was the first to give a theoretical justification to this idea by providing an asymptotic convergence result when the discrete tree structure gets closer and closer to the actual noise stochastic process.

In this section, we first briefly recall the formulation of the approximate problem over a scenario tree, which is substituted for the original problem (7.9). We then consider the optimality conditions of this approximate problem and give brief indications on how to solve them. Finally, we come back to our benchmark problem and show some of the corresponding results. The main purpose is here to illustrate the difficulty in achieving the last step recalled hereabove, namely to obtain implementable feedback functions.

The initial step of the whole methodology is to obtain a scenario tree, a topic already broached in Sect. 6.3.1. The starting point may be a collection of scenarios such as those depicted in Fig. 7.2, from which one must construct a tree as illustrated by Fig. 7.5. This approach is followed e.g. in [61, 71, 72]. Alternatively, one may use various techniques to derive scenario trees directly by appropriate sampling or quantization techniques (see [12, 111, 114, 115]) or by further reducing an already given scenario tree [73]. We do not thoroughly investigate this issue here because our main concern is to illustrate the difficulties already discussed in Sect. 6.3.2. Some indications are, however, provided in the context of our benchmark problem in Sect. 7.4.5 hereafter.

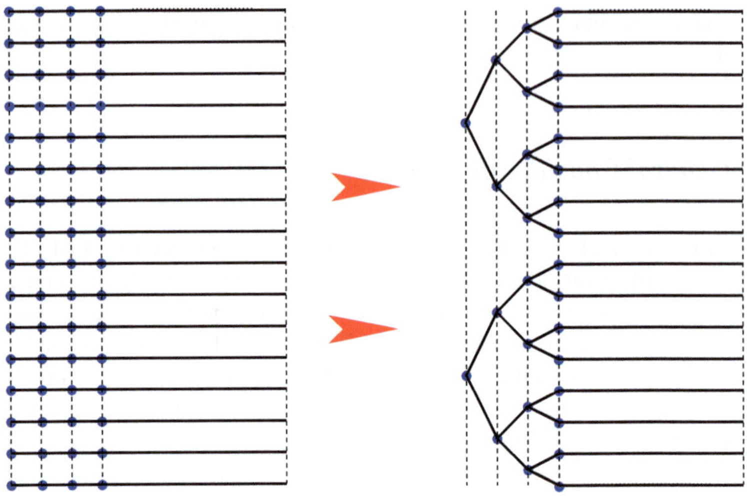

Fig. 7.5 From scenarios to a scenario tree

7.4.2 Formulation of the Problem over a Scenario Tree

Notations

We follow the main stream of [33] with some differences. Although a tree is a two-dimensional structure, it is more convenient to number nodes with a single index ν, e.g. starting at time stage 0 from top to bottom, and then proceed with increasing values of t, still from top to bottom. This index set is denoted $\mathcal{N}$.

We need to recover the value of t for a given value of ν. To that purpose, we introduce the mapping $\theta : \mathcal{N} \to \{0, \dots, T\}$ which plays the part of the "clock" function. Therefore, $\theta^{-1}(t)$ is the subset of nodes located at stage t. Observe that $\theta^{-1}(0)$ is not necessarily reduced to a singleton (see the right-hand side of Fig. 7.5): it represents the samples of $\boldsymbol{X}_0 = \boldsymbol{W}_0$. Hence, since the root of the tree is not unique, we should rather speak of a "forest" rather than of a "tree" but we still keep on using the terminology "tree" which is more common.

The topology of the tree is described by the "father" function $\mathfrak{f} : \mathcal{N} \to \mathcal{N}$: $\mathfrak{f}(\nu)$ is the *unique* node which precedes node ν at stage $\theta(\nu) - 1$. Hence $\theta\big(\mathfrak{f}(\nu)\big) = \theta(\nu) - 1$. In fact, $\mathfrak{f}$ is undefined over the subset $\theta^{-1}(0)$ of roots, and it ranges in $\mathcal{N} \setminus \theta^{-1}(T)$ since the "leaves" $\theta^{-1}(T)$ have no "sons". "Sons" of a given node ν are the nodes that follow this node ν at stage $\theta(\nu) + 1$, namely $\mathfrak{f}^{-1}(\nu)$. Of course, this subset is empty for nodes which are leaves of the tree (nodes at stage T).

The knowledge of the set $\mathcal{N}$ of nodes, and of the father function $\mathfrak{f}$, completely characterizes the topology of the tree. For example, the whole past history of a node is obtained by iterating function $\mathfrak{f}$. Similarly, the subtree hanging at a node is obtained by iterating the set-valued mapping $\mathfrak{f}^{-1}$.

In addition to the topology, other data and variables are required, and they are attached to the nodes of the tree. Every node ν of the tree carries a sampled noise value w_ν of the random vector $\mathbf{W}_{\theta(\nu)}$. Depending on how the tree has been derived from the underlying noise stochastic process, nodes at any time stage t (those belonging to $\theta^{-1}(t)$) may be equiprobable, or they may have different probability weights. We introduce a probability function $\pi : \mathcal{N} \to (0, 1]$ satisfying the following conditions:

$$\pi(\nu) = \sum_{\mu \in \mathfrak{f}^{-1}(\nu)} \pi(\mu), \quad \forall \nu \in \mathcal{N} \setminus \theta^{-1}(T), \tag{7.13a}$$

$$\sum_{\nu \in \theta^{-1}(T)} \pi(\nu) = 1. \tag{7.13b}$$

Observe that the combination of both conditions above implies that property (7.13b) holds true for all $t = 0, \ldots, T$, and not only for T. Moreover, according to (7.13a), the numbers $\{\pi(\mu)/\pi(\nu)\}_{\mu \in \mathfrak{f}^{-1}(\nu)}$ form also a discrete probability distribution, which is indeed a *conditional* probability distribution (of next outcomes) knowing that node ν has been reached.

Remark 7.3 The technique of scenario trees does not require a priori that the noise process be a white noise. Hence, the above mentioned conditional probability distribution may differ when jumping from one node to another at the same time stage t (and a fortiori when t differs). ◊

Finally, attached to every node $\nu \in \mathcal{N} \setminus \theta^{-1}(T)$ is a control variable u_ν, and for $\nu \in \mathcal{N}$ is a "state" variable x_ν. Those variables are the unknowns of the following problem.

Formulation

The following optimization problem is a transposition of Problem (7.9) over the tree. Indeed, if one assumes that the noise process exactly follows the discrete probability distribution depicted by the scenario tree, then Problem (7.9) boils down to

$$x_\nu = w_\nu, \qquad \forall \nu \in \theta^{-1}(0), \tag{7.14a}$$

$$x_\nu = f_{\theta(\mathfrak{f}(\nu))}\big(x_{\mathfrak{f}(\nu)}, u_{\mathfrak{f}(\nu)}, w_\nu\big), \qquad \forall \nu \in \mathcal{N} \setminus \theta^{-1}(0), \tag{7.14b}$$

$$u_\nu \in C_{\theta(\nu)}, \qquad \forall \nu \in \mathcal{N} \setminus \theta^{-1}(T), \tag{7.14c}$$

$$\min \Bigg(\sum_{\nu \in \mathcal{N} \setminus \theta^{-1}(0)} \pi(\nu) L_{\theta(\mathfrak{f}(\nu))}\big(x_{\mathfrak{f}(\nu)}, u_{\mathfrak{f}(\nu)}, w_\nu\big) + \sum_{\nu \in \theta^{-1}(T)} \pi(\nu) K(x_\nu) \Bigg). \tag{7.14d}$$

In this transposition, there is no explicit equivalent to the non anticipativity constraint (7.9c), since this constraint is now encoded in the tree structure.

7.4.3 Optimality Conditions and Resolution

To obtain the optimality conditions for Problem (7.14), we follow the methodology already used in Sect. 5.4, which consists in forming a Lagrangian associated with the problem and using its derivatives w.r.t. the primal and dual variables to derive the stationary conditions, or the gradient w.r.t. the control variables according to the technique explained in Sect. 5.4.1.

In dualizing Eq. (7.14a) and (7.14b) to form the Lagrangian, we introduce the weights $\pi(\nu)$: this trick amounts to scaling the corresponding dual variables λ_ν, which corresponds to a mere change of variables aimed at making their probabilistic interpretation smarter:

$$\begin{aligned}\ell(u, x; \lambda) = &\sum_{\nu\in\theta^{-1}(T)} \pi(\nu)K(x_\nu) + \\ &\sum_{\nu\in\mathcal{N}\setminus\theta^{-1}(0)} \pi(\nu)\Big(L_{\theta(\mathfrak{f}(\nu))}\big(x_{\mathfrak{f}(\nu)}, u_{\mathfrak{f}(\nu)}, w_\nu\big) + \lambda_\nu^\top\Big(f_{\theta(\mathfrak{f}(\nu))}\big(x_{\mathfrak{f}(\nu)}, u_{\mathfrak{f}(\nu)}, w_\nu\big) - x_\nu\Big)\Big) \\ &+ \sum_{\nu\in\theta^{-1}(0)} \pi(\nu)\lambda_\nu^\top(w_\nu - x_\nu). \end{aligned} \tag{7.15}$$

The stationarity of this Lagrangian gradients w.r.t. the λ_ν leads back to (7.14a) and (7.14b) (up to the weighting factor $\pi(\nu)$). The stationarity of the gradients w.r.t. the x_ν's provides, as usual, backward co-state equations (analogous to (5.18)):

$$\lambda_\nu = \nabla K(x_\nu), \qquad \nu \in \theta^{-1}(T), \tag{7.16a}$$

$$\begin{aligned}\lambda_\nu = \frac{1}{\pi(\nu)} &\sum_{\mu\in\mathfrak{f}^{-1}(\nu)} \pi(\mu)\Big(\nabla_x f_{\theta(\nu)}\big(x_\nu, u_\nu, w_\mu\big)\lambda_\mu \\ &+ \nabla_x L_{\theta(\nu)}\big(x_\nu, u_\nu, w_\mu\big)\Big), \qquad \nu \in \mathcal{N} \setminus \theta^{-1}(T). \end{aligned} \tag{7.16b}$$

Finally, the gradients w.r.t. the u_ν's yield the gradients of the cost (7.14d) (denoted $J(u)$) w.r.t. the controls:

$$\begin{aligned}\nabla_{u_\nu} J(u) = &\sum_{\mu\in\mathfrak{f}^{-1}(\nu)} \pi(\mu)\Big(\nabla_u f_{\theta(\nu)}\big(x_\nu, u_\nu, w_\mu\big)\lambda_\mu \\ &+ \nabla_u L_{\theta(\nu)}\big(x_\nu, u_\nu, w_\mu\big)\Big). \end{aligned} \tag{7.17}$$

Remark 7.4 The expression (7.17) is somewhat different by nature from (5.19): the latter corresponds to the cost gradient w.r.t. control evaluated along a particular noise scenario (hence the notation $\nabla_{u_t} j(u, w)$); the former corresponds to the gradient of the cost function (7.14d) which already contains an operation of expectation w.r.t. noise (hence the notation $\nabla_{u_\nu} J(u)$). However, since the derivative is w.r.t. u_ν

which is the control variable acting at stage $\theta(\nu)$ along scenarios which go through node ν, only the subtree hanging at this node is involved in this expectation operation (which is thus rather akin to a *conditional* expectation). Indeed, in formula (7.17), one may observe the weights $\pi(\mu)$ for the nodes $\mu \in \mathfrak{f}^{-1}(\nu)$ following immediately node ν in the tree; those weights "shape" a conditional probability up to the normalizing factor $\pi(\nu)$ (see (7.13a) and the comments below that formula); as for the other nodes at the next stages in the future, they are indirectly present in (7.17) through the co-state variables.

As a matter of fact, if we now observe the "geometry" of the co-state equations (7.16), they start at the leaves of the tree (see (7.16a)) and then progress backwards by successive conditional expectations which are apparent in the right-hand side of (7.16b). Therefore, these co-state equations are comparable to the *adapted* co-state equations encountered in (5.22c) or (5.25c) rather than to the *non adapted* version (5.20d). $\lozenge$

The gradient expressions (7.17) may serve to write optimality conditions according to (5.9), or to solve the optimization problem (7.14) using a projected gradient algorithm by iterating the following steps:

- guess initial control values $\{u_\nu^{(0)}\}_{\nu \in \mathcal{N} \setminus \theta^{-1}(T)}$ with $u_\nu^{(0)} \in C_{\theta(\nu)}$;
- at iteration k, knowing the $u_\nu^{(k)}$, integrate the state equations (7.14a) and (7.14b) forwards along the tree from roots to leaves, which yields the $x_\nu^{(k)}$;
- then, integrate the co-state equations (7.16) backwards from leaves to roots, which yields the $\lambda_\nu^{(k)}$;
- with all those values, form the gradient expressions (7.17) and perform a gradient step projected onto the feasible subsets $C_{\theta(\nu)}$ to update the $u_\nu^{(k)}$ into $u_\nu^{(k+1)}$.

7.4.4 About Feedback Synthesis

Once Problem (7.14) has been solved, three variable values are attached to each node ν of the tree, namely u_ν, x_ν and λ_ν. Recall that our final goal is to obtain a feedback function γ_t at each time stage t which tells us which control value must be used as a function of observations available at time t. In the formulation (7.9), according to (7.9c), the observations available at time t are the whole past history of the noise process. As we noticed already, the scenario tree technique does not require particular assumptions about the noise process, except that it is observed in a causal manner.

Only a few noise past histories are considered at every time stage t (as many as the number of elements in $\theta^{-1}(t)$): they are the elements of a grid which must then serve for the final phase of interpolation-regression of the feedback functions. However, this operation must be achieved in a space of increasing dimension as t grows, which makes it rather difficult. A possibility is to attempt to reduce this dimension, at the price, possibly, of some loss in performance, by trying to limit the arguments of the feedback functions to the most significant observations (which may be determined for example by data analysis techniques).

This difficulty is alleviated in the Markovian case since we know from theory that the optimal performance can be achieved by using only *state* feedback instead of the whole past noise history. This is, in particular, the case for our benchmark problem of Sect. 7.2. Therefore, in the Markovian case, one may examine the clouds of dots $\{(x_\nu, u_\nu)\}_{\nu \in \theta^{-1}(t)}$ for each value of t and use those clouds of dots as the grids to derive the feedback functions γ_t by interpolation-regression. For our benchmark problem, we skip this last stage of interpolation and limit ourselves to the observation of the clouds at some time stages. This is enough to draw interesting conclusions.

7.4.5 Results Obtained for the Benchmark Problem

In order to formulate Problem (7.14) for our benchmark problem, we must first obtain a scenario tree from the initial data described in Sect. 7.2.2. As indicated there, 200 noise scenarios are available: each one is made up of an initial storage value x_0^i, $i = 1, \ldots, 200$, followed, at the next time stages $t = 1, \ldots, 24$, by a pair of values (a_t^i, d_t^i) for water inflows and electricity demands as depicted in Fig. 7.2. Out of those 200 scenarios, Barty [12] derived a scenario tree by following these steps:

- for $t = 0$, by considering the collection of values $\{x_0^i\}_{i=1,\ldots,200}$, and by using Lloyd's algorithm described at the end of Sect. 6.1.2, a quantization of this sample set into two cells (subsets) and two corresponding centroids is obtained;
- for $t = 1$, the same procedure is applied *separately* to the two previous subsets of scenarios: for each one, the two-dimensional vector values (a_1^i, d_1^i) are taken into account and each subset is divided again into two subsets by optimal quantization with help of Lloyd's algorithm; therefore, 4 subsets are now obtained, which are issued from 2 "fathers", namely the two subsets of stage $t = 0$;
- this process is repeated at the next time stages by dividing each of the previously obtained subsets into two subsets, based on the current two-dimensional vector values (a_t^i, d_t^i);
- so doing, one builds up a binary[1] "tree" (indeed, two such trees with two distinct roots at level $t = 0$), but since $2^8 = 256$, already at $t = 7$, the 200 distinct scenarios are exhausted, and the tree cannot keep on its branching rate of 2; therefore, beyond that time stage, each node is followed by a single "son", that is, the future becomes *conditionally deterministic*.

Once this scenario tree has been defined, Problem 7.14 can be numerically solved based on the other numerical data described in Sect. 7.2.2 by using the algorithm described at the end of Sect. 7.4.3. As explained earlier, we examine the clouds of

[1] "Binary" means that each node is followed by two sons, or, otherwise stated, the branching rate is equal to 2.

dots (x_ν, u_ν) for $\nu \in \theta^{-1}(t)$ and for certain values of t. Figure 7.6 shows these results for $t \in \{0, 12, 23\}$ together with the optimal curves estimated by DP and already depicted in Fig. 7.4.

Several comments are in order here. First of all, the number of nodes at every time stage t (the cardinality of $\theta^{-1}(t)$) determines the number of available dots in the cloud at the corresponding time stage t. The tree starts with 2 nodes at $t = 0$, hence the two dots in the first plot of Fig. 7.6; the next two plots shown in this figure have 200 dots each. Unfortunately, and this is our second comment, a larger number of particles available as t increases does not mean better conditions for feedback synthesis. Indeed, each particle (x_ν, u_ν) provides (through the value u_ν) an *estimation* of the value of the feedback function at x_ν. Hence, as for any stochastic estimation, the quality of this estimation is characterized by its potential *bias* and also by its *variance*. This is summarized by the Mean Quadratic Error (MQE—see (6.3)), which is equal to the square of the bias plus the variance.

If we observe the first plot of Fig. 7.6 corresponding to $t = 0$, the two dots of this plot are rather close to the actual feedback curve. This is an indication that the MQE is rather small at this time stage. This may be explained by the fact that, for each particle, the pending future in the tree is rich enough to provide an adequate sample of the noise process. Therefore, at the early stages of the tree, the problem is not so much with the quality of estimation but rather with the small number of particles available: with two dots in dimension 1, one can hardly obtain anything else than a *linear* feedback function guess, whereas the true function is obviously nonlinear.

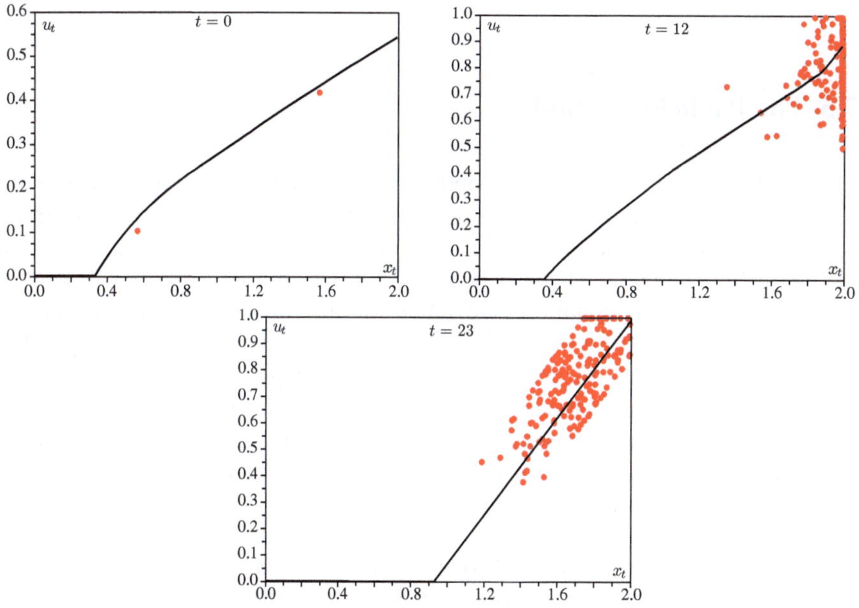

Fig. 7.6 Scenario tree: optimal pairs (x, u) at three time instants

The other two plots show 200 dots, but the task of estimating the feedback function is nevertheless not made easier: now the problem is with the average deviation from the true curve. Remember that when the number of particles reaches 200, which was the initial number of noise scenarios available, necessarily, the future of each such particle is *deterministic*, that is, it consists of a *single* scenario. Therefore, we are in the situation of producing a stochastic estimation based on a single sample, which obviously leads to a very high variance. We also refer the reader back to Sect. 6.3.2 in which variance issues have been discussed in connection with the construction of scenario trees for white noise processes.

In this section on the scenario tree technique, we have illustrated the factors that make this technique not well suited for the numerical estimation of optimal feedback functions for SOC problems, especially when the time horizon exceeds a few stages. Indeed, even with a minimal (binary) rate of branching, the number of scenarios grows exponentially with the time horizon, unless those scenarios become deterministic after a few stages (that is, if they stop branching), which, however, has a disastrous effect on the quality of the results. Using large deviation techniques, Shapiro [139] showed that the size of the scenario tree required to achieve a specified accuracy in terms of the optimal cost value grows exponentially with the time horizon. Girardeau [70] obtained similar results by numerical experiments while studying the MQE between the optimal feedback function and that derived by a piecewise constant interpolation-regression technique from the particles computed by the scenario tree technique.

The rest of this chapter is devoted to the study of another numerical method that overcomes these drawbacks.

7.5 The Particle Method

In this section, we begin with the optimality conditions obtained in Chap. 5 for SOC problems, and we derive from them numerical methods of resolution. Here, we limit ourselves to the Markovian situation of Sect. 5.5 and, more precisely, to the optimality conditions obtained in Sect. 5.5.4, which used a functional point of view. These optimality conditions are, among all the variants examined in Chap. 5, those which are the closest to the DP approach. In particular, in contrast to the other sets of conditions, they do not involve *conditional* expectations, but only simple expectations which are classically approximated by straightforward Monte Carlo arithmetic means. For this reason, the sophisticated techniques for the approximation of conditional expectations, which provide only *biased* estimates (see [86]), are avoided. But this comes at the expense of using a *functional* representation of co-state variables, as explained in Sect. 5.5.4, and this technique requires tools similar to those used for the approximation of conditional expectations (e.g. Nadaraya-Watson kernel regression [107, 151] mentioned in Sect. 7.3.2). This is not surprising since conditional expectations are, in fact, *functions* of the conditioning variables.

The reader interested in numerical methods derived from the other sets of optimality conditions developed in Chap. 5 may refer to Dallagi's PhD thesis [48]. In this section, we rather try to illustrate the similarities and differences of the method proposed, referred to as the "particle method", with the DP technique. The most significant difference lies in the fact that the particle method automatically constructs a discretization grid in the state space which reflects the optimal state probability distribution. This happens precisely because "state particles" are repeatedly simulated while the algorithm iteratively improves the feedback control. This may be a decisive advantage over the DP approach in the case when the optimal state distribution does not fill up the whole state space domain but tends to concentrate in particular narrow regions at some time stages, as happens with our benchmark problem (this is apparent in Fig. 7.6 especially in the second and third plots). As a matter of fact, the DP technique requires a systematic a priori discretization of the whole state domain which cannot take the final optimal distribution into account.

7.5.1 Algorithm

Our purpose is to solve the optimality conditions (5.26) numerically, using N independent and identically distributed noise processes $\{\boldsymbol{W}_t^i\}_{t=0,\dots,T}^{i=1,\dots,N}$ which are supposed to be samples of the noise process $\boldsymbol{W}$. A realization of $\boldsymbol{W}_t^i$ is denoted w_t^i.

Unlike the scenario tree technique, there is no need to derive a tree structure here. That is, those initial scenarios are used as they are. As for the comparison with the DP numerical approach described in Sect. 7.3.3, there is a common ingredient, namely, the need to appeal to some numerical representation of certain functions of the state variable; but there is also a noticeable difference, namely that the discretization of the state space is, here, "self-constructive" and adapted to the optimal solution of the problem.

The algorithm considered here is iterative and it is akin to a projected gradient algorithm. Let (k) be the superscript indexing all manipulated entities at the current iteration k. For reasons on which we come back later on (see Remark 7.7 hereafter), as iterations proceed, feedback laws appearing in (5.26) need to be represented only by particles $\{u_t^{i,(k)}\}_{t=0,\dots,T-1}^{i=1,\dots,N}$ corresponding to the N noise scenarios; these control particles are improved up to their final values $u_t^{i,(\infty)}$ and the "solution" delivered by the algorithm, namely the feedback laws $\{\gamma_t^{(\infty)}\}_{t=0,\dots,T-1}$, are synthesized using the final values of the control particles.

Building Adaptive Grids in the State Space

At iteration k, knowing the control particles, the state particles $\{x_t^{i,(k)}\}_{t=0,\dots,T}^{i=1,\dots,N}$ are generated by integration of (5.26a) and (5.26b) along the N noise scenarios: for $i = 1, \dots, N$,

$$x_0^{i,(k)} = w_0^i, \tag{7.18a}$$

and for $t = 0, \ldots, T-1$,

$$x_{t+1}^{i,(k)} = f_t\Big(x_t^{i,(k)}, u_t^{i,(k)}, w_{t+1}^i\Big), \tag{7.18b}$$

which yields the grids

$$\boldsymbol{x}_t^{(k)} = \big\{x_t^{i,(k)}\big\}_{i=1,\ldots,N}. \tag{7.18c}$$

Functional Co-State Backward Recursion

Consider now stage t of the backward recursion (5.26e). In the right-hand side of this equation, expectations of functions w.r.t. $\boldsymbol{W}_{t+1}$ are classically approximated by the arithmetic average of these functions evaluated at the samples $\{w_{t+1}^i\}_{i=1,\ldots,N}$. So doing, even if we want only evaluate an approximation of the trace of the function Λ_t at the points of the grid $\boldsymbol{x}_t^{(k)}$, we need an approximation $\Lambda_{t+1}^{(k)}$ of the *whole* function Λ_{t+1}, and not only its evaluations on the grid $\boldsymbol{x}_{t+1}^{(k)}$. The right-hand side of (5.26e) is approximated by

$$\frac{1}{N}\sum_{j=1}^{N}\Bigg(\nabla_x L_t\Big(x_t^{i,(k)}, u_t^{i,(k)}, w_{t+1}^j\Big) + \nabla_x f_t\Big(x_t^{i,(k)}, u_t^{i,(k)}, w_{t+1}^j\Big)\Lambda_{t+1}^{(k)}\Big(f_t\Big(x_t^{i,(k)}, u_t^{i,(k)}, w_{t+1}^j\Big)\Big)\Bigg).$$

This expression shows that $\Lambda_{t+1}^{(k)}$ must be evaluated at

$$f_t\Big(x_t^{i,(k)}, u_t^{i,(k)}, w_{t+1}^j\Big) \tag{7.19}$$

which, for $j \neq i$, is *not* a point of the grid $\boldsymbol{x}_{t+1}^{(k)}$. This is where an interpolation-regression operation is needed to produce an approximation $\Lambda_{t+1}^{(k)}$ of Λ_{t+1}.

To that purpose, for every t, we need an estimation $\{l_t^{i,(k)}\}_{i=1,\ldots,N}$ of the trace of Λ_t on the grid $\boldsymbol{x}_t^{(k)}$. It is constructed in a backward recursion, together with interpolated functions $\Lambda_t^{(k)}$ as follows:

$$\Lambda_T^{(k)} = \nabla K, \tag{7.20a}$$

for $t = 1, \ldots, T-1$ and $i = 1, \ldots, N$,

$$l_t^{i,(k)} = \frac{1}{N} \sum_{j=1}^{N} \Big(\nabla_x L_t \big(x_t^{i,(k)}, u_t^{i,(k)}, w_{t+1}^j \big)$$

$$+ \nabla_x f_t \big(x_t^{i,(k)}, u_t^{i,(k)}, w_{t+1}^j \big) \Lambda_{t+1}^{(k)} \Big(f_t \big(x_t^{i,(k)}, u_t^{i,(k)}, w_{t+1}^j \big) \Big) \Big), \quad (7.20b)$$

$$\boldsymbol{l}_t^{(k)} = \{ l_t^{i,(k)} \}_{i=1,\dots,N}, \quad (7.20c)$$

$$\Lambda_t^{(k)} = \mathfrak{R}_{\mathbb{X}_t} \big(\boldsymbol{x}_t^{(k)}, \boldsymbol{l}_t^{(k)} \big). \quad (7.20d)$$

Updating the Control Particles

We now examine how the control particles can be improved for the next iteration of the algorithm. The left-hand side member of (5.26c) represents the gradient of the cost function w.r.t. the control used at stage t when the system is in state x. Therefore, using the grid $\boldsymbol{x}_t^{(k)}$, the control particle $u_t^{i,(k)}$ is updated using a gradient step (with stepsize $\varepsilon^{(k)}$) projected on the feasible set C_t: for $t = 0, \dots, T-1$ and $i = 1, \dots, N$,

$$u_t^{i,(k+1)} = \text{proj}_{C_t} \Big(u_t^{i,(k)} - \varepsilon^{(k)} \frac{1}{N} \sum_{j=1}^{N} \Big(\nabla_u L_t \big(x_t^{i,(k)}, u_t^{i,(k)}, w_{t+1}^j \big)$$

$$+ \nabla_u f_t \big(x_t^{i,(k)}, u_t^{i,(k)}, w_{t+1}^j \big) \Lambda_{t+1}^{(k)} \Big(f_t \big(x^{i,(k)}, u_t^{i,(k)}, w_{t+1}^j \big) \Big) \Big) \Big). \quad (7.21)$$

Termination

Assuming the convergence of the iterations driven by index k and the stabilization of state and control particle values at

$$\boldsymbol{x}_t^{(\infty)} = \{ x_t^{i,(\infty)} \}_{t=0,\dots,T}^{i=1,\dots,N} \quad \text{and} \quad \boldsymbol{u}_t^{(\infty)} = \{ u_t^{i,(\infty)} \}_{t=0,\dots,T-1}^{i=1,\dots,N}$$

respectively, one uses those values as an approximate trace of the feedback function optimal solution to build up an approximation of this solution by interpolation-regression: for $t = 0, \dots, T-1$,

$$\gamma_t^{(\infty)} = \mathfrak{R}_{\mathbb{U}_t} \big(\boldsymbol{x}_t^{(\infty)}, \boldsymbol{u}_t^{(\infty)} \big). \quad (7.22)$$

Remark 7.5 The fact that the particles $u_t^{i,(\infty)}$ belong to C_t does not guarantee that the interpolated functions $\gamma_t^{(\infty)}$ assume values that are confined to C_t. This depends on the technique used for interpolation-regression. This is ensured for example if only convex combinations of the particles are used to interpolate (yielding piecewise linear feedback functions) and if C_t is a convex subset. Observe that, in this particular case, the values of the particles $u_t^{i,(\infty)}$ remain those of the interpolated feedback functions $\gamma_t^{(\infty)}$ at the particles $x_t^{i,(\infty)}$ (refer back to the discussion leading to (7.11) in Sect. 7.3.2). ◊

Summarizing
We summarize the whole algorithm.

Algorithm 7.6

Initialization: Set $k = 0$ and guess initial control particles $u_t^{i,(0)}$ for $t = 0, \ldots, T-1$ and $i = 1, \ldots, N$.
Iteration k:

1. Compute the state grids $\boldsymbol{x}_t^{(k)}$ using the forward recursion (7.18).
2. Compute the co-state functions $\Lambda_t^{(k)}$ using the backward recursion (7.20).
3. Update the control particles to $u_t^{i,(k+1)}$ for $t = 0, \ldots, T-1$ and $i = 1, \ldots, N$ using (7.21).
4. Iterate with $k + 1 \leftarrow k$ or stop if stationarity is (almost) achieved.

Termination: With the limit values $\boldsymbol{x}_t^{(\infty)}$ and $\boldsymbol{u}_t^{(\infty)}$, build up feedback functions $\phi_t^{(\infty)}$ for $t = 0, \ldots, T-1$, according to (7.22).

7.5.2 Results Obtained for the Benchmark Problem and Comments

We return to the problem of Sect. 7.2 and show the results obtained for this problem using Algorithm 7.6. To make the comparison of these results with those obtained with the scenario tree technique (see Sect. 7.4.5) easier, in Fig. 7.7 we provide the same type of plots as those shown in Fig. 7.6. That is, we limit ourselves to showing the clouds of dots with coordinates $\left(x_t^{i,(\infty)}, u_t^{i,(\infty)}\right)$ for $i = 1, \ldots, 200$ and $t = 0, 12, 23$. In particular, we did not achieve the termination step of Algorithm 7.6, namely the feedback synthesis, and we also show, as in Fig. 7.6, the "exact" solution curves provided by the DP approach (see Fig. 7.4).

When comparing Figs. 7.6 and 7.7, several observations can be pointed out. First of all, there are now 200 dots for every plot corresponding to any time instant for $t = 0$ to 24, unlike the scenario tree technique which yields $\max(2^{t+1}, 200)$ dots at time t. This is, of course, a first advantage of the particle method, especially for the earlier values of t.

Moreover, even when both approaches provide 200 particles on which to base the reconstruction of a feedback function, which is the case for $t = 12$ and 23, it is clear that those particles are much closer to the true curves (provided by DP) with the particle method than with the scenario tree technique. We already explained why the latter approach is handicapped, since, when t increases, the pending subtree at any node of level t tends to be reduced to a single scenario. Obviously, the particle method does not suffer the same drawback and the three plots shown in Fig. 7.7 suggest that the MQE (Mean Quadratic Error) is more or less the same (and here rather moderate with 200 scenarios) from the beginning to the end of the time horizon. At

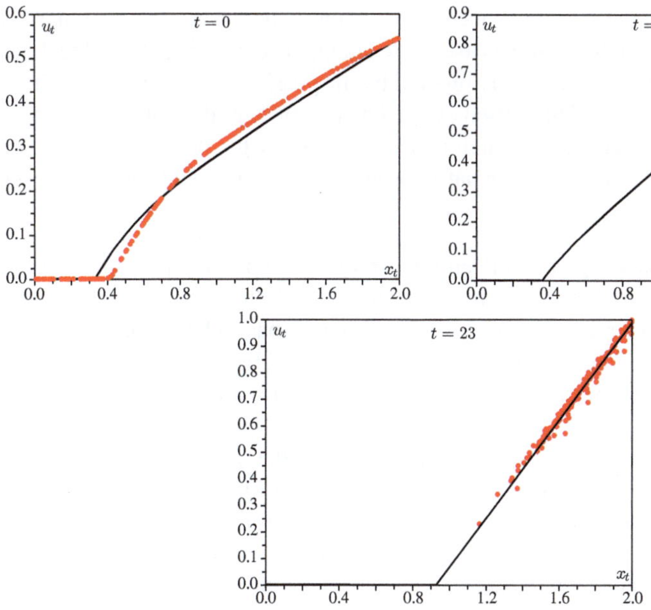

Fig. 7.7 Scenario tree: optimal pairs (x, u) at three time instants

this moment, there are no theoretical results to support this claim, but more experimental results available in Girardeau's PhD thesis [70] lead to the same impression which is also intuitively reasonable when examining the type of computations done in Algorithm 7.6.

However, one may object that even if the average error in the locations of particles seems rather moderate, this does not mean that the *whole* feedback functions can be accurately estimated because the particles are not evenly distributed over the whole state domain, namely the segment [0, 2] here. This is the case for $t = 0$ since $\boldsymbol{X}_0$ has a uniform distribution over this segment which is thus also reflected by the sample distribution. But for $t = 12$ and $t = 23$, only parts of the state domain are covered by the particles. Indeed, remember that those parts of the state domain are the ones visited by the final grids $\boldsymbol{x}_t^{(\infty)}$. Those grids are not designed a priori by the user, as in the case of the DP resolution, but they are automatically produced by the algorithm itself. They, in fact, reflect the optimal state distribution of the problem under consideration.

This feature, which may sound like a drawback of the method, may also be considered an advantage. Indeed, in the DP approach, in which one has no a priori idea of the optimal state distribution, the natural way to proceed is to attempt to cover the whole possible state domain as evenly as possible with the discretization grid. But as we have discussed already, this approach is the source of the "curse of dimensionality" because it requires a number of particles which increases exponentially with the state vector dimension. In the present example, it turns out that, in the optimal behavior of the system, the state concentrates at some time stages in a very narrow part of the state space, so that it seems useless to attempt to compute optimal feedback functions in regions which are mostly ignored by the possible state trajectories. If we knew that a priori, we would have used our given amount of grid points (200 here) to more finely discretize the region which is really visited by the flow of optimal state trajectories. In some sense, this is what is progressively achieved by the proposed iterative algorithm.

Remark 7.7 In Remark 7.2, we noticed that, in the backward recursion to numerically solve the DP equation, an approximation of the Bellman function V_t is only needed to perform this recursion because this function has to be evaluated outside the discretization grid. On the contrary, the feedback function γ_t is used only at the grid points so that its reconstruction by interpolation-regression is necessary only to deliver a final answer.

In some sense, we have benefited in Algorithm 7.6 of the same kind of observation in which Λ_t plays the role of V_t (the relation between those two functions was commented at the end of Sect. 5.5.4). Since the feedback function is used only at the points of the grid, we have only been able to handle its trace of the grid all along the iterations of the algorithm and defer the feedback synthesis operation only when convergence has been achieved. ◊

7.6 Conclusion

This chapter was devoted to translate the optimality conditions (5.26) into a numerical algorithm. We obtained some results for our benchmark problem of Sect. 7.2 and compared them with those obtained by the scenario tree technique and by DP. The optimality conditions (5.26) concern SOC problems that fit in the Markovian setting (see Sect. 5.5.1) as does also the DP approach, whereas the scenario tree technique can potentially address more general SOC problems. We refer the reader to [48] for the description of numerical algorithms exploiting the other optimality conditions studied in Chap. 5 which concern those more general problems.

The philosophy of the approach leading to Algorithm 7.6 follows the path "optimize then discretize", a feature also shared by the DP approach, whereas the scenario tree technique follows the alternative path "discretize then optimize". In the latter, the crucial informational constraint must find a translation at the first discretization step, hence the tree structure that must be given to scenarios. As we have discussed, this tree structure has important negative consequences regarding the quality of Monte Carlo estimators it can produce, and this has been illustrated by the results shown in Fig. 7.6.

In the alternative path "optimize then discretize", the informational constraint is handled at the first stage of writing down optimality conditions. This may practically translate into using a functional point of view (which is the case in DP) or by using conditional expectations as shown in Chap. 5. However, in the Markovian case, those conditional expectations boil down to simple expectations, which makes their numerical approximation by the standard Monte Carlo approach far easier.

The DP approach is recursive in time, whereas, the particle method is also iterative and closer to the spirit of a variational (gradient-like) algorithm. In both methods, some functional approximation must be used and the arguments of the manipulated functions belong to the state space. However, one distinctive feature is that DP must a priori explore the whole state space, which leads to the exponential growth of the number of discretization points needed in that space with the dimension of this space (the "curse of dimensionality"), whereas the particle method incorporates an automatic construction of the discretization grid in this space whose cardinality is connected with the size of the Monte Carlo sample set considered (i.e. the number N of particles). This does not mean that the dimension of the state space plays no role, and it is likely to have an influence on the interpolation-regression operations required by the method. Theoretical results about the rate of convergence of the method with the parameter N are still missing, but the reader may consult [70] for some experimental results. However, the fact that the particle method is able to construct a grid in the state space which is adapted to the optimal state distribution, as illustrated by our benchmark problem, should be considered as an advantage, if not a definite answer, to the curse of dimensionality.

As mentioned earlier, in front of this challenging issue of the curse of dimensionality, several other methods have been proposed in the literature to tackle the DP equation numerical resolution. A comparison of all these methods with the particle method proposed in this chapter is out of the scope of this book and still to be done. Some referees claimed strong relations between the particle method and either the so-called Stochastic Dual Dynamic Programming (SDDP) technique [113, 116] or the Approximate Dynamic Programming (ADP) [126]. Inevitably, as long as all those methods address the same problem resolution, namely the SOC problem, there are obviously possible connections and some common features between them. But as far as we understand SDDP and ADP, we do not see very tight relations with the particle method. Both SDDP and ADP focus on approximations of the Bellman function whereas the starting point of the particle method is more on the side of Pontryagin like optimality conditions. In addition, SDDP assumes convexity of the Bellman function in order to build outer approximations by supporting hyperplanes whereas ADP is often based on an initial guess of a parametric class of functions to which, hopefully, the exact Bellman function "almost" belongs. No such a priori assumption or guess is needed with the particle method.

Part IV
Convergence Analysis

Chapter 8
Convergence Issues in Stochastic Optimization

8.1 Introduction

In stochastic programming, the decision maker makes decisions which are subject to partial observations. A typical example is in multi-stage decision problems when the decision maker is constrained to use *non-anticipative* strategies, i.e. admissible strategies are constrained to be adapted to the filtration generated by the observations. This leads to constraints on admissible controls which are called measurability constraints (see Sect. 1.2.1). Investigating the dependence of optimal strategies and optimal values on measurability constraints is of deep importance for both theory and applications. Moreover, if we are concerned with numerical solutions of stochastic problems, approximations of random variables and subfields involved in the problem are also considered. We are thus led to examine optimal strategies and optimal values when both random variable approximation and measurability constraints approximation are considered. For some concepts used in the text without definitions, the reader is referred to classical Analysis textbooks such as [2] and Appendix A for basics in optimization.

More precisely, let $(\Omega, \mathcal{A}, \mathbb{P})$ be a probability space, and consider problems formulated as in (1.3),

$$\min_{\boldsymbol{U} \preceq \mathcal{G}} \mathbb{E}\big(j(\boldsymbol{U}, \boldsymbol{W})\big), \tag{8.1}$$

with a measurability constraint specified by a fixed subfield $\mathcal{G} \subset \mathcal{A}$. As exposed in Sect. 1.2.2, we are in a case of static information structure. Moreover, we will consider problems with admissible controls in subspaces of L^p spaces with $1 \leq p < +\infty$.

Using a sequence $\{\boldsymbol{W}^{(n)}\}_{n\in\mathbb{N}}$ of approximations of the random variable $\boldsymbol{W}$ and a sequence $\{\mathcal{G}^{(n)}\}_{n\in\mathbb{N}}$ of subfields of $\mathcal{A}$ as approximations of the subfield $\mathcal{G}$, we are led to solve a sequence of approximated problems:

$$\min_{\boldsymbol{U} \preceq \mathcal{G}^{(n)}} \mathbb{E}\big(j(\boldsymbol{U}, \boldsymbol{W}^{(n)})\big).$$

P. Carpentier et al., *Stochastic Multi-Stage Optimization*, Probability Theory and Stochastic Modelling 75, DOI 10.1007/978-3-319-18138-7_8

Using the sequence of functions

$$F^{(n)}(\boldsymbol{U}) := \mathbb{E}\big(j(\boldsymbol{U}, \boldsymbol{W}^{(n)})\big) + \chi_{L^0(\Omega, \mathcal{G}^{(n)}, \mathbb{P}; \mathbb{U})}(\boldsymbol{U}),$$

where $L^0(\Omega, \mathcal{G}^{(n)}, \mathbb{P}; \mathbb{U})$ is the space of $\mathcal{G}^{(n)}$-measurable functions and where χ_A is the characteristic function of the set A (see Eq. A.2), we can rewrite the sequence of optimization problems as:

$$\min_{\boldsymbol{U} \in L^p(\Omega, \mathcal{A}, \mathbb{P}; \mathbb{U})} F^{(n)}(\boldsymbol{U}).$$

In the following, we give sufficient conditions under which the sequence $\{F^{(n)}\}_{n \in \mathbb{N}}$ Mosco converges to F in $L^p(\Omega, \mathcal{A}, \mathbb{P}; \mathbb{U})$ where the function F is defined by:

$$F(\boldsymbol{U}) := \mathbb{E}\big(j(\boldsymbol{U}, \boldsymbol{W})\big) + \chi_{L^0(\Omega, \mathcal{G}, \mathbb{P}; \mathbb{U})}(\boldsymbol{U}).$$

Mosco convergence ensures epi-convergence of the sequence of functions $F^{(n)}$ in both the strong and weak topologies. As developed in [6], epi-convergence of a sequence of functions $\{f^{(n)}\}_{n \in \mathbb{N}}$, together with compactness assumptions, makes it possible to approximate the infimal value and the set of minimizers of the epi-limit of the sequence $\{f^{(n)}\}_{n \in \mathbb{N}}$ [6, Theorems 1.10 and 2.11]. Related results can be found in [4, 163] where the dependence of the optimal strategies and optimal values upon subfield variation is examined. An additional constraint on the control $\boldsymbol{U}$ under the form $\mathbb{E}(\boldsymbol{U}) \in C$, where C is a convex subset of $\mathbb{U}$, is also considered in both papers. In the context of open-loop stochastic optimization, approximations of optimization problems through random variables approximation have been widely studied [23, 62, 101, 112, 130]. Note also that mixing random variable approximations and measurability constraint approximations are considered in [12, 110], with specific assumptions which link the random variable approximations to the measurability constraint approximations.

This chapter is organized as follows. In Sect. 8.2, we introduce convergence notions. First, for sequences of functions taking values in the extended real numbers, we introduce epi- and Mosco convergence; then, we introduce the strong convergence of sequences of subfields. In Sect. 8.3, we recall a set of definitions and theorems on multifunctions, integrands and upper integrals. These are the theoretical tools for manipulating cost functions given by expectations. A definition of the conditional expectation of normal integrands is also presented, and the section ends up with a general theorem on interchange of minimization and integration. We state the mathematical framework for the precise statement of problem (8.1) and justify the interchange operation (6.6). In Sect. 8.4 we recall some classical applications of epi-convergence to the solution of open-loop optimization problems. Section 8.5 is devoted to a convergence theorem accompanied with examples and discussions about related works.

8.2 Convergence Notions

We present epi-convergence and Mosco convergence of sequences of functions, as well as convergence of sequences of subfields.

8.2.1 Epi-Convergence and Mosco Convergence

We now define and give some properties of *epi-convergence* for functions taking values in $\mathbb{R} \cup \{+\infty\}$, or more generally in the extended real numbers $\overline{\mathbb{R}} := \mathbb{R} \cup \{+\infty\} \cup \{-\infty\}$. Its main importance comes from the fact that epi-convergence of a sequence of functions $\{f^{(n)}\}_{n\in\mathbb{N}}$ to f is what is needed to ensure the convergence of the minimizers of $\{f^{(n)}\}_{n\in\mathbb{N}}$ to the minimizers of f. For a precise and complete study of the properties of epi-convergence, we refer the reader to [6, 135].

Definition 8.1 Let $(\mathbb{S}, \rho)$ be a topological space whose topology is denoted here by ρ and $\{f^{(n)}\}_{n\in\mathbb{N}}$ be a sequence of functions $f^{(n)} : \mathbb{S} \to \overline{\mathbb{R}}$. We denote by $V(z)$ the family of neighborhoods of $z \in \mathbb{S}$ relative to topology ρ. The following functions taking values in $\overline{\mathbb{R}}$ are said to be, respectively, the *ρ-epigraphical lower limit* and the *ρ-epigraphical upper limit* of the sequence $\{f^{(n)}\}_{n\in\mathbb{N}}$:

$$(\rho\text{-li}_{\mathrm{e}} f^{(n)})(z) := \sup_{\mathcal{V}\in V(z)} \liminf_{n} \inf_{\overline{z}\in\mathcal{V}} f^{(n)}(\overline{z}),$$

$$(\rho\text{-ls}_{\mathrm{e}} f^{(n)})(z) := \sup_{\mathcal{V}\in V(z)} \limsup_{n} \inf_{\overline{z}\in\mathcal{V}} f^{(n)}(\overline{z}).$$

It follows from the definitions that we have $(\rho\text{-li}_{\mathrm{e}} f^{(n)})(z) \leq (\rho\text{-ls}_{\mathrm{e}} f^{(n)})(z)$. Moreover, using [6, Theorem 2.1], we note that both functions $\rho\text{-li}_{\mathrm{e}} f^{(n)}$ and $\rho\text{-ls}_{\mathrm{e}} f^{(n)}$ are ρ-lower semicontinuous.

Definition 8.2 Let $(\mathbb{S}, \rho)$ be a topological space. A sequence of functions $\{f^{(n)}\}_{n\in\mathbb{N}}$ defined on $\mathbb{S}$ is said to be *ρ-epi-convergent at point* $z \in \mathbb{S}$ when the upper and lower epigraphical limits are equal:

$$(\rho\text{-li}_{\mathrm{e}} f^{(n)})(z) = (\rho\text{-ls}_{\mathrm{e}} f^{(n)})(z).$$

The common value is called the *ρ-epigraphical limit* of the sequence $\{f^{(n)}\}_{n\in\mathbb{N}}$ at point z and is denoted by $\rho\text{-lim}_{\mathrm{e}} f^{(n)}(z)$. When the sequence is epi-convergent for all $z \in \mathbb{S}$, it is said to be *epi-convergent*.

Moreover, let us notice that if $\mathbb{S}$ is a first countable topological space (which means that each point in $\mathbb{S}$ has a countable neighbourhood basis), we have the following easier sequential characterizations of the ρ-epigraphical lower and upper limits [6, Theorem 1.13]. This sequential characterization of epi-convergence can be used in metric spaces which are first countable topological spaces.

Proposition 8.3 *Let $\mathbb{S}$ be a metric space whose topology is denoted by ρ and $\{f^{(n)}\}_{n\in\mathbb{N}}$ be a sequence of functions $f^{(n)} : \mathbb{S} \to \overline{\mathbb{R}}$. For every $z \in \mathbb{S}$, the following holds true:*

$$(\rho\text{-li}_e f^{(n)})(z) \\ = \min\left\{\liminf_n f^{(n)}(z^{(n)}) \;\middle|\; \{z^{(n)}\} \textit{ is such that } z = \rho\text{-lim}\, z^{(n)}\right\}, \tag{8.2}$$

$$(\rho\text{-ls}_e f^{(n)})(z) \\ = \min\left\{\limsup_n f^{(n)}(z^{(n)}) \;\middle|\; \{z^{(n)}\} \textit{ is such that } z = \rho\text{-lim}\, z^{(n)}\right\}. \tag{8.3}$$

The right-hand side of (8.2) (resp. of (8.3)) is also called the sequential ρ-epi-lower limit (resp. -upper limit). For a fixed $z \in \mathbb{S}$, Eqs. (8.2) and (8.3) can be used to check ρ-epi-convergence at point z as follows (see [6, Proposition 1.14]).

Proposition 8.4 *Let $\mathbb{S}$ be a metric space whose topology is denoted by ρ and $\{f^{(n)}\}_{n\in\mathbb{N}}$ a sequence of functions $f^{(n)} : \mathbb{S} \to \overline{\mathbb{R}}$. For every $z \in \mathbb{S}$, the statement $f(z) = \rho\text{-lim}_e f^{(n)}(z)$ is equivalent to the following two conditions:*

(i) there exists a sequence $\{z^{(n)}\}_{n\in\mathbb{N}}$ in $\mathbb{S}$ with $z = \rho\text{-lim}_n z^{(n)}$ such that

$$\limsup_n f^{(n)}(z^{(n)}) \leq f(z),$$

(ii) for any sequence $\{z^{(n)}\}_{n\in\mathbb{N}}$ in $\mathbb{S}$ such that $z = \rho\text{-lim}_n z^{(n)}$, then

$$\liminf_n f^{(n)}(z^{(n)}) \geq f(z).$$

Remark 8.5 Epi-convergence of functions is related to the Painlevé-Kuratowski convergence of sets as explained now. For a sequence $\{S^{(n)}\}_{n\in\mathbb{N}}$ of closed nonempty subsets of $\mathbb{S}$, we define the ρ-lower limit denoted by ρ-li and the ρ-upper limit denoted by ρ-ls as follows

$$\rho\text{-li}\, S^{(n)} := \left\{z \in \mathbb{S} \;\middle|\; z = \rho\text{-lim}\, z^{(n)},\, z^{(n)} \in S^{(n)}\right\}, \tag{8.4}$$

$$\rho\text{-ls}\, S^{(n)} := \left\{z \in \mathbb{S} \;\middle|\; \exists\{n(k)\}_{k\in\mathbb{N}},\, z = \rho\text{-lim}\, z^{(k)},\, z^{(k)} \in S^{(n(k))}\right\}, \tag{8.5}$$

where $\{S^{(n(k))}\}_{k\in\mathbb{N}}$ is a subsequence of $\{S^{(k)}\}_{k\in\mathbb{N}}$. This means that ρ-li $S^{(n)}$ is the set composed of limit values of converging sequences $\{z^{(n)}\}_{n\in\mathbb{N}}$ with $z^{(n)} \in S^{(n)}$, whereas ρ-ls $S^{(n)}$ is the set composed of cluster points of sequences $\{z^{(n)}\}_{n\in\mathbb{N}}$ with $z^{(n)} \in S^{(n)}$. Indeed we have ρ-li $S^{(n)} \subset \rho$-ls $S^{(n)}$ and, when the two sets are identical, the sequence $\{S^{(n)}\}_{n\in\mathbb{N}}$ is said to converge in the Painlevé-Kuratowski sense to $S = \rho\text{-li}\, S^{(n)} = \rho\text{-ls}\, S^{(n)}$. The Painlevé-Kuratowski limit of the sequence $\{S^{(n)}\}_{n\in\mathbb{N}}$ is denoted by PK-lim $S^{(n)}$. The epi-convergence of a sequence of l.s.c. proper functions $\{f^{(n)}\}_{n\in\mathbb{N}}$ is equivalent to the convergence of their epigraphs in the Painlevé-Kuratowski sense [6, Theorem 1.39]. ◊

When the topological $\mathbb{S}$ is equipped with two topologies ρ and σ, a new convergence notion known as Mosco convergence is defined as follows.

Definition 8.6 Let $\mathbb{S}$ be a topological space with two topologies ρ and σ and $\{f^{(n)}\}_{n\in\mathbb{N}}$ be a sequence of functions $f^{(n)} : \mathbb{S} \to \overline{\mathbb{R}}$. The sequence $\{f^{(n)}\}_{n\in\mathbb{N}}$ *Mosco converges* at point z when the sequence epi-converges for both topologies and the two limits are the same:

$$\rho\text{-lim}_{\mathrm{e}}\, f^{(n)}(z) = \sigma\text{-lim}_{\mathrm{e}}\, f^{(n)}(z).$$

The common value is denoted by $M(\rho, \sigma)\text{-lim}\, f^{(n)}(z)$.

We recall that the family of all topologies on $\mathbb{S}$ is partially ordered by set inclusion. That is, if every ρ-open set is also σ-open, then we say that σ is stronger than ρ which is denoted by $\rho \preceq \sigma$ (or $\rho \subset \sigma$). When $\mathbb{S}$ is a metric space with two topologies ρ and σ, such that the σ topology is stronger than the ρ topology, the characterization of Mosco convergence by Proposition 8.4 applied to both topologies can be simplified as follows.

Proposition 8.7 *Let $\mathbb{S}$ be a topological space with two first countable topologies $\rho \preceq \sigma$, and $\{f^{(n)}\}_{n\in\mathbb{N}}$ be a sequence of functions $f^{(n)} : \mathbb{S} \to \overline{\mathbb{R}}$. For every $z \in \mathbb{S}$, the statement $f(z) = M(\rho, \sigma)\text{-lim} f^{(n)}(z)$ is equivalent to the following two conditions:*

(i) there exists a sequence $\{z^{(n)}\}_{n\in\mathbb{N}}$ in $\mathbb{S}$ with $z = \sigma\text{-lim} z^{(n)}$ such that

$$\limsup_n f^{(n)}(z^{(n)}) \le f(z);$$

(ii) for any sequence $\{z^{(n)}\}_{n\in\mathbb{N}}$ such that $z = \rho\text{-lim} z^{(n)}$, then

$$\liminf_n f^{(n)}(z^{(n)}) \ge f(z).$$

Note that Condition 8.7-(i) implies Condition 8.4-(i) for both topologies ρ and σ since the sequence $\{z^{(n)}\}_{n\in\mathbb{N}}$ converging for topology σ is also converging for topology ρ with the same limit. The same argument shows that Condition 8.7-(ii) also implies Condition 8.4-(ii) for both topologies ρ and σ. We use this criteria when σ and ρ are, respectively, the strong (s-) and weak (w-) topologies of function spaces [2, Chap. 2]. In that case, the weak topology is not first countable, but the notion which turns out to be more interesting for the weak topology is the sequential epi-convergence. Thus, Mosco convergence proved using Proposition 8.7 gives epi-convergence for the strong topology and sequential epi-convergence for the weak topology.

The link between epi-convergence of a sequence of functions and the convergence of sequence of minimizers is described now. We first need to extend the notion of minimizer of a function by introducing the ϵ-minimizer notion as follows:

$$\epsilon\text{- arg min } f := \left\{ z \in \mathbb{S} \;\middle|\; f(z) \leq \sup\left(-\frac{1}{\epsilon};\ \inf_{z \in \mathbb{S}} f(z) + \epsilon \right) \right\}.$$

This definition is valid even when the expression $\inf_{z \in \mathbb{S}} f(z)$ is equal to $-\infty$.

For a given sequence of positive numbers $\{\epsilon^{(n)}\}_{n \in \mathbb{N}}$ converging to zero, when a sequence of functions $\{f^{(n)}\}_{n \in \mathbb{N}}$ epi-converges to the function f, then any cluster point of sequences $\{z^{(n)}\}_{n \in \mathbb{N}}$, with $z^{(n)} \in \epsilon^{(n)}\text{- arg min } f^{(n)}$ is element of $\arg\min f$ [6, Proposition 2.9, Corollary 2.10, Theorem 2.11].

Theorem 8.8 *Suppose that the sequence of extended real-valued functions $\{f^{(n)}\}_{n \in \mathbb{N}}$ epi-converges to the function f. Then, the following relation holds true:*

$$\limsup_n \left(\inf_{z \in \mathbb{S}} f^{(n)}(z) \right) \leq \inf_{z \in \mathbb{S}} f(z), \tag{8.6}$$

and, for every sequence $\{\epsilon^{(n)}\}_{n \in \mathbb{N}}$, $\epsilon^{(n)} > 0$ converging to zero,

$$\rho\text{-ls}\left(\epsilon^{(n)}\text{- arg min } f^{(n)}\right) \subset \arg\min f. \tag{8.7}$$

Moreover, if the set $\rho\text{-ls}(\epsilon^{(n)}\text{- arg min } f^{(n)})$ is not empty for a given sequence $\{\epsilon^{(n)}\}_{n \in \mathbb{N}}$, then inequality (8.6) turns out to be an equality with a standard limit:

$$\lim_{n \to +\infty} \left(\inf_{z \in \mathbb{S}} f^{(n)}(z) \right) = \inf_{z \in \mathbb{S}} f(z). \tag{8.8}$$

The property that the set $\rho\text{-ls}(\epsilon^{(n)}\text{- arg min } f^{(n)})$ is nonempty is called an *inf-compactness property*. This is the minimal assertion that is needed in order to derive, from epi-convergence of a sequence $\{f^{(n)}\}_{n \in \mathbb{N}}$, the convergence of the infimum sequence $\{\inf_{z \in \mathbb{S}} f^{(n)}(z)\}_{n \in \mathbb{N}}$ and the existence of a minimum for the limit problem (note that under the inf-compactness property, $\arg\min f$ is not empty since it contains a nonempty subset using Eq. (8.7)). These results are rephrased in the next theorem [6, Theorem 2.11].

Theorem 8.9 *Let $\mathbb{S}$ be a metric space whose topology is denoted by ρ and $\{f^{(n)}\}_{n \in \mathbb{N}}$ be a sequence of functions $f^{(n)} : \mathbb{S} \to \overline{\mathbb{R}}$ assumed to ρ-epi-converge to the function f. Then, the following three statements are equivalent:*

(i) $\lim_{n \to \infty} \left(\inf_{z \in \mathbb{S}} f^{(n)}(z) \right) = \inf_{z \in \mathbb{S}} f(z)$ *and* $\arg\min f \neq \emptyset$;

(ii) *there exist a sequence* $\{\epsilon^{(n)}\}_{n\in\mathbb{N}}$ *which converges to zero and a* ρ*-relatively compact sequence* $\{z^{(n)}\}_{n\in\mathbb{N}}$ *in* $\mathbb{S}$ *for which* $z^{(n)} \in \epsilon^{(n)}\text{-}\arg\min f^{(n)}$*, for all* $n \in \mathbb{N}$*;*

(iii) *there exist a sequence* $\{\epsilon^{(n)}\}_{n\in\mathbb{N}}$ *which converges to zero and a nonempty* ρ*-relatively compact set* K *in the space* $\mathbb{S}$ *such that:*

$$\inf_{z\in K} f^{(n)}(z) \leq \sup\left\{-\frac{1}{\epsilon^{(n)}};\ \inf_{z\in\mathbb{S}} f^{(n)}(z) + \epsilon^{(n)}\right\},\quad \forall n \in \mathbb{N}.$$

Epi-convergence proofs frequently use the Fenchel transform (see Sect. A.1.3) because it is bicontinuous for the Mosco convergence on the set of closed convex proper functions taking values in $\mathbb{R}\cup\{+\infty\}$ (see Sect. A.1.1 for convex proper definition). This comes from the fact that, by Fenchel duality, sequential weak and strong topologies are exchanged and upper and lower limit are exchanged. The bicontinuity is shown by the following theorem [6, Theorem 3.18] (see reflexive Banach space in [2, p. 232]).

Theorem 8.10 *Let* $\mathbb{U}$ *be a reflexive Banach space,* f *be a closed convex proper function taking values in* $\mathbb{R}\cup\{+\infty\}$ *and* $\{f^{(n)}\}_{n\in\mathbb{N}}$ *be a sequence of closed convex proper functions taking values in* $\mathbb{R}\cup\{+\infty\}$*. Then the following two statements are equivalent:*

(i) *the sequence* $\{f^{(n)}\}_{n\in\mathbb{N}}$ *Mosco converges to* f *on* $\mathbb{U}$ $(M(w,s))$*;*

(ii) *the sequence* $\{f^{(n)\star}\}_{n\in\mathbb{N}}$ *Mosco converges to* $f^\star$ *on* $\mathbb{U}^*$ $(M(w,s))$*,*

where the star superscript denotes Fenchel transform for functions and dual topological space for spaces.

8.2.2 Convergence of Subfields

This paragraph is devoted to information discretization and convergence of subfields. This viewpoint is not so common in Probability Theory as compared to the noise discretization—related to the convergence of measures and random variables—which is somewhat conventional.

Let $(\Omega, \mathcal{A}, \mathbb{P})$ be a given probability space, we recall some results about the set $\mathcal{A}^\diamond$ of subfields of $\mathcal{A}$ (see [43, 95] for further details).

We draw the following definition from Neveu [109].

Definition 8.11 The sequence $\{\mathcal{G}^{(n)}\}_{n\in\mathbb{N}}$ of subfields of $\mathcal{A}$ is said to converge to $\mathcal{G}$ strongly if $\mathbb{E}(\mathbf{1}_A \mid \mathcal{G}^{(n)})$ converges to $\mathbb{E}(\mathbf{1}_A \mid \mathcal{G})$ in probability (see Sect. B.3.3) for any $A \in \mathcal{A}$.

Recalling the definition $\|g\|_1 = \int |g(\omega)|\,\mathrm{d}\mathbb{P}(\omega)$ of the L^1 norm, Kudo proves in [95, Theorem 2.1] that it is equivalent to the convergence of $\left\|\mathbb{E}(f \mid \mathcal{G}^{(n)})\right\|_1$ to $\|\mathbb{E}(f \mid \mathcal{G})\|_1$ for any bounded $\mathcal{A}$-measurable function f.

Moreover, in [95] Kudo introduces the notions of upper and lower limits of the sequence $\{\mathcal{G}^{(n)}\}_{n\in\mathbb{N}}$, that we recall here.

Definition 8.12 Let a sequence $\{\mathcal{G}^{(n)}\}_{n\in\mathbb{N}}$ of subfields of $\mathcal{A}$ be given. A subfield is called the *upper limit* (resp. *lower limit*) of $\{\mathcal{G}^{(n)}\}_{n\in\mathbb{N}}$, and is denoted by $\mathcal{G}^{\sharp}$ (resp. $\mathcal{G}^{\flat}$) if it is the minimal (resp. maximal) subfield among subfields $\mathcal{G}$ of $\mathcal{A}$ such that,

$$\limsup_{n\to+\infty} \|\mathbb{E}(f \mid \mathcal{G}^{(n)})\|_1 \leq \|\mathbb{E}(f \mid \mathcal{G})\|_1,$$

$$\left(\text{resp.} \liminf_{n\to+\infty} \|\mathbb{E}(f \mid \mathcal{G}^{(n)})\|_1 \geq \|\mathbb{E}(f \mid \mathcal{G})\|_1\right),$$

for every bounded $\mathcal{A}$-measurable function f.

In this definition, the order relation between subfields of a given σ-field $\mathcal{A}$ is induced by the inclusion relation. The existence of minimal and maximal subfields stated in the previous definition is proved in [95].

Definition 8.13 Let a sequence $\{\mathcal{G}^{(n)}\}_{n\in\mathbb{N}}$ of subfields of $\mathcal{A}$ be given. When the upper limit $\mathcal{G}^{\sharp}$ and the lower limit $\mathcal{G}^{\flat}$ of the sequence of subfields $\{\mathcal{G}^{(n)}\}_{n\in\mathbb{N}}$ are equal, the sequence $\{\mathcal{G}^{(n)}\}_{n\in\mathbb{N}}$ is said to *Kudo converge* to the subfield $\mathcal{G}^{\infty}$ ($\mathcal{G}^{(n)} \to \mathcal{G}^{\infty}$) where $\mathcal{G}^{\infty}$ denotes the common value of the upper and lower limits.

Kudo convergence is equivalent to the strong convergence, as defined in Definition 8.11 and as proved in [95, Theorem 3.4]. Following the terminology given by Neveu and Kudo [95], this convergence is known as the *strong convergence* topology on $\mathcal{A}^{\diamondsuit}$. But, although being termed as "strong", this topology actually corresponds to a pointwise convergence notion. There exist "stronger" convergence notions for subfields, such as the *uniform convergence* defined by Boylan [29]. Note that the strong convergence topology on $\mathcal{A}^{\diamondsuit}$ is also proved to be a metrizable topology [43].

Another important characterization of the strong convergence on $\mathcal{A}^{\diamondsuit}$ is the fact that it is the coarsest topology such that the conditional expectation is continuous in $L^1(\Omega, \mathcal{A}, \mathbb{P}; \mathbb{R})$ (see Sect. B.2.1) w.r.t. the conditioning subfield. And, as stated in Proposition 8.14, the property remains true when considering $L^p(\Omega, \mathcal{A}, \mathbb{P}; \mathbb{U})$ spaces for $p \in [1, +\infty)$ endowed with the strong or weak topology [118, Theorem 2.2].

Proposition 8.14 *With* $1 \leq p < +\infty$, *we have that* $\mathcal{G}^{(n)} \to \mathcal{G}^{\infty}$ *if, and only if,* $\mathbb{E}(f \mid \mathcal{G}^{(n)})$ *converges to* $\mathbb{E}(f \mid \mathcal{G}^{\infty})$ *for every* $f \in L^p(\Omega, \mathcal{A}, \mathbb{P}; \mathbb{R})$, *where* $L^p(\Omega, \mathcal{A}, \mathbb{P}; \mathbb{R})$ *is endowed with the strong or weak topology.*

Note that the case $p = +\infty$ cannot be treated with the strong topology [117]. When concerned with discretization or numerical approximations, we may have to manipulate random variables taking a finite number of possible values. The subfields generated by such variables are generated by finite partitions and are called atomic σ-fields (see Definition 3.26). Thus, we need to have a topology for which the limit of a sequence of subfields generated by a finite partition is large enough to cover any possible subfield of $\mathcal{A}$. Such a requirement is fulfilled by the strong convergence as shown by the next proposition [43].

Proposition 8.15 *The set of subfields of $\mathcal{A}$ generated by finite partitions of Ω is dense in $\mathcal{A}^{\Diamond}$ equipped with the strong convergence topology.*

In Chap. 6, quantization was introduced and a measurability constraint given by the subfield $\sigma(\boldsymbol{Y})$ generated by the random variable $\boldsymbol{Y}$, was replaced through quantization by the sequence of subfields $\{\sigma(\boldsymbol{Y}^{(n)})\}_{n\in\mathbb{N}}$ with $\boldsymbol{Y}^{(n)} = Q^{(n)} \circ \boldsymbol{Y}$. The next proposition shows that sufficient conditions for strong convergence of the sequence of such subfields are the convergence in probability of the sequence $\{\boldsymbol{Y}^{(n)}\}_{n\in\mathbb{N}}$ to $\boldsymbol{Y}$ and the fact that all the σ-fields generated by the sequence of random variables are included in the limit σ-field $\sigma(\boldsymbol{Y})$ [12].

Proposition 8.16 *Let $\{\boldsymbol{Y}^{(n)}\}_{n\in\mathbb{N}}$ be a sequence of random variables converging in probability to a random variable $\boldsymbol{Y}$. If $\sigma(\boldsymbol{Y}^{(n)}) \subset \sigma(\boldsymbol{Y})$ for all $n \in \mathbb{N}$, then the sequence of σ-fields $\{\sigma(\boldsymbol{Y}^{(n)})\}_{n\in\mathbb{N}}$ strongly converges to $\sigma(\boldsymbol{Y})$.*

8.3 Operations on Integrands

This section is devoted to multifunctions and integrands and, more precisely, to normal integrands. Normal integrands are used as cost functions in stochastic optimization. In Sects. 8.3.4 and 8.3.5, we present the assumptions which have to be made in order to define the conditional expectation of normal integrands w.r.t. a given subfield and to interchange expectation and minimization.

In the next sections, we use the following notations. Let $(\Omega, \mathcal{A}, \mathbb{P})$ be a complete probability space (see Sect. B.1.2) and $\mathbb{U}$ be a topological space with its Borel σ-field $\mathcal{B}^{o}_{\mathbb{U}}$ (see Sect. B.6.1). We recall that $L^0(\Omega, \mathcal{A}, \mathbb{P}; \mathbb{U})$ is the space of measurable functions on $(\Omega, \mathcal{A})$ taking values in the measurable space $(\mathbb{U}, \mathcal{B}^{o}_{\mathbb{U}})$. If $\mathbb{U}$ is a normed space, we denote by $L^p(\Omega, \mathcal{A}, \mathbb{P}; \mathbb{U})$ (for $1 \leq p \leq +\infty$) the subspace of L^0 of functions f such that $\omega \mapsto \|f(\omega)\|_{\mathbb{U}}$ belongs to $L^p(\Omega, \mathcal{A}, \mathbb{P}; \mathbb{R})$ (see Sect. B.2.1 for further details).

8.3.1 Multifunctions

A *multifunction*, or *set-valued mapping*, F denoted by $F : \Omega \rightrightarrows \mathbb{U}$ is a map from Ω to the space $2^{\mathbb{U}}$ of all subsets of $\mathbb{U}$. The *domain* of F is the subset of Ω defined by

$$\operatorname{dom} F := \{\omega \in \Omega \mid F(\omega) \neq \emptyset\}.$$

A *selection* of a multifunction F is a function $s : \Omega \to \mathbb{U}$ such that for all $\omega \in \operatorname{dom} F$, $s(\omega) \in F(\omega)$. We denote by $\operatorname{Sel}(F)$ the set of all the selections of F.

Definition 8.17 A multifunction F is said to be *measurable* if, for every open set O in $\mathbb{U}$, we have $F^{-}(O) \in \mathcal{A}$, where $F^{-}(O) := \{\omega \in \Omega \mid F(\omega) \cap O \neq \emptyset\}$.

A *measurable selection* of F is a selection of F which is $\mathcal{A}$-measurable. We denote by $L^0_F(\Omega, \mathcal{A}, \mathbb{P}; \mathbb{U})$ the set of $\mathcal{A}$-measurable selections of F,

$$L^0_F(\Omega, \mathcal{A}, \mathbb{P}; \mathbb{U}) := L^0(\Omega, \mathcal{A}, \mathbb{P}; \mathbb{U}) \cap \mathrm{Sel}(F),$$

and by $L^p_F(\Omega, \mathcal{A}, \mathbb{P}; \mathbb{U})$ (for $1 \leq p \leq +\infty$) the set of all the $\mathcal{A}$-measurable selections of F which belong to the space $L^p(\Omega, \mathcal{A}, \mathbb{P}; \mathbb{U})$:

$$L^p_F(\Omega, \mathcal{A}, \mathbb{P}; \mathbb{U}) := L^p(\Omega, \mathcal{A}, \mathbb{P}; \mathbb{U}) \cap \mathrm{Sel}(F).$$

Definition 8.18 A nonempty, closed-valued and measurable multifunction is called a *random closed set*.

The importance of random closed sets defined in [76, 77] arises from the following theorem.

Theorem 8.19 *Let F be a random closed set. Then, the set of measurable selections $L^0_F(\Omega, \mathcal{A}, \mathbb{P}; \mathbb{U})$ is not empty.*

The proof is, in fact, valid for a closed-valued and measurable multifunction as proved in [132, Corollary 1, p. 221].

Note, also, that the graph of a random closed set $F : \Omega \rightrightarrows \mathbb{U}$,

$$\mathrm{Gr}(F) := \{(\omega, u) \in \Omega \times \mathbb{U} \mid u \in F(\omega)\}, \tag{8.9}$$

is $\mathcal{A} \otimes \mathcal{B}^o_{\mathbb{U}}$-measurable.

Remark 8.20 We have made the assumption that the σ-field $\mathcal{A}$ is complete w.r.t. the underlying probability space. Under this assumption, and when the space $\mathbb{U}$ is a separable and complete metric space, all the reasonable definitions of measurability for multifunctions are equivalent. More precise interrelations can be found in [34], [132, Theorem 1, p. 219] and [133]. ◇

8.3.2 Integrands

Integrands and associated multifunctions are defined in [135, p. 661] as follows.

Definition 8.21 A function $f : \mathbb{U} \times \Omega \to \overline{\mathbb{R}}$ such that $f(u, \omega)$ is measurable w.r.t. ω for every $u \in \mathbb{U}$ is called an *integrand*.[1]

Note that it is important to consider maps which may possibly assume infinite values because it enables efficient representation of constraints. In the optimization framework, two natural multifunctions are associated with an integrand f:

[1] In order to maintain consistency with all previous chapters, we have interverted the traditional order $f(\omega, u)$ of arguments of an integrand.

- the *epigraphical multifunction* associated with an integrand f is the multifunction

$$S_f : \omega \mapsto \operatorname{epi} f(\cdot, \omega) = \{(u, \alpha) \in \mathbb{U} \times \mathbb{R} \mid f(u, \omega) \le \alpha\};$$

- the *domain mapping* D_f associated with an integrand f is the multifunction

$$D_f : \omega \mapsto \operatorname{dom} f(\cdot, \omega) = \{u \in \mathbb{U} \mid f(u, \omega) < +\infty\}.$$

We define now normal integrands as in [135, Definition 14.27, p. 661].

Definition 8.22 A *normal integrand* f is an integrand with an associated epigraphical multifunction S_f which is measurable and closed-valued.

We also speak of $\mathcal{A}$-normal integrands when the σ-field used for measurability needs to be specified. Moreover, if, for almost every $\omega \in \Omega$, the function $f(\cdot, \omega)$ satisfies a property P, then the integrand f is said to be a P-integrand. For example, f is a convex integrand if it is an integrand and $f(\cdot, \omega)$ is convex for almost every $\omega \in \Omega$. With the previous conventions, the epigraphical multifunction $\operatorname{epi} f(\cdot, \omega)$ associated with a proper (see Sect. A.1.1) normal integrand f is a random closed set, so that a normal integrand is lower semicontinuous (l.s.c., see Definition A.2). If an integrand $f : \mathbb{U} \times \Omega \to \overline{\mathbb{R}}$ is normal, then it is $(\mathcal{B}^o_{\mathbb{U}} \otimes \mathcal{A})$-measurable. A lower semicontinuous and $(\mathcal{B}^o_{\mathbb{U}} \otimes \mathcal{A})$-measurable function is a normal integrand when the σ-field $\mathcal{A}$ is complete [132].

As already mentioned, normal integrands are $(\mathcal{B}^o_{\mathbb{U}} \otimes \mathcal{A})$-measurable and thus, when composed with a measurable control, they yield measurable functions. Normal integrands take their importance from that property [132, p. 221] and [135, Proposition 14.28].

Proposition 8.23 *Consider $f : \mathbb{U} \times \Omega \to \overline{\mathbb{R}}$ a normal integrand and $\boldsymbol{U} : \Omega \to \mathbb{U}$ a random variable in $L^0(\Omega, \mathcal{A}, \mathbb{P}; \mathbb{U})$. Then, the mapping $\omega \mapsto f(\boldsymbol{U}(\omega), \omega)$ is measurable.*

Fenchel transform is extended naturally to $\mathcal{A}$-normal integrands by considering the Fenchel transform for each value of ω as defined now.

Definition 8.24 Denote by $\mathbb{U}^\star$ the topological dual space of $\mathbb{U}$, and by $\langle \cdot \, , \cdot \rangle$ the pairing product. Let $f : \mathbb{U} \times \Omega \to \overline{\mathbb{R}}$ be an $\mathcal{A}$-normal integrand. The Fenchel transform of f is the mapping $f^\star : \mathbb{U}^\star \times \Omega \to \overline{\mathbb{R}}$ defined by

$$\forall (u^\star, \omega) \in \mathbb{U}^\star \times \Omega, \; f^\star(u^\star, \omega) = \sup_{u \in \mathbb{U}} \big(\langle u \, , u^\star \rangle - f(u, \omega) \big). \tag{8.10}$$

This extension is shown to be consistent with the notion of $\mathcal{A}$-normal integrand by the next proposition.

Proposition 8.25 *Let $f : \mathbb{U} \times \Omega \to \overline{\mathbb{R}}$ be an $\mathcal{A}$-normal integrand. Then, its conjugate $f^\star : \mathbb{U}^\star \times \Omega \to \overline{\mathbb{R}}$ is an $\mathcal{A}$-normal integrand.*

Proposition 8.25 can be found in [135, Theorem 14.50] when $\mathbb{U} := \mathbb{R}^n$ and in [34, 132] or [74, 75] (without the completeness assumption on the probability space) when $\mathbb{U}$ is a separable Banach space.

8.3.3 Upper Integral

As normal integrands take values in $\overline{\mathbb{R}}$, we want to define integrals of measurable functions in such a way that they are adapted to functions that may possibly assume infinite values on measurable sets with non null measure. For that purpose, the upper integral is introduced [135].

Definition 8.26 Given $\varphi \in L^0(\Omega, \mathcal{A}, \mathbb{P}; \mathbb{R})$, we denote by I_φ or $\int_{\mathbb{U}}^\star \varphi(\omega)\,\mathrm{d}\mathbb{P}(\omega)$ the *upper integral* of φ defined by

$$I_\varphi := \inf\left\{ \int_\Omega \boldsymbol{U}(\omega)\,\mathrm{d}\mathbb{P}(\omega) \;\middle|\; \boldsymbol{U} \in L^1(\Omega, \mathcal{A}, \mathbb{P}; \mathbb{R}) \text{ and } \varphi \le \boldsymbol{U} \quad \mathbb{P}\text{-a.s.}\right\}.$$

Note that the previous definition conforms to the formula

$$I_\varphi = \int_\Omega \max\left\{\varphi(\omega), 0\right\}\mathrm{d}\mathbb{P}(\omega) + \int_\Omega \min\left\{\varphi(\omega), 0\right\}\mathrm{d}\mathbb{P}(\omega),$$

if the conventions $\infty + (-\infty) = (-\infty) + \infty = +\infty$ and $0.\infty = \infty.0 = 0$ are adopted [135, p. 675].

Definition 8.27 The *integral function* associated with a normal integrand $f : \mathbb{U} \times \Omega \to \overline{\mathbb{R}}$ is the functional $I_f : L^0(\Omega, \mathcal{A}, \mathbb{P}; \mathbb{U}) \to \overline{\mathbb{R}}$ defined by

$$I_f(\boldsymbol{U}) := I_{f(\boldsymbol{U})} = \int_\Omega^\star f(\boldsymbol{U}(\omega), \omega)\,\mathrm{d}\mathbb{P}(\omega), \tag{8.11}$$

where $f(\boldsymbol{U})$ is a shorthand notation for the random variable $\omega \mapsto f\big(\boldsymbol{U}(\omega), \omega\big)$.

The integral function is well defined for any normal integrand f and measurable random variable $\boldsymbol{U}$ [135, Proposition 14.56]. Furthermore, we have that

$$I_f(\boldsymbol{U}) < +\infty \Rightarrow \boldsymbol{U}(\omega) \in D_f(\omega) \quad \mathbb{P}\text{-a.s.}.$$

Remark 8.28 Suppose that F is a random closed set used to define a.s. constraints on random variables through the admissible set $\mathcal{U}^{\mathrm{ad}} := \left\{\boldsymbol{U} \;\middle|\; \boldsymbol{U} \in L^0_F(\Omega, \mathcal{A}, \mathbb{P}; \mathbb{U})\right\}$. Then, it is proved in [18, p. 144] that the characteristic function of the graph of the random closed set F defined by $f(u, \omega) = \chi_{F(\omega)}(u)$, where χ_A is the characteristic function of the set A (see Eq. A.2), is a normal integrand. Moreover, for all random variables $\boldsymbol{U} : \Omega \to \mathbb{U}$ we have that:

$$I_f(\boldsymbol{U}) = \chi_{\mathcal{U}^{\mathrm{ad}}}(\boldsymbol{U}). \qquad \Diamond$$

8.3.4 Conditional Expectation of a Normal Integrand

To define the conditional expectation of a normal integrand with respect to $\mathcal{G}$, a $\mathbb{P}$-complete subfield of $\mathcal{A}$ (see Sect. B.1), we introduce the notion of $\mathcal{A}$-*quasi integrable* integrand [146, Proposition 12].

Definition 8.29 An integrand $f : \mathbb{U} \times \Omega \to \overline{\mathbb{R}}$ is said to be $\mathcal{A}$-*quasi integrable* if there exists a sequence $\{\alpha^{(n)}\}_{n \in \mathbb{N}}$ with $\alpha^{(n)} \in L^1(\Omega, \mathcal{A}, \mathbb{P}; \mathbb{R})$ and such that

$$\inf_{\|u\| \le n} f(u, \omega) \ge \alpha^{(n)}(\omega), \quad \forall n \in \mathbb{N} \;\; \mathbb{P}\text{-a.s.}.$$

Assuming quasi integrability, it is possible to define the conditional expectation of an $\mathcal{A}$-normal integrand (see [146, Proposition 12]).

Theorem 8.30 *Let $\mathcal{G}$ be a $\mathbb{P}$-complete subfield of $\mathcal{A}$ and $f : \mathbb{U} \times \Omega \to \overline{\mathbb{R}}$ be an $\mathcal{A}$-normal integrand which is $\mathcal{A}$-quasi integrable. Then, there exists a $\mathcal{B}^{\mathrm{o}}_{\mathbb{U}} \otimes \mathcal{G}$-measurable mapping $g_{\mathcal{G}} : \mathbb{U} \times \Omega \to \overline{\mathbb{R}}$ such that*

(i) for all $B \in \mathcal{G}$ and all $\mathcal{G}$-measurable random variable $\boldsymbol{U}$, we have that

$$\int_B f\big(\boldsymbol{U}(\omega), \omega\big)\, \mathrm{d}\mathbb{P}(\omega) = \int_B g_{\mathcal{G}}\big(\boldsymbol{U}(\omega), \omega\big)\big)\, \mathrm{d}\mathbb{P}(\omega);$$

(ii) the mapping $g_{\mathcal{G}}$ is unique $\mathbb{P}$-a.s., is a $\mathcal{G}$-normal integrand and is $\mathcal{G}$-quasi integrable;

(iii) moreover, if f is a convex integrand then $g_{\mathcal{G}}$ is also a convex integrand.

The $\mathcal{G}$-normal integrand $g_{\mathcal{G}}$ is called the conditional expectation *of f w.r.t. the σ-field $\mathcal{G}$ and is denoted by $\mathbb{E}(f \mid \mathcal{G})$.*

The proof of (i) and (ii) can be found in [146, Proposition 12] and the proof of (iii) in [146, Proposition 15]. A proof of (i) and (ii) under weaker assumptions is given in [36] and a survey can be found in [143].

Remark 8.31 The Fenchel transform of a normal integrand f can be used to check the $\mathcal{A}$-quasi integrability of f using the fact that, for all $(u, u^\star) \in \mathbb{U} \times \mathbb{U}^\star$, we have that $f(u, \omega) \ge \langle u^\star, u \rangle - f^\star(u^\star, \omega)$. As an example, if there exists $\boldsymbol{U}^\star \in L^0(\Omega, \mathcal{A}, \mathbb{P}; \mathbb{U}^\star)$ such that $\|\boldsymbol{U}^\star\|$ and $f^\star\big(\boldsymbol{U}^\star(\cdot), \cdot\big)$ are $\mathcal{A}$-integrable, then f is $\mathcal{A}$-quasi integrable [146, Proposition 4]. The previous condition is satisfied if there exists $\boldsymbol{U}^\star \in L^p(\Omega, \mathcal{A}, \mathbb{P}; \mathbb{U}^\star)$ with $1 \le p < +\infty$ such that $\omega \mapsto f^\star(\boldsymbol{U}^\star(\omega), \omega)$ is integrable. $\Diamond$

8.3.5 Interchange of Minimization and Integration

In Sect. 3.5.2, the interchange of minimization and integration in a problem with measurability constraints has already been stated in a finite context. Our aim here is to extend this result to a more general setting.

Let $(\Omega, \mathcal{A}, \mathbb{P})$ be a probability space or more generally a measurable space with a σ-finite positive measure $\mathbb{P}$—see Sect. B.1.2—and $(\mathbb{U}, \mathcal{B}^{\text{o}}_{\mathbb{U}})$ be a separable Banach space with its Borel σ-field.

We first introduce two technical definitions which are used for defining the constraint set of a minimization problem. The second one is only relevant when considering a σ-finite positive measure. We recall that Ω is said to have a σ-finite measure when it is a countable union of measurable sets with finite measure. The countable set of measurable sets covering Ω is called a σ-finite covering of Ω.

Definition 8.32 A subset $\mathbb{G}$ of $L^0(\Omega, \mathcal{A}, \mathbb{P}; \mathbb{U})$ is said to be *decomposable* if, for all $A \in \mathcal{A}$ and all f and g in $\mathbb{G}$, the function $f\mathbf{1}_A + g\mathbf{1}_{A^c}$ is in $\mathbb{G}$.

Definition 8.33 Let $\mathbb{G}$ and $\mathbb{H}$ be two decomposable sets in $L^0(\Omega, \mathcal{A}, \mathbb{P}; \mathbb{U})$. The set $\mathbb{G}$ is said to be *rich* in $\mathbb{H}$ if $\mathbb{G} \subset \mathbb{H}$ and if, for each $f \in \mathbb{H}$, there exists a σ-finite covering $\{\Omega^{(n)}\}_{n\in\mathbb{N}}$ of Ω and a sequence $\{h^{(n)}\}_{n\in\mathbb{N}}$ of elements of $\mathbb{G}$ such that, for all $n \in \mathbb{N}$, $f = h^{(n)}$ on $\Omega^{(n)}$.

The interchange of minimization and integration is stated in [28, Corollary 3.9] and in [135, Theorem 14.60, p. 677].

Theorem 8.34 *Suppose that the σ-field $\mathcal{A}$ is complete. Let C be a multifunction with nonempty values and with an $\mathcal{A} \otimes \mathcal{B}^{\text{o}}_{\mathbb{U}}$-measurable graph and let f be a $\mathcal{B}^{\text{o}}_{\mathbb{U}} \otimes \mathcal{A}$-measurable integrand. If $\mathbb{G}_C$ is a decomposable subset which is rich in $L^0_C(\Omega, \mathcal{A}, \mathbb{P}; \mathbb{U})$, we have that*

$$\inf_{U\in\mathbb{G}_C} I_f(U) = \int^{\star}_{\Omega} \inf_{u\in C(\omega)} f(u,\omega)\,\mathrm{d}\mathbb{P}(\omega), \tag{8.12}$$

provided that the left-hand side be not equal to $+\infty$.

Example 8.35 Suppose that $\mathbb{G}_C = L^p(\Omega, \mathcal{A}, \mathbb{P}; \mathbb{U})$ $(p \geq 1)$ and consider the constant multifunction $C : \Omega \rightrightarrows \mathbb{U}$ such that $C(\omega) = \mathbb{U}$. The set $\mathbb{G}_C$ is decomposable and rich in $L^0_C(\Omega, \mathcal{A}, \mathbb{P}; \mathbb{U}) = L^0(\Omega, \mathcal{A}, \mathbb{P}; \mathbb{U})$. Indeed, given $\boldsymbol{Y} \in L^p(\Omega, \mathcal{A}, \mathbb{P}; \mathbb{U})$ and using a σ-finite covering $\{\Omega^{(n)}\}_{n\in\mathbb{N}}$, we can build a new σ-finite covering $\{\overline{\Omega}^{(n)}\}_{n\in\mathbb{N}}$ by

$$\overline{\Omega}^{(n)} := \Omega^{(n)} \cap \big\{\omega \in \Omega \mid \|\boldsymbol{Y}(\omega)\| \leq n\big\},$$

and a sequence $\{\boldsymbol{G}^{(n)}\}_{n\in\mathbb{N}}$ of elements of $\mathbb{G}_C$ defined by $\boldsymbol{G}^{(n)} = \boldsymbol{Y}\mathbf{1}_{\overline{\Omega}^{(n)}}$ such that for all $n \in \mathbb{N}$, $\boldsymbol{Y} = \boldsymbol{G}^{(n)}$ on $\overline{\Omega}^{(n)}$. Thus, we obtain

$$\inf_{U \in L^p(\Omega, \mathcal{A}, \mathbb{P}; \mathbb{U})} I_f(U) = \int_\Omega^\star \inf_{u \in \mathbb{U}} f(u, \omega) \, d\mathbb{P}(\omega), \tag{8.13}$$

provided that the left-hand side be not equal to $+\infty$.

The minimization of I_f over $L^p(\Omega, \mathcal{A}, \mathbb{P}; \mathbb{U})$ is therefore reduced to a pointwise minimization over $\mathbb{U}$. △

Note that if the multifunction C is a random closed set, then C has a measurable graph in $\Omega \times \mathbb{U}$, and if the integrand f is a normal integrand, it is $\mathcal{B}^o_{\mathbb{U}} \otimes \mathcal{A}$-measurable. Moreover, the right-hand side of Eq. (8.12) is properly defined since, when f is a normal integrand and C is a multifunction with a measurable graph, the function $\inf_{u \in C(\omega)} f(u, \omega)$ is $\mathcal{A}$-measurable. Indeed, the integrand $g(u, \omega) = f(u, \omega) + \chi_{C(\omega)}(u)$ is an $\mathcal{A}$-normal integrand given the assumptions made about f and C by [135, Example 14.32]. Then, the $\mathcal{A}$-measurability of $\inf_{u \in \mathbb{U}} g(u, \omega)$ follows from [135, Theorem 14.37] when $\mathbb{U} = \mathbb{R}^n$ and from [34, III.39] in a more general setting.

Theorem 8.34 is useful to take into account pointwise constraints on the decision variable U. We now propose an extension of the interchange Theorem 8.34 to take into account measurability constraints. More precisely, letting $\mathcal{G}$ be a subfield of $\mathcal{A}$, we consider the interchange Theorem 8.34 with the additional constraint $U \preceq \mathcal{G}$, also denoted by $U \in L^0(\Omega, \mathcal{G}, \mathbb{P}; \mathbb{U})$.

Corollary 8.36 *Suppose that the subfield $\mathcal{G}$ is $\mathbb{P}$-complete. Let $C : \Omega \rightrightarrows \mathbb{U}$ be a multifunction with nonempty values and with a $\mathcal{G} \otimes \mathcal{B}^o_{\mathbb{U}}$-measurable graph, and $f : \mathbb{U} \times \Omega \to \overline{\mathbb{R}}$ be a normal integrand which is $\mathcal{A}$-quasi integrable. If $\mathbb{G}_C$ is a decomposable subset which is rich in $L^0_C(\Omega, \mathcal{A}, \mathbb{P}; \mathbb{U})$, then we have that*

$$\inf_{U \in \mathbb{G}_C \cap L^0(\Omega, \mathcal{G}, \mathbb{P}; \mathbb{U})} I_f(U) = \int_\Omega^\star \inf_{u \in C(\omega)} \mathbb{E}(f \mid \mathcal{G})(u, \omega) \, d\mathbb{P}(\omega), \tag{8.14}$$

provided that the left-hand side be not equal to $+\infty$.

Proof When the random variable U is $\mathcal{G}$-measurable, we can use property (i) of Theorem 8.30, to obtain that

$$I_{\mathbb{E}(f \mid \mathcal{G})}(U) = I_f(U). \tag{8.15}$$

Thus, replacing f by its conditional expectation w.r.t. the subfield $\mathcal{G}$ in the left-hand side of Eq. (8.14) does not change the optimization problem defined by (8.14).

If the multifunction C with nonempty values has a $\mathcal{G} \otimes \mathcal{B}^o_{\mathbb{U}}$-measurable graph and $\mathbb{G}_C$ is $\mathcal{A}$-decomposable and rich in $L^0_C(\Omega, \mathcal{A}, \mathbb{P}; \mathbb{U})$, then it is straightforward to see that $\mathbb{G}_C \cap L^0(\Omega, \mathcal{G}, \mathbb{P}; \mathbb{U})$ is $\mathcal{G}$-decomposable. Moreover $\mathbb{G}_C \cap L^0(\Omega, \mathcal{G}, \mathbb{P}; \mathbb{U})$ is rich in $L^0_C(\Omega, \mathcal{A}, \mathbb{P}; \mathbb{U}) \cap L^0(\Omega, \mathcal{G}, \mathbb{P}; \mathbb{U})$ and remains rich in $L^0_C(\Omega, \mathcal{G}, \mathbb{P}; \mathbb{U})$. Thus, we can use Theorem 8.34 applied to the probability space $(\Omega, \mathcal{G}, \mathbb{P})$ to derive the result. □

In Corollary 8.36, the multifunction C is assumed to have a $\mathcal{G} \otimes \mathcal{B}^o_{\mathbb{U}}$-measurable graph (see Eq. (8.9)). Note, however, that instead of using the multifunction C as a pointwise constraint, we could add to the objective function f the integrand $\chi_{C(\omega)}(u)$ where χ_A is the characteristic function of the set A.

Thus, when the multifunction C has a $\mathcal{A} \otimes \mathcal{B}^o_{\mathbb{U}}$-measurable graph, using Remark 8.28, Corollary 8.36 can be applied to the new integrand $\tilde{f}(u,\omega) := f(u,\omega) + \chi_{C(\omega)}(u)$ and no explicit $\mathbb{G}_C$ constraints, yielding the following equality:

$$\inf_{U \in L^0(\Omega,\mathcal{G},\mathbb{P};\mathbb{U})} I_{\tilde{f}}(U) = \int_{\Omega}^{\star} \big(\inf_{u \in \mathbb{U}} \mathbb{E}(f \mid \mathcal{G})(u,\omega) + \mathbb{E}(\chi_C \mid \mathcal{G})(u,\omega) \big) \, d\mathbb{P}(\omega).$$

Now, if we add the assumption that the multifunction C has a $\mathcal{G} \otimes \mathcal{B}^o_{\mathbb{U}}$-measurable graph, we have that $\mathbb{E}(\chi_C \mid \mathcal{G})(u,\omega) = \chi_C(u,\omega)$ and we recover Eq. (8.14).

Remark 8.37 As developed in [147], it is possible to define the conditional expectation of the multifunction C denoted by $\mathbb{E}(C \mid \mathcal{G})$ which can be characterized as the unique closed-valued multifunction with $\mathcal{G} \otimes \mathcal{B}^o_{\mathbb{U}}$-measurable graph such that $\chi_{\mathbb{E}(C|\mathcal{G})} = \mathbb{E}(\chi_C \mid \mathcal{G})$. Using this conditional expectation, the integrand $\mathbb{E}(\chi_C \mid \mathcal{G})(u,\omega)$ can be turned into the constraint $u \in \chi_{\mathbb{E}(C|\mathcal{G})}$ and we obtain that

$$\inf_{U \in L^0(\Omega,\mathcal{G},\mathbb{P};\mathbb{U})} I_{\tilde{f}}(U) = \int_{\Omega}^{\star} \Big(\inf_{u \in \mathbb{E}(C|\mathcal{G})} \mathbb{E}(f \mid \mathcal{G})(u,\omega) \Big) \, d\mathbb{P}(\omega). \qquad \Diamond$$

Using Theorem 8.34, we see that the optimization problem (8.12) can be reduced at least formally to a pointwise problem. The optimization problem is to be solved for each value of $\omega \in \Omega$. Recovering a $\mathcal{A}$-measurable minimizer for problem (8.12) is done via measurable selection theorems [135, Theorem 14.37 and Example 14.32]. When considering Corollary 8.36, the situation is the same except that the measurable selection theorem is applied to the minimization problem

$$\min_{u \in C(\omega)} \mathbb{E}(f \mid \mathcal{G})(u,\omega),$$

where the integrand $\mathbb{E}(f \mid \mathcal{G})$ is a $\mathcal{G}$-normal integrand and C is a multifunction with a $\mathcal{G} \otimes \mathcal{B}^o_{\mathbb{U}}$-measurable graph. Selection theorems give $\mathcal{G}$-measurable minimizers in that case.

When the integrand f is a $\mathcal{G}$-measurable convex integrand, it is worth noting that if there exists a minimizer for problem (8.13), then there also exists a $\mathcal{G}$-measurable minimizer for the same problem. Thus, imposing that the constraint set be $\mathcal{G}$-measurable in order to obtain a $\mathcal{G}$-measurable minimizer is not necessary when the involved integrand f is a $\mathcal{G}$-measurable convex integrand. A $\mathcal{G}$-measurable minimizer is obtained thanks to the following proposition.

Proposition 8.38 *Let $\mathcal{G}$ be a subfield of $\mathcal{A}$ and $g(u,\omega)$ be a convex normal integrand which is $\mathcal{G}$-measurable. For $U \in L^p(\Omega,\mathcal{A},\mathbb{P};\mathbb{U})$ $(p \geq 1)$, we have that*

$$I_g(\boldsymbol{U}) \geq I_g(\mathbb{E}(\boldsymbol{U} \mid \mathcal{G})).$$

The proof consists in noting that $\boldsymbol{U}$ is also in $L^1(\Omega, \mathcal{A}, \mathbb{P}; \mathbb{U})$ and in applying [4, Proposition 3.2] which is reminiscent of Jensen's inequality.

8.4 Application to Open-Loop Optimization Problems

We give here a classical example of epi-convergence [94] which relies on the previous bicontinuity Theorem 8.10. The stochastic program to solve is an open-loop problem

$$\min_{u \in \mathbb{U}} \mathbb{E}\big(j(u, \cdot)\big) = \min_{u \in \mathbb{U}} \int_{\Omega}^{\star} j(u, \omega) \, \mathrm{d}\mathbb{P}(\omega) \tag{8.16}$$

where $\mathbb{U} = \mathbb{R}^n$ is equipped with its Borel σ-field $\mathcal{B}^{\mathrm{o}}_{\mathbb{U}}$ and j is a convex random l.s.c. function $j : \mathbb{U} \times \Omega \to \overline{\mathbb{R}}$ defined on the $\mathbb{P}$-complete probability space $(\Omega, \mathcal{A}, \mathbb{P})$.

This problem was described in detail in Chap. 2. One possible approach to numerically solve problem (8.16) consists in replacing the expectation to be minimized by a Monte Carlo approximation. This method, presented in Sect. 2.5.3 is called sample average approximation (SAA). Issues of convergence in the SAA framework involves epi-convergence or, more precisely, epi-consistency to be defined below. We briefly recall here epi-convergence formulation of SAA and give a convergence theorem.

We approximate the probability measure $\mathbb{P}$ by empirical measures derived from independent random variables $\{\boldsymbol{W}^n\}_{n \in \mathbb{N}}$. Thus, a family of approximations of problem (8.16) indexed by the number of random observations can be considered:

$$\min_{u \in \mathbb{U}} J^{(n)}(u) := \min_{u \in \mathbb{U}} \frac{1}{n} \sum_{i=1}^{n} j(u, \boldsymbol{W}^i). \tag{8.17}$$

Note that, now, each approximated problem is a random problem where randomness arises from the sample space carrying an infinite number of i.i.d. random variables (see Sect. B.7). We have to study almost sure epi-convergence of the sequence of random functions $\{J^{(n)}\}_{n \in \mathbb{N}}$ to the initial objective function

$$J : u \in \mathbb{U} \mapsto \mathbb{E}\big(j(u, \cdot)\big). \tag{8.18}$$

This is known as epi-consistency [94] and is defined as follows.

Definition 8.39 A sequence $\{h^{(n)}\}_{n \in \mathbb{N}}$ of random l.s.c. functions is *epi-consistent* if there is a (necessarily) l.s.c. function h such that $\{h^{(n)}\}_{n \in \mathbb{N}}$ epi-converges to h with probability one.

Following [94, Theorem 2.3], we give here assumptions for epi-consistence involving convex normal integrands given (see also [5, 62]).

Theorem 8.40 *Let $j : \mathbb{U} \times \Omega \to \overline{\mathbb{R}}$ be a convex normal integrand. We assume that there exists a point $\overline{u} \in \mathbb{U}$ for which we have that $j(\overline{u}, \cdot) \in L^1(\Omega, \mathcal{A}, \mathbb{P}; \mathbb{R})$ and a random variable $\overline{\boldsymbol{U}}^\star : \Omega \to \mathbb{U}^\star$ in $L^1(\Omega, \mathcal{A}, \mathbb{P}; \mathbb{U}^\star)$ for which we have that $j^\star\big(\overline{\boldsymbol{U}}^\star(\cdot), \cdot\big) \in L^1(\Omega, \mathcal{A}, \mathbb{P}; \mathbb{R})$ where $j^\star$ is the Fenchel transform of the normal integrand j. Then, the function J defined by Eq. (8.18) is proper, convex and l.s.c. and the sequence $\{J^{(n)}\}_{n\in\mathbb{N}}$ is epi-consistent with limit J.*

8.5 Application to Closed-Loop Optimization Problems

The previous Sects. 8.2 and 8.3 have provided the main notions which make it possible now to prove a convergence theorem for approximations of closed-loop optimization problems such as those studied throughout this book.

Let $(\Omega, \mathcal{A}, \mathbb{P})$ be a complete probability space, and let $(\mathbb{W}, \mathcal{B}^{\text{o}}_{\mathbb{W}})$ and $(\mathbb{U}, \mathcal{B}^{\text{o}}_{\mathbb{U}})$ be $\mathbb{R}^n$ and $\mathbb{R}^m$ with their associated Borel σ-fields. Given a random variable $\boldsymbol{W}$ with values in $\mathbb{W}$ and a subfield $\mathcal{G}$ of $\mathcal{A}$, which, respectively, represent the noise and the observation, we are concerned with the stochastic optimization problem (8.1) which is written again here as follows:

$$\min_{\boldsymbol{U} \in L^p(\Omega, \mathcal{G}, \mathbb{P}; \mathbb{U})} \mathbb{E}\big(j(\boldsymbol{U}, \boldsymbol{W})\big). \tag{8.19}$$

We now study approximate versions of problem (8.19), where the noise $\boldsymbol{W}$ is discretized using a sequence of random variables $\{\boldsymbol{W}^{(n)}\}_{n\in\mathbb{N}}$ and where the subfield $\mathcal{G}$ is approximated by a sequence of subfields $\{\mathcal{G}^{(n)}\}_{n\in\mathbb{N}}$. Thanks to the material introduced in Sects. 8.2 and 8.3, we are able to formulate and prove a convergence theorem when both the noise and the measurability constraint converge. We compare and discuss our result with other related works in the literature. Finally, we apply our convergence result to the examples analyzed in Sects. 1.3.1 and 1.4.2.

8.5.1 Main Convergence Theorem

Using a sequence of random variables $\{\boldsymbol{W}^{(n)}\}_{n\in\mathbb{N}}$ taking values in $(\mathbb{W}, \mathcal{B}^{\text{o}}_{\mathbb{W}})$ which approximates (technical assumptions given later) the given random variable $\boldsymbol{W}$, we define a sequence of integrands $\{f^{(n)}\}_{n\in\mathbb{N}}$ by:

$$f^{(n)}(u, \omega) = j(u, \boldsymbol{W}^{(n)}(\omega)). \tag{8.20}$$

Now, the measurability constraint imposed by the subfield $\mathcal{G}$ of $\mathcal{A}$ is approximated by means of a sequence of subfields $\{\mathcal{G}^{(n)}\}_{n\in\mathbb{N}}$. Using the sequence $f^{(n)} : \mathbb{U} \times \Omega \to \overline{\mathbb{R}}$

of integrands defined for $n \in \mathbb{N}$ by Eq. (8.20) and the sequence $\{\mathcal{G}^{(n)}\}_{n\in\mathbb{N}}$ of subfields of $\mathcal{A}$, we define a sequence of functions $F^{(n)} : L^0(\Omega, \mathcal{A}, \mathbb{P}; \mathbb{U}) \to \overline{\mathbb{R}}$ as follows:

$$F^{(n)}(\boldsymbol{U}) := \begin{cases} \int_{\Omega}^{\star} f^{(n)}\big(\boldsymbol{U}(\omega), \omega\big)\, \mathrm{d}\mathbb{P}(\omega) & \text{if } \boldsymbol{U} \text{ is } \mathcal{G}^{(n)}\text{-measurable,} \\ +\infty & \text{otherwise.} \end{cases} \tag{8.21}$$

In the same way, we define a function $F : L^0(\Omega, \mathcal{A}, \mathbb{P}; \mathbb{U}) \to \overline{\mathbb{R}}$ associated with the integrand f as follows:

$$F(\boldsymbol{U}) := \begin{cases} \int_{\Omega}^{\star} f\big(\boldsymbol{U}(\omega), \omega\big)\, \mathrm{d}\mathbb{P}(\omega) & \text{if } \boldsymbol{U} \text{ is } \mathcal{G}\text{-measurable,} \\ +\infty & \text{otherwise.} \end{cases} \tag{8.22}$$

We therefore have two discretization schemes. One scheme is used for noise discretization ($\{\boldsymbol{W}^{(n)}\}_{n\in\mathbb{N}}$), and it gives rise to a sequence of integrands ($\{f^{(n)}\}_{n\in\mathbb{N}}$). The second discretization scheme is applied to the measurability constraint, and it gives rise to a sequence of subfields ($\{\mathcal{G}^{(n)}\}_{n\in\mathbb{N}}$). When the sequence $\{f^{(n)}\}_{n\in\mathbb{N}}$ of integrands defined by Eq. (8.20) are normal integrands which are $\mathcal{A}$-quasi integrable, we consider also the sequence of normal integrands $g^{(n)} : \mathbb{U} \times \Omega \to \overline{\mathbb{R}}$ for $n \in \mathbb{N}$ defined as follows:

$$g^{(n)}(u, \omega) = \mathbb{E}\big(f^{(n)}(u, \cdot) \mid \mathcal{G}^{(n)}\big)(\omega), \tag{8.23}$$

as well as the normal integrand g defined by:

$$g(u, \omega) = \mathbb{E}\big(f(u, \cdot) \mid \mathcal{G}\big)(\omega). \tag{8.24}$$

To completely define the optimization problems that we consider, we have to choose a functional space for the controls. In the following, p is a fixed real number $1 \leq p < +\infty$ and control variables are restricted to belong to the functional space $L^p(\Omega, \mathcal{A}, \mathbb{P}; \mathbb{U})$.

With all these notations, we consider the sequence of optimization problems

$$\min_{\boldsymbol{U} \in L^p(\Omega, \mathcal{A}, \mathbb{P}; \mathbb{U})} F^{(n)}(\boldsymbol{U}), \quad \forall n \in \mathbb{N}, \tag{8.25}$$

and the original problem (8.1) restated as follows:

$$\min_{\boldsymbol{U} \in L^p(\Omega, \mathcal{A}, \mathbb{P}; \mathbb{U})} F(\boldsymbol{U}). \tag{8.26}$$

When considering problems reformulated as (8.25) and (8.26), convergence of the discretization scheme can be investigated through the Mosco convergence of the sequence $\{F^{(n)}\}_{n\in\mathbb{N}}$ as seen in Sect. 8.2. We first gather some technical assumptions.

Assumption 8.41

(i) The function $j : \mathbb{U} \times \mathbb{W} \to \overline{\mathbb{R}}$ is l.s.c. as a function of (u, w) and proper;
(ii) $j(\cdot, w)$ is a convex function for all $w \in \mathbb{W}$;
(iii) $j(u, w) \geq 0$ for all $(u, w) \in \mathbb{U} \times \mathbb{W}$;
(iv) there exists a sequence $\{\boldsymbol{U}^{(n)}\}_{n\in\mathbb{N}}$ with $\boldsymbol{U}^{(n)} \in L^p(\Omega, \mathcal{A}, \mathbb{P}; \mathbb{U})$ and functions $\boldsymbol{k}$ and $\boldsymbol{k}_0$ in $L^p(\Omega, \mathcal{A}, \mathbb{P}; \mathbb{R})$ such that $\|\boldsymbol{U}^{(n)}(\omega)\|_{\mathbb{U}} \leq \boldsymbol{k}(\omega)$ $\mathbb{P}$-a.s. and

$$j(\boldsymbol{U}^{(n)}, \boldsymbol{W}^{(n)}) \leq \boldsymbol{k}_0;$$

(v) the random variable $\boldsymbol{k}_0$ given by Assumption 8.41-(iv) is such that there exists $\boldsymbol{k}_1$ in $L^p(\Omega, \mathcal{A}, \mathbb{P}; \mathbb{R})$ such that we have that $\mathbb{E}(\boldsymbol{k}_0 \mid \mathcal{G}^{(n)}) \leq \boldsymbol{k}_1$ for all $n \in \mathbb{N}$.

The main convergence theorem is stated now, with a short proof depending on a set of propositions which are postponed to Sect. 8.5.5.

Theorem 8.42 *Let the function $j : \mathbb{U} \times \mathbb{W} \to \overline{\mathbb{R}}$ be given together with a random variable $\boldsymbol{W}$ and a subfield $\mathcal{G}$ of $\mathcal{A}$. We consider the function $f : \mathbb{U} \times \Omega \to \overline{\mathbb{R}}$ given by $f(u, \omega) = j\big(u, \boldsymbol{W}(\omega)\big)$ and the function $g : \mathbb{U} \times \Omega \to \overline{\mathbb{R}}$ given by Eq. (8.24). For a sequence $\{\boldsymbol{W}^{(n)}\}_{n\in\mathbb{N}}$ of random variables and a sequence $\{\mathcal{G}^{(n)}\}_{n\in\mathbb{N}}$ of subfields of $\mathcal{A}$, we define two sequences of integrands $\{f^{(n)}\}_{n\in\mathbb{N}}$ by (8.20) and $\{g^{(n)}\}_{n\in\mathbb{N}}$ by (8.23). Suppose that Assumptions 8.41 are satisfied together with the following assumptions:*

(i) the sequence of subfields $\{\mathcal{G}^{(n)}\}_{n\in\mathbb{N}}$ Kudo converges to $\mathcal{G}$;
(ii) the sequence $\{\boldsymbol{W}^{(n)}\}_{n\in\mathbb{N}}$ converges to $\boldsymbol{W}$ in probability;
(iii) for almost every $\omega \in \Omega$ and each $u \in \operatorname{dom} g(\cdot, \omega)$, we have that

$$g(u, \omega) \geq \big(\text{s-ls}_{\text{e}} g^{(n)}\big)(u, \omega);$$

(iv) the probability $\mathbb{P}$ is non-atomic.

Then, the sequence $\{F^{(n)}\}_{n\in\mathbb{N}}$ given by (8.21) Mosco converges to F given by (8.22) in $L^p(\Omega, \mathcal{A}, \mathbb{P}; \mathbb{U})$.

Proof First, we use the fact that Assumption 8.41-(iii) ensures Assumption (iii) of Proposition 8.56. Invoking Assumptions 8.42-(iii) and 8.41-(i), (iv), (v), we can use Proposition 8.56 to conclude that, for every function $\boldsymbol{U}$ in $L^p(\Omega, \mathcal{A}, \mathbb{P}; \mathbb{U})$, we have that $I_g(\boldsymbol{U}) \geq \big(\text{s-ls}_{\text{e}} I_{g^{(n)}}\big)(\boldsymbol{U})$. Second, by Assumption 8.41-(iv) there exists $\boldsymbol{U} \in L^p(\Omega, \mathcal{A}, \mathbb{P}; \mathbb{U})$ such that $I_f(\boldsymbol{U})$ is finite. This fact, together with Assumptions 8.41-(i), (ii), (iii) and 8.42-(ii), yields, by Proposition 8.60, that for every function $\boldsymbol{U}$ in $L^p(\Omega, \mathcal{A}, \mathbb{P}; \mathbb{U})$, we have that $I_f(\boldsymbol{U}) \leq \big(\text{w-li}_{\text{e}} I_{f^{(n)}}\big)(\boldsymbol{U})$. It is also noted that Assumption 8.41-(i), which is stronger than requiring that the integrand j is normal integrand, is requested in Proposition 8.60. The conclusion then follows from Proposition 8.51. □

8.5.2 Revisiting a Basic Example in Static Information

We illustrate Theorem 8.42 with the help of a specialized version of the example presented in Sect. 1.3.1. Our aim here is to illustrate the fact that σ-field discretization and random variable discretization can be done independently, thus avoiding pitfalls described in Sect. 6.2.2. Here $\Omega = [-1, 1]$ with its Borel σ-field and $\mathbb{P}$ is the uniform distribution on Ω. The random variable $\boldsymbol{W}$ is the identity function on Ω and thus $\boldsymbol{W}$ is a uniformly distributed random variable. We consider the random variable $\boldsymbol{Y} : \Omega \to \Omega$ defined by $\boldsymbol{Y}(\omega) = |\omega|$ and the subfield $\mathcal{G} = \sigma(\boldsymbol{Y})$ generated by the random variable $\boldsymbol{Y}$. Thus, $\mathcal{G}$-measurable functions are even functions. We consider the stochastic optimization problem (8.19) where the integrand j is defined by $j(u, w) = (u - w)^2$:

$$\min_{\boldsymbol{U} \in L^p(\Omega, \mathcal{G}, \mathbb{P}; \mathbb{R})} \mathbb{E}\Big(\big(\boldsymbol{U} - \boldsymbol{W}\big)^2\Big). \tag{8.27}$$

This problem is a specialized version of problem 1.4 for which $\boldsymbol{Y} = h(\boldsymbol{W})$ and $h(w) = |w|$.

Checking the technical assumptions 8.41 is left to the reader. We pay attention to the last three assumptions of Theorem 8.42 related to the discretization process.

Optimal Value and Optimal Solution

It was reviewed in Sect. 1.3.1 that the optimal solution of Problem (8.27) is given by the conditional expectation of $\boldsymbol{W}$ w.r.t. $\mathcal{G}$. Moreover, we can here explicitely compute conditional expectations w.r.t. the subfield $\mathcal{G}$ using [12, Lemma IV.32]. For a given random variable $\boldsymbol{H}$, we have that:

$$\mathbb{E}(\boldsymbol{H} \mid \mathcal{G}) = \frac{\boldsymbol{H} + \overline{\boldsymbol{H}}}{2} \quad \text{with} \quad \overline{\boldsymbol{H}}(\omega) := \boldsymbol{H}(-\omega). \tag{8.28}$$

The optimal solution is thus given by $\boldsymbol{U}^\sharp = (\boldsymbol{W} + \overline{\boldsymbol{W}})/2 \equiv 0$ (since $\boldsymbol{W} = \mathrm{I}_\Omega$) and the optimal cost is $1/3$.

Approximation Schemes

We approximate the subfield $\mathcal{G} = \sigma(\boldsymbol{Y})$ by the sequence $\{\mathcal{G}^{(n)}\}_{n\in\mathbb{N}}$ where

$$\mathcal{G}^{(n)} = \sigma\big(Q^{(n)}(\boldsymbol{Y})\big), \tag{8.29}$$

and where $\{Q^{(n)}\}_{n\in\mathbb{N}}$ is a sequence of piecewise constant functions $Q^{(n)} : [0, 1] \to [0, 1]$ defined as follows:

$$Q^{(n)}(y) = \sum_{k=1}^{n} \frac{k}{n} \mathbf{1}_{I^{k,(n)}}(y) \quad \text{and} \quad I^{k,(n)} = \left(\frac{k-1}{n}, \frac{k}{n}\right]. \tag{8.30}$$

The function $Q^{(n)}$ is a piecewise approximation of the identity function on [0, 1]. We approximate the noises using a sequence of random variables $\{W^{(n)}\}_{n\in\mathbb{N}}$ and we suppose that $\{W^{(n)}\}_{n\in\mathbb{N}}$ converges in L^2 to the random variable W (see Sect. 6.1 for the theory of quantization). We will also suppose that the discretizations of the noise and of the subfield are asymptotically linked by

$$\mathbb{E}\big((W^{(n)})^k \mid \mathcal{G}^{(n)}\big) \xrightarrow{\text{a.s.}} \mathbb{E}\big((W)^k \mid \mathcal{G}\big) \quad \text{for } k \in \{1, 2\} \tag{8.31}$$

The approximation scheme consists of replacing $\mathcal{G}$ and W by their discretized versions in problem (8.27):

$$\min_{U \preceq \mathcal{G}^{(n)}} \mathbb{E}\big(j(U, W^{(n)})\big). \tag{8.32}$$

Using the definition (8.29) of the σ-field $\mathcal{G}^{(n)}$, it is easy to check that admissible solutions for the discretized problem (8.32) are piecewise constant functions of the absolute value of the noise (see Proposition 3.46):

$$\{U \mid U \preceq \mathcal{G}^{(n)}\} = \Big\{U \;\Big|\; U = \sum_{k=1}^{n} u_k \mathbf{1}_{I^{k,(n)}}\big(|W|\big), \quad (u_1, \dots, u_n) \in \mathbb{R}^n\Big\}. \tag{8.33}$$

For a given random variable H, the conditional expectation w.r.t. the σ-field $\mathcal{G}^{(n)}$ can be explicitely computed. Using Definition B.5, we compute the argmin of $\mathbb{E}\left(\|H - U\|^2\right)$ for U given by Eq. 8.33, and we get:

$$\mathbb{E}(H \mid \mathcal{G}^{(n)}) = n \sum_{k=1}^{n} \mathbf{1}_{I^{k,(n)}}\big(|W|\big) \int_{I^{k,(n)}} \frac{H(\omega) + \overline{H}(\omega)}{2} \, \mathrm{d}\omega, \tag{8.34}$$

where the factor n comes from $1/n = \int_{I^{k,(n)}} \mathrm{d}\omega$. We therefore obtain an explicit optimal solution $U^{\sharp,(n)}$ of the discretized problem (8.32) given by:

$$U^{\sharp,(n)} = \mathbb{E}(W^{(n)} \mid \mathcal{G}^{(n)}) \tag{8.35a}$$

$$= n \sum_{k=1}^{n} \mathbf{1}_{I^{k,(n)}}\big(|W|\big) \int_{I^{k,(n)}} \frac{W^{(n)}(\omega) + \overline{W}^{(n)}(\omega)}{2} \, \mathrm{d}\omega. \tag{8.35b}$$

Kudo Convergence (Assumption (i) of Theorem 8.42)

We prove that $\{\mathcal{G}^{(n)}\}_{n\in\mathbb{N}}$ Kudo converges to $\mathcal{G}$. Consider $\{Y^{(n)}\}_{n\in\mathbb{N}}$ a sequence of random variables defined by $Y^{(n)} = Q^{(n)} \circ Y$. It is easy to check that the sequence of random variables $\{Y^{(n)}\}_{n\in\mathbb{N}}$ converges almost surely to the random variable Y and thus also converges in probability (see Sect. B.3.3). Moreover, a $\mathcal{G}^{(n)}$-measurable function is even and is therefore $\sigma(\mathcal{G})$-measurable. Thus, we have that $\sigma(\mathcal{G}^{(n)}) \subset \sigma(\mathcal{G})$. We obtain strong convergence of the sequence of subfields $\{\mathcal{G}^{(n)}\}_{n\in\mathbb{N}}$ by Proposition 8.16, ensuring Assumption 8.42-(i).

Noise Convergence (Assumption (ii) of Theorem 8.42)
If the sequence of approximated noises $\{\boldsymbol{W}^{(n)}\}_{n\in\mathbb{N}}$ converges in L^2, then it also converges in probability to $\boldsymbol{W}$ (see Sects. B.3.3 and B.3.5).

Assumption (iii) of Theorem 8.42
Using Eq. (8.3) with a constant sequence $\{u^{(n)}\}_{n\in\mathbb{N}}$ having value $u \in \mathbb{R}$, we obtain that $\big((\text{s-ls}_{\text{e}} g^{(n)})\big)(u,\omega) \leq \limsup_n g^{(n)}(u,\omega)$ $\mathbb{P}$-a.s.. Thus, in order to ensure Assumption 8.42-(iii), it is sufficient to prove that, for each fixed value of $u \in \mathbb{R}$, we have that:

$$\lim_{n\mapsto\infty} \mathbb{E}\big((u-\boldsymbol{W}^{(n)})^2 \mid \mathcal{G}^{(n)}\big) = \mathbb{E}\big((u-\boldsymbol{W})^2 \mid \mathcal{G}\big) \quad \mathbb{P}\text{-a.s.}.$$

This is indeed the case under the assumption given by Eq. (8.31).

Remark 8.43 Note that, using very similar lines, we can give a direct proof of the almost sure convergence of the optimal control $\boldsymbol{U}^{\sharp,(n)}$ in (8.35) to the optimal control $\boldsymbol{U}^{\sharp} = 0$ of problem (8.27). ◊

8.5.3 Discussion About Related Works

In Theorem 8.42, we have proposed an approximation scheme in which the discretization of the noise and the discretization of the informational constraint are done separately and which converges towards the optimal cost of the original problem, and so does the argmin. It is interesting to compare this approach with other approaches also taking into account the whole discretization process (noise and information) for stochastic optimal control problems.

Barty's Approach
In his PhD thesis [12], Barty proves the convergence of a discretization scheme for problem (8.1). The result he provides [12, Theorem IV.28] makes use of the strong convergence of σ-fields as in Theorem 8.42 and makes use of the convergence in distribution for the noise approximation. His approach involves two consecutive steps.

(i) The σ-field $\mathcal{G}$ is approximated by a sequence $\{\mathcal{G}^{(k)}\}_{k\in\mathbb{N}}$ of σ-fields. It is assumed that, for all $k \in \mathbb{N}$, we have that $\mathcal{G}^{(k)} \subset \mathcal{G}$ and each σ-field $\mathcal{G}^{(k)}$ is generated by a finite partition. Note that each σ-field $\mathcal{G}^{(k)}$ is a partition field (see Sect. 3.3.2) and the finite partition which generates $\mathcal{G}^{(k)}$ is the partition $\mathsf{part}(\mathcal{G}^{(k)})$ composed of the atoms of the partition field $\mathcal{G}^{(k)}$ (see Proposition 3.18). The problem (8.1) is approximated by the following sequence of optimization problems:

$$V^{(k)} = \min_{U \preceq \mathcal{G}^{(k)}} \mathbb{E}\big(j(U, W)\big). \tag{8.36}$$

The optimal value $V^{(k)}$ of problem (8.36) converges with k towards the optimal value of problem (8.1) as $\mathcal{G}^{(k)}$ strongly converges to $\mathcal{G}$ [12, Theorem IV.21]. In practical examples, Barty considers the case when the σ-field $\mathcal{G}$ is generated by the random variable $h(W)$, for a given function h, and a σ-field discretization is obtained by considering quantization approximation $\mathcal{G}^{(k)} = \sigma\big(Q^{(k)} \circ h(W)\big)$ (we thus have that $\mathcal{G}^{(k)} \subset \mathcal{G}$). The sequence $\{Q^{(k)} \circ h(W)\}_{k \in \mathbb{N}}$ of random variables is assumed to converge in probability to the random variable $h(W)$ (which ensures the strong convergence of σ-fields by Proposition 8.16), and the mappings $Q^{(k)}$ are assumed to take values in a finite set to provide σ-field generated by partitions. Note also that a $\mathcal{G}^{(k)}$-measurable random variable U is constant over each subset building up $\mathsf{part}(\mathcal{G}^{(k)})$: such a random variable U is characterized by a n_k-uple (where $n_k \in \mathbb{N}$ is the number of subset building up $\mathsf{part}(\mathcal{G}^{(k)})$) $\big(u_1, \dots, u_{n_k}\big) \in \mathbb{U}^{n_k}$, and the minimization in (8.36) is thus performed over the finite dimensional space $\mathbb{U}^{n_k}$.

(ii) For a given index k, the random variable W is approximated by a finitely valued random variable $W^{(n)}$ and problem (8.36) is replaced by:

$$V^{(n,k)} = \min_{U \preceq \mathcal{G}^{(k)}} \mathbb{E}\big(j(U, W^{(n)})\big). \tag{8.37}$$

The optimal value $V^{(n,k)}$ of problem (8.37) converges with n towards the optimal value $V^{(k)}$ of problem (8.36) as $W^{(n)}$ converges in distribution toward W [12, Theorem IV.26]. Note that this step only involves open-loop problems, which are approximated using the traditional Monte Carlo approach (see Sects. 8.4 and 2.5.3) and, as pointed out later in Remark 8.46, the convergence in distribution of the noises proves insufficient for closed-loop problems.

In this approach, the global discretization error $\big|V - V^{(n,k)}\big|$ is bounded from above by the sum of the term $\big|V - V^{(k)}\big|$, which accounts for an *information structure discretization error*, and the term $\big|V^{(k)} - V^{(k,n)}\big|$, which accounts for a *mean computation discretization error*.

Combining the two steps, we can find an increasing function $k : \mathbb{N} \to \mathbb{N}$ such that $V^{(n,k(n))}$ converges to V as n goes to infinity [12, Corollary IV.29]. In addition to the fact that the assumptions on the cost function were more stringent in Barty's approach than in Theorem 8.42, the provided results are convergence results of the value function and not convergence results of the control through epi-convergence.

Pennanen's Approach

An approach inspired by Pennanen's work [110] was described in Sect. 6.2.2. In this approach, the observation $\boldsymbol{Y}$ was assumed to be given as a function of the noise $\boldsymbol{W}$:

$$\boldsymbol{Y} = h(\boldsymbol{W}).$$

The discretization process is then mainly driven by noise approximations. The random variable $\boldsymbol{W}$ is approximated by a sequence of finitely valued random variables $\{\boldsymbol{W}^{(n)}\}_{n\in\mathbb{N}}$.

Then the σ-field $\mathcal{G} = \sigma\big(h(\boldsymbol{W})\big)$ is approximated by a sequence $\{\mathcal{G}^{(n)}\}_{n\in\mathbb{N}}$ of σ-fields given by $\mathcal{G}^{(n)} = \sigma\big(h(\boldsymbol{W}^{(n)})\big)$ (see (6.8) and (6.9)). The σ-field discretization is thus derived from the noise approximation. An additional requirement is, however, necessary: it is expressed as $h(\boldsymbol{W}^{(n)}) \preceq h(\boldsymbol{W})$ (see (6.10)), and it can be equivalently stated as $\mathcal{G}^{(n)} \subset \mathcal{G}$ for all $n \in \mathbb{N}$. As discussed in Sect. 6.2.2, this last requirement is important since an admissible solution of the discretized problem is also an admissible solution for the initial problem (8.1). Moreover, as far as convergence is concerned, this requirement has an implication in terms of σ-field convergence as shown by Proposition 8.16. Problem (8.1) is thus approximated by the following sequence of optimization problems:

$$V^{(n)} := \min_{\boldsymbol{U}\preceq\mathcal{G}^{(n)}} \mathbb{E}\big(j(\boldsymbol{U}, \boldsymbol{W}^{(n)})\big). \tag{8.38}$$

The approximation of the σ-field is intimately related to the approximation of the random variable in Pennanen's approach, requiring the additional Assumption (6.10), whereas these two approximations are designed *separately* in our approach in Sect. 8.5.1. An epi-convergence theorem when following Pennanen's approach is given in [110].

8.5.4 Revisiting the Example of Sect. 1.4.2

In this section, we revisit the example presented in Sect. 1.4.2. The optimization problem (1.8) can be reformulated as

$$\min_{\boldsymbol{U}\preceq\mathcal{G}} \mathbb{E}\big(j(\boldsymbol{U}, \boldsymbol{W})\big) \quad \text{with} \quad \mathcal{G} = \sigma\big(h(\boldsymbol{W})\big), \tag{8.39}$$

with $j\big(u, (w_0, w_1)\big) = \varepsilon u^2 + (w_0 + u + w_1)^2$ and $h\big((w_0, w_1)\big) = w_0$. It was shown that the optimal cost of problem (8.39) is given by Eq. (1.10), that we recall here

$$\min_{\boldsymbol{U}\preceq\mathcal{G}} \mathbb{E}\big(j(\boldsymbol{U}, \boldsymbol{W})\big) = \frac{1}{3}\frac{1+2\varepsilon}{1+\varepsilon} \approx \frac{1}{3}, \tag{8.40}$$

and the optimal control is given by:

$$U = -\frac{W_0}{(1+\epsilon)}. \tag{8.41}$$

Monte Carlo Sampling Fails

It was also shown in Sect. 1.4.2 that, with a naive Monte Carlo discretization method, one may end up with an admissible control for Problem (1.10) giving a cost of about $2/3$! This was demonstrated by numerical simulations since it was difficult to mathematically study the admissible control defined by Eq. (1.15) because of the randomness of the a_i's, which delimit the strips in Fig. 1.3. We now propose another discretization method, which is also based on Monte Carlo discretization, leading to the same cost of about $2/3$, but for which we are able to prove that the discretized noise converges in distribution, but not in probability, to the original noise as required in Theorem 8.42. A noticeable difference with the previous scheme is that, now, the boundaries of the strips (see Fig. 8.1) are deterministic.

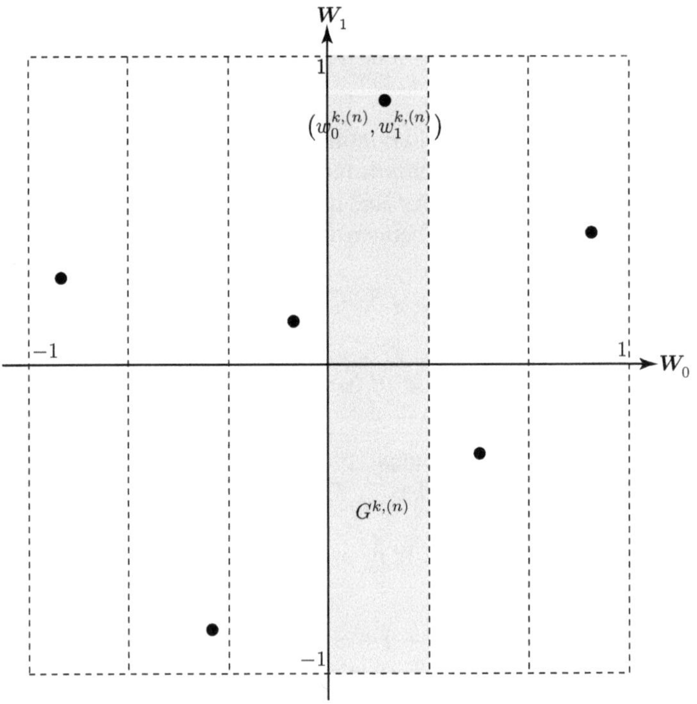

Fig. 8.1 Partition of $[-1, 1]^2$ and associated sample

Noise Discretization

A possible probability space associated with problem (8.39) is $\big(\Omega, \mathcal{B}^{o}_{\Omega}, \mathbb{P}\big)$, with $\Omega = [-1, 1]^2$ and where $\mathcal{B}^{o}_{\Omega}$ is the Borel σ-field on $[-1, 1]^2$ and $\mathbb{P}$ is the product of two uniform probability distributions on $[-1, 1]$. The random variables $\boldsymbol{W}_0$ and $\boldsymbol{W}_1$ are the two components of the identity application $\mathrm{I}_{[-1,1]^2}$ on $[-1, 1]^2$, the real valued control variable $\boldsymbol{U}$ being defined on $[-1, 1]^2$.

Let $\{\zeta^n\}_{n\in\mathbb{N}}$ be a sequence of elements in $[-1, 1]^2$, with $\zeta^n = \big(\zeta_1^n, \zeta_2^n\big)$, and let μ^n be the empirical probability distribution function associated with $(\zeta^1, \dots, \zeta^n)$

$$\mu^n := \frac{1}{n}\sum_{k=1}^{n} \delta_{\zeta^k},$$

where δ denotes the Dirac measure. We assume that the sequence $\{\mu^n\}_{n\in\mathbb{N}^*}$ of empirical probability distributions narrowly converges to the probability measure $\mathbb{P}$ (see Sect. B.3.4).

Remark 8.44 Such a sequence $\{\zeta^n\}_{n\in\mathbb{N}^*}$ is usually obtained as the realization of an infinite Monte carlo sample $\{\boldsymbol{\zeta}^n\}_{n\in\mathbb{N}^*}$ of i.i.d. random variables on $[-1, 1]^2$ with distribution $\mathbb{P}$. The narrow convergence assumption is then, almost surely, a consequence of the Glivenko-Cantelli Theorem [22]. ◊

For $n \in \mathbb{N}^\star$ and for any $k \in \{1, \dots, n\}$, we define

$$\big(w_0^{k,(n)}, w_1^{k,(n)}\big) = \left(\frac{2k-1}{n} - 1 + \frac{\zeta_1^k}{n}, \zeta_2^k\right) \tag{8.42}$$

and

$$\overline{G}^{k,(n)} = \left(\frac{2k-2}{n} - 1, \frac{2k}{n} - 1\right], \quad G^{k,(n)} = \overline{G}^{k,(n)} \times [-1, 1]. \tag{8.43}$$

By construction, $\big(G^{1,(n)}, \dots, G^{n,(n)}\big)$ is a partition of $[-1, 1]^2$, made of vertical strips as displayed in Fig. 8.1, and $\big(w_0^{k,(n)}, w_1^{k,(n)}\big) \in G^{k,(n)}$ for all $k \in \{1, \dots, n\}$.

We now derive a noise discretization scheme following the quantization approach of Chap. 6. Let $Q^{(n)} : [-1, 1]^2 \to [-1, 1]^2$ be the piecewise constant function taking value $\big(w_0^{k,(n)}, w_1^{k,(n)}\big)$ on the vertical strip $G^{k,(n)}$ for $k \in \{1, \dots, n\}$, that is:

$$Q^{(n)}\big(w_0, w_1\big) = \sum_{k=1}^{n} \big(w_0^{k,(n)}, w_1^{k,(n)}\big)\mathbf{1}_{G^{k,(n)}}\big(w_0, w_1\big).$$

We define the sequence $\big\{\big(\boldsymbol{W}_0^{(n)}, \boldsymbol{W}_1^{(n)}\big)\big\}_{n\in\mathbb{N}^*}$ of random variables by

$$\big(\boldsymbol{W}_0^{(n)}, \boldsymbol{W}_1^{(n)}\big) = Q^{(n)}\big(\boldsymbol{W}_0, \boldsymbol{W}_1\big). \tag{8.44}$$

According to this definition, the discretized random variable $\big(\mathbf{W}_0^{(n)}, \mathbf{W}_1^{(n)}\big)$ is constant over each subset $G^{k,(n)}$.

Lemma 8.45 *The sequence* $\big\{\big(\mathbf{W}_0^{(n)}, \mathbf{W}_1^{(n)}\big)\big\}_{n\in\mathbb{N}^*}$ converges in distribution *to* $\big(\mathbf{W}_0, \mathbf{W}_1\big)$ *as* $n \to +\infty$.

Proof Consider the joint cumulative distribution function $D^{(n)}$ of $\big(\mathbf{W}_0^{(n)}, \mathbf{W}_1^{(n)}\big)$:

$$D^{(n)}\big(w_0, w_1\big) = \mathbb{P}\Big(\mathbf{W}_0^{(n)} \leq w_0, \mathbf{W}_1^{(n)} \leq w_1\Big)$$
$$= \frac{1}{n}\sum_{k=1}^{n} \mathbf{1}_{[-1,w_0]\times[-1,w_1]}\big(w_0^{k,(n)}, w_1^{k,(n)}\big).$$

For a given $w_0 \in [-1, 1]$ and $n \in \mathbb{N}^*$, let k_0 be the index such that $w_0 \in \overline{G}^{k_0,(n)}$ (see (8.43)) and let ν_0 be equal to 0 if $w_0 \leq w_0^{k_0,(n)}$ and equal to 1 otherwise. Then, we have that

$$\begin{aligned} D^{(n)}\big(w_0, w_1\big) &= \frac{1}{n}\sum_{k=1}^{k_0-1} \mathbf{1}_{[-1,w_1]}\big(w_1^{k,(n)}\big) + \frac{\nu_0}{n}\mathbf{1}_{[-1,w_1]}\big(w_1^{k_0,(n)}\big) \\ &= \frac{k_0-1}{n}\left(\frac{1}{k_0-1}\sum_{k=1}^{k_0-1} \mathbf{1}_{[-1,w_1]}\big(w_1^{k,(n)}\big)\right) \\ &\quad + \frac{\nu_0}{n}\mathbf{1}_{[-1,w_1]}\big(w_1^{k_0,(n)}\big). \end{aligned} \tag{8.45}$$

Using the definition (8.43) of $\overline{G}^{k_0,(n)}$, we have that

$$\frac{k_0-1}{n} < \frac{w_0+1}{2} \leq \frac{k_0}{n}.$$

Thus, the index k_0 goes to infinity as n goes to infinity (for any $w_0 > -1$) and $(k_0-1)/n$ converges to $(w_0+1)/2$. The last term of Eq. (8.45) converges to zero since ν_0 is bounded by the constant 1 and, noting that $w_1^{k,(n)} = \zeta_2^k$, the term inside parentheses converges to the cumulative distribution of the uniform probability distribution on $[-1, 1]$. We thus conclude that $D^{(n)}\big(w_0, w_1\big)$ converges to $D\big(w_0, w_1\big) = (1+w_0)\,(1+w_1)/4$, the distribution function of $\mathbb{P}$, the uniform probability on the square $[-1, 1]^2$. □

Remark 8.46 Carrying on Remark 8.44, the sequence $\big\{\big(\mathbf{W}_0^{(n)}, \mathbf{W}_1^{(n)}\big)\big\}_{n\in\mathbb{N}^*}$ is usually a sequence of random variables based on the realization $\{\zeta^n\}_{n\in\mathbb{N}^*}$ of an i.i.d. sample. From Lemma 8.45, the sequence of associated probability distributions converges in distribution to $\mathbb{P}$. Note that this is precisely the condition required in [62] in order to ensure convergence when discretizing an open-loop stochastic optimization problem. However, the convergence in distribution proves insufficient for closed-loop problems as it was stated in Theorem 8.42. $\lozenge$

Information

Since $\boldsymbol{W}_0$ is the first component of $\mathrm{I}_{[-1,1]^2}$, the subfield $\sigma(\boldsymbol{W}_0)$ of $\mathcal{B}^{\mathrm{o}}_{[-1,1]^2}$ generated by the random variable $\boldsymbol{W}_0$ is the product

$$\mathcal{G} = \mathcal{B}^{\mathrm{o}}_{[-1,1]} \otimes \{\emptyset, [-1,1]\}.$$

For a given $n \in \mathbb{N}^\star$, we approximate $\mathcal{G}$ by the σ-field $\mathcal{G}^{(n)}$ generated by the partition $\left(G^{1,(n)}, \dots, G^{n,(n)}\right)$:

$$\mathcal{G}^{(n)} = \sigma\left(G^{1,(n)}, \dots, G^{n,(n)}\right). \tag{8.46}$$

From the definition of the subsets $G^{k,(n)}$, the inclusion $\mathcal{G}^{(n)} \subset \mathcal{G}$ holds true. As was already pointed out in Sect. 8.5.4, we note that the approximated information constraint "$\boldsymbol{U}$ is $\mathcal{G}^{(n)}$-measurable" is equivalent to "$\boldsymbol{U}$ is constant over each subset $G^{k,(n)}$", that is constant on each vertical strip of Fig. 8.1. Such a control variable $\boldsymbol{U}$ is thus parameterized by the values $u^{k,(n)}$ taken on each subset $G^{k,(n)}$:

$$\boldsymbol{U}\left(w_0, w_1\right) = \sum_{k=1}^{n} u^{k,(n)} \mathbf{1}_{G^{k,(n)}}\left(w_0, w_1\right). \tag{8.47}$$

Notice that $\mathbf{1}_{G^{k,(n)}}\left(w_0, w_1\right)$ does not depend actually upon w_1, and, therefore, $\boldsymbol{U}\left(w_0, w_1\right)$ depends only upon w_0.

Lemma 8.47 *The sequence* $\left\{\mathcal{G}^{(n)}\right\}_{n\in\mathbb{N}^*}$ *of σ-fields* strongly converges *to $\mathcal{G}$ as $n \to +\infty$.*

Proof Using Eq. (8.44), we obtain that the random variable $\boldsymbol{W}_0^{(n)}$ is constant over each subset $G^{k,(n)}$ with value $w_0^{k,(n)}$. Moreover, since the values $w_0^{k,(n)}$, $k = 1, \dots, n$, defined by (8.42) are all different, we obtain that $\mathcal{G}^{(n)} = \sigma\left(\boldsymbol{W}_0^{(n)}\right)$. Following Proposition 8.16, it is sufficient to show that $\boldsymbol{W}_0^{(n)} \longrightarrow \boldsymbol{W}_0$ in probability. This last claim follows from the definition of $w_0^{k,(n)}$ in Eq. (8.42). $\square$

Approximated Solution

Approximating problem (8.39) consists of replacing $\mathcal{G}$ and $\left(\boldsymbol{W}_0, \boldsymbol{W}_1\right)$ by their discretized versions $\mathcal{G}^{(n)}$ and $\left(\boldsymbol{W}_0^{(n)}, \boldsymbol{W}_1^{(n)}\right)$:

$$\min_{\boldsymbol{U} \preceq \mathcal{G}^{(n)}} \mathbb{E}\big(j(\boldsymbol{U}, \boldsymbol{W}^{(n)})\big). \tag{8.48}$$

The resulting function to be minimized is constant over each $G^{k,(n)}$, so that the approximated problem specializes as follows

$$\min_{\left(u^{1,(n)},\dots,u^{n,(n)}\right) \in \mathbb{R}^n} \sum_{k=1}^{n} \mathbb{P}\big(G^{k,(n)}\big) \left(\epsilon\big(u^{k,(n)}\big)^2 + \big(w_0^{k,(n)} + u^{k,(n)} + w_1^{k,(n)}\big)^2\right). \tag{8.49}$$

This optimization problem is deterministic, and we can get an explicit solution. Indeed, since $\mathbb{P}\big(G^{k,(n)}\big) > 0$ (equal to $1/n$), problem (8.49) splits up into n independent subproblems, namely

$$\min_{u^{k,(n)} \in \mathbb{R}} \epsilon\big(u^{k,(n)}\big)^2 + \big(w_0^{k,(n)} + u^{k,(n)} + w_1^{k,(n)}\big)^2. \tag{8.50}$$

The optimal solution of this quadratic minimization problem is

$$\widehat{u}^{k,(n)} = -\frac{w_0^{k,(n)} + w_1^{k,(n)}}{1+\epsilon}, \tag{8.51}$$

and the associated implementable control variable $\widehat{\boldsymbol{U}}^{(n)}$ is, according to (8.47),

$$\widehat{\boldsymbol{U}}^{(n)}\big(w_0, w_1\big) = -\sum_{k=1}^{n} \frac{w_0^{k,(n)} + w_1^{k,(n)}}{1+\epsilon} \mathbf{1}_{G^{k,(n)}}\big(w_0, w_1\big). \tag{8.52}$$

By construction, the approximated feedback $\widehat{\boldsymbol{U}}^{(n)}$ is $\mathcal{G}^{(n)}$-measurable. Since $\mathcal{G}^{(n)} \subset \mathcal{G}$, we conclude that $\widehat{\boldsymbol{U}}^{(n)}$ satisfies the measurability constraint of problem (8.39) as, of course, also does $\boldsymbol{U}^\sharp$ given by Eq. (8.41). We compare the cost $\widehat{J}^{(n)}$ induced by $\widehat{\boldsymbol{U}}^{(n)}$, namely

$$\widehat{J}^{(n)} = \mathbb{E}\big(\epsilon \widehat{\boldsymbol{U}}^{(n),2} + (\boldsymbol{W}_0 + \widehat{\boldsymbol{U}}^{(n)} + \boldsymbol{W}_1)^2\big), \tag{8.53}$$

to the true optimal cost $J^\sharp$ (see (8.40)) in order to evaluate the quality of the approximation, that is the optimality loss induced by $\widehat{\boldsymbol{U}}^{(n)}$ w.r.t. $\boldsymbol{U}^\sharp$.

Lemma 8.48 *The sequence* $\big\{\widehat{J}^{(n)}\big\}_{n \in \mathbb{N}}$ *in* (8.53) *is such that* $\widehat{J}^{(n)} \xrightarrow{a.s.} \dfrac{2}{3}$ *as* n *goes to infinity.*

Proof We have that

$$\widehat{J}^{(n)} = \int_{[-1,1]^2} \Big(\epsilon\big(\widehat{\boldsymbol{U}}^{(n)}(w_0, w_1)\big)^2 + \big(w_0 + \widehat{\boldsymbol{U}}^{(n)}(w_0, w_1) + w_1\big)^2\Big)\, \mathbb{P}\big(\mathrm{d}w_0\, \mathrm{d}w_1\big)$$

$$= \sum_{k=1}^{n} \int_{G^{k,(n)}} \left(\epsilon\left(\frac{w_0^{k,(n)} + w_1^{k,(n)}}{1+\epsilon}\right)^2 + \left(w_0 + w_1 - \frac{w_0^{k,(n)} + w_1^{k,(n)}}{1+\epsilon}\right)^2\right) \mathbb{P}\big(\mathrm{d}w_0\, \mathrm{d}w_1\big).$$

Remembering that $\mathbb{P}\big(G^{k,(n)}\big) = 1/n$ and expanding the last quadratic term in the previous expression leads to

$$\widehat{J}^{(n)} = \frac{2}{3} + \frac{1}{n}\sum_{k=1}^{n} \frac{\big(w_0^{k,(n)} + w_1^{k,(n)}\big)^2}{1+\epsilon} - 2\sum_{k=1}^{n}\left(\frac{w_0^{k,(n)} + w_1^{k,(n)}}{1+\epsilon}\right)\int_{G^{k,(n)}} (w_0 + w_1)\, \mathbb{P}\big(\mathrm{d}w_0\, \mathrm{d}w_1\big)$$

$$= \frac{2}{3} + \frac{1}{n}\sum_{k=1}^{n} \frac{\big(w_0^{k,(n)} + w_1^{k,(n)}\big)^2}{1+\epsilon} - \frac{2}{n}\sum_{k=1}^{n}\left(\frac{2k-1}{n} - 1\right)\left(\frac{w_0^{k,(n)} + w_1^{k,(n)}}{1+\epsilon}\right).$$

Using the convergence in distribution of $\big(\boldsymbol{W}_0^{(n)}, \boldsymbol{W}_1^{(n)}\big)_{n\in\mathbb{N}^*}$ to $\big(\boldsymbol{W}_0, \boldsymbol{W}_1\big)$ as $n \to +\infty$ (see Lemma 8.45), and the inequalities $\left|w_0^{k,(n)} - \big(\frac{2k-1}{n} - 1\big)\right| \leq \frac{1}{n}$ for every k, we obtain that the sum of the last two terms in the previous equality goes to zero as n goes to infinity, hence the result. □

From the expression (8.40) of the optimal cost $J^\sharp$ of problem (8.39), we deduce that the following inequality holds true for any $\epsilon > 0$:

$$\lim_{n\to+\infty} \widehat{J}^{(n)} > J^\sharp.$$

Remark 8.49 It is easy to prove that the optimal cost $\widetilde{J}^{(n)}$ of problem (8.49) is such that:

$$\lim_{n\to+\infty} \widetilde{J}^{(n)} = \frac{2}{3}\left(\frac{\epsilon}{1+\epsilon}\right).$$

Once again, this limit is different from $J^\sharp$ and goes to zero as ϵ goes to zero. Hence, this "apparent" cost is very optimistic, whereas the real cost induced by the solution of the approximated problem (which is respectful of the information constraint) is indeed very bad. This is another illustration that problem (8.49) is not a valid approximation of problem (8.39). Although the *criterion* in (8.49) looks like a good approximation of the one in (8.39), the two corresponding *optimization problems* are definitely different. The minimization in problem (8.49) yields a solution $\widehat{u}^{k,(n)} = -(w_0^{k,(n)} + w_1^{k,(n)})/(1+\epsilon)$ depending on both $w_0^{k,(n)}$ and $w_1^{k,(n)}$. The computation of $\widetilde{J}^{(n)}$ therefore corresponds to the numerical integration of (8.39) using the feedback law $\widetilde{\boldsymbol{U}}(w_0, w_1) = -(w_0 + w_1)/(1+\epsilon)$, which is not $\sigma(\boldsymbol{W}_0)$-measurable, hence the gap with $J^\sharp$. ◊

As in Sect. 1.4.2, we thus conclude that the discretization scheme (8.48)—with $\mathcal{G}^{(n)}$ defined by Eq. (8.46) and $\boldsymbol{W}^{(n)}$ defined by Eq. (8.44)—fails to asymptotically provide the optimal solution of problem (8.39).

What Has Gone Wrong

Using Proposition 8.36, we obtain the following equivalent form of problem (8.39)

$$\mathbb{E}\Big(\min_{u\in\mathbb{R}} \mathbb{E}\big(\epsilon u^2 + (\boldsymbol{W}_0 + u + \boldsymbol{W}_1)^2 \mid \boldsymbol{W}_0\big)\Big). \tag{8.54}$$

Equation (8.50) should have been interpreted as a discretized version of the inner conditional expectation appearing in Eq. (8.54) when $\boldsymbol{W}_0$ is equal to $w_0^{k,(n)}$. But, it appears that, in the discretization scheme we have devised in (8.44)–(8.47) and (8.48) and leading to Eq. (8.50), the conditional expectation is approximated using a sample of $\boldsymbol{W}_1$ of length one (this fact remains true, even when n goes to infinity) which is namely a really poor way of evaluating this conditional expectation.

The discretization scheme defined by (8.44)–(8.47) and (8.48) is such that each subproblem derived from (8.49) is optimized using a *unique* sample of the noise random variable.

If we revisit the assumptions of Theorem 8.42, we realize that Assumption 8.42-(ii) is not satisfied: the noise convergence notion provided by the example is significantly weaker that the one required by the theorem. Let us show that the convergence of the sequence $\{(\boldsymbol{W}_0^{(n)}, \boldsymbol{W}_1^{(n)})\}_{n\in\mathbb{N}}$ towards $(\boldsymbol{W}_0, \boldsymbol{W}_1)$ *does not hold in probability*.

Indeed, consider some number $\tau > 0$, consider the norm $\|(w_0, w_1)\| = \max\{|w_0|, |w_1|\}$ on $[-1,1]^2$, and let $A^{(n)}$ be the subset of $[-1,1]^2$ defined by:

$$A^{(n)} = \Big\{w \in [-1,1]^2 \;\Big|\; \Big\|\big(\boldsymbol{W}_0^{(n)}, \boldsymbol{W}_1^{(n)}\big)(w) - \big(\boldsymbol{W}_0, \boldsymbol{W}_1\big)(w)\Big\| \leq \tau\Big\}.$$

Since $(\boldsymbol{W}_0, \boldsymbol{W}_1) = \mathrm{I}_{[-1,1]^2}$ and since $(\boldsymbol{W}_0^{(n)}, \boldsymbol{W}_1^{(n)})$ is constant over each $G^{k,(n)}$, the subset $A^{(n)}$ can be expressed as the disjoint union of n subsets $A^{k,(n)}$, with

$$\begin{aligned}A^{k,(n)} &= A^{(n)} \cap G^{k,(n)} \\ &= \Big\{w \in G^{k,(n)} \;\Big|\; \max\Big\{\Big|w_0^{k,(n)} - w_0\Big|, \Big|w_1^{k,(n)} - w_1\Big|\Big\} \leq \tau\Big\}.\end{aligned}$$

From the definitions of $G^{k,(n)}$ and of $A^{(n)}$, the subset $A^{k,(n)}$ is included in a $\frac{2}{n} \times 2\tau$ rectangle. We thus obtain $\mathbb{P}\big(A^{k,(n)}\big) \leq \frac{\tau}{n}$, and then $\mathbb{P}\big(A^{(n)}\big) \leq \tau$ by summation. This proves that

$$\mathbb{P}\left(\left\|\left(\boldsymbol{W}_0^{(n)}, \boldsymbol{W}_1^{(n)}\right) - \left(\boldsymbol{W}_0, \boldsymbol{W}_1\right)\right\| > \tau\right) \geq 1 - \tau \quad \forall n \in \mathbb{N}.$$

Therefore, this prevents convergence in probability.

As in Sect. 1.4.2, we thus conclude that the discretization scheme (8.48)—with $\mathcal{G}^{(n)}$ defined by Eq. (8.46) and $\boldsymbol{W}^{(n)}$ defined by Eq. (8.44)—fails to asymptotically provide the optimal solution of problem (8.39).

What has gone wrong in our discretization example (8.48) is now clear: although the discretizations of the noise and the information are a priori unrelated, we have chosen to bind them in a very specific way; however, with this particular binding, one of the assumptions of Theorem 8.42, namely the convergence in distribution of the noise discretization is not fulfilled.

A Convergent Discretization Scheme

We ultimately illustrate how a direct application of Theorem 8.42, and thus the use of a stronger convergence notion for the noise, leads to a positive convergence result for the optimization problem (8.39). We do not change the information discretization defined by Eq. (8.46) since it satisfies strong convergence. The information discretization leads to the vertical strips $G^{k,(n)}$. In order to discretize the noise $\boldsymbol{W}$, we appeal now to the theory of quantization developed in Chap. 6, and introduce the Voronoi cells $C^{k,(n)}$ around the centroids $(w_0^{k,(n)}, w_1^{k,(n)})$ (see Fig. 8.2). The discretized random variable $\big(\boldsymbol{W}_0^{(n)}, \boldsymbol{W}_1^{(n)}\big)$ is, accordingly, defined by

$$\left(\boldsymbol{W}_0^{(n)}, \boldsymbol{W}_1^{(n)}\right)(w_0, w_1) = \sum_{k=1}^{n} \left(w_0^{k,(n)}, w_1^{k,(n)}\right)\mathbf{1}_{C^{k,(n)}}(w_0, w_1). \tag{8.55}$$

The optimal Voronoi tessellation is based on the L^2 norm. We suppose that the diameters of the cells go to zero as n goes to infinity, so that the discretized noise $\{(\boldsymbol{W}_0^{(n)}, \boldsymbol{W}_1^{(n)})\}_{n\in\mathbb{N}}$ converges in $L^2\big([-1,1]^2, \mathcal{B}^{\mathrm{o}}_{[-1,1]^2}, \mathbb{P}; [-1,1]^2\big)$ norm and therefore converges in probability (see Sect. B.3.5).

The problem to be solved as an approximation of Problem (8.39) is given by (8.55), (8.47) and (8.48), and it again splits into n open-loop subproblems. Denoting by $\pi_{k,l} = \mathbb{P}(G^{k,(n)} \cap C^{l,(n)})$ the probability weight of the subset $G^{k,(n)} \cap C^{l,(n)}$, the k-th subproblem writes

$$\min_{u^{k,(n)}\in\mathbb{R}} \sum_{l=1}^{n} \pi_{k,l} \left(\epsilon\big(u^{k,(n)}\big)^2 + \big(w_0^{l,(n)} + u^{k,(n)} + w_1^{l,(n)}\big)^2\right). \tag{8.56}$$

Note that the discretization scheme proposed here is similar to the first one used in Sect. 6.2.3.

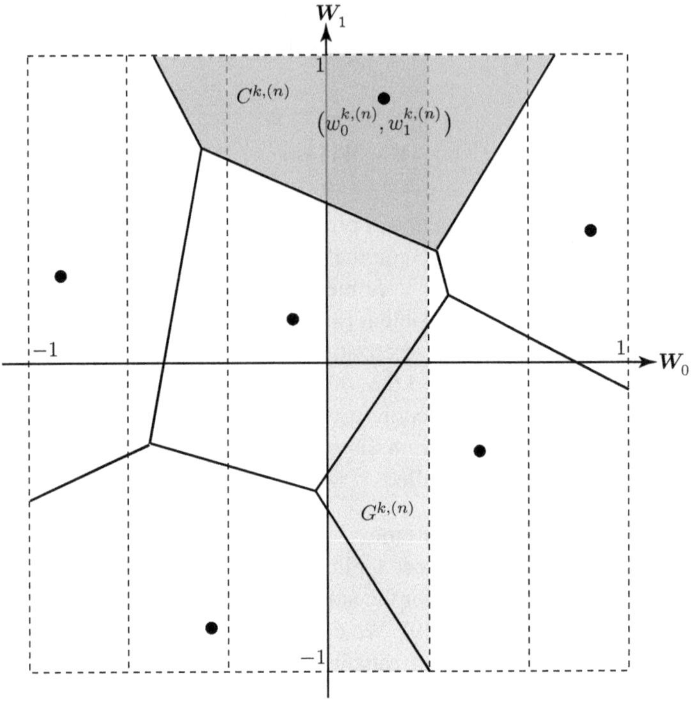

Fig. 8.2 Noise discretization induced by Voronoi tessellation

As n goes to infinity for a fixed value of k the width of any strip $G^{k,(n)}$ goes to zero whereas its height remains equal to 1, and the diameter of any cell $C^{l,(n)}$ goes to zero. Thus, we intuitively expect that for any k, the number of nonempty subsets $\{G^{k,(n)} \cap C^{l,(n)}\}_{l=1,\dots,n}$ goes to infinity as n goes to infinity. Therefore, each optimal value $\widehat{u}^{k,(n)}$ is computed using a large (in fact, asymptotically infinite) number of samples $(w_0^{l,(n)}, w_1^{l,(n)})$. This drastic difference with the approximation scheme illustrated by Fig. 8.1, where each optimal value $\widehat{u}^{k,(n)}$ is computed using one sample, explains the success of the approximation (8.56).

Remark 8.50 A distinctive feature of the discretization scheme (8.56) is that the same sample $(w_0^{l,(n)}, w_1^{l,(n)})$ may enter in the computation of several control values $\widehat{u}^{k,(n)}$. This observation shows that such a discretization scheme cannot be represented by a scenario tree. In fact, in a tree, a control value is attached to a node and the subtree hanging at that node is not shared by any other node at the same level (time instant) of the tree. ◊

We end up this section with remarks on the information discretization. In problem (8.39), the information σ-field of the original problem is generated by a function of the noise $\mathcal{G} = \sigma\big(h(\boldsymbol{W})\big)$. We have used an information discretization based on the strips $G^{k,(n)}$, that is,

$$\mathcal{G}^{(n)} = \sigma(G^{1,(n)}, \ldots, G^{n,(n)}).$$

We first come back to the naive Monte Carlo non convergent scheme used for solving problem (8.39). In this scheme, defined by Eqs. (8.44), (8.47) and (8.48), we have that $\mathcal{G}^{(n)} = \sigma(h(\boldsymbol{W}^{(n)}))$ and we have that $\mathcal{G}^{(n)} \subset \mathcal{G}$. Thus, Pennanen's Assumption (6.10) is satisfied by the naive Monte Carlo non convergent discretization scheme.

We now consider the convergent discretization scheme (8.55), (8.47) and (8.48) which leads to Eq. (8.56). Since the random variable $h(\boldsymbol{W}_0^{(n)}, \boldsymbol{W}_1^{(n)})$ is constant over the cells $\{C^{k,(n)}\}_{k\in 1,\ldots,n}$, we note that the subfield $\widetilde{\mathcal{G}}^{(n)} = \sigma\big(h(\boldsymbol{W}_0^{(n)}, \boldsymbol{W}_1^{(n)})\big)$ generated by the noise discretization is such that

$$\widetilde{\mathcal{G}}^{(n)} \subset \sigma\big(C^{1,(n)}, \ldots, C^{n,(n)}\big). \tag{8.57}$$

Moreover, if we assume that the first coordinates of the centroids $(w_0^{k,(n)}, w_1^{k,(n)})$ are all distinct, which is likely to be the case in practice, then the previous inclusion turns out to be an equality. Using Eq. (8.57), it is clear that $\widetilde{\mathcal{G}}^{(n)} \not\subset \mathcal{G}$. Moreover, the sequence of σ-fields $\{\widetilde{\mathcal{G}}^{(n)}\}_{n\in\mathbb{N}}$ Kudo converges to the σ-field $\mathcal{A}$ rather than to the σ-field $\mathcal{G}$. Thus, the subfield $\widetilde{\mathcal{G}}^{(n)}$ deduced from the noise discretization is not a good candidate for approximating problem (8.39). As it was already discussed in Sect. 6.2.3, we cannot obtain a quantization of the observation using $h(\boldsymbol{W}_0^{(n)}, \boldsymbol{W}_1^{(n)})$; we need to use more involved schemes as described in Sect. 6.2.3.

8.5.5 Companion Propositions to Theorem 8.42

Here, we give a set of propositions and theorems used in the proof of Theorem 8.42.

We recall that the notation $F^{(n)}(\boldsymbol{U})$ (see (8.21)) stands for:

$$F^{(n)}(\boldsymbol{U}) := \begin{cases} \displaystyle\int_{\Omega}^{\star} f^{(n)}\Big(\boldsymbol{U}(\omega), \omega\Big)\, \mathrm{d}\mathbb{P}(\omega) & \text{if } \boldsymbol{U} \text{ is } \mathcal{G}^{(n)}\text{-measurable,} \\ +\infty & \text{otherwise.} \end{cases} \tag{8.58}$$

For a sequence $\{f^{(n)}\}_{n\in\mathbb{N}}$ of $\mathcal{A}$-quasi integrable integrands, the sequence $\{g^{(n)}\}_{n\in\mathbb{N}}$ defined by Eq. (8.23) is well defined and each $g^{(n)}$ is a $\mathcal{G}^{(n)}$-normal integrand by Theorem 8.30.

In order to prove that the sequence $\{F^{(n)}\}_{n\in\mathbb{N}}$ Mosco converges to F in $L^p(\Omega, \mathcal{A}, \mathbb{P}; \mathbb{U})$, we use Proposition 8.7. We first start by proving that the Assumption 8.7-(i) for the sequence $\{F^{(n)}\}_{n\in\mathbb{N}}$ can be deduced from a similar assumption on the sequence $\{I_{g^{(n)}}\}_{n\in\mathbb{N}}$ and, similarly, the Assumption 8.7-(ii) for the sequence $\{F^{(n)}\}_{n\in\mathbb{N}}$ can be deduced from a similar assumption on the sequence $\{I_{f^{(n)}}\}_{n\in\mathbb{N}}$. More precisely, we have the following proposition, the proof of which relies on two postponed Propositions 8.53 and 8.54.

Proposition 8.51 *Let $\{f^{(n)}\}_{n\in\mathbb{N}}$ be a sequence of convex normal integrands and f be a convex normal integrand. Suppose that the sequence of σ-fields $\{\mathcal{G}^{(n)}\}_{n\in\mathbb{N}}$ converges in the Kudo sense to $\mathcal{G}$. For any random variable $\boldsymbol{U} \in L^p(\Omega, \mathcal{A}, \mathbb{P}; \mathbb{U})$ with $1 \leq p < +\infty$, if*

$$\left(\text{s-ls}_e I_{g^{(n)}}\right)(\boldsymbol{U}) \leq I_g(\boldsymbol{U}) \text{ and } I_f(\boldsymbol{U}) \leq \left(\text{w-li}_e I_{f^{(n)}}\right)(\boldsymbol{U}), \tag{8.59}$$

then the sequence $\{F^{(n)}\}_{n\in\mathbb{N}}$ Mosco converges to F in $L^p(\Omega, \mathcal{A}, \mathbb{P}; \mathbb{U})$.

Proof The proof easily follows from Propositions 8.53 and 8.54. □

We now introduce a lemma that states that, when restricted to $\mathcal{G}^{(n)}$-measurable arguments, the three functions $I_{g^{(n)}}$, $I_{f^{(n)}}$ and $F^{(n)}$ are identical.

Lemma 8.52 *Let $\{f^{(n)}\}_{n\in\mathbb{N}}$ be a sequence of $\mathcal{A}$-normal integrands fulfilling conditions of Theorem 8.30. We consider the sequence $\{I_{f^{(n)}}\}_{n\in\mathbb{N}}$ of integral functions. If $\boldsymbol{U} \in L^0(\Omega, \mathcal{A}, \mathbb{P}; \mathbb{U})$ is $\mathcal{G}^{(n)}$-measurable, then we have that $I_{f^{(n)}}(\boldsymbol{U}) = I_{g^{(n)}}(\boldsymbol{U}) = F^{(n)}(\boldsymbol{U})$, where $g^{(n)}$ and $F^{(n)}$ are defined by (8.23) and (8.21).*

Proof If $\boldsymbol{U}$ is $\mathcal{G}^{(n)}$-measurable, then $I_{f^{(n)}}(\boldsymbol{U}) = F^{(n)}(\boldsymbol{U})$ by definition of $F^{(n)}$. Using Theorem 8.30, we obtain that $I_{f^{(n)}}(\boldsymbol{U}) = I_{g^{(n)}}(\boldsymbol{U})$. □

Proposition 8.53, given now, links Assumption 8.7-(ii) for the sequence of extended functions $\{F^{(n)}\}_{n\in\mathbb{N}}$ to Assumption 8.7-(ii) for the sequence of integrands $\{f^{(n)}\}_{n\in\mathbb{N}}$.

Proposition 8.53 *Let $\{f^{(n)}\}_{n\in\mathbb{N}}$ be a sequence of normal integrands and let $\{\mathcal{G}^{(n)}\}_{n\in\mathbb{N}}$ be a sequence of subfields which converges in the Kudo sense to $\mathcal{G}$. For a given random variable $\boldsymbol{U} \in L^p(\Omega, \mathcal{A}, \mathbb{P}; \mathbb{U})$, we consider the sequence of integral functions $\left\{I_{f^{(n)}}(\boldsymbol{U})\right\}_{n\in\mathbb{N}}$. Let f be a normal integrand such that $I_f(\boldsymbol{U}) \leq \left(\tau\text{-li}_e I_{f^{(n)}}\right)(\boldsymbol{U})$. Then we have that $F(\boldsymbol{U}) \leq \left(\tau\text{-li}_e F^{(n)}\right)(\boldsymbol{U})$ where the functions $\{F^{(n)}\}_{n\in\mathbb{N}}$ and F are defined by Eqs. (8.58) and (8.22) and τ can be the strong or weak topology in $L^p(\Omega, \mathcal{A}, \mathbb{P}; \mathbb{U})$ with $1 \leq p < +\infty$.*

Proof The result is obvious if $\left(\tau\text{-li}_e F^{(n)}\right)(\boldsymbol{U}) = +\infty$. Consider a fixed real number $\epsilon > 0$. If $\left(\tau\text{-li}_e F^{(n)}\right)(\boldsymbol{U}) < +\infty$, we can find a sequence of random variables $\{\boldsymbol{U}^{(n)}\}_{n\in\mathbb{N}}$ which τ-converges to $\boldsymbol{U}$ and a subsequence $\nu(n)$ for which the sequence $\{F^{(\nu(n))}(\boldsymbol{U}^{(\nu(n))})\}_{n\in\mathbb{N}}$ has a limit when n converges to $+\infty$ satisfying:

$$\lim_{n\to\infty} F^{(\nu(n))}(\boldsymbol{U}^{(\nu(n))}) \leq \left(\tau\text{-li}_e F^{(n)}\right)(\boldsymbol{U}) + \epsilon.$$

We start by proving that we can assume that $\boldsymbol{U}^{(\nu(n))}$ is $\mathcal{G}^{(\nu(n))}$-measurable. Indeed, if there is a subsequence $\{\beta(n)\}$ of $\{\nu(n)\}$ for which $\boldsymbol{U}^{(\beta(n))}$ is not $\mathcal{G}^{(\beta(n))}$-measurable, the result is obvious since we would have for all $n \in \mathbb{N}$ that $F^{(\beta(n))}(\boldsymbol{U}^{(\beta(n))}) = +\infty$ (see (8.58)). Assuming that $\boldsymbol{U}^{(\nu(n))}$ is $\mathcal{G}^{(\nu(n))}$-measurable, we have that

$F^{(\nu(n))}(\boldsymbol{U}^{(\nu(n))}) = I_{f^{(\nu(n))}}(\boldsymbol{U}^{(\nu(n))})$ and consequently $I_f(\boldsymbol{U}) \leq \big(\tau\text{-li}_e F^{(n)}\big)(\boldsymbol{U}) + \epsilon$. Using [118, Theorem 2.3], the conditional expectation $\mathbb{E}\big(\boldsymbol{U}^{(\nu(n))} \mid \mathcal{G}^{(\nu(n))}\big)$ τ-converges to $\mathbb{E}(\boldsymbol{U} \mid \mathcal{G})$, the result being true for the strong or weak topology of $L^p(\Omega, \mathcal{A}, \mathbb{P}; \mathbb{U})$ when $1 \leq p < +\infty$. Since $\boldsymbol{U}^{(\nu(n))}$ is $\mathcal{G}^{(\nu(n))}$-measurable, we have that $\boldsymbol{U}^{(\nu(n))} = \mathbb{E}\big(\boldsymbol{U}^{(\nu(n))} \mid \mathcal{G}^{(\nu(n))}\big)$ and we obtain that $\boldsymbol{U}$ is $\mathcal{G}$-measurable. Using this last assertion we obtain that $I_f(\boldsymbol{U}) = F(\boldsymbol{U})$. □

The Proposition 8.54, given now, links Assumption 8.7-(i) for the sequence of extended functions $\{F^{(n)}\}_{n\in\mathbb{N}}$ to Assumption 8.7-(i) for the sequence of integrands $\{g^{(n)}\}_{n\in\mathbb{N}}$.

Proposition 8.54 *Let $\{f^{(n)}\}_{n\in\mathbb{N}}$ be a sequence of convex normal integrands and f be a convex normal integrand. Suppose that the sequence of σ-fields $\{\mathcal{G}^{(n)}\}_{n\in\mathbb{N}}$ converges in the Kudo sense to $\mathcal{G}$. For any random variable $\boldsymbol{U} \in L^p(\Omega, \mathcal{A}, \mathbb{P}; \mathbb{U})$ with $1 \leq p < +\infty$, if the random variable $\boldsymbol{U}$ satisfies the inequality $\big(\tau\text{-ls}_e I_{g^{(n)}}\big)(\boldsymbol{U}) \leq I_g(\boldsymbol{U})$, then we have that $\big(\tau\text{-ls}_e F^{(n)}\big)(\boldsymbol{U}) \leq F(\boldsymbol{U})$.*

Proof The only case to consider is when $\boldsymbol{U}$ is $\mathcal{G}$-measurable. Since the random variable $\boldsymbol{U}$ satisfies

$$\big(\tau\text{-ls}_e I_{g^{(n)}}\big)(\boldsymbol{U}) \leq I_g(\boldsymbol{U}),$$

we can find a sequence $\{\boldsymbol{U}^{(n)}\}_{n\in\mathbb{N}}$ of random variables in $L^p(\Omega, \mathcal{A}, \mathbb{P}; \mathbb{U})$ such that $\limsup_{n\to\infty} I_{g^{(n)}}(\boldsymbol{U}^{(n)}) \leq I_g(\boldsymbol{U})$. Since $\boldsymbol{U}$ is $\mathcal{G}$-measurable, we have, by Lemma 8.52, that $I_g(\boldsymbol{U}) = I_f(\boldsymbol{U}) = F(\boldsymbol{U})$. Let $\boldsymbol{V}^{(n)} = \mathbb{E}(\boldsymbol{U}^{(n)} \mid \mathcal{G}^{(n)})$; the sequence $\{\boldsymbol{V}^{(n)}\}_{n\in\mathbb{N}}$ converges to $\mathbb{E}(\boldsymbol{U} \mid \mathcal{G}) = \boldsymbol{U}$ in $L^p(\Omega, \mathcal{A}, \mathbb{P}; \mathbb{U})$. Using again Lemma 8.52 and Proposition 8.38, we have that $F^{(n)}(\boldsymbol{V}^{(n)}) = I_{g^{(n)}}(\boldsymbol{V}^{(n)}) \leq I_{g^{(n)}}(\boldsymbol{U}^{(n)})$ and thus:

$$\limsup_{n\mapsto\infty} F^{(n)}(\boldsymbol{V}^{(n)}) \leq \limsup_{n\mapsto\infty} I_{g^{(n)}}(\boldsymbol{U}^{(n)}) \leq I_g(\boldsymbol{U}) = F(\boldsymbol{U}). \tag{8.60}$$

This ends the proof. □

Inequality for the s-Epigraphical Upper Limit

We have established in Proposition 8.51 that sufficient conditions for the Mosco convergence of the sequence $\{F^{(n)}\}_{n\in\mathbb{N}}$ to F are given by Eq. (8.59). In this paragraph, we provide results in order to derive the first inequality of Eq. (8.59), that is, $\big(\text{s-ls}_e I_{g^{(n)}}\big)(\boldsymbol{U}) \leq I_g(\boldsymbol{U})$ where the sequence of integrands $\{g^{(n)}\}_{n\in\mathbb{N}}$ is given by (8.23). The proof is based on [44, Theorem 4.1] in which the space $\mathbb{U}$ may be assumed to be a separable Banach space. In the special case when $\mathbb{U} = \mathbb{R}^n$, similar results were proved in [88]. We first recall a key result [44, Theorem 4.1].

Theorem 8.55 *Let $\mathbb{U}$ be a separable Banach space and $\{\zeta^{(n)}\}_{n\in\mathbb{N}}$ be a sequence of proper normal integrands ($\zeta^{(n)} : \mathbb{U} \times \Omega \to (-\infty, +\infty]$) and $\zeta : \mathbb{U} \times \Omega \to (-\infty, +\infty]$ be a proper integrand and let p, q be numbers with $1 \leq p < +\infty$, and $p^{-1} + q^{-1} = 1$, satisfying the following assumptions:*

(i) *for almost every* $\omega \in \Omega$ *and each* $u \in \operatorname{dom} \zeta(\cdot, \omega)$, *we have that*

$$\zeta(u, \omega) \geq \left(\text{s-ls}_{\text{e}} \zeta^{(n)}\right)(u, \omega);$$

(ii) *there exist a sequence* $\{\boldsymbol{U}^{(n)}\}_{n \in \mathbb{N}}$ *in* $L^p(\Omega, \mathcal{A}, \mathbb{P}; \mathbb{U})$ *and functions* $\boldsymbol{k}$ *and* $\boldsymbol{k}_0$ *in* $L^p(\Omega, \mathcal{A}, \mathbb{P}; \mathbb{R})$ *such that for each* n, $\|\boldsymbol{U}^{(n)}(\omega)\| \leq \boldsymbol{k}(\omega)$ *and*

$$\zeta^{(n)}(\boldsymbol{U}^{(n)}(\omega), \omega) \leq \boldsymbol{k}_0(\omega) \text{ a.s.};$$

(iii) *for almost every* $\omega \in \Omega$, *each* $u \in \mathbb{U}$ *and each* $n \geq 1$, $\zeta^{(n)}(u, \omega) \geq -h(\omega)\|u\| - h_0(\omega)$, *where* h *and* h_0 *belong to* $L^q(\Omega, \mathcal{A}, \mathbb{P}; \mathbb{R})$ *with* $h(\omega) > 0$ *a.s.*

Then, for every function $\boldsymbol{U}$ *in* $L^p(\Omega, \mathcal{A}, \mathbb{P}; \mathbb{U})$, *we have that* $\left(\text{s-ls}_{\text{e}} I_{\zeta^{(n)}}\right)(\boldsymbol{U}) \leq I_\zeta(\boldsymbol{U})$.

We now relate general results of Theorem 8.55 to the problem formulation used in Theorem 8.42 in which the integrands $f^{(n)}$ are given by $f^{(n)}(u, \omega) = j\left(u, \boldsymbol{W}^{(n)}(\omega)\right)$. In the next proposition, $\{\mathcal{G}^{(n)}\}_{n \in \mathbb{N}}$ is a given sequence of σ-fields which are used to define a sequence $\{g^{(n)}\}_{n \in \mathbb{N}}$ of functions from a sequence $\{f^{(n)}\}_{n \in \mathbb{N}}$ by Eq. (8.23). In a similar way, $\mathcal{G}$ is a given σ-field used to define integrand g from integrand f.

Proposition 8.56 *For a sequence* $\{\boldsymbol{W}^{(n)}\}_{n \in \mathbb{N}}$ *of random variables, we define a sequence* $\{f^{(n)}\}_{n \in \mathbb{N}}$ *by* $f^{(n)}(u, \omega) = j(u, \boldsymbol{W}^{(n)}(\omega))$ *for all* $n \in \mathbb{N}$. *In a similar way, for a given random variable* $\boldsymbol{W}$, *we define the integrand* $f(u, \omega) = j\left(u, \boldsymbol{W}(\omega)\right)$. *Let* $\{g^{(n)}\}_{n \in \mathbb{N}}$ *be the sequence of integrands defined by (8.23) and* g *be the integrand* $g(u, \omega) = \mathbb{E}\left(f(u, \cdot) \mid \mathcal{G}\right)(\omega)$. *Suppose that*

(i) *for almost every* $\omega \in \Omega$ *and each* $u \in \operatorname{dom} g(\cdot, \omega)$, *we have that*

$$g(u, \omega) \geq \left(\text{s-ls}_{\text{e}} g^{(n)}\right)(u, \omega);$$

(ii) $j : \mathbb{U} \times \mathbb{W} \to \overline{\mathbb{R}}$ *is l.s.c. as a function of* (u, w) *and proper;*

(iii) *there exists a measurable function* $h(w)$ *such that* $j(u, w) \geq h(w)$ *and such that* $h(\boldsymbol{W}^{(n)}) \in L^q(\Omega, \mathcal{A}, \mathbb{P}; \mathbb{R})$ *and* $\mathbb{E}\left(h(\boldsymbol{W}^{(n)}) \mid \mathcal{G}^{(n)}\right) \geq \boldsymbol{G}$ *with* $\boldsymbol{G} \in L^q(\Omega, \mathcal{A}, \mathbb{P}; \mathbb{R})$;

(iv) *there exists a sequence* $\{\boldsymbol{U}^{(n)}\}_{n \in \mathbb{N}}$ *with* $\boldsymbol{U}^{(n)} \in L^p(\Omega, \mathcal{A}, \mathbb{P}; \mathbb{U})$ *and functions* $\boldsymbol{k}$ *and* $\boldsymbol{k}_0$ *in* $L^p(\Omega, \mathcal{A}, \mathbb{P}; \mathbb{R})$ *such that* $\|\boldsymbol{U}^{(n)}\|_{\mathbb{U}} \leq \boldsymbol{k}$ *and*

$$j(\boldsymbol{U}^{(n)}, \boldsymbol{W}^{(n)}) \leq \boldsymbol{k}_0;$$

(v) k_0 *given by* (iv) *is such that* $\mathbb{E}(\boldsymbol{k}_0 \mid \mathcal{G}^{(n)}) \leq \boldsymbol{k}_1$ *with* $\boldsymbol{k}_1$ *in* $L^p(\Omega, \mathcal{A}, \mathbb{P}; \mathbb{R})$.

Then for every random variable $\boldsymbol{U}$ *in* $L^p(\Omega, \mathcal{A}, \mathbb{P}; \mathbb{U})$, *we have that* $I_g(\boldsymbol{U}) \geq \left(\text{s-ls}_{\text{e}} I_{g^{(n)}}\right)(\boldsymbol{U})$.

Proof The proof consists of applying Theorem 8.55 to the sequence $\{g^{(n)}\}_{n\in\mathbb{N}}$ and the integrand g. The first step is to check that the sequence $\{g^{(n)}\}_{n\in\mathbb{N}}$ and the integrand g are proper normal integrands. Using [135, Proposition 14.45], the sequence of integrands $\{f^{(n)}\}_{n\in\mathbb{N}}$ is a sequence of normal integrands if j is a normal integrand and the $\boldsymbol{W}^{(n)}$ are measurable functions. The function j is a normal integrand since it does not explicitely depend on the noise ω and it is is l.s.c. as a function of (u, w) [135, Example 14.31]. The sequence $\{f^{(n)}\}_{n\in\mathbb{N}}$ is also proper using Assumption 8.56-(ii). Consequently, the sequence $\{g^{(n)}\}_{n\in\mathbb{N}}$ is well defined. Using the same lines, we also obtain that the integrand g is well defined.

Using Assumption 8.56-(iii) and Theorem 8.30, the integrands g and $\{g^{(n)}\}_{n\in\mathbb{N}}$ are normal integrands, which are also proper using again Assumption 8.56-(ii). Now,

8.55-(i) is just given by 8.56-(i);
8.55-(ii) is easily deduced from Assumption 8.56-(iv) and (v) with $\boldsymbol{k}_0$ replaced by $\boldsymbol{k}_1$ in (b);
8.55-(iii) follows from the property that, using Assumption 8.56-(iii) we have that $f^{(n)}(u,\omega) \geq -\|u\|_{\mathbb{U}} + h(\boldsymbol{W}^{(n)})$ and with $h_1 = 1$ and $h^{(n),2} = \mathbb{E}\big(h(\boldsymbol{W}^{(n)}) \mid \mathcal{G}^{(n)}\big)$, we obtain that

$$g^{(n)}(u,\omega) \geq -h_1(\omega)\|u\| + h^{(n),2}(\omega) \geq -h_1(\omega)\|u\| + \boldsymbol{G}(\omega),$$

where h_1 and $h^{(n),2}$ belong to $L^q(\Omega, \mathcal{A}, \mathbb{P}; \mathbb{U})$ with $h_1(\omega) > 0$ a.s.; this gives 8.55-(iii) for the $\{g^{(n)}\}_{n\in\mathbb{N}}$ sequence. □

Remark 8.57 Under the following additional assumption to Proposition 8.56

(ii)′ $j(., w)$ is a convex function for all $w \in \mathbb{W}$,

we obtain that $\{f^{(n)}\}_{n\in\mathbb{N}}$ is a sequence of convex normal integrands and, using Assumption 8.56-(iii) and [146, Proposition 15], $g^{(n)}$ is also convex. ◊

Remark 8.58 The key condition to satisfy in Proposition 8.56 is 8.56-(i) and we list here some sufficient conditions on the sequence $\{\boldsymbol{W}^{(n)}\}_{n\in\mathbb{N}}$, and on the subfields $\{\mathcal{G}^{(n)}\}_{n\in\mathbb{N}}$ to fulfill it.

- We suppose that the sequence of σ-fields $\{\mathcal{G}^{(n)}\}_{n\in\mathbb{N}}$ is increasing and Kudo converges to the σ-field $\mathcal{G}$ ($\mathcal{G}^{(n)} \uparrow \mathcal{G}$). The sequence of noises is assumed to be a constant sequence ($\boldsymbol{W}^{(n)} = \boldsymbol{W}$ for all $n \in \mathbb{N}$). If for each fixed $u \in \operatorname{dom} g(\cdot,\omega)$ and any $n \in \mathbb{N}$ the random variable $f^{(n)}(u,\cdot) = j(u, \boldsymbol{W}(\cdot))$ is in $L^1(\Omega, \mathcal{A}, \mathbb{P}; \mathbb{R})$, then the sequence $\{g^{(n)}(u,\cdot)\}_{n\in\mathbb{N}}$ is a martingale for each fixed value of u which converges a.s. ($\sup_n \mathbb{E}\big(|g^{(n)}(u,\omega)|\big) = \mathbb{E}\big(|f(u,\omega)|\big) < +\infty$) to $g(u,\omega)$. Then we easily get, using $u^{(n)} = u$, that $g(u,\omega) \geq \big(\text{s-ls}_{\text{e}}\, g^{(n)}\big)(u,\omega)$.
- We suppose that the sequence of σ-fields $\{\mathcal{G}^{(n)}\}_{n\in\mathbb{N}}$ is increasing and Kudo converges to the σ-field $\mathcal{G}$ ($\mathcal{G}^{(n)} \uparrow \mathcal{G}$). For all $n \in \mathbb{N}$, we assume that $f^{(n)}(u,\omega) \leq f(u,\omega)$ and, for each fixed $u \in \operatorname{dom} g(\cdot,\omega)$, the random variable $f(u,\cdot)$ is

in $L^1(\Omega, \mathcal{A}, \mathbb{P}; \mathbb{R})$ and $\mathbb{E}\big(f^{(n)}(u,\omega) \mid \mathcal{G}^{(n)}\big) \leq \mathbb{E}\big(f(u,\omega) \mid \mathcal{G}^{(n)}\big)$. Then, for $u \in \operatorname{dom} g(\cdot, \omega)$, the sequence of random variables $\{\mathbb{E}\big(f(u,\omega) \mid \mathcal{G}^{(n)}\big)\}_{n\in\mathbb{N}}$ converges a.s. to $g(u,\omega) = \mathbb{E}\big(f(u,\omega) \mid \mathcal{G}\big)$. Thus, we obtain that

$$\limsup_{n\mapsto\infty} g^{(n)}(u,\omega) \leq g(u,\omega)\ \mathbb{P}\text{-a.s.}\,.$$

- We suppose that the sequence of σ-fields $\{\mathcal{G}^{(n)}\}_{n\in\mathbb{N}}$ is increasing and Kudo converges to the σ-field $\mathcal{G}$ and, for each fixed u, $j(u, \boldsymbol{W}^{(n)})$ converges to $j(u, \boldsymbol{W})$ in $L^1(\Omega, \mathcal{A}, \mathbb{P}; \mathbb{R})$. Then, for each fixed u, the sequence $\{g^{(n)}\}_{n\in\mathbb{N}}$ is a $\{\mathcal{G}^{(n)}\}_{n\in\mathbb{N}}$-adapted sequence of random variables which converges in $L^1(\Omega, \mathcal{A}, \mathbb{P}; \mathbb{R})$ by Proposition 8.14. It is thus uniformly integrable (see Sect. B.2.5). Suppose now that the sequence of random variables $\{g^{(n)}\}_{n\in\mathbb{N}}$ is a martingale in the limit, namely, for all $m \in \mathbb{N}$,
$$\lim_{n\to\infty} \mathbb{E}(g^{(n)} \mid \mathcal{G}^{(m)}) - g^{(m)} = 0 \quad \mathbb{P}\text{-a.s.}.$$

  Then, by [106, Theorem 2], $\{g^{(n)}\}_{n\in\mathbb{N}}$ converges $\mathbb{P}$-a.s. to
$$g(u,\omega) = \mathbb{E}\big(f(u,\omega) \mid \mathcal{G}\big).$$

  This proves Assumption 8.56-(i), namely $g(u,\omega) \geq \big(\text{s-ls}_{\text{e}} g^{(n)}\big)(u,\omega)$. ◊

Inequality for the w-Epigraphical Lower Limit

In this paragraph, we provide sufficient conditions in order to derive the second inequality of Eq. (8.59), that is, $I_f(\boldsymbol{U}) \leq \big(\text{w-li}_{\text{e}} I_{f^{(n)}}\big)(\boldsymbol{U})$. We show that convergence in probability is sufficient to obtain this inequality. The proof is based on [87, Theorem 1]. We assume that $\mathbb{P}$ is a finite positive non-atomic (see Definition B.2) complete measure on the space Ω. The Ioffe theorem applies to extended-real-valued functions $\zeta : \Omega \times \mathbb{W} \times \mathbb{U} \to \overline{\mathbb{R}}$ such that $\zeta(\omega, \boldsymbol{W}(\omega), \boldsymbol{U}(\omega))$ is measurable for any measurable mappings $\boldsymbol{W} \in L^0(\Omega, \mathcal{A}; \mathbb{W})$ and $\boldsymbol{U} \in L^0(\Omega, \mathcal{A}; \mathbb{U})$. It gives necessary and sufficient conditions for the integral function (see Definition 8.27) $I_\zeta : \mathbb{L} \times \mathbb{M} \to \overline{\mathbb{R}}$, defined by

$$I_\zeta(\boldsymbol{W}, \boldsymbol{U}) = \int_\Omega^\star \zeta\big(\omega, \boldsymbol{W}(\omega), \boldsymbol{U}(\omega)\big)\, \mathrm{d}\mathbb{P}(\omega), \tag{8.61}$$

to be sequentially lower semicontinuous on $\mathbb{L} \times \mathbb{M}$, where $\mathbb{L}$ and $\mathbb{M}$ are two linear topological subspaces of $L^0(\Omega, \mathcal{A}, \mathbb{P}; \mathbb{W})$ and $L^0(\Omega, \mathcal{A}, \mathbb{P}; \mathbb{U})$. The two subspaces $\mathbb{L}$ and $\mathbb{M}$ must satisfy two technical assumptions denoted by (H_1) and (H_2) in the original paper. We do not review here these two assumptions, but recall that assumptions (H_1) and (H_2) are fulfilled when $\mathbb{L}$ and $\mathbb{M}$ are $L^p(\Omega, \mathcal{A}, \mathbb{P})$ spaces (with $1 \leq p \leq +\infty$) with strong or weak topologies and if we assume that the topology in $\mathbb{L}$ is not weaker than the topology of convergence in measure.

Before recalling [87, Theorem 1], we need the following definition. A function ζ satisfies the *lower compactness property* on $\mathbb{L} \times \mathbb{M}$ if any sequence $\{\zeta^-(\omega, \boldsymbol{W}^{(k)}(\omega),$

$U^{(k)}(\omega))\}_{k\in\mathbb{N}}$ (here $f^- = \min(f, 0)$) is weakly precompact in L^1 whenever the sequence $\{W^{(k)}\}_{k\in\mathbb{N}}$ converges in $\mathbb{L}$, the sequence $\{U^{(k)}\}_{k\in\mathbb{N}}$ converges in $\mathbb{M}$ and there exists a real number a such that $I_\zeta(W^{(k)}, U^{(k)}) \leq a < +\infty$ for all k. Note that, when the integrand ζ is a positive function, then it satisfies the *lower compactness property* [87, Theorem 3].

Theorem 8.59 *Let $(\Omega, \mathcal{A}, \mathbb{P})$ be a probability space and suppose that $\mathbb{P}$ is non-atomic. Let $\mathbb{L}$ and $\mathbb{M}$ be two linear topological subspaces of $L^0(\Omega, \mathcal{A}, \mathbb{P}; \mathbb{W})$ and $L^0(\Omega, \mathcal{A}, \mathbb{P}; \mathbb{U})$ satisfying Ioffe assumptions (H_1) and (H_2). Assume that $\zeta(\omega, w, u)$ is $\mathcal{A} \times \mathcal{B}^{o}_{\mathbb{W}\times\mathbb{U}}$-measurable, l.s.c. in (w, u) and convex in u. In order that I_ζ be lower semicontinuous on $\mathbb{L}\times\mathbb{M}$ and everywhere on $\mathbb{L}\times\mathbb{M}$ greater than $-\infty$, it is necessary and (if I_ζ is finite for at least one point in $\mathbb{L} \times \mathbb{M}$) sufficient that ζ satisfies the lower compactness property.*

We propose now sufficient conditions to ensure that $I_f(U) \leq \big(\text{w-li}_{\text{e}} I_{f^{(n)}}\big)(U)$. The proof mainly consists in showing that we can use Theorem 8.59.

Proposition 8.60 *For a given random variable W and a sequence of random variables $\{W^{(n)}\}_{n\in\mathbb{N}}$, we define a sequence $\{f^{(n)}\}_{n\in\mathbb{N}}$ of integrands as in (8.20) by $f^{(n)}(u, \omega) = j\big(u, W^{(n)}(\omega)\big)$ for all $n \in \mathbb{N}$ and an integrand $f(u, \omega) = j\big(u, W(\omega)\big)$. Suppose that, for $1 \leq p < +\infty$, we have the following properties*

(i) the function $j : \mathbb{U} \times \mathbb{W} \to \overline{\mathbb{R}}$ is l.s.c. as a function of (u, w) and proper;
(ii) $j(\cdot, w)$ is a convex function for all $w \in \mathbb{W}$;
(iii) $j(u, w) \geq 0$ for all $(u, w) \in \mathbb{U} \times \mathbb{W}$;
(iv) there exists $U \in L^p(\Omega, \mathcal{A}, \mathbb{P}; \mathbb{U})$ such that $I_f(U)$ is finite;
(v) the sequence $\{W^{(n)}\}_{n\in\mathbb{N}}$ of random variables converges in probability to the random variable W.

Then, for every function U in $L^p(\Omega, \mathcal{A}, \mathbb{P}; \mathbb{U})$, $I_f(U) \leq \big(\text{w-li}_{\text{e}} I_{f^{(n)}}\big)(U)$.

Proof As in the proof of Proposition 8.56, the sequence $\{f^{(n)}\}_{n\in\mathbb{N}}$ is composed of normal integrands and thus has the requested measurability property. The result then follows directly from Theorem 8.59 [87, Theorem 1] applied to the homogeneous integrands $f(\omega, w, u) = j(u, w)$. □

Remark 8.61 It is possible to weaken Assumption 8.60-(iii), but it restricts the values of p for which the conclusion of Proposition 8.60 is valid. For example, replacing Assumption 8.60-(iii) by using the following assumption in the homogeneous case [87, Theorem 5]

$$j(u, w) \geq -c(\|w\| + \|u\|) + b,$$

with $(c, b) \in \mathbb{R}^2$ and assuming that $\{W^{(n)}\}$ converges in $L^1(\Omega, \mathcal{A}, \mathbb{P}; \mathbb{R}^l)$, we obtain that

$$I_f(U) \leq \big(\text{w-li}_{\text{e}} I_{f^{(n)}}\big)(U)$$

for U in $L^1(\Omega, \mathcal{A}, \mathbb{P}; \mathbb{U})$ only. ◇

8.6 Conclusion

In Chap. 1, we presented case studies and puzzles around discretization of closed loop stochastic optimization problems. In Chap. 3, we introduced basic ways to handle information, and we examined its impact on stochastic optimization problems in Chap. 4. Approximation schemes respecting information constraints were presented in Chap. 6. In the present chapter, we are finally able to prove a convergence theorem for appropriate notions of noise and information convergence in closed-loop stochastic optimization problems. The main Theorem 8.42 is based on epi-convergence results, which take into account constraints defined by characteristic functions of sets.

Part V
Multi-Agent Systems

Chapter 9
Multi-Agent Decision Problems

9.1 Introduction

The notion of a system carries the idea of interconnected parts that have interactions. In this chapter, we present some methods for analyzing systems comprised of several agents, making decisions in an order that is not fixed in advance.

In Sect. 9.2, we expose the so-called Witsenhausen intrinsic model, with its high level of generality on how information is taken into account, and we examine the notions of solvability and of causality in Sect. 9.3. With these notions, we can pursue the discussion about the concept of "state", already approached in Sects. 1.2.1 and 4.4.

In Sect. 9.4, we provide a unified framework to define and study four binary relations between agents: (1) precedence, (2) subsystem, (3) information-memory relations (these three relations are scattered in the literature), and we add (4) the decision-memory relation.

Equipped with the four binary relations between agents, we display a typology of systems in Sect. 9.5, among which we distinguish the sequential and the partially nested ones. In a sequential system, agents are ordered in advance, and each agent is influenced, at most, by the "previous" agents. Such systems have been discussed in Sect. 4.4. In a partially nested system, whenever an agent is a predecessor of another agent, all information available to the former is available to the latter. We return to the notion of "signaling", previously discussed in Sect. 1.3.2. Whenever an agent is a predecessor of another agent, the former can, by means of her decisions, send a "signal" (information transmission) to the latter. However, this potential for signaling is useless for partially nested systems, for which the latter already knows what the former knows. This is why Linear-Quadratic-Gaussian (LQG) stochastic optimization problems display optimal linear strategies under partially nested information patterns [81, 83], whereas they display optimal tricky nonlinear strategies under non-nested patterns [155], as this is the case in the Witsenhausen counterexample, considered in Sect. 1.3.3 and discussed in Sect. 4.2. We also provide examples of systems composed of two agents and a summary table of systems.

P. Carpentier et al., *Stochastic Multi-Stage Optimization*, Probability Theory and Stochastic Modelling 75, DOI 10.1007/978-3-319-18138-7_9

In Sect. 9.6, we examine stochastic control issues. In the typology of systems discussed above, classical ones are those for which agents are ordered in advance, each agent is influenced at most by the "previous" agents and passes on her information to the "next" agent (in particular, classical systems are both sequential and partially nested). The application of Dynamic Programming to classical stochastic control problems is made possible by the following property, quoted from Witsenhausen in [160] (and whose essential role was pointed out by Chernoff in his unpublished 1963 memo *Backward induction in dynamic programming* and by Striebel in [142]): "If an observer of a stochastic control system observes both the decision taken by an agent in the system and the data that was available for this decision, then the conclusions that the observer can draw do not depend on the functional relation (policy, control law) used by this agent to reach his decision." In [160], Witsenhausen considers the case of general information structures. We recall Witsenhausen's notion of policy independence of conditional expectations and the prominent conditions under which it holds true. Thanks to the precedence, subsystem, information-memory and decision-memory relations, we discuss the Witsenhausen's conditions making it possible to decompose an optimization problem into subproblems.

This chapter heavily relies on notions presented at Chap. 3.

9.2 Witsenhausen Intrinsic Model

The so-called *Witsenhausen intrinsic model* is introduced in [159]. The model consists of a finite set of agents, of a collection of decision (or control) sets, with a corresponding collection of σ-fields, and of a single sample set (universe) equipped with a σ-field, and representing uncertainties, or states of Nature. This model does not suppose any temporal ordering of decisions.

9.2.1 The Extensive Space of Decisions and States of Nature

Let A be a finite set, whose elements are called *agents*.

Definitions
Each agent $\alpha \in A$ is supposed to make one decision $u_\alpha \in \mathbb{U}_\alpha$, where $\mathbb{U}_\alpha$ is the control set for agent α, equipped with a σ-field $\mathcal{U}_\alpha$.

Example 9.1 When $A = \{0, \ldots, T-1\}$, this framework includes the (discrete time) dynamic case by considering that a decision maker making one decision at each period $t = 0, \ldots, T-1$ is in fact made up of several agents, one for each period. △

We define the *decision space* as the product set

$$\mathbb{U}_A := \prod_{\beta \in A} \mathbb{U}_\beta, \tag{9.1}$$

equipped with the product *decision field* (see Remark 3.25)

$$\mathcal{U}_A := \bigotimes_{\beta \in A} \mathcal{U}_\beta. \tag{9.2}$$

Let Ω be a measurable set, equipped with σ-field $\mathcal{F}$,[1] which represents all uncertainties: any $\omega \in \Omega$ is called a *state of Nature*, or a *sample path*.

The *history space* is the product space

$$\mathbb{H} := \mathbb{U}_A \times \Omega = \prod_{\beta \in A} \mathbb{U}_\beta \times \Omega, \tag{9.3}$$

equipped with the product *history field*

$$\mathcal{H} := \mathcal{U}_A \otimes \mathcal{F} = \bigotimes_{\beta \in A} \mathcal{U}_\beta \otimes \mathcal{F}. \tag{9.4}$$

Example 9.2 Let us illustrate the above definitions in the simple case of two agents, each of them having two possible decision values, and of states of Nature reduced to two elements. Thus, we consider the case $A = \{a, b\}$, $\mathbb{U}_a = \{a_1, a_2\}$, $\mathbb{U}_b = \{b_1, b_2\}$, together with $\mathcal{U}_a = 2^{\mathbb{U}_a}$ and $\mathcal{U}_b = 2^{\mathbb{U}_b}$ the complete σ-fields (see Sect. 3.3.2), and with $\Omega = \{\omega_-, \omega_+\}$ and $\mathcal{F} = 2^{\Omega}$. We have that $\mathbb{H} = \mathbb{U}_a \times \mathbb{U}_b \times \Omega = \{a_1, a_2\} \times \{b_1, b_2\} \times \{\omega_-, \omega_+\}$ and $\mathcal{H} = \mathcal{U}_a \otimes \mathcal{U}_b \otimes \mathcal{F} = 2^{\mathbb{H}}$. This is illustrated in Fig. 9.1, where the σ-field $\mathcal{H}$ is the complete partition generated by the singletons (see Sect. 3.3.2). △

Cylindric Extensions and Decision Subfields

Consider now a subset B of A. Let us define the *cylindric extension* of $\bigotimes_{\beta \in B} \mathcal{U}_\beta$ to the decision field $\mathcal{U}_A$ defined in (9.2) by:

$$\mathcal{U}_B := \bigotimes_{\beta \in B} \mathcal{U}_\beta \otimes \bigotimes_{\beta \notin B} \{\emptyset, \mathbb{U}_\beta\} \subset \mathcal{U}_A. \tag{9.5}$$

[1] We avoid the notation $\mathcal{A}$ for the state of Nature σ-field, because there might be a visual confusion with the set A of agents.

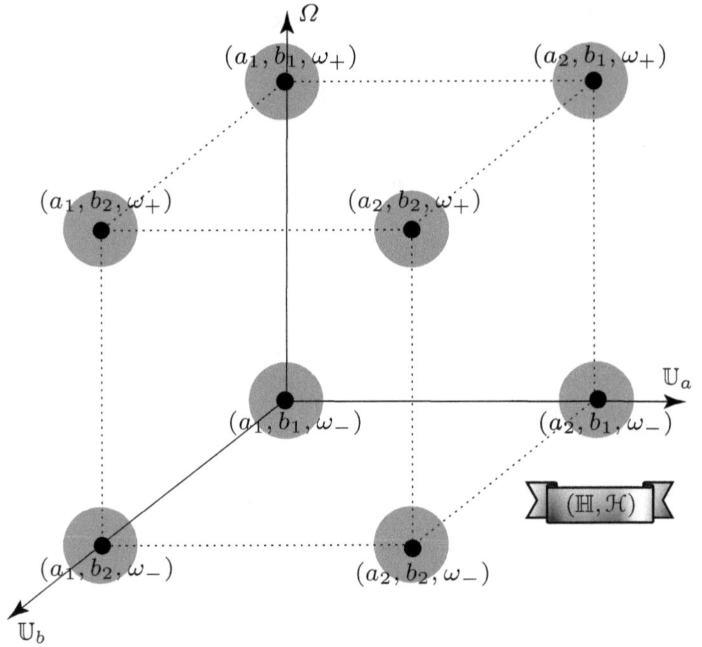

Fig. 9.1 The extensive space $\mathbb{H}$ and its finest partition $\mathcal{H} = 2^{\mathbb{H}}$

Any element in $\mathcal{U}_B$ is of the form $K \times \prod_{\beta \notin B} \mathbb{U}_\beta$, where $K \in \bigotimes_{\beta \in B} \mathcal{U}_\beta$. For all subsets $B \subset A$ and $C \subset A$, the following properties hold true

$$\mathcal{U}_{B \cap C} = \mathcal{U}_B \wedge \mathcal{U}_C \text{ and } \mathcal{U}_{B \cup C} = \mathcal{U}_B \vee \mathcal{U}_C, \tag{9.6}$$

where the operators $\wedge$ and $\vee$ between σ-fields have been introduced in Proposition 3.15 for π-fields, and extended to σ-Fields in Sect. 3.3.3.

The field $\mathcal{U}_B$ in (9.5) is a subfield of the decision field $\mathcal{U}_A$ defined in (9.2). We now consider and define another *cylindric extension* of $\bigotimes_{\beta \in B} \mathcal{U}_\beta$ to the history field $\mathcal{H}$ defined in (9.4) by:

$$\mathcal{D}_B := \mathcal{U}_B \otimes \{\emptyset, \Omega\} = \bigotimes_{\beta \in B} \mathcal{U}_\beta \otimes \bigotimes_{\beta \notin B} \{\emptyset, \mathbb{U}_\beta\} \otimes \{\emptyset, \Omega\} \subset \mathcal{U}_A \otimes \mathcal{F} = \mathcal{H}. \tag{9.7}$$

The subfield $\mathcal{D}_B$ of $\mathcal{H}$ represents the information provided by the decisions of the agents in B. When B is reduced to a singleton $\{\alpha\}$, we set

$$\mathcal{D}_\alpha := \mathcal{D}_{\{\alpha\}}, \tag{9.8}$$

to alleviate the notations, and we call $\mathcal{D}_\alpha$ the *decision subfield* of agent α. Any element in $\mathcal{D}_B$ is of the form $K \times \prod_{\beta \notin B} \mathbb{U}_\beta \times \Omega$, where $K \in \bigotimes_{\beta \in B} \mathcal{U}_\beta$. The relations (9.6) also hold true with $\mathcal{U}$ replaced by $\mathcal{D}$.

9.2.2 Information Fields and Policies

Information Fields

The *information field* of agent $\alpha \in A$ is a σ-field

$$\mathcal{I}_\alpha \subset \mathcal{H}. \tag{9.9}$$

In this way, the information of agent α may depend upon the states of Nature and all agents decisions (including herself in case of "self-information", see Definition 9.5 right below). An example is given in Fig. 9.2 where the information field is described by the associated partition.

Definition 9.3 A *stochastic control system* (shortly *system*) is a collection consisting of a finite set A of agents, states of Nature $(\Omega, \mathcal{F})$, control sets, fields and information fields $\{\mathbb{U}_\alpha, \mathcal{U}_\alpha, \mathcal{I}_\alpha\}_{\alpha \in A}$.

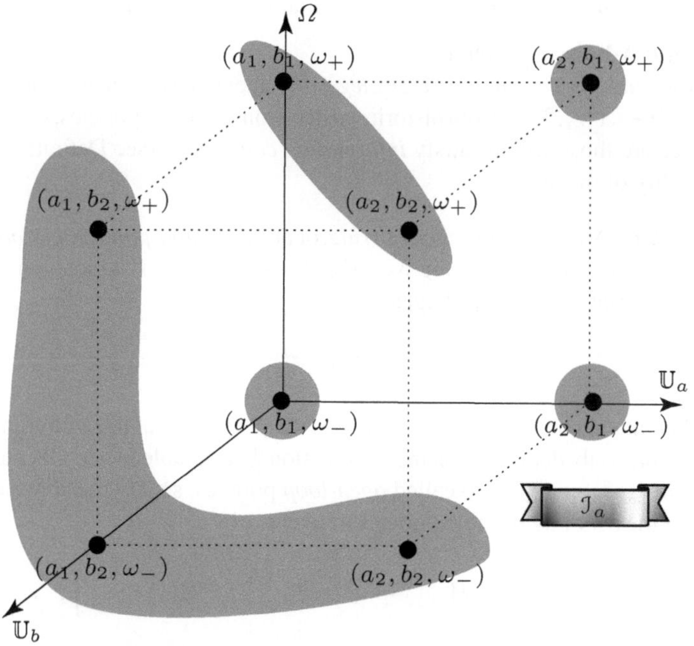

Fig. 9.2 An example of information field given by a partition

Example 9.4 For instance, in the description of a sequential optimal stochastic control problem as exposed in Sect. 4.5.1, we recognize a system with agents $A = \{0, \dots, T-1\}$. △

To illustrate the use of information fields, consider two agents α and β. The condition $\mathcal{I}_\beta \subset \mathcal{I}_\alpha$ mathematically expresses that what agent β *knows* is passed on to agent α. The condition $\mathcal{D}_\beta \subset \mathcal{I}_\alpha$ expresses that what *does* agent β is passed on to agent α.

Define the *information* $\mathcal{I}_B \subset \mathcal{H}$ *of the subset* $B \subset A$ *of agents* by

$$\mathcal{I}_B := \bigvee_{\beta \in B} \mathcal{I}_\beta. \tag{9.10}$$

With the above notations, we can express the property that an agent information cannot depend on her own decision.

Definition 9.5 The *absence of self-information* is the property that

$$\mathcal{I}_\alpha \subset \mathcal{U}_{A \setminus \{\alpha\}} \otimes \mathcal{F}, \tag{9.11}$$

for all agent $\alpha \in A$.

In practice, we either impose such a restriction or it is the consequence of stronger assumptions (causality, solvability or sequentiality) to be discussed in Sect. 9.3.

Policies and Admissible Policies

A *decision rule*, or a *policy*, or a *strategy*, for agent α is a measurable mapping $\lambda_\alpha : (\mathbb{H}, \mathcal{H}) \to (\mathbb{U}_\alpha, \mathcal{U}_\alpha)$, from histories to decisions. Among policies, of paramount importance are those which satisfy *information constraints* (see Definition 3.44 for measurability of mappings).

Definition 9.6 An *admissible decision rule*, or an *admissible policy*, or an *admissible strategy*, for agent α is a mapping $\lambda_\alpha : (\mathbb{H}, \mathcal{H}) \to (\mathbb{U}_\alpha, \mathcal{U}_\alpha)$ which is measurable w.r.t. the information field $\mathcal{I}_\alpha$ of agent α:

$$\lambda_\alpha^{-1}(\mathcal{U}_\alpha) \subset \mathcal{I}_\alpha. \tag{9.12}$$

The mathematical condition (9.12) expresses the property that the admissible policy of agent α may only depend upon the information $\mathcal{I}_\alpha$ available to her. Of course, constant mappings $\mathbb{H} \to \mathbb{U}_\alpha$, also called *open-loop* policies, satisfy the above property. We denote the set of admissible policies of agent α by

$$\Lambda_\alpha^{\mathrm{ad}} := \left\{ \lambda_\alpha : (\mathbb{H}, \mathcal{H}) \to (\mathbb{U}_\alpha, \mathcal{U}_\alpha) \;\middle|\; \lambda_\alpha^{-1}(\mathcal{U}_\alpha) \subset \mathcal{I}_\alpha \right\}, \tag{9.13}$$

and the set of admissible policies of all agents is

$$\Lambda_A^{\text{ad}} := \prod_{\alpha \in A} \Lambda_\alpha^{\text{ad}}. \tag{9.14}$$

Remark 9.7 Assume that $\mathcal{U}_\alpha$ contains singletons. In the absence of self-information (see Definition 9.5), any admissible policy λ has the property that, for any agent α, the domain $\mathbb{H}$ of λ_α can, in fact, be restricted to $\mathbb{U}_{A\setminus\{\alpha\}} \times \Omega$. Indeed, on the one hand, we have that $\mathcal{I}_\alpha \subset \mathcal{U}_{A\setminus\{\alpha\}} \otimes \mathcal{F}$. On the other hand, we have that $\lambda_\alpha^{-1}(\mathcal{U}_\alpha) \subset \mathcal{I}_\alpha$ since λ is an admissible policy. Hence, $\lambda_\alpha^{-1}(\mathcal{U}_\alpha) \subset \mathcal{U}_{A\setminus\{\alpha\}} \otimes \mathcal{F}$, and we conclude that the value of $\lambda_\alpha(u_\alpha, \{u_\beta\}_{\beta\neq\alpha}, \omega)$ does not depend on u_α (we assumed that $\mathcal{U}_\alpha$ contains singletons). $\diamondsuit$

Remark 9.8 Notice that information may be given by mappings: agent α learns about the history $h \in \mathbb{H}$ through a mapping (a signal) $\boldsymbol{Y}_\alpha : \mathbb{H} \to \mathbb{Y}_\alpha$, where $(\mathbb{Y}_\alpha, \mathcal{Y}_\alpha)$ is some measurable space. Assuming that $\boldsymbol{Y}_\alpha$ is measurable from $(\mathbb{H}, \mathcal{H})$ to $(\mathbb{Y}_\alpha, \mathcal{Y}_\alpha)$, the connection with the "field" approach is given by the σ-field generated by the mapping:

$$\mathcal{I}_\alpha = \boldsymbol{Y}_\alpha^{-1}(\mathcal{Y}_\alpha) = \sigma(\boldsymbol{Y}_\alpha).$$

Though not all σ-fields are given by signals, as discussed in [9, 10], the signal/mapping approach to information is widespread; the Witsenhausen counterexample presented in Sect. 4.2 is an example. $\diamondsuit$

Example 9.9 Let us formulate the classical information pattern of Sect. 4.3.2 and the Witsenhausen counterexample information pattern of Sect. 4.2 in the above Witsenhausen intrinsic model framework. Following the notations of Sect. 4.2, we set $A = \{0, 1\}$, $\Omega = \mathbb{X}_0 \times \mathbb{W}_1 = \mathbb{R}^2$ with typical element $\omega = (x_0, w_1)$ and σ-field $\mathcal{F} = \mathcal{X}_0 \otimes \mathcal{W}_1 = \mathcal{B}^{\mathrm{o}}_{\mathbb{R}^2}$, and $(\mathbb{U}_0, \mathcal{U}_0) = (\mathbb{R}, \mathcal{B}^{\mathrm{o}}_{\mathbb{R}}) = (\mathbb{U}_1, \mathcal{U}_1)$. We introduce the history space $\mathbb{H} = \mathbb{U}_0 \times \mathbb{U}_1 \times \Omega$ and, following (4.3), the coordinate mappings $\boldsymbol{U}_0(u_0, u_1, x_0, w_1) = u_0$, $\boldsymbol{X}_0(u_0, u_1, x_0, w_1) = x_0$, $\boldsymbol{W}_1(u_0, u_1, x_0, w_1) = w_1$, and two signals $\boldsymbol{Y}_0 = \boldsymbol{X}_0$ and $\boldsymbol{Y}_1 = \boldsymbol{X}_0 + \boldsymbol{U}_0 + \boldsymbol{W}_1$. The classical pattern of Sect. 4.3.2 corresponds to $\mathcal{I}_0 = \sigma(\boldsymbol{Y}_0) = \{\emptyset, \mathbb{U}_0\} \otimes \{\emptyset, \mathbb{U}_1\} \otimes \mathcal{X}_0 \otimes \{\emptyset, \mathbb{W}_1\}$ and $\mathcal{I}_1 = \sigma(\boldsymbol{Y}_0, \boldsymbol{U}_0, \boldsymbol{Y}_1) = \sigma(\boldsymbol{X}_0, \boldsymbol{U}_0, \boldsymbol{W}_1) = \mathcal{U}_0 \otimes \{\emptyset, \mathbb{U}_1\} \otimes \mathcal{X}_0 \otimes \mathcal{W}_1$. The Witsenhausen counterexample measurability constraints (4.6) correspond to $\mathcal{I}_0 = \sigma(\boldsymbol{Y}_0)$ and $\mathcal{I}_1 = \sigma(\boldsymbol{Y}_1) = \sigma(\boldsymbol{X}_0 + \boldsymbol{U}_0 + \boldsymbol{W}_1)$. $\triangle$

9.3 Causality and Solvability for Stochastic Control Systems

In the Witsenhausen intrinsic model, agents make decisions in an order which is not fixed in advance. Once the agents have selected their policies, Witsenhausen wonders whether it is possible for the agents to play. Briefly speaking, solvability is the property that, for each state of Nature, their decisions are uniquely determined by their policies. In a causal system, agents are ordered, one playing after the other with available information depending only on agents acting earlier, but the order may depend upon the history.

9.3.1 Solvability and Solvability/Measurability

Consider a collection $\lambda = \{\lambda_\alpha\}_{\alpha \in A} \in \Lambda_A^{\mathrm{ad}}$ of admissible policies: by (9.13) and (9.14), the policy $\lambda_\alpha : \mathbb{U}_A \times \Omega \to \mathbb{U}_\alpha$ of agent α is measurable w.r.t. the information field $\mathfrak{I}_\alpha$. Thus, agent α makes a decision according to the information she has on the state of Nature ω and on all the decisions $\{u_\beta\}_{\beta \in A}$ (including, a priori, her own decision!). The problem is:

Is it possible for the agents to play? And will their decisions be uniquely determined by the rules?

Mathematically, the problem is to find, for any $\omega \in \Omega$, solutions $u \in \mathbb{U}_A$ (depending upon ω) satisfying the implicit equations

$$u = \lambda(u, \omega) \tag{9.15}$$

or, equivalently,

$$u_\alpha = \lambda_\alpha(\{u_\beta\}_{\beta \in A}, \omega), \ \forall \alpha \in A. \tag{9.16}$$

For example, for a collection of open-loop policies defined in Sect. 9.2.2, Eq. (9.15) has a unique solution $u_\alpha = \lambda_\alpha$, by identifying the mapping λ_α with its constant value.

Existence and uniqueness of the solutions of (9.15) is related to information patterns. For instance, consider an information structure with two agents α and β, and displaying the absence of self-information (see Definition 9.5). Assuming that $\mathcal{U}_\alpha$ and $\mathcal{U}_\beta$ contain singletons, by Remark 9.7 and by (9.12), policies have the form $\lambda_\alpha(u, \omega) = \widetilde{\lambda}_\alpha(u_\beta, \omega)$ and $\lambda_\beta(u, \omega) = \widetilde{\lambda}_\beta(u_\alpha, \omega)$. Equation (9.15) is now $u_\alpha = \widetilde{\lambda}_\alpha(u_\beta, \omega)$ and $u_\beta = \widetilde{\lambda}_\beta(u_\alpha, \omega)$, which may display zero solutions (deadlock), one solution (solvability) or multiple solutions (undeterminacy). For sequential optimal stochastic control problem as exposed in Sect. 4.5.1, existence and uniqueness of the solutions of (9.15) is automatically ensured by an explicit inductive procedure (4.60).

Definition 9.10 The *solvability property* holds true when, for any collection $\lambda \in \Lambda_A^{\text{ad}}$ of admissible policies, and any state of Nature $\omega \in \Omega$, there exists one, and only one, decision $u \in \mathbb{U}_A$ satisfying (9.15). Denoting $M_\lambda(\omega)$ this unique $u \in \mathbb{U}_A$, we obtain a mapping $M_\lambda : \Omega \to \mathbb{U}_A$. The *solvability/measurability* property holds true when, in addition, the mapping $M_\lambda : \Omega \to \mathbb{U}_A$ is measurable from $(\Omega, \mathcal{F})$ to $(\mathbb{U}_A, \mathcal{U}_A)$.

Equation (9.15) defines a *multifunction* or *set-valued mapping* $\Omega \rightrightarrows \mathbb{U}_A$ (see definition in Sect. 8.3.1) that associates with every $\omega \in \Omega$ the (possibly empty) set of solutions $u \in \mathbb{U}_A$. The *solvability property* is equivalent to the property that this multifunction is a function with *domain* Ω.

Definition 9.11 Suppose that the solvability property holds true. Thanks to the mapping M_λ, we define the *solution map* $S_\lambda : \Omega \to \mathbb{H}$ by

$$S_\lambda(\omega) := \big(M_\lambda(\omega), \omega\big), \quad \forall \omega \in \Omega, \tag{9.17}$$

that is,

$$(u, \omega) = S_\lambda(\omega) \iff u = \lambda(u, \omega), \quad \forall (u, \omega) \in \mathbb{U}_A \times \Omega. \tag{9.18}$$

We include ω in the image of $S_\lambda(\omega)$ to map the universe Ω towards the history space $\mathbb{H}$ and to interpret $S_\lambda(\omega)$ as a "state trajectory". Indeed, the mapping S_λ yields all the history generated by the state of Nature ω and by the admissible policy λ. It is a generalization of the state trajectory mapping for discrete time dynamical control systems (see the state map in Definition 4.6 for state models, and the solution map defined in Sect. 4.5.1 for sequential optimal stochastic control problems).

Remark 9.12 Given a collection $\lambda \in \Lambda_A^{\text{ad}}$ of admissible policies, we introduce:

$$\mathcal{F}_\alpha^\lambda := S_\lambda^{-1}(\mathcal{I}_\alpha) \subset \mathcal{F}. \tag{9.19}$$

Let us insist upon the fact that $\mathcal{F}_\alpha^\lambda$ is a subfield of $\mathcal{F}$, the σ-field attached to the universe Ω, whereas $\mathcal{I}_\alpha$ is a subfield of the history field $\mathcal{H}$. Moreover, $\mathcal{F}_\alpha^\lambda$ depends upon the whole policy λ: the policies of other agents, and not only the policy of agent α, can reveal more or less information as to the state of Nature. We come back to this issue in Sect. 9.6. ◊

Witsenhausen highlights a connection between solvability and absence of self-information. We refer the reader to [159] for a proof of the following proposition.

Proposition 9.13 *Assume that $\mathcal{U}_\alpha$ contains singletons, for all agent α in A. Solvability implies absence of self-information.*

9.3.2 Causality

The subtle notion of *causality* is treated with care in [156] (see also [3, 83, 159]). The idea is that, in a causal system, agents are ordered, one playing after the other with available information depending only on agents acting earlier, but the order may depend upon the history. As Witsenhausen says, the "main difficulty is one of notation".

Let $\mathbb{S}$ denote the set of total orderings of agents in A, that is, injective mappings from $\{1, \ldots, n\}$ to A, where $n = \text{card}A$. For $k \in \{1, \ldots, n\}$, let $\mathbb{S}_k$ denote the set of k-orderings, that is, injective mappings from $\{1, \ldots, k\}$ to A (thus $\mathbb{S} = \mathbb{S}_n$). There is a natural mapping ψ_k from $\mathbb{S}$ to $\mathbb{S}_k$, the restriction of any ordering of A to the domain set $\{1, \ldots, k\}$.

Define a *history-ordering* as a mapping $\varphi : \mathbb{H} \to \mathbb{S}$ from histories towards orderings: along each history $h \in \mathbb{H}$, the agents are ordered by $\sigma = \varphi(h) \in \mathbb{S}$, so that agent $\sigma(1)$ is the first, $\sigma(2)$ is the second, etc. With any $k \in \{1, \ldots, n\}$ and k-ordering $\sigma \in \mathbb{S}_k$, associate the set

$$\mathbb{H}_{k,\sigma} = \{h \in \mathbb{H} \mid \psi_k(\varphi(h)) = \sigma\} \tag{9.20}$$

of histories along which $\varphi(h)$ begins with $(\alpha_1, \ldots, \alpha_k) = (\sigma(1), \ldots, \sigma(k))$: along any history $h \in \mathbb{H}_{k,\sigma}$, the first k agents are known and given by $(\sigma(1), \ldots, \sigma(k))$.

Formally, a system is *causal* if there exists (at least one) history-ordering φ from $\mathbb{H}$ towards $\mathbb{S}$, with the property that for any $k \in \{1, \ldots, n\}$ and $\sigma \in \mathbb{S}_k$, the set $\mathbb{H}_{k,\sigma}$ satisfies

$$\mathbb{H}_{k,\sigma} \cap G \in \mathcal{U}_{\{\sigma(1),\ldots,\sigma(k-1)\}} \otimes \mathcal{F}, \ \forall G \in \mathcal{I}_{\sigma(k)}. \tag{9.21}$$

In other words, when the first k agents are known and given by $(\sigma(1), \ldots, \sigma(k))$, the information $\mathcal{I}_{\sigma(k)}$ of the agent $\sigma(k)$ with rank k depends at most on the decisions of agents $\sigma(1)$, …, $\sigma(k)$ with rank stricly less than k.

In Sect. 9.5.1, so-called sequential systems are defined as those for which the causality condition holds true with a constant history-ordering mapping $\varphi : \mathbb{H} \to \mathbb{S}$. In Sect. 9.5.2, we detail the analysis of a non causal system (deadlock).

9.3.3 Solvability, Causality and "State"

We refer the reader to [156] for the proof of the following proposition.

Proposition 9.14 *Causality implies (recursive) solvability with a measurable solution map.*

In [156], to prove that causality implies recursive solvability, Witsenhausen defines a sequence of subfields $\mathcal{H}^{(k)}$ of the history field $\mathcal{H}$ and interprets this sequence as the information available to an umpire after she has seen k agents make their decisions.

He proves that the umpire fields $\mathcal{H}^{(k)}$ are nested ($\mathcal{H}^{(k)} \subset \mathcal{H}^{(k+1)}$), meaning that the umpire enjoys perfect recall. Witsenhausen also introduces maps corresponding to the idea of state transition equations, with the σ-field $\mathcal{H}^{(k)}$ representing the state of knowledge of the umpire.

9.4 Four Binary Relations Between Agents

We provide a unified framework to define and study three binary relations between agents, which are found in the literature. The precedence and information-memory relations were introduced by Ho and Chu in [81, 83] for the multi-agent LQG problem. Their representation as graphs forms the *information structure diagram*. Witsenhausen in [159] presented a fairly general theory, introducing, in particular, the subsystem relation (additional results can be found in [13] and other references are [82, 161]). We add a fourth relation: the decision-memory relation. We make use of notions on binary relations reviewed in Sect. 3.2.1.

9.4.1 The Precedence Relation $\mathfrak{P}$

The precedence[2] binary relation identifies the agents whose decisions indeed influence the observations of a given agent.

For a given agent α, let us consider the set $P_\alpha \subset 2^A$ of subsets $B \subset A$ such that $\mathcal{I}_\alpha \subset \mathcal{U}_B \otimes \mathcal{F}$. Any $B \in P_\alpha$ contains agents whose decisions affect the information $\mathcal{I}_\alpha$ available to agent α. We have that $A \in P_\alpha$ because $\mathcal{I}_\alpha \subset \mathcal{U}_A \otimes \mathcal{F}$. By (9.6), the set P_α is stable under intersection. This motivates the following definition.

Definition 9.15 *([13])* Let $\langle\alpha\rangle_{\mathfrak{P}} \subset A$ be the intersection of subsets $B \subset A$ such that $\mathcal{I}_\alpha \subset \mathcal{U}_B \otimes \mathcal{F}$. We define a *precedence* binary relation $\mathfrak{P}$ on A by

$$\beta \mathfrak{P} \alpha \iff \beta \in \langle\alpha\rangle_{\mathfrak{P}}, \tag{9.22}$$

and we say that β is a *predecessor* of α.

In other words, the decisions of any *predecessor* of an agent *affect* the information of this agent: any agent is influenced by his predecessors (when they exist, because $\langle\alpha\rangle_{\mathfrak{P}}$ might be empty).

Remark 9.16 Assume that $\mathcal{U}_\beta$ contains singletons, for all $\beta \in A$. Being inspired by Remark 9.7, we can notice that any admissible policy λ has the property that, for any agent α, the domain $\mathbb{H}$ of λ_α can, in fact, be restricted to $\mathbb{U}_{\beta\in\langle\alpha\rangle_{\mathfrak{P}}} \times \Omega$. ♢

[2]Our definition of precedence is directly inspired by the one that Ho and Chu give in their 1974 paper [83], and not by the one they provide in their 1972 paper [81] (which rather relates to the subsystem relation to be discussed in Sect. 9.4.2).

By construction, the subset of agents $\langle\alpha\rangle_{\mathfrak{P}}$ is the smallest subset $B \subset A$ such that $\mathfrak{I}_\alpha \subset \mathfrak{U}_B \otimes \mathfrak{F}$. In other words, $\langle\alpha\rangle_{\mathfrak{P}}$ is characterized by:

$$\mathfrak{I}_\alpha \subset \mathfrak{U}_{\langle\alpha\rangle_{\mathfrak{P}}} \otimes \mathfrak{F} \text{ and } \big(\mathfrak{I}_\alpha \subset \mathfrak{U}_B \otimes \mathfrak{F} \Rightarrow \langle\alpha\rangle_{\mathfrak{P}} \subset B\big) . \tag{9.23}$$

Example 9.17 In Example 9.9, we have that $\langle 0\rangle_{\mathfrak{P}} = \emptyset$ and $\langle 1\rangle_{\mathfrak{P}} = \{0\}$, both for the classical pattern of Sect. 4.3.2 and for the Witsenhausen counterexample information pattern. △

Remark 9.18 Whenever $\langle\alpha\rangle_{\mathfrak{P}} \neq \emptyset$, there is a potential for *signaling*, that is for information transmission. Indeed, any agent β in $\langle\alpha\rangle_{\mathfrak{P}}$ influences the information $\mathfrak{I}_\alpha$ upon which agent α bases her decisions. Therefore, whenever agent β is a predecessor of agent α, the former can, by means of her decisions, send a "signal" to the latter. In case $\langle\alpha\rangle_{\mathfrak{P}} = \emptyset$, the decisions of agent α depend, at most, on the state of Nature, and there is no room for signaling. ◊

We use the following definitions.

Definition 9.19 For any $B \subset A$, we introduce the following subset of agents:

$$\langle B\rangle_{\mathfrak{P}} := \bigcup_{\beta\in B} \langle\beta\rangle_{\mathfrak{P}}, \ \langle B\rangle_{\mathfrak{P}}^0 := B \text{ and } \langle B\rangle_{\mathfrak{P}}^{n+1} := \langle\langle B\rangle_{\mathfrak{P}}^n\rangle_{\mathfrak{P}}, \ \forall n \in \mathbb{N}. \tag{9.24}$$

When B is a singleton $\{\alpha\}$, we denote $\langle\alpha\rangle_{\mathfrak{P}}^n$ for $\langle\{\alpha\}\rangle_{\mathfrak{P}}^n$.

The proof of the following proposition is a straightforward application of Definitions 9.15 and 9.19. We leave it to the reader.

Proposition 9.20 *Let B and C be two subsets of the set A of agents. We have that*

$$\mathfrak{I}_C \subset \mathfrak{U}_B \otimes \mathfrak{F} \iff \langle C\rangle_{\mathfrak{P}} \subset B, \tag{9.25}$$

and that, if agents in C pass on their information to agents in B, then those agents who influence C also influence B:

$$\mathfrak{I}_C \subset \mathfrak{I}_B \Rightarrow \langle C\rangle_{\mathfrak{P}} \subset \langle B\rangle_{\mathfrak{P}}. \tag{9.26}$$

With the precedence relation $\mathfrak{P}$, and especially with the complementary relation $\mathfrak{P}^c$ defined in (3.20), we can easily express that a system displays absence of self-information (see Definition 9.5).

Proposition 9.21 *The following assertions are equivalent.*

1. *A system is without self-information.*
2. *No agent can be a predecessor of herself:* $\alpha \notin \langle\alpha\rangle_{\mathfrak{P}}, \ \forall\alpha \in A$.
3. *The diagonal* Δ_A *of* $\mathfrak{P}$ *is not in* $\mathfrak{P}$*:* $(\alpha, \alpha) \notin \mathfrak{P}, \ \forall\alpha \in A$.
4. *The complementary relation* $\mathfrak{P}^c$ *is reflexive:* $\alpha\, \mathfrak{P}^c\, \alpha, \ \forall\alpha \in A$.

Proof

1 $\iff$ 2. By Definition 9.5, a system is without self-information if, and only if, $\mathfrak{I}_\alpha \subset \mathfrak{U}_{A\setminus\{\alpha\}} \otimes \mathfrak{F}$, for all $\alpha \in A$. By (9.25) with $C = \{\alpha\}$ and $B = A\setminus\{\alpha\}$, this is equivalent to $\langle\alpha\rangle_{\mathfrak{P}} \subset A\setminus\{\alpha\}$, hence to $\alpha \notin \langle\alpha\rangle_{\mathfrak{P}}$.

2 $\iff$ 3. By Definition 9.15 of the precedence relation $\mathfrak{P}$, and especially by (9.22), $\alpha \notin \langle\alpha\rangle_{\mathfrak{P}}$ is equivalent to $(\alpha, \alpha) \notin \mathfrak{P}$, that is, $\Delta_A \cap \mathfrak{P} = \emptyset$ by definition (3.2) of the diagonal relation.

3 $\iff$ 4. By the definition (3.20) of the complementary relation $\mathfrak{P}^c$, we have that $\alpha \mathfrak{P}^c \alpha \iff \alpha \notin \langle\alpha\rangle_{\mathfrak{P}}$. Therefore, item 3 is equivalent to the property that $\mathfrak{P}^c$ is reflexive. □

9.4.2 The Subsystem Relation $\mathfrak{S}$

Witsenhausen defines and studies subsystems in [159].

Definition 9.22 (*[159]*) A nonempty subset B of agents in A is a *subsystem* if the information field $\mathfrak{I}_B$ defined in (9.10) at most depends on the decisions of the agents in B:

$$\mathfrak{I}_B \subset \mathfrak{U}_B \otimes \mathfrak{F} .$$

Thus, the information received by agents in B depends upon states of Nature and decisions of members of B only.

Example 9.23 In Example 9.9, $\{0\}$ and $\{0, 1\}$ are the only subsystems. △

As a consequence of (9.6), Witsenhausen notices that subsystems form the closed sets of a topology (after considering the empty set $\emptyset$ as a subsystem).

Recall that a *topology* of A is a set $\mathfrak{T} \subset 2^A$ of subsets of A, containing $\emptyset$ and A, and which is stable (closed) under union and under finite intersection. The elements of $\mathfrak{T}$ are called *open sets*, whereas the complementary of an open set in A is a *closed set*. A subset B is said to be *connected* if it cannot be written as the union of two disjoint nonempty open sets. The maximal connected subsets (for the inclusion relation order) are called the *connected components* of the space; they form a partition of A.

Remark 9.24 Here, since A is a finite set, a topology of A is a set $\mathfrak{T} \subset 2^A$ containing $\emptyset$ and A, and which is stable under finite union and under finite intersection. Therefore, if $\mathfrak{T}$ is a topology, so is $\mathfrak{T}^c = \{B^c, \ B \in \mathfrak{T}\}$, where B^c is the complement of the subset B in A. This is why, closed and open sets satisfy the same axioms. ◇

In the proposition that follows, Witsenhausen emphasizes the role of closed sets.

Proposition 9.25 ([159]) *The subsystems are stable under the intersection and union operations, and thus form the closed sets of a topology $\mathfrak{T}$ on A.*

As Witsenhausen says, connected components of the topological space $(A, \mathfrak{T})$ are dynamically decoupled subsystems; a static coupling remains through the common dependence upon the states of Nature.

Definition 9.26 *([159])* The closure $\overline{B}$, for the topology $\mathfrak{T}$, of a subset $B \subset A$ is the smallest subsystem containing B; it is called the *subsystem generated by* B. The *subsystem generated by agent* α is the closure $\overline{\{\alpha\}}$ of the singleton $\{\alpha\}$. The corresponding *subsystem* binary relation $\mathfrak{S}$ between agents is defined by:

$$\beta \,\mathfrak{S}\, \alpha \iff \beta \in \overline{\{\alpha\}}, \quad \forall(\alpha, \beta) \in A^2. \tag{9.27}$$

In other words, $\beta \,\mathfrak{S}\, \alpha$ means that agent β belongs to the subsystem generated by agent α or, equivalently, that the subsystem generated by agent α contains that generated by agent β. Indeed, by the properties of a topological closure, we have that

$$\beta \,\mathfrak{S}\, \alpha \iff \overline{\{\beta\}} \subset \overline{\{\alpha\}}, \quad \forall(\alpha, \beta) \in A^2. \tag{9.28}$$

Notice also that a subset $B \subset A$ is a subsystem if, and only if, it coincides with the generated subsystem:

$$B \text{ is a subsytem} \iff B = \overline{B}. \tag{9.29}$$

Proposition 9.27 ([159]) *The subsystem relation $\mathfrak{S}$ is a pre-order, namely it is reflexive and transitive.*

Proof The subsystem relation $\mathfrak{S}$ is reflexive since, in a topology, any element $\alpha \in A$ belongs to the closure $\overline{\{\alpha\}}$ (see also (9.28) with $\beta = \alpha$).

The relation $\mathfrak{S}$ is also transitive. Indeed, let agents α, β and δ be such that $\alpha\mathfrak{S}\beta$ and $\beta\mathfrak{S}\delta$, that is $\alpha \in \overline{\{\beta\}}$ and $\beta \in \overline{\{\delta\}}$. From $\beta \in \overline{\{\delta\}}$, we deduce that $\overline{\{\beta\}} \subset \overline{\{\delta\}}$ and thus $\alpha \in \overline{\{\delta\}}$, that is, $\alpha \,\mathfrak{S}\, \delta$ by (9.27). □

Remark 9.28 The relation $\mathfrak{S}$ is not necessarily anti-symmetric since $\overline{\{\beta\}} = \overline{\{\alpha\}}$ may occur with $\alpha \neq \beta$. In Sect. 9.5.2, the reader can find an example of non causal system (deadlock) such that $\mathfrak{S}$ is not anti-symmetric. ◇

The following Proposition 9.29 and Theorem 9.30 describe connections between the subsystem relation $\mathfrak{S}$ and the precedence binary relation $\mathfrak{P}$.

Proposition 9.29 *Let B be a subset of the set A of agents, and α be an agent in A.*

1. *A subset $B \subset A$ is a subsystem iff $\langle B \rangle_{\mathfrak{P}} \subset B$, that is, iff the predecessors of agents in B belong to B:*

$$B \textit{ is a subsytem} \iff \overline{B} = B \iff \langle B \rangle_{\mathfrak{P}} \subset B. \tag{9.30}$$

2. *For any $\alpha \in A$, the subsystem generated by agent α is the union of α and of all her iterated predecessors (see Definition 9.19):*

$$\overline{\{\alpha\}} = \bigcup_{n \in \mathbb{N}} \langle \alpha \rangle_{\mathfrak{P}}^{n}. \tag{9.31}$$

Proof

1. We have already noticed in (9.29) that B is a subsytem iff $\overline{B} = B$. For the rest, by Definition 9.22, a subset $B \subset A$ is a subsystem iff $\mathfrak{I}_B \subset \mathfrak{U}_B \otimes \mathfrak{F}$, and thus iff $\langle B \rangle_{\mathfrak{P}} \subset B$ by (9.25).
2. First, we prove by induction that $\langle \alpha \rangle_{\mathfrak{P}}^{n} \subset \overline{\{\alpha\}}$ for $n \in \mathbb{N}$. This holds true for $n = 0$ since $\alpha \in \overline{\{\alpha\}}$. Assuming $\langle \alpha \rangle_{\mathfrak{P}}^{n} \subset \overline{\{\alpha\}}$, we deduce that $\langle \alpha \rangle_{\mathfrak{P}}^{n+1} \subset \langle \overline{\{\alpha\}} \rangle_{\mathfrak{P}}$ by definition (see (9.24)). Now, since $\overline{\{\alpha\}}$ is a subsystem, we have that $\langle \overline{\{\alpha\}} \rangle_{\mathfrak{P}} \subset \overline{\{\alpha\}}$ by (9.30). Thus $\langle \alpha \rangle_{\mathfrak{P}}^{n+1} \subset \overline{\{\alpha\}}$ and the induction is proven. We deduce that $\bigcup_{n \in \mathbb{N}} \langle \alpha \rangle_{\mathfrak{P}}^{n} \subset \overline{\{\alpha\}}$.
   Second, $\bigcup_{n \in \mathbb{N}} \langle \alpha \rangle_{\mathfrak{P}}^{n}$ is a subsystem since $\langle \bigcup_{n \in \mathbb{N}} \langle \alpha \rangle_{\mathfrak{P}}^{n} \rangle_{\mathfrak{P}} = \bigcup_{n \in \mathbb{N}} \langle \alpha \rangle_{\mathfrak{P}}^{n+1} \subset \bigcup_{n \in \mathbb{N}} \langle \alpha \rangle_{\mathfrak{P}}^{n}$. To sum up, $\bigcup_{n \in \mathbb{N}} \langle \alpha \rangle_{\mathfrak{P}}^{n}$ is a subsystem containing α and contained in $\overline{\{\alpha\}}$: it is thus equal to $\overline{\{\alpha\}}$.

This ends the proof. □

We now show that the precedence relation $\mathfrak{P}$ is included in the subsystem relation $\mathfrak{S}$, and the subsystem relation $\mathfrak{S}$ is the reflexive and transitive closure $\mathfrak{P}^*$ of the precedence relation $\mathfrak{P}$, as defined in (3.17). Recall that $\mathfrak{P}^{\infty}$ is the transitive closure of the relation $\mathfrak{P}$, as in (3.11).

Theorem 9.30 *The following sequence of inclusions and equalities holds true:*

$$\mathfrak{P} \subset \mathfrak{P}^{\infty} \subset \mathfrak{S} = \mathfrak{P}^* = \mathfrak{P}^{\infty} \cup \Delta_A. \tag{9.32}$$

Proof We have that $\mathfrak{P} \subset \mathfrak{S}$, since $\beta \in \langle \alpha \rangle_{\mathfrak{P}} \Rightarrow \beta \in \overline{\{\alpha\}}$ by (9.31).

By Proposition 9.27, $\mathfrak{S}$ is transitive and thus it is equal to its transitive closure $\mathfrak{S}^{\infty}$. Thus, $\mathfrak{P} \subset \mathfrak{S} \Rightarrow \mathfrak{P}^{\infty} \subset \mathfrak{S}^{\infty} = \mathfrak{S}$. Now, by the definition (3.11) of $\mathfrak{P}^{\infty}$, the identity (9.31) means that $\mathfrak{P}^{\infty} \cup \Delta_A = \mathfrak{S}$. We conclude with the equality (3.17). □

Here is a proposition which proves important for partially nested systems to be discussed in Sect. 9.5.3.

Proposition 9.31 *If the precedence binary relation $\mathfrak{P}$ is transitive, then*

1. *the following inclusions and equalities hold true*

$$\mathfrak{P} = \mathfrak{P}^{\infty} \subset \mathfrak{S} = \mathfrak{P} \cup \Delta_A;$$

2. *for any subset B of agents in A, the set $\langle B \rangle_{\mathfrak{P}}$ of her predecessors is a subsystem;*
3. *for any agent $\alpha \in A$, the subsystem $\overline{\{\alpha\}}$ generated by agent α is the union of α and of her predecessors: $\overline{\{\alpha\}} = \langle \alpha \rangle_{\mathfrak{P}} \cup \{\alpha\}$;*
4. *if the system is without self-information, the subset $\overline{\{\alpha\}} \setminus \{\alpha\} = \langle \alpha \rangle_{\mathfrak{P}}$ of agents is a subsystem for any agent $\alpha \in A$.*

Proof

1. Since $\mathfrak{P}$ is transitive, then $\mathfrak{P} = \mathfrak{P}^{\infty}$. The inclusion $\mathfrak{P}^{\infty} \subset \mathfrak{S} = \mathfrak{P} \cup \Delta_A$ is a straightforward consequence of (9.32).
2. Since $\mathfrak{P}$ is transitive, we have that $\mathfrak{P}^2 \subset \mathfrak{P}$, and thus $\langle B \rangle_{\mathfrak{P}}^2 \subset \langle B \rangle_{\mathfrak{P}}$. We conclude by (9.30).
3. The equality $\mathfrak{S} = \mathfrak{P} \cup \Delta_A$ is equivalent to saying that, for any $\alpha \in A$, $\overline{\{\alpha\}} = \langle \alpha \rangle_{\mathfrak{P}} \cup \{\alpha\}$.
4. The subset $\overline{\{\alpha\}} \setminus \{\alpha\} = \langle \alpha \rangle_{\mathfrak{P}} \setminus \{\alpha\}$ is a subsystem by item 2, and $\langle \alpha \rangle_{\mathfrak{P}} = \langle \alpha \rangle_{\mathfrak{P}} \setminus \{\alpha\}$ when the system is without self-information (see item 4 in Proposition 9.21). □

9.4.3 The Information-Memory Relation $\mathfrak{M}$

The following definition of information-memory is inspired by [81] and is generalized in [13].

Definition 9.32 *([13])* With any agent $\alpha \in A$, we associate the subset $\langle \alpha \rangle_{\mathfrak{M}}$ of agents who pass on their information to α:

$$\langle \alpha \rangle_{\mathfrak{M}} := \left\{ \beta \in A \;\middle|\; \mathfrak{I}_\beta \subset \mathfrak{I}_\alpha \right\}. \tag{9.33}$$

We define an *information-memory* binary relation $\mathfrak{M}$ on A by

$$\beta \,\mathfrak{M}\, \alpha \iff \beta \in \langle \alpha \rangle_{\mathfrak{M}} \iff \mathfrak{I}_\beta \subset \mathfrak{I}_\alpha, \;\; \forall (\alpha, \beta) \in A^2. \tag{9.34}$$

When $\beta \,\mathfrak{M}\, \alpha$, we say that agent β information is *remembered by* or *passed on to* agent α, or that the information of agent β is embedded in the information of agent α. When agent β belongs to $\langle \alpha \rangle_{\mathfrak{M}}$, the information available to β is also available to agent α.

By construction, the subset of agents $\langle \alpha \rangle_{\mathfrak{M}}$ is the largest subset $B \subset A$ such that $\mathfrak{I}_B \subset \mathfrak{I}_\alpha$. Therefore, we have that

$$B \subset \langle \alpha \rangle_{\mathfrak{M}} \iff \mathfrak{I}_B \subset \mathfrak{I}_\alpha, \tag{9.35}$$

and, in particular, that

$$\alpha \in \langle \alpha \rangle_{\mathfrak{M}} \text{ and } \mathfrak{I}_{\langle \alpha \rangle_{\mathfrak{M}}} = \mathfrak{I}_\alpha. \tag{9.36}$$

Example 9.33 In Example 9.9, we have that $\langle 0 \rangle_{\mathfrak{M}} = \{0\}$ and $\langle 1 \rangle_{\mathfrak{M}} = \{0, 1\}$ in the classical pattern, whereas $\langle 1 \rangle_{\mathfrak{M}} = \{1\}$ in the Witsenhausen counterexample. △

Proposition 9.34 ([13]) *The* information-memory *relation $\mathfrak{M}$ is a pre-order, namely $\mathfrak{M}$ is reflexive and transitive.*

Proof The relation $\mathfrak{M}$ is reflexive since $\alpha \in \langle \alpha \rangle_{\mathfrak{M}}$ by construction. It is transitive by (9.34). □

The following Proposition 9.35 and Corollary 9.36 describe connections between the information-memory relation $\mathfrak{M}$ and the precedence binary relation $\mathfrak{P}$. For any $B \subset A$, let us introduce the set of agents

$$\langle B \rangle_{\mathfrak{M}} := \bigcup_{\beta \in B} \langle \beta \rangle_{\mathfrak{M}}, \tag{9.37}$$

who pass on their information to at least one agent in B.

Proposition 9.35 ([13]) *Let B and C be two subsets of the set A of agents. We have that*

$$\mathfrak{P}\mathfrak{M} \subset \mathfrak{P}, \tag{9.38}$$

and that, if a set C of agents pass on their information to at least one agent in B, then the predecessors of C are predecessors of B:

$$C \subset \langle B \rangle_{\mathfrak{M}} \Rightarrow \langle C \rangle_{\mathfrak{P}} \subset \langle B \rangle_{\mathfrak{P}}. \tag{9.39}$$

Proof We first show that $\mathfrak{I}_{\langle B \rangle_{\mathfrak{M}}} = \mathfrak{I}_B$. Indeed, we have that

$$\begin{aligned}
\mathfrak{I}_{\langle B \rangle_{\mathfrak{M}}} &= \bigvee_{\gamma \in \langle B \rangle_{\mathfrak{M}}} \mathfrak{I}_\gamma \text{ by (9.10)} \\
&= \bigvee_{\beta \in B} \bigvee_{\gamma \in \langle \beta \rangle_{\mathfrak{M}}} \mathfrak{I}_\gamma \text{ by (9.37)} \\
&= \bigvee_{\beta \in B} \mathfrak{I}_{\langle \beta \rangle_{\mathfrak{M}}} \text{ by (9.10)} \\
&= \bigvee_{\beta \in B} \mathfrak{I}_\beta \text{ by (9.36)} \\
&= \mathfrak{I}_B \text{ by (9.10).}
\end{aligned}$$

By (9.26), we deduce that

$$\langle \langle B \rangle_{\mathfrak{M}} \rangle_{\mathfrak{P}} = \langle B \rangle_{\mathfrak{P}}. \tag{9.40}$$

Then, (9.39) follows from the monotonocity of $\langle\ \rangle_{\mathfrak{P}}$ for the inclusion (see Definition 9.19):

$$C \subset \langle B\rangle_{\mathfrak{M}} \Rightarrow \langle C\rangle_{\mathfrak{P}} \subset \langle\langle B\rangle_{\mathfrak{M}}\rangle_{\mathfrak{P}} = \langle B\rangle_{\mathfrak{P}}.$$

Regarding (9.38), consider three agents α, β and γ such that $\beta\,\mathfrak{P}\,\gamma$ and $\gamma\,\mathfrak{M}\,\alpha$. By (9.22) and (9.34), we have that $\beta \in \langle\gamma\rangle_{\mathfrak{P}}$ and $\gamma \in \langle\alpha\rangle_{\mathfrak{M}}$. By (9.39) with $C = \{\gamma\}$ and $B = \{\alpha\}$, we deduce that $\langle\gamma\rangle_{\mathfrak{P}} \subset \langle\alpha\rangle_{\mathfrak{P}}$. As a consequence, $\beta \in \langle\alpha\rangle_{\mathfrak{P}}$, that is, $\beta\,\mathfrak{P}\,\alpha$ by (9.22). Thus, by the definition of the composition relation introduced in Sect. 3.2.1, we conclude that $\mathfrak{P}\mathfrak{M} \subset \mathfrak{P}$. □

The converse (3.9) of the precedence relation $\mathfrak{P}$ is the *successor relation* $\mathfrak{P}^{-1}$, characterized by

$$\beta\,\mathfrak{P}^{-1}\,\alpha \iff \alpha\,\mathfrak{P}\,\beta. \tag{9.41}$$

Quite naturally, β is a successor of α iff α is a predecessor of β. The following corollary states that, in absence of self-information, information-memory and successor are mutually exclusive relations in the sense that if β succeeds α, then the information of agent β cannot be passed on to agent α. In other words, if the information of agent β is passed on to agent α and that α influences β, then agent α is auto-influenced (that is, self-information holds true).

Corollary 9.36 *In absence of self-information, we have that* $\mathfrak{M} \cap \mathfrak{P}^{-1} = \emptyset$ *or, equivalently, that*

$$\beta\,\mathfrak{P}^{-1}\,\alpha \Rightarrow \neg(\beta\,\mathfrak{M}\,\alpha) \iff \beta\,\mathfrak{M}^{c}\,\alpha, \tag{9.42}$$

where the complementary operator ^{c} *is defined in (3.20),*

Proof By Proposition 9.21, a system without self-information is one for which $\alpha \notin \langle\alpha\rangle_{\mathfrak{P}}$, whatever the agent $\alpha \in A$. Consider two agents α and β in A. We have that

$$\begin{aligned}
\beta\,\mathfrak{M}\,\alpha &\Rightarrow \beta \in \langle\alpha\rangle_{\mathfrak{M}} \text{ by Definition 9.32}\\
&\Rightarrow \langle\alpha\rangle_{\mathfrak{P}} \supset \langle\beta\rangle_{\mathfrak{P}} \text{ by Proposition 9.35}\\
&\Rightarrow \alpha \notin \langle\beta\rangle_{\mathfrak{P}} \text{ since } \alpha \notin \langle\alpha\rangle_{\mathfrak{P}}\\
&\Rightarrow \neg(\alpha\,\mathfrak{P}\,\beta) \text{ by (9.22), which defines } \mathfrak{P}\\
&\Rightarrow \neg(\beta\,\mathfrak{P}^{-1}\,\alpha) \text{ by (9.41).}
\end{aligned}$$

This completes the proof, because $\beta\,\mathfrak{M}\,\alpha \Rightarrow \neg(\beta\,\mathfrak{P}^{-1}\,\alpha)$ is equivalent to (9.42), and to $\mathfrak{M} \cap \mathfrak{P}^{-1} = \emptyset$. □

9.4.4 The Decision-Memory Relation $\mathfrak{D}$

The following decision-memory relation is not, to our knowledge, found in the literature. It proves useful to express strictly classical and strictly quasiclassical systems in Sect. 9.5.1, and to formulate appropriate assumptions for policy independence of conditional expectations in Sect. 9.6.

Definition 9.37 With any agent $\alpha \in A$, we associate

$$\langle\alpha\rangle_{\mathfrak{D}} = \left\{\beta \in A \;\middle|\; \mathcal{D}_\beta \subset \mathcal{I}_\alpha\right\}, \tag{9.43}$$

the subset of agents β whose decision is passed on to α, where the decision subfield $\mathcal{D}_\beta$ is defined in (9.7).

We define a *decision-memory* binary relation $\mathfrak{D}$ on A by

$$\beta \,\mathfrak{D}\, \alpha \iff \beta \in \langle\alpha\rangle_{\mathfrak{D}} \iff \mathcal{D}_\beta \subset \mathcal{I}_\alpha, \;\; \forall(\alpha, \beta) \in A^2. \tag{9.44}$$

When $\beta \,\mathfrak{D}\, \alpha$, we say that agent β decision is *remembered by* or *passed on to* agent α, or that the decision of agent β is embedded in the information of agent α. By (9.43) and (9.7), we have that

$$\mathcal{D}_{\langle\alpha\rangle_{\mathfrak{D}}} \subset \mathcal{I}_\alpha. \tag{9.45}$$

By comparing the definition of the decision-memory relation $\mathfrak{D}$ with that of the precedence relation $\mathfrak{P}$ in Definition 9.15, and by the definition (9.7) of $\mathcal{D}_B$, we see that

$$\mathcal{D}_{\langle\alpha\rangle_{\mathfrak{D}}} = \mathcal{U}_{\langle\alpha\rangle_{\mathfrak{D}}} \otimes \{\emptyset, \Omega\} \subset \mathcal{I}_\alpha \subset \mathcal{U}_{\langle\alpha\rangle_{\mathfrak{P}}} \otimes \mathcal{F}. \tag{9.46}$$

Therefore, we conclude that

$$\langle\alpha\rangle_{\mathfrak{D}} \subset \langle\alpha\rangle_{\mathfrak{P}}, \;\; \forall\alpha \in A, \tag{9.47}$$

or, equivalently, that

$$\mathfrak{D} \subset \mathfrak{P}. \tag{9.48}$$

Remark 9.38 If $\beta \,\mathfrak{D}\, \alpha$, the decision made by agent β decision is passed on to agent α and, by the fact that $\mathfrak{D} \subset \mathfrak{P}$, β is a predecessor of α. However, the agent β can be a predecessor of α, but his influence may happen without passing on his decision to α. For instance, in the Witsenhausen counterexample discussed in Sect. 4.2, agent $t = 1$ is influenced by agent $t = 0$, but this is made through the observation $\boldsymbol{X}_0 + \boldsymbol{U}_0 + \boldsymbol{W}_1$, and not by passing the first control $\boldsymbol{U}_0$. The following example is another illustration, where $\langle\alpha\rangle_{\mathfrak{D}} \neq \langle\alpha\rangle_{\mathfrak{P}}$ for at least one agent α. ◇

Example 9.39 In Example 9.9, we have that $\langle 0\rangle_{\mathfrak{P}} = \emptyset$, $\langle 1\rangle_{\mathfrak{P}} = \{0\}$, $\langle 0\rangle_{\mathfrak{D}} = \emptyset$ and $\langle 1\rangle_{\mathfrak{D}} = \{0\}$ in the classical pattern. In the Witsenhausen counterexample, we have that $\langle 0\rangle_{\mathfrak{P}} = \emptyset$, $\langle 1\rangle_{\mathfrak{P}} = \{0\}$, $\langle 0\rangle_{\mathfrak{D}} = \emptyset$ and $\langle 1\rangle_{\mathfrak{D}} = \emptyset$. △

9.5 A Typology of Stochastic Control Systems

Equipped with the four binary relations between agents, exposed in Sect. 9.4, we display a typology of systems in Sect. 9.5.1. Then, we provide examples of systems composed of two agents in Sect. 9.5.2. We focus on partially nested and on sequential systems in Sect. 9.5.3, and end with a summary table of systems in Sect. 9.5.4.

9.5.1 A Typology of Systems

We gather here definitions found in the literature.

A *station* [157] is a subset of agents such that the set of information fields of these agents is totally ordered under inclusion (i.e., nested). In other words, a subset B of agents in A is a station iff the information-memory relation $\mathfrak{M}$ induces a total order on B (i.e. consists of a chain of length $m = \operatorname{card} B$) iff there exists an ordering $(\alpha_1, \ldots, \alpha_m)$ of B such that

$$\mathfrak{I}_{\alpha_1} \subset \cdots \subset \mathfrak{I}_{\alpha_k} \subset \mathfrak{I}_{\alpha_{k+1}} \subset \cdots \subset \mathfrak{I}_{\alpha_m}, \tag{9.49}$$

or, equivalently, that $\alpha_k \in \langle \alpha_{k+1} \rangle_{\mathfrak{M}}$, for $k = 1, \ldots, m$.

An agent, or a subset of agents, enjoys *perfect recall* when the subsystem generated (see Definition 9.26) forms a station.

A *static team* [159] is a subset B of A such that $\langle B \rangle_{\mathfrak{P}} = \emptyset$ or, equivalently, that agents in B have no predecessors (see (9.25)). When the whole set A of agents is a static team, any agent $\alpha \in A$ has no predecessor: $\langle \alpha \rangle_{\mathfrak{P}} = \emptyset$, $\forall \alpha \in A$. Equivalently, the precedence relation $\mathfrak{P}$ is empty or, equivalently, the subsystem relation $\mathfrak{S}$ is reduced to the equality relation Δ_A: the system exhibits no dynamic relations between agents (see Fig. 9.3 for an illustration). A system is *static* if the set A of agents is a static team.

A system is *monic* [159] iff it is reduced to a singleton that is a static team: $A = \{\alpha\}$ and $\mathfrak{I}_\alpha \subset \{\emptyset, \mathbb{U}_\alpha\} \otimes \mathfrak{F}$.

A system is said to be *sequential* [159] iff there exists an ordering $(\alpha_1, \ldots, \alpha_n)$ of A such that each agent α_k is influenced at most by the "previous" agents $\alpha_1, \ldots, \alpha_{k-1}$ that is,

$$\langle \alpha_1 \rangle_{\mathfrak{P}} = \emptyset \text{ and } \langle \alpha_k \rangle_{\mathfrak{P}} \subset \{\alpha_1, \ldots, \alpha_{k-1}\}, \ \forall k = 2, \ldots, n. \tag{9.50}$$

Equivalently, there exists an ordering $(\alpha_1, \ldots, \alpha_n)$ of A such that $\alpha_i \, \mathfrak{P} \, \alpha_j \Rightarrow i < j$ that is, an ordering strictly compatible with $\mathfrak{P}$ (see Sect. 3.2.1).

A system is *partially nested* [81, 83] iff the precedence relation $\mathfrak{P}$ is included in the information-memory relation $\mathfrak{M}$ that is, $\mathfrak{P} \subset \mathfrak{M}$, namely $\langle \alpha \rangle_{\mathfrak{P}} \subset \langle \alpha \rangle_{\mathfrak{M}}$ for any agent $\alpha \in A$. In a partially nested system, any agent knows what her predecessors know or, in other words, any predecessor of a given agent passes her information to

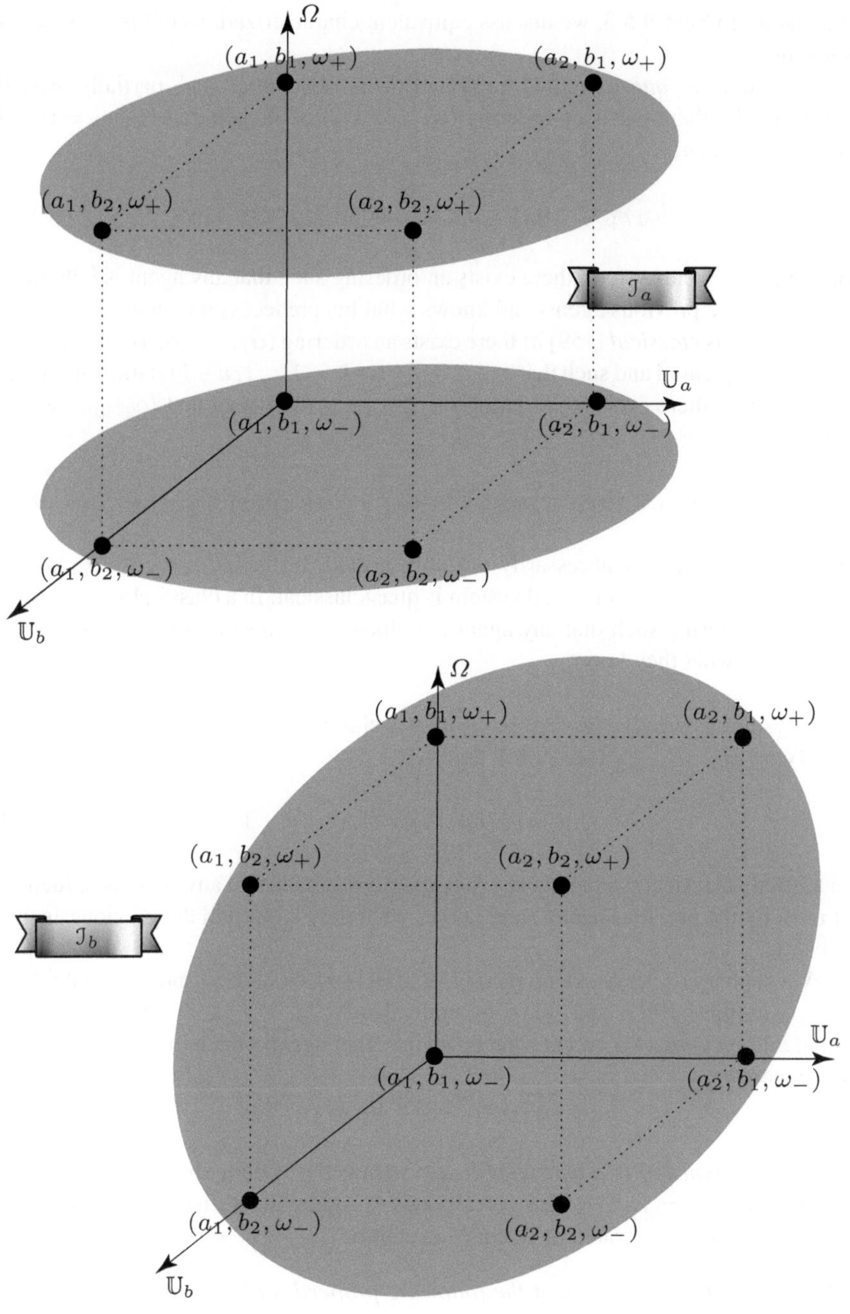

Fig. 9.3 A static team

that agent. In Sect. 9.5.3, we discuss equivalent characterizations of partially nested systems.

A system is *quasiclassical* [159] iff it is sequential and partially nested. Equivalently, there exists an ordering $(\alpha_1, \ldots, \alpha_n)$ of A such that $\langle\alpha_1\rangle_{\mathfrak{P}} = \emptyset$ and, for $k = 2, \ldots, n$,

$$\langle\alpha_k\rangle_{\mathfrak{P}} \subset \{\alpha_1, \ldots, \alpha_{k-1}\} \text{ and } \langle\alpha_k\rangle_{\mathfrak{P}} \subset \langle\alpha_k\rangle_{\mathfrak{M}}. \tag{9.51}$$

In a quasiclassical system, there exists an ordering such that any agent is influenced at most by the previous agents and knows what his predecessors know.

A system is *classical* [159] iff there exists an ordering $(\alpha_1, \ldots, \alpha_n)$ of A for which it is both sequential and such that $\mathfrak{I}_{\alpha_k} \subset \mathfrak{I}_{\alpha_{k+1}}$ for $k = 1, \ldots, n-1$ (station property). Equivalently, there exists an ordering $(\alpha_1, \ldots, \alpha_n)$ of A such that $\langle\alpha_1\rangle_{\mathfrak{P}} = \emptyset$ and, for $k = 2, \ldots, n$,

$$\langle\alpha_k\rangle_{\mathfrak{P}} \subset \{\alpha_1, \ldots, \alpha_{k-1}\} \subset \{\alpha_1, \ldots, \alpha_{k-1}, \alpha_k\} \subset \langle\alpha_k\rangle_{\mathfrak{M}}. \tag{9.52}$$

A classical system is necessarily partially nested, because $\langle\alpha_k\rangle_{\mathfrak{P}} \subset \langle\alpha_k\rangle_{\mathfrak{M}}$ for $k = 1, \ldots, n$; hence, a classical system is quasiclassical. In a classical system, there exists an ordering such that any agent is influenced at most by the previous agents and knows what they know.

A system is *strictly classical* [159] iff it is classical for an ordering $(\alpha_1, \ldots, \alpha_n)$ satisfying $\mathfrak{D}_{\alpha_k} \subset \mathfrak{I}_{\alpha_{k+1}}$ (see (9.8)), that is,

$$\alpha_k \in \langle\alpha_{k+1}\rangle_{\mathfrak{D}}, \ \forall k = 1, \ldots, n-1. \tag{9.53}$$

In a strictly classical system, there exists an ordering such that any agent is influenced at most by the previous agents, and knows what they know and the previous agent's decision.

A system is *strictly quasiclassical* [159] iff it is quasiclassical and such that $\beta \,\mathfrak{S}\, \alpha$, $\beta \neq \alpha$ implies $\beta \,\mathfrak{M}\, \alpha$ and $\beta \,\mathfrak{D}\, \alpha$ or, equivalently, $\mathfrak{I}_\beta \vee \mathfrak{D}_\beta \subset \mathfrak{I}_\alpha$. In other words, agent α knows what know and what do all the other agents which form the subsystem she generates:

$$\beta \in \overline{\{\alpha\}} \backslash \{\alpha\} \Rightarrow \beta \in \langle\alpha\rangle_{\mathfrak{M}} \cap \langle\alpha\rangle_{\mathfrak{D}}. \tag{9.54}$$

The *strict expansion of a quasiclassical system* is the system obtained by replacing $\mathfrak{I}_\alpha$ by $\mathfrak{I}_\alpha \vee \bigvee_{\beta \in \overline{\{\alpha\}} \backslash \{\alpha\}} \mathfrak{D}_\beta$. This corresponds to adding the decisions of those agents in the subsystem generated by agent α, except α.

Theorem 9.40 ([159]) *Any of the following properties of a system—monic, static team, strictly classical, classical, strictly quasiclassical, quasiclassical, sequential, causal, solvable, without self-information—is shared by all its subsystems.*

9.5.2 Examples of Systems with Two Agents

We now illustrate the typology in Sect. 9.5.1, and also causal and non causal systems as in Sect. 9.3, with systems composed of two agents (see Example 9.2 for the notations).

Static Team
The set of agents $A = \{a, b\}$ forms a static team iff

$$\mathfrak{I}_a \subset \{\emptyset, \mathbb{U}_a\} \otimes \{\emptyset, \mathbb{U}_b\} \otimes \mathfrak{F}, \ \ \mathfrak{I}_b \subset \{\emptyset, \mathbb{U}_a\} \otimes \{\emptyset, \mathbb{U}_b\} \otimes \mathfrak{F}.$$

There are no interdecisions between agents, just a dependence upon states of Nature. Figure 9.3 is an illustration where agent b knows nothing, whereas agent a knows only the state of Nature, corresponding to $\mathfrak{I}_b = \{\emptyset, \mathbb{H}\}$, $\mathfrak{I}_a = \{\emptyset, \mathbb{U}_a\} \otimes \{\emptyset, \mathbb{U}_b\} \otimes \mathfrak{F}$. The precedence relation is $\mathfrak{P} = \emptyset$ ($\langle a \rangle_{\mathfrak{P}} = \langle b \rangle_{\mathfrak{P}} = \emptyset$); the subsystem relation is $\mathfrak{S} = \{(a, a), (b, b)\}$ ($\overline{\{a\}} = \{a\}$, $\overline{\{b\}} = \{b\}$); in the specific case corresponding to Fig. 9.3, the information-memory relation is $\mathfrak{M} = \{(a, a), (b, b), (b, a)\}$ ($\langle a \rangle_{\mathfrak{M}} = \{a\}$, $\langle b \rangle_{\mathfrak{M}} = \{a, b\}$); the decision-memory relation is $\mathfrak{D} = \emptyset$ ($\langle a \rangle_{\mathfrak{D}} = \langle b \rangle_{\mathfrak{D}} = \emptyset$).

Station
The set of agents $A = \{a, b\}$ forms a station iff

$$\mathfrak{I}_a \subset \mathfrak{I}_b \text{ or } \mathfrak{I}_b \subset \mathfrak{I}_a.$$

Example of Classical System
The set of agents $A = \{a, b\}$ with information fields given by

$$\mathfrak{I}_a = \{\emptyset, \mathbb{U}_a\} \otimes \{\emptyset, \mathbb{U}_b\} \otimes \mathfrak{F}, \ \ \mathfrak{I}_b = \mathfrak{U}_a \otimes \{\emptyset, \mathbb{U}_b\} \otimes \mathfrak{F},$$

corresponding to Fig. 9.4, forms a classical system. Indeed, first, the system is sequential where a precedes b ($\langle a \rangle_{\mathfrak{P}} = \emptyset$ and $a \in \langle b \rangle_{\mathfrak{P}}$): agent a observes the state of Nature and makes her decision accordingly; agent b observes both agent a's decision and the state of Nature and makes her decision based on that information. Second, one has that $\mathfrak{I}_a \subset \mathfrak{I}_b$ ($a \in \langle b \rangle_{\mathfrak{M}}$), which may be interpreted in different ways. One may say that agent a communicates her own information to agent b. If agent a is an individual at time $t = 0$, whereas agent b is the same individual at time $t = 1$, one may say that the information is not forgotten with time (memory of past knowledge). The precedence relation is $\mathfrak{P} = \{(a, b)\}$ ($\langle a \rangle_{\mathfrak{P}} = \emptyset$, $\langle b \rangle_{\mathfrak{P}} = \{a\}$); the subsystem relation is $\mathfrak{S} = \{(a, a), (b, b), (a, b)\}$ ($\overline{\{a\}} = \{a\}$, $\overline{\{b\}} = \{a, b\}$); the information-memory relation is $\mathfrak{M} = \{(a, a), (b, b), (a, b)\}$ ($\langle a \rangle_{\mathfrak{M}} = \{a\}$, $\langle b \rangle_{\mathfrak{M}} = \{a, b\}$); the decision-memory relation is $\mathfrak{D} = \{(a, b)\}$ ($\langle a \rangle_{\mathfrak{D}} = \emptyset$, $\langle b \rangle_{\mathfrak{D}} = \{a\}$).

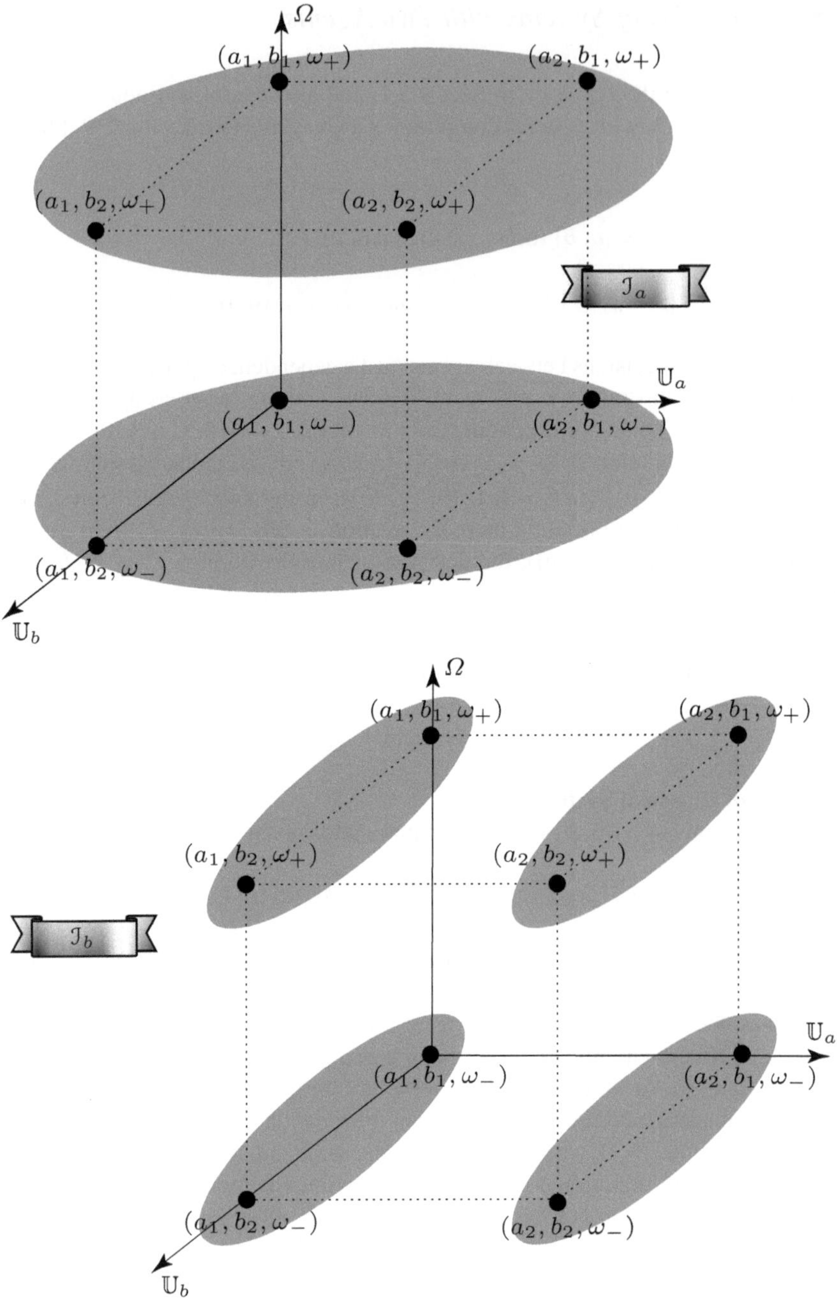

Fig. 9.4 A classical system

Example of Sequential but Not Classical System
The set of agents $A = \{a, b\}$ with information fields given by

$$\mathfrak{I}_a = \{\emptyset, \mathbb{U}_a\} \otimes \{\emptyset, \mathbb{U}_b\} \otimes \mathfrak{F}, \ \mathfrak{I}_b = \mathfrak{U}_a \otimes \{\emptyset, \mathbb{U}_b\} \otimes \{\emptyset, \Omega\}$$

forms a sequential system which is not classical (see Fig. 9.5). Indeed, agent a precedes agent b ($\langle a\rangle_{\mathfrak{P}} = \emptyset$ and $a \in \langle b\rangle_{\mathfrak{P}}$), but $\mathfrak{I}_a$ and $\mathfrak{I}_b$ are not comparable: agent a observes only the state of Nature, whereas agent b observes only agent a's decision. The precedence relation is $\mathfrak{P} = \{(a, b)\}$ ($\langle a\rangle_{\mathfrak{P}} = \emptyset$, $\langle b\rangle_{\mathfrak{P}} = \{a\}$); the subsystem relation is $\mathfrak{S} = \{(a, a), (b, b), (a, b)\}$ ($\overline{\{a\}} = \{a\}$, $\overline{\{b\}} = \{a, b\}$); the information-memory relation is $\mathfrak{M} = \{(a, a), (b, b)\}$ ($\langle a\rangle_{\mathfrak{M}} = \{a\}$, $\langle b\rangle_{\mathfrak{M}} = \{b\}$); the decision-memory relation is $\mathfrak{D} = \{(a, b)\}$ ($\langle a\rangle_{\mathfrak{D}} = \emptyset$, $\langle b\rangle_{\mathfrak{D}} = \{a\}$).

Example of Non Causal System (Deadlock)
The set of agents $A = \{a, b\}$ with information fields given by

$$\mathfrak{I}_a = \{\emptyset, \mathbb{U}_a\} \otimes \mathfrak{U}_b \otimes \{\emptyset, \Omega\}, \ \mathfrak{I}_b = \mathfrak{U}_a \otimes \{\emptyset, \mathbb{U}_b\} \otimes \{\emptyset, \Omega\}$$

corresponds to Fig. 9.6. Such a system displays a *deadlock* situation [3] where the decision process in (9.15) may have no solution, one solution, or multiple solutions. Agent a observes agent b's decision, whereas agent b observes agent a's decision. Thus, agent a precedes agent b ($a \in \langle b\rangle_{\mathfrak{P}}$) and agent b precedes agent a ($b \in \langle a\rangle_{\mathfrak{P}}$).

This system is not causal (hence, not sequential). Indeed, following the definition of causality given in Sect. 9.3.2, consider a history-ordering mapping φ from $\mathbb{H}$ towards $\{(a, b), (b, a)\}$ and denote by $\mathbb{H}_a = \varphi^{-1}\big((a, b)\big)$ those histories for which agent a is the first, and similarly for $\mathbb{H}_b$. For causality to hold true, the trace of the information field $\mathfrak{I}_a$ of agent a on $\mathbb{H}_a$ must belong to $\{\emptyset, \mathbb{U}_a\} \otimes \{\emptyset, \mathbb{U}_b\} \otimes \mathfrak{F}$. Now, the atoms of the partition field $\mathfrak{I}_a$ are cylinders with base in $\mathbb{U}_b$ (facing vertical disks in Fig. 9.6). Hence, the intersection of $\mathbb{H}_a$ with facing vertical disks must be composed of the atoms of $\{\emptyset, \mathbb{U}_a\} \otimes \{\emptyset, \mathbb{U}_b\} \otimes \mathfrak{F}$, which are horizontal disks. This is impossible. The same reasoning applies to $\mathbb{H}_b$. No history-ordering mapping φ satisfying the conditions exposed in Sect. 9.3.2 exists: the system is not causal.

The precedence relation is $\mathfrak{P} = \{(a, b), (b, a)\}$ ($\langle a\rangle_{\mathfrak{P}} = \{b\}$, $\langle b\rangle_{\mathfrak{P}} = \{a\}$); the subsystem relation is $\mathfrak{S} = \{(a, a), (b, b), (a, b), (b, a)\}$ ($\overline{\{a\}} = \overline{\{b\}} = \{a, b\}$); the information-memory relation is $\mathfrak{M} = \{(a, a), (b, b)\}$ ($\langle a\rangle_{\mathfrak{M}} = \{a\}$, $\langle b\rangle_{\mathfrak{M}} = \{b\}$); the decision-memory relation is $\mathfrak{D} = \{(a, b), (b, a)\}$ ($\langle a\rangle_{\mathfrak{D}} = \{b\}$, $\langle b\rangle_{\mathfrak{D}} = \{a\}$).

Remark 9.41 Notice that the system is non sequential but that, however, the precedence relation $\mathfrak{P}$ is not empty. ◊

Example of Causal but Non Sequential System
The set of agents $A = \{a, b\}$ with information fields given by

$$\mathfrak{I}_a = \sigma(\{a_1, a_2\} \times \{b_1, b_2\} \times \{\omega_+\}, \{a_1, a_2\} \times \{b_1\} \times \{\omega_-\}),$$
$$\mathfrak{I}_b = \sigma(\{a_1, a_2\} \times \{b_1, b_2\} \times \{\omega_-\}, \{a_1\} \times \{b_1, b_2\} \times \{\omega_+\})$$

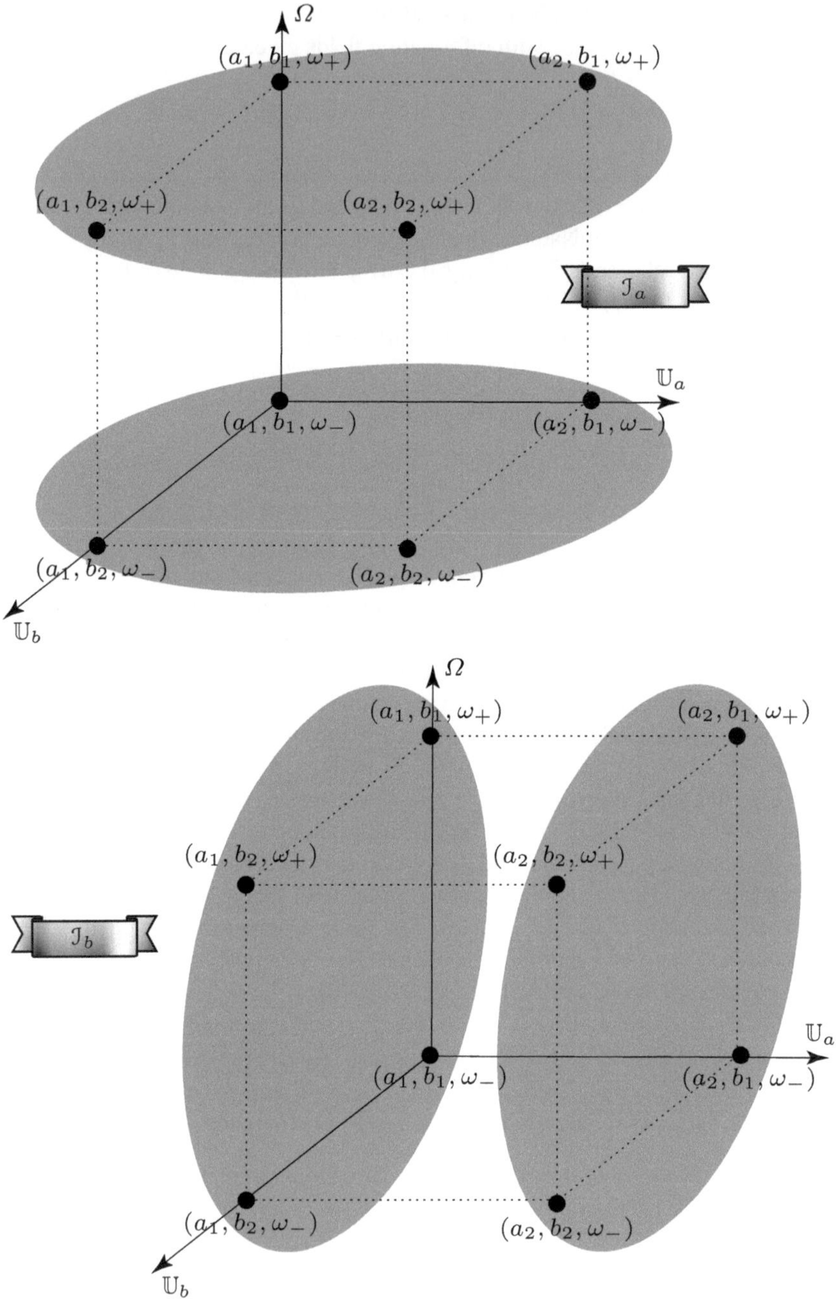

Fig. 9.5 A sequential but not classical system

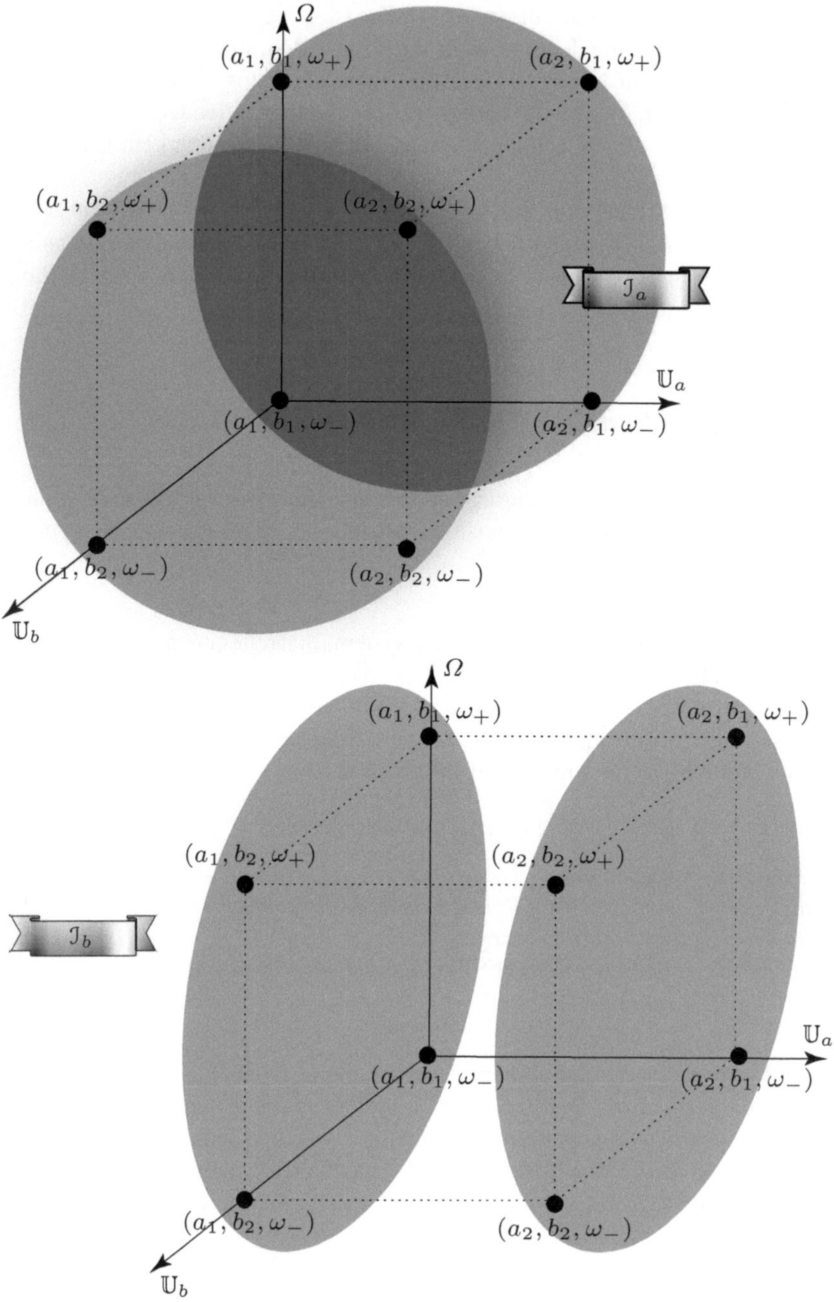

Fig. 9.6 An information structure with deadlock

corresponds to Fig. 9.7. When the state of Nature is ω_+, agent a only sees ω_+, whereas agent b sees ω_+ and the decision of a: thus a acts first, then b. The reverse holds true when the state of Nature is ω_-. Thus, there are history-ordering mappings φ from $\mathbb{H}$ towards $\{(a, b), (b, a)\}$ (see Sect. 9.3.2), but they differ according to history ($\varphi((u_a, u_b, \omega_+)) = (a, b)$ and $\varphi((u_a, u_b, \omega_-)) = (b, a)$): the system is causal but not sequential.

The precedence relation is $\mathfrak{P} = \{\emptyset\}$ ($\langle a\rangle_{\mathfrak{P}} = \langle b\rangle_{\mathfrak{P}} = \emptyset$), the subsystem relation is $\mathfrak{S} = \{(a, a), (b, b)\}$ ($\overline{\{a\}} = \{a\}$, $\overline{\{b\}} = \{b\}$); the information-memory relation is $\mathfrak{M} = \{(a, a), (b, b)\}$ ($\langle a\rangle_{\mathfrak{M}} = \{a\}$, $\langle b\rangle_{\mathfrak{M}} = \{b\}$); the decision-memory relation is $\mathfrak{D} = \emptyset$ ($\langle a\rangle_{\mathfrak{D}} = \langle b\rangle_{\mathfrak{D}} = \emptyset$).

9.5.3 Partially Nested and Sequential Systems

Here, we first provide a characterization of sequential systems. Then, we return to the so-called partially nested systems. We conclude with a result connecting partially nested and sequential systems.

Sequential Systems

We provide a characterization of sequential systems that relies upon notions on binary relations recalled in Sect. 3.2.1.

The equivalence of the first two assertions in the following Theorem 9.42 is due to Witsenhausen in [159]. The others are new. They prove useful for characterizing partially nested systems without self-information.

Theorem 9.42 *The following assertions are equivalent:*

1. *the system is sequential;*
2. *the system is without self-information and the subsystem pre-order $\mathfrak{S}$ is an order*[3]*;*
3. *the precedence relation $\mathfrak{P}$ is* acyclic[4]*;*
4. *the complementary relation $(\mathfrak{P}^\infty)^c$ of the transitive closure $\mathfrak{P}^\infty$ of the precedence relation $\mathfrak{P}$ is reflexive.*

Proof The proof is a straightforward transcription of Proposition 3.1 with $\mathfrak{R} = \mathfrak{P}$.

For this, recall from Sect. 9.5.1 that a system is sequential iff there exists an ordering of A strictly compatible with $\mathfrak{P}$. Also, a system is without self-information iff $\mathfrak{P}^c$ is reflexive (see Proposition 9.21). At last, the pre-order $\mathfrak{S}$ is an order iff $\mathfrak{S} = \mathfrak{P}^*$ is an order. □

[3]That is, $\mathfrak{S}$ is antisymmetric: $\forall(\alpha, \beta) \in A^2, \overline{\{\alpha\}} = \overline{\{\beta\}} \Rightarrow \alpha = \beta$ or, in other words, if two agents generate the same subsystem, they must be equal.

[4]That is, $\alpha \notin \langle\alpha\rangle_{\mathfrak{P}}^k, \forall\alpha \in A, \forall k \geq 1$.

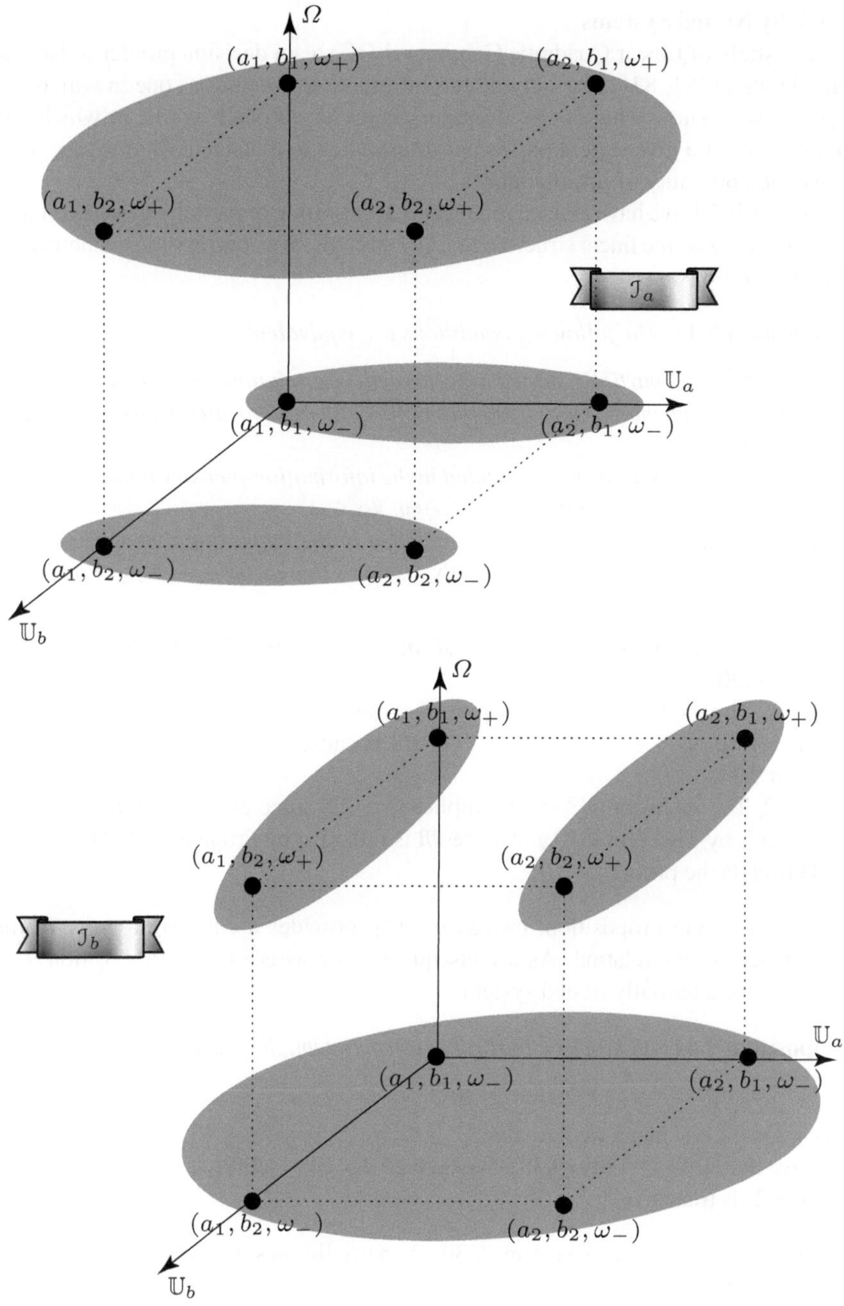

Fig. 9.7 A causal but not sequential system

Partially Nested Systems

In their study of Linear Quadratic Gaussian (LQG) team decision problems, Ho and Chu define in [81, 83] a partially nested information structure as one in which any agent always know what her predecessors know or, in other words, in which any predecessor of a given agent passes her information to that agent. In that case, they prove the optimality of linear controls.

In Sect. 9.5.1, we have given a more general definition of partially nested systems, not depending on the linear structure of LQG systems. We now provide mathematical characterizations.

Proposition 9.43 *The following conditions are equivalent.*

1. *The system is partially nested (the precedence relation $\mathfrak{P}$ is included in the information-memory relation $\mathfrak{M}$, that is, $\mathfrak{P} \subset \mathfrak{M}$ or, equivalently, $\langle\alpha\rangle_{\mathfrak{P}} \subset \langle\alpha\rangle_{\mathfrak{M}}$, $\forall \alpha \in A$).*
2. *The subsystem relation $\mathfrak{S}$ is included in the information-memory relation $\mathfrak{M}$, that is, $\mathfrak{S} \subset \mathfrak{M}$ or, equivalently, $\overline{\{\alpha\}} \subset \langle\alpha\rangle_{\mathfrak{M}}$, $\forall \alpha \in A$.*
3. *The transitive closure $\mathfrak{P}^{\infty}$ of $\mathfrak{P}$ is included in the information-memory relation $\mathfrak{M}$, that is, $\mathfrak{P}^{\infty} \subset \mathfrak{M}$ or, equivalently, $\langle\alpha\rangle_{\mathfrak{P}}^{k} \subset \langle\alpha\rangle_{\mathfrak{M}}$, $\forall \alpha \in A$, $\forall k \geq 1$.*

Proof 2 $\Rightarrow$ 1. The inclusion $\mathfrak{S} \subset \mathfrak{M}$ implies $\mathfrak{P} \subset \mathfrak{M}$, since $\mathfrak{P} \subset \mathfrak{S}$ by Theorem 9.30.

1 $\Rightarrow$ 3. The inclusion $\mathfrak{P} \subset \mathfrak{M}$ implies that transitive closures satisfy $\mathfrak{P}^{\infty} \subset \mathfrak{M}^{\infty}$; therefore, $\mathfrak{P}^{\infty} \subset \mathfrak{M}$ since $\mathfrak{M}^{\infty} = \mathfrak{M}$ because $\mathfrak{M}$ is transitive, by Proposition 9.34.

3 $\Rightarrow$ 2. The inclusion $\mathfrak{P}^{\infty} \subset \mathfrak{M}$ implies $\mathfrak{S} \subset \mathfrak{M}$ since $\mathfrak{S}$ is the reflexive closure of $\mathfrak{P}^{\infty}$ by Theorem 9.30 and since $\mathfrak{M}$ is reflexive by Proposition 9.34.

This ends the proof. □

The following proposition, proved in [13], provides a curious property of the precedence binary relation. As a consequence, the assertions of Proposition 9.31 hold true for a partially nested system.

Proposition 9.44 ([13]) *For a partially nested system, the precedence relation $\mathfrak{P}$ is transitive.*

Proof On the one hand, we have that $\mathfrak{P} \subset \mathfrak{M}$ by Proposition 9.43. On the other hand, we have that $\mathfrak{P}\mathfrak{M} \subset \mathfrak{P}$ by (9.38). We deduce that $\mathfrak{P}^2 \subset \mathfrak{P}\mathfrak{M} \subset \mathfrak{P}$: the precedence relation $\mathfrak{P}$ is transitive. □

As a consequence of Theorem 9.30, a partially nested system satisfies the following inclusions and equalities:

$$\mathfrak{P} = \mathfrak{P}^{\infty} \subset \mathfrak{S} = \mathfrak{P} \cup \Delta_A \subset \mathfrak{M}. \tag{9.55}$$

We conclude with a result connecting partially nested and sequential systems.

Theorem 9.45 *A partially nested system without self-information is sequential. As a consequence, a causal partially nested system is sequential (hence is quasiclassical).*

Proof On the one hand, by Proposition 9.44, a partially nested system satisfies $\mathfrak{P} = \mathfrak{P}^{\infty}$. On the other hand, by Proposition 9.21, a system without self-information is such that $\mathfrak{P}^{c}$ is reflexive. Thus, $(\mathfrak{P}^{\infty})^{c}$ is reflexive and we conclude by Theorem 9.42 that the system is sequential. Hence, by the definition introduced in Sect. 9.5.1, a partially nested system without self-information is necessarily quasiclassical. □

9.5.4 Summary Table

We have presented four binary relations between agents, namely the precedence, subsystem, information-memory and decision-memory relations. They provide a way to classify systems, be they sequential or not.

Now, we summarize prominent properties of the precedence, subsystem, information-memory and decision-memory binary relations in Table 9.1. For instance,

Table 9.1 Summary of properties of the precedence, subsystem, information-memory and decision-memory binary relations between agents

		Relations between agents					
		Precedence		Subsystem		Information-memory	Decision-memory
		$\mathfrak{P}$		$\mathfrak{S}$		$\mathfrak{M}$	$\mathfrak{D}$
Properties		$\mathfrak{P}^{*} = \mathfrak{S}$		Pre-order		Pre-order	$\mathfrak{D} \subset \mathfrak{P}$
No self-information	$\Longleftrightarrow$	$\mathfrak{P}^{c}$ reflexive					
No self-information	$\Rightarrow$	$\mathfrak{M} \cap \mathfrak{P}^{-1} = \emptyset$					
Station	$\Longleftrightarrow$					Total pre-order	
Static team	$\Longleftrightarrow$	$\mathfrak{P} = \emptyset$					
Static team	$\Longleftrightarrow$			$\mathfrak{S} = \Delta_{A}$			
Sequential	$\Longleftrightarrow$	Acyclic					
Sequential	$\Longleftrightarrow$	$(\mathfrak{P}^{\infty})^{c}$ reflexive					
Sequential	$\Longleftrightarrow$	$\mathfrak{P}^{c}$ reflexive	and	Order			
Sequential	$\Longleftrightarrow$	$\mathfrak{P}^{c}$ reflexive	and	Antisym.			
Quasiclassical	$\Longleftrightarrow$	Acyclic			and	$\mathfrak{M} \supset \mathfrak{S} \supset \mathfrak{P}$	
Quasiclassical	$\Longleftrightarrow$	$(\mathfrak{P}^{\infty})^{c}$ reflexive			and	$\mathfrak{M} \supset \mathfrak{S} \supset \mathfrak{P}$	
Classical	$\Rightarrow$	Acyclic				Total pre-order	
Partially nested	$\Longleftrightarrow$					$\mathfrak{M} \supset \mathfrak{S} \supset \mathfrak{P}$	
Partially nested	$\Rightarrow$	$\mathfrak{P} = \mathfrak{P}^{\infty}$					

a system is quasiclassical iff $\mathfrak{S}$ is a pre-order and $\mathfrak{M} \supset \mathfrak{S} \supset \mathfrak{P}$; a system without self-information is such that $\mathfrak{M} \cap \mathfrak{P}^{-1} = \emptyset$.

9.6 Policy Independence of Conditional Expectations and Dynamic Programming

We now examine optimization issues. In the Witsenhausen intrinsic model introduced in Sect. 9.2, we suppose that the universe Ω (with the σ-field $\mathcal{F}$) is equipped with a probability $\mathbb{P}$. Supposing that the solvability/measurability property holds true (see Definition 9.10), the solution map $S_\lambda : \Omega \to \mathbb{H}$, from the universe Ω to the history space $\mathbb{H}$, can be defined (see Definition 9.11) and is measurable, for any collection $\lambda \in \Lambda_A^{\mathrm{ad}}$ of admissible policies (see (9.13)). Given a measurable and bounded[5] criterion $\tilde{j} : \mathbb{H} \to \mathbb{R}$, the optimization problem we consider is

$$\min_{\lambda \in \Lambda_A^{\mathrm{ad}}} \mathbb{E}(\tilde{j} \circ S_\lambda), \tag{9.56}$$

where the expectation $\mathbb{E}$ is taken w.r.t. $\mathbb{P}$ on $(\Omega, \mathcal{F})$.

In Sect. 9.6.1, we recall Witsenhausen's notion of policy independence of conditional expectations and the prominent conditions under which it holds true. Thanks to the precedence, subsystem, information-memory and decision-memory relations exposed in Sect. 9.4, we discuss in Sect. 9.6.2 the Witsenhausen's conditions making it possible to decompose an optimization problem into subproblems.

9.6.1 Policy Independence of Conditional Expectations

In [160], Witsenhausen considers an observer whose information is characterized by a subfield $\mathcal{G}$ of the history field $\mathcal{H}$. Once the agents play according to a collection λ of admissible policies, the solution map $S_\lambda : \Omega \to \mathbb{H}$ (supposed to exist) generates a subfield $S_\lambda^{-1}(\mathcal{G})$ of the σ-field $\mathcal{F}$ on the universe Ω (see Remark 9.12). In general, conditional expectations w.r.t. $S_\lambda^{-1}(\mathcal{G})$ indeed depend on the collection λ of admissible policies. The opposite situation when it does not is of particular interest. Following [160], we define policy independence of conditional expectations as follows.

Definition 9.46 Suppose that the solvability/measurability property holds true (see Definition 9.10), and recall that $S_\lambda : \Omega \to \mathbb{H}$ denotes the solution map introduced in Definition 9.11.

---

[5] Bounded is for the sake of simplicity, so that integrability with respect to the probability $\mathbb{P}$ is easily established. A nonnegative criterion would also do.

Consider a bounded[6] measurable function $\tilde{j} : (\mathbb{H}, \mathcal{H}) \to \mathbb{R}$, and a subfield $\mathcal{G}$ of the history field $\mathcal{H}$. The *policy independence of conditional expectations property* (PICE) holds true if there exists a $\mathcal{G}$-measurable function $J : \mathbb{H} \to \mathbb{R}$ such that, for any collection $\lambda \in \Lambda_A^{\mathrm{ad}}$ of admissible policies, one has that

$$\mathbb{E}\big(\tilde{j} \circ S_\lambda \mid S_\lambda^{-1}(\mathcal{G})\big) = J \circ S_\lambda, \quad \mathbb{P}\text{-a.s.}. \tag{9.57}$$

The crucial requirement in PICE is that the function $J : \mathbb{H} \to \mathbb{R}$ does not depend on the collection $\lambda \in \Lambda_A^{\mathrm{ad}}$ of policies.

In [160], Witsenhausen provides conditions—in particular, that the observer knows what the agents know and do—which ensure policy independence of conditional expectations.

Theorem 9.47 ([160]) *Assume that*

- *either the decision sets $\{\mathbb{U}_\alpha\}_{\alpha \in A}$ are countable or the probability $\mathbb{P}$ has countable support;*
- *for all $\alpha \in A$, the diagonal of $\mathbb{U}_\alpha \times \mathbb{U}_\alpha$ belongs to $\mathcal{U}_\alpha \otimes \mathcal{U}_\alpha$;*
- *for all $\alpha \in A$, $\mathcal{I}_\alpha \subset \mathcal{G}$ (the observer knows what the agents know) and $\mathcal{D}_\alpha \subset \mathcal{G}$ (the observer knows what the agents do).*

Then, policy independence of conditional expectations holds true.

Proof We just give a sketch of the proof in the case where the decision sets $\{\mathbb{U}_\alpha\}_{\alpha \in A}$ are countable. In this case, one can display a probability on $\mathbb{U}_A$ assigning positive measure to each singleton. By product with the probability $\mathbb{P}$ on $(\Omega, \mathcal{F})$, we obtain a product probability $\mathbb{Q}$ on $(\mathbb{H}, \mathcal{H})$, independent of the collection λ of admissible policies. Witsenhausen shows that the image $\mathbb{P} \circ S_\lambda^{-1}$ of the probability $\mathbb{P}$ on $\mathbb{H}$ by the solution map $S_\lambda : \Omega \to \mathbb{H}$ has a density $\boldsymbol{T}_\lambda$ w.r.t. $\mathbb{Q}$: $\mathbb{P} \circ S_\lambda^{-1} = \boldsymbol{T}_\lambda \mathbb{Q}$ (see Sect. B.2.4).

We now show that, if the density $\boldsymbol{T}_\lambda : \mathbb{H} \to \mathbb{R}_+$ is $\mathcal{G}$-measurable, the function $J = \mathbb{E}_{\mathbb{Q}}(\tilde{j} \mid \mathcal{G})$ satisfies (9.57). Indeed, by (B.23), one has

$$\mathbb{E}_{\mathbb{P}}\big(\tilde{j} \circ S_\lambda \mid S_\lambda^{-1}(\mathcal{G})\big) = \mathbb{E}_{\mathbb{P} \circ S_\lambda^{-1}}\big(\tilde{j} \mid \mathcal{G}\big) \circ S_\lambda \;\; \mathbb{P}\text{-a.s.}.$$

Then, by (B.22), one obtains

$$\mathbb{E}_{\mathbb{P} \circ S_\lambda^{-1}}(\tilde{j} \mid \mathcal{G}) = \mathbb{E}_{\boldsymbol{T}_\lambda \mathbb{Q}}(\tilde{j} \mid \mathcal{G}) = \frac{\mathbb{E}_{\mathbb{Q}}\left(\boldsymbol{T}_\lambda \tilde{j} \mid \mathcal{G}\right)}{\mathbb{E}_{\mathbb{Q}}\left(\boldsymbol{T}_\lambda \mid \mathcal{G}\right)} = \mathbb{E}_{\mathbb{Q}}\left(\tilde{j} \mid \mathcal{G}\right) \;\; \mathbb{Q}\text{-a.s.}, \tag{9.58}$$

since the term $\boldsymbol{T}_\lambda$ factors out, being $\mathcal{G}$-measurable.

Now, when is the density $\boldsymbol{T}_\lambda$ $\mathcal{G}$-measurable? It happens that the support of the density $\boldsymbol{T}_\lambda$ is the graph $G_\lambda \subset \mathbb{H}$ of the solution map S_λ. Since, for each $\alpha \in A$, the diagonal of $\mathbb{U}_\alpha \times \mathbb{U}_\alpha$ belongs to $\mathcal{U}_\alpha \otimes \mathcal{U}_\alpha$, then the graph G_λ belongs to

[6]See Footnote 5 in p. 286.

$\bigvee_{\alpha \in A} (\mathcal{I}_\alpha \vee \mathcal{D}_\alpha)$. Therefore, by assumption G_λ belongs to $\mathcal{G}$. In the same way, Witsenhausen proves that $\boldsymbol{T}_\lambda$ is $\mathcal{G}$-measurable.

Notice that the ingredients of the proof are that the image $\mathbb{P} \circ S_\lambda^{-1}$ of the probability $\mathbb{P}$ on $\mathbb{H}$ by the solution map $S_\lambda : \Omega \to \mathbb{H}$ has a density $\boldsymbol{T}_\lambda$ w.r.t. a fixed probability $\mathbb{Q}$ on $\mathbb{H}$, and that this density $\boldsymbol{T}_\lambda$ is $\mathcal{G}$-measurable. □

Remark 9.48 In the Witsenhausen counterexample, discussed in Sect. 4.2 and in Sect. 4.3.5, consider that the decision maker (DM) at time $t = 1$ is the observer of the agent identified with the DM at time $t = 0$. By (4.35), we see that the observer does not know what the agent $t = 0$ knows (neither does he know what agent $t = 0$ does). Policy independence of conditional expectations does not hold true, as can be illustrated in Remark 4.2. Indeed, in Eq. 4.18, the policy $\boldsymbol{U}_0{}'$ explicitly appears in the term $C_{[\boldsymbol{U}_0{}']}$ in the expression of the conditional expectation $\mathrm{Var}\left(\boldsymbol{W}_1 \mid \boldsymbol{U}_0{}' + \boldsymbol{W}_1\right)(\omega) = C_{[\boldsymbol{U}_0{}']}\left(\boldsymbol{U}_0{}'(\omega) + \boldsymbol{W}_1(\omega)\right)$, $\mathbb{P}$-a.s.. We refer the reader back to the discussion at p. xv of the preamble on notations, regarding the notation $\left[\boldsymbol{U}_0{}'\right]$. ◊

9.6.2 Application to Decomposition by Dynamic Programming

In the introduction of this chapter, we discussed the importance of precedence and information-memory relations to enlighten the phenomena of "signaling". Now, we are going to see that precedence, subsystem, information-memory and decision-memory relations are useful tools to delineate conditions under which an optimization problem can be decomposed into subproblems. In the last part of [160], Witsenhausen sketches a decomposition procedure that we try to detail below.

We still consider the optimization problem (9.56) introduced at the beginning of Sect. 9.6. We assume that $\mathcal{U}_\alpha$ contains singletons, for all agents $\alpha \in A$.

Application of PICE

We consider a causal stochastic control system (see Sect. 9.3.2), with a focal agent $\alpha \in A$ such that

- $A \backslash \{\alpha\}$ is a subsystem (see Definition 9.22),
- policy independence of conditional expectations property holds true for $\mathcal{G} = \mathcal{I}_\alpha$ (see Definition 9.46).

Under technical assumptions to be found in Theorem 9.47, the above assumptions are the consequence of the following ones:

- $A \backslash \{\alpha\}$ is a subsystem: $\langle A \backslash \{\alpha\} \rangle_{\mathfrak{P}} \subset A \backslash \{\alpha\}$,
- the agent α knows what the other agents know: $A \backslash \{\alpha\} \subset \langle \alpha \rangle_{\mathfrak{M}}$,
- the agent α knows what the agents do: $A \backslash \{\alpha\} \subset \langle \alpha \rangle_{\mathfrak{D}}$.

Then, we hint at how these assumptions are building blocks for a decomposition procedure.

By Definition 9.46, since policy independence of conditional expectations property holds true for $\mathcal{G} = \mathcal{I}_\alpha$, there exists an $\mathcal{I}_\alpha$-measurable function $\tilde{j}_\alpha$ such that, for any collection $\lambda \in \Lambda_A^{\text{ad}}$ of admissible policies:

$$\mathbb{E}\big(\tilde{j} \circ S_\lambda \mid S_\lambda^{-1}(\mathcal{I}_\alpha)\big) = \tilde{j}_\alpha \circ S_\lambda, \ \mathbb{P}\text{-a.s.}. \tag{9.59}$$

Therefore, since $\mathbb{E}(\tilde{j} \circ S_\lambda) = \mathbb{E}\Big(\mathbb{E}\big(\tilde{j} \circ S_\lambda \mid S_\lambda^{-1}(\mathcal{I}_\alpha)\big)\Big)$ by elementary properties of the conditional expectation recalled in Sect. B.4, the optimization problem (9.56) now becomes

$$\min_{\lambda \in \Lambda_A^{\text{ad}}} \mathbb{E}(\tilde{j}_\alpha \circ S_\lambda). \tag{9.60}$$

We need the following lemma that relates subsystems and decomposition of the solution map.

Lemma 9.49 *Suppose that causality, hence the solvability property, holds true (see Definition 9.10). Let $\alpha \in A$ be such that $A \backslash \{\alpha\}$ is a subsystem. Consider a collection $\lambda \in \Lambda_A^{\text{ad}}$ of admissible policies. Then,*

1. *when restricted to $A \backslash \{\alpha\}$, the Eq. (9.16) define a mapping $S_{\lambda_{A \backslash \{\alpha\}}} : \Omega \to \mathbb{U}_{A \backslash \{\alpha\}} \times \Omega$, which is the solution map on $A \backslash \{\alpha\}$;*
2. *the policy λ_α may be restricted to the domain $\mathbb{U}_{A \backslash \{\alpha\}} \times \Omega$, instead of $\mathbb{U}_A \times \Omega$;*
3. *the solution map S_λ may be written as*

$$S_\lambda = (\lambda_\alpha \circ S_{\lambda_{A \backslash \{\alpha\}}}, S_{\lambda_{A \backslash \{\alpha\}}}). \tag{9.61}$$

Proof

1. Among Equations (9.16), consider the subset of equalities

$$u_\delta = \lambda_\delta(\{u_\beta\}_{\beta \in A}, \omega), \ \forall \delta \in A \backslash \{\alpha\}. \tag{9.62}$$

   Since $A \backslash \{\alpha\}$ is a subsystem, the policies λ_δ are $\mathcal{U}(A \backslash \{\alpha\})$-measurable for $\delta \in A \backslash \{\alpha\}$, hence they do depend only upon the decisions u_β for $\beta \in A \backslash \{\alpha\}$ (by the same reasoning as in Remark 9.7, using that $\mathcal{U}_\alpha$ contains singletons, for all agents $\alpha \in A$). Therefore, since the solvability property hold true, this subset of equalities has, for each $\omega \in \Omega$, a unique solution which belongs to $\mathbb{U}_{A \backslash \{\alpha\}}$. This defines a mapping $S_{\lambda_{A \backslash \{\alpha\}}} : \Omega \to \mathbb{U}_{A \backslash \{\alpha\}} \times \Omega$, which is the solution map on $A \backslash \{\alpha\}$.
2. Since causality implies absence of self-information, we apply the observation of Remark 9.7.

3. To solve (9.16) and obtain thus the solution map S_λ, we first solve the subset of equalities bearing only on ω and on the decisions u_β for $\beta \in A\backslash\{\alpha\}$, since $A\backslash\{\alpha\}$ is a subsystem, as proved by item 1. The solution is $\{u_\beta\}_{\beta\neq\alpha} = S_{\lambda_{A\backslash\{\alpha\}}}(\omega)$. Then, in (9.16), there only remains the equation
$$u_\alpha = \lambda_\alpha(u_\alpha, \{u_\beta\}_{\beta\neq\alpha}, \omega) = \lambda_\alpha(\{u_\beta\}_{\beta\neq\alpha}, \omega),$$
where the last equality comes from the property, seen at the previous item, that the domain of the policy λ_α may be restricted to $\mathbb{U}_{A\backslash\{\alpha\}} \times \Omega$. Equation (9.61) is derived.

This ends the proof. □

As a consequence of Lemma 9.49, by (9.59) and (9.61), one obtains that
$$\tilde{j}_\alpha \circ S_\lambda = \tilde{j}_\alpha(\lambda_\alpha \circ S_{\lambda_{A\backslash\{\alpha\}}}, S_{\lambda_{A\backslash\{\alpha\}}}) = \big(\tilde{j}_\alpha(\lambda_\alpha(\cdot), \cdot)\big) \circ S_{\lambda_{A\backslash\{\alpha\}}}. \tag{9.63}$$

Now, suppose that there exists $\lambda_\alpha^\sharp : \prod_{\beta\neq\alpha} \mathbb{U}_\beta \times \Omega \to \mathbb{U}_\alpha$ such that
$$\tilde{j}_\alpha(\lambda_\alpha^\sharp(\{u_\beta\}_{\beta\neq\alpha}, \omega), \{u_\beta\}_{\beta\neq\alpha}, \omega) \leq \tilde{j}_\alpha(u_\alpha, \{u_\beta\}_{\beta\neq\alpha}, \omega),$$
$$\forall (u_\alpha, \omega) \in \mathbb{U}_\alpha \times \Omega. \tag{9.64}$$

This amounts to saying that $\inf_{u_\alpha\in\mathbb{U}_\alpha} \tilde{j}_\alpha(u_\alpha, \{u_\beta\}_{\beta\neq\alpha}, \omega)$ is achieved at $\lambda_\alpha^\sharp(\{u_\beta\}_{\beta\neq\alpha}, \omega)$ for each $(\{u_\beta\}_{\beta\neq\alpha}, \omega)$. The mapping $\lambda_\alpha^\sharp : \prod_{\beta\neq\alpha}\mathbb{U}_\beta \times \Omega \to \mathbb{U}_\alpha$ is extended to a policy $\lambda_\alpha^\sharp : \mathbb{H} \to \mathbb{U}_\alpha$ by
$$\lambda_\alpha^\sharp(u_\alpha, \{u_\beta\}_{\beta\neq\alpha}, \omega) = \lambda_\alpha^\sharp(\{u_\beta\}_{\beta\neq\alpha}, \omega).$$

Since the function $\tilde{j}_\alpha$ is $\mathfrak{I}_\alpha$-measurable, it is likely (but technical assumptions may be required) that the policy $\lambda_\alpha^\sharp : \mathbb{H} \to \mathbb{U}_\alpha$ is also $\mathfrak{I}_\alpha$-measurable, as the arg min of a $\mathfrak{I}_\alpha$-measurable function (though this is a delicate issue, discussed in particular in [21], and related to measurable selections as discussed in Sect. 8.3).

By combining (9.63) and (9.64), we obtain that
$$\big(\tilde{j}_\alpha(\lambda_\alpha^\sharp(\cdot), \cdot)\big) \circ S_{\lambda_{A\backslash\{\alpha\}}} \leq \big(\tilde{j}_\alpha(\lambda_\alpha(\cdot), \cdot)\big) \circ S_{\lambda_{A\backslash\{\alpha\}}} = \tilde{j}_\alpha \circ S_\lambda. \tag{9.65}$$

Therefore, by elementary properties of the conditional expectation recalled at Sect. B.4, we have that
$$\mathbb{E}\Big(\big(\tilde{j}_\alpha(\lambda_\alpha^\sharp(\cdot), \cdot)\big) \circ S_{\lambda_{A\backslash\{\alpha\}}}\Big) \leq \mathbb{E}(\tilde{j}_\alpha \circ S_\lambda). \tag{9.66}$$

By the policy independence of conditional expectations property, the function $\tilde{j}_\alpha$ does not depend upon any collection $\lambda \in \Lambda^{\mathrm{ad}}_{A\setminus\{\alpha\}}$ of policies. Thus, the optimization problem (9.60) can be replaced by

$$\min_{\lambda \in \Lambda^{\mathrm{ad}}_{A\setminus\{\alpha\}}} \mathbb{E}\Big(\big(\tilde{j}_\alpha(\lambda^\sharp_\alpha(\cdot), \cdot)\big) \circ S_{\lambda_{A\setminus\{\alpha\}}}\Big). \tag{9.67}$$

Hence, at the end of the process, the original optimization problem (9.56) has been transformed into the optimization problem (9.67): the set A of agents is reduced to $A\setminus\{\alpha\}$ and the criterion $\tilde{j}$ is replaced by $\tilde{j}_\alpha\big(\lambda^\sharp_\alpha(\cdot), \cdot\big)$. Equation (9.64) relating $\tilde{j}_\alpha\big(\lambda^\sharp_\alpha(\cdot), \cdot\big)$ to $\tilde{j}$ can be rewritten as

$$\tilde{j}_\alpha(\lambda^\sharp_\alpha(\{u_\beta\}_{\beta\neq\alpha}, \omega), \{u_\beta\}_{\beta\neq\alpha}, \omega) = \min_{u_\alpha \in \mathbb{U}_\alpha} \tilde{j}_\alpha(u_\alpha, \{u_\beta\}_{\beta\neq\alpha}, \omega). \tag{9.68}$$

It is the analogue of the Dynamic Programming (DP) equation (4.57) between value functions (see Sects. 4.4.5 and 4.5.4): In the comparison with the classical DP equation, the variable $(\{u_\beta\}_{\beta\neq\alpha}, \omega)$ plays the role of the state (see the discussion in Sect. 4.5.3).

A Decomposition Principle

In the last part of [160], Witsenhausen sketches a decomposition procedure related to dynamic programming (see Sect. 4.4) as follows.

1. Suppose that the set A of agents is the disjoint union of $\{\alpha\}$, B and C.
2. Suppose the policies $\{\lambda_\gamma\}_{\gamma\in C}$ of the agents in C are fixed and known. Then the system reduces to $A' = \{\alpha\} \cup B$ with an information structure and cost function obtainable using $\{\lambda_\gamma\}_{\gamma\in C}$.[7]
3. Suppose that, in the reduced system, the agents in B form a subsystem of A' ($\langle B\rangle'_{\mathfrak{P}} \subset B$), and that agent α knows everything that the agents in B know ($B \subset \langle\alpha\rangle'_{\mathfrak{M}}$) and make ($B \subset \langle\alpha\rangle'_{\mathfrak{D}}$).

Remark 9.50 In the state model of Sect. 4.4, the set A of agents is the set of times $\{0, \dots, T\}$. We consider the information structure given by

$$\mathfrak{I}_t = \sigma\big(u_0, u_1, \dots, u_{t-1}, x_0, w_1, w_2, \dots, w_t\big), \tag{9.69}$$

where $\big(u_0, u_1, \dots, u_{t-1}, x_0, w_1, w_2, \dots, w_t\big)$ are here identified with the corresponding coordinate mappings on $\prod_{s=0}^{T-1} \mathbb{U}_s \times \mathbb{X}_0 \times \prod_{s=1}^{T} \mathbb{W}_s$. We suppose that

[7] For this, map the reduced history space $\mathbb{U}_{A'} \times \Omega$ into the full history space $\mathbb{U}_A \times \Omega$ by associating with each reduced history $(u_{A'}, \omega) \in \mathbb{U}_{A'} \times \Omega$ the image $S_{\bar{\lambda}}(\omega)$ of the solution map, where the policy $\bar{\lambda}$ is given by the constant mapping $\bar{\lambda}_\beta \equiv u_\beta$ for $\beta \in A'$, and by the policies $\bar{\lambda}_\beta = \lambda_\beta$ for $\beta \in C$. Then, the reciprocal images of the information fields $\mathfrak{I}_\beta$ for $\beta \in A'$ provide information fields $\mathfrak{I}'_\beta$ over the reduced history space $\mathbb{U}_{A'} \times \Omega$.

$\prod_{s=1}^{T} \mathbb{W}_s$ is equipped with a product probability (see Sect. B.1.4), so that the coordinate mappings are independent random variables. We set

$$\overline{x}_t = \big(u_0, u_1, \ldots, u_{t-1}, x_0, w_1, w_2, \ldots, w_t\big), \tag{9.70}$$

so that we have a trivial dynamical equation

$$\begin{aligned} \overline{x}_{t+1} = \big(u_0, u_1, \ldots, u_{t-1}, u_t, x_0, w_1, w_2, \ldots, w_t, w_{t+1}\big) \\ = f_t\big(\overline{x}_t, u_t, w_{t+1}\big). \end{aligned} \tag{9.71}$$

The set C of agents corresponds to the times $\{t+1, \ldots, T\}$ and the fixed and known policies $\{\lambda_\gamma\}_{\gamma \in C}$ correspond to the feedbacks already computed backward from the horizon T to time $t+1$ thanks to the Dynamic Programming (DP) equation (4.57b). Agent α corresponds to time t. The agents in B correspond to the previous times $\{0, \ldots, t-1\}$. By the DP equation, the system indeed reduces to the times $\{0, \ldots, t\}$, corresponding to $A' = \{\alpha\} \cup B$, with a cost function obtainable using the optimal feedbacks yet computed, that is, the cost-to-go (4.54) or value function (4.56) at time $t+1$. In the reduced system, the times $\{0, \ldots, t-1\}$ form a subsystem of $\{0, \ldots, t\}$ by (9.69), which expresses sequential causality. The times $\{0, \ldots, t-1\}$ are remembered at time t (the information fields in (9.69) are nested). The actions made at times $\{0, \ldots, t-1\}$ are known at time t by (9.69). The system is stricly quasiclassical as defined in Sect. 9.5.1. ◊

The main result in [160] is that if an optimal $\lambda_\alpha^\sharp$ can be found that solves (9.68), it is optimal for any fixed choice of the policies of the agents in B by Theorem 9.47. Witsenhausen concludes that, "adopting $\lambda_\alpha^\sharp$, agent α moves into set C, leaving the smaller set B for further optimization". He points out that no special relationship is required between the information subfields $\mathfrak{I}_\beta$ and the decision subfields $\mathfrak{D}_\beta$. Thus, the conditions

$$\langle B \rangle_{\mathfrak{P}} \subset B, \quad B \subset \langle \alpha \rangle_{\mathfrak{M}} \text{ and } B \subset \langle \alpha \rangle_{\mathfrak{D}} \tag{9.72}$$

are the appropriate ingredients to break a multi-agent optimization problem on $\{\alpha\} \cup B$ into smaller problems on $\{\alpha\}$ and on B.

9.7 Conclusion

In this chapter, we have examined general systems comprising several agents, making decisions in an order which is not fixed in advance. We have presented the Witsenhausen intrinsic model and four binary relations between agents, which make it possible to define a typology of systems. When optimization is the issue, Witsenhausen's notion of policy independence of conditional expectations is a key to obtain a Dynamic Programming-like decomposition of a stochastic optimization problem into subproblems.

Chapter 10
Dual Effect for Multi-Agent Stochastic Input-Output Systems

10.1 Introduction

In stochastic optimal control, a key issue is that "solutions" are searched for in terms of "closed-loop control laws" over available information and, as a consequence, a major potential difficulty is the fact that present control may affect future available information. This is known as the "dual effect" of control, and has been discussed in Sects. 1.1.3, 1.2.1, 1.3.2 and 4.2.3. Following [13], we will characterize the maximal set of closed-loop control laws containing open-loop laws and for which the information provided by observations closed with such a feedback remains fixed.

For this purpose, we consider in Sect. 10.2 the following variant of the Witsenhausen intrinsic model in Sect. 9.2. A *multi-agent stochastic input-output system*, in short MASIOS, is a multi-agent stochastic control system as in Sect. 9.2 where the information of an agent is described by an observation mapping (a signal), and where measurability is w.r.t. (complete) partition fields and not to σ-fields (see Sects. 3.3.2 and 3.4.2). In parallel to the discussion on the precedence and information-memory relations for multi-agent stochastic control systems in Sect. 9.4, we introduce their counterparts for MASIOS, as well as a typology of MASIOS.

The counterpart of a policy in the Witsenhausen intrinsic model in Sect. 9.2 is a control law, that is, a random variable defined on the universe Ω. An admissible control law for a focal agent is one that is measurable w.r.t. the agent closed-loop observation after control (of all agents). A collection of control laws (indexed by the set of agents) induces a partition of the universe Ω. No open-loop dual effect holds true when all constant control laws induce the same fixed partition. This is the object of Sect. 10.3.

Thanks to the typology of MASIOS introduced in Sect. 10.2, we characterize in Sect. 10.3.3 classes of control laws for which the induced partition coincides with this fixed partition. Therefore, if we restrict a stochastic optimization problem to such *no dual effect* control laws, the discretization of the control laws domain can be made in advance.

P. Carpentier et al., *Stochastic Multi-Stage Optimization*, Probability Theory and Stochastic Modelling 75, DOI 10.1007/978-3-319-18138-7_10

10.2 Multi-Agent Stochastic Input-Output Systems (MASIOS)

We introduce a *multi-agent stochastic input-output system*, which is a multi-agent stochastic control system as in Sect. 9.2, but where the information of an agent is described by an observation mapping (a signal), and where measurability is w.r.t. (complete) partition fields and not to σ-fields. We provide state models, especially linear ones, as examples inducing MASIOS. In parallel to the discussion on the precedence and information-memory relations for multi-agent stochastic control systems in Sect. 9.4, we introduce their counterparts for MASIOS, as well as a typology of MASIOS.

10.2.1 *Definition of Multi-Agent Stochastic Input-Output Systems*

Let A be a finite set representing agents. Each agent $\alpha \in A$ is supposed to make only one decision $u_\alpha \in \mathbb{U}_\alpha$, where $\mathbb{U}_\alpha$ is the control set for agent α, equipped with the complete π-field $\mathcal{U}_\alpha = 2^{\mathbb{U}_\alpha}$. Let Ω (*universe* or *sample space*) be a measurable set, with the complete π-field $\mathcal{F} = 2^{\Omega}$, which represents all uncertainties: any $\omega \in \Omega$ is called a *state of Nature*.

Remark 10.1 We adopt the same formalism as in Chap. 9, but for measurability which, here, is w.r.t. (complete) partition fields and not to σ-fields. We refer the reader to Sect. 3.3.2 for details. This option makes statements more compact and proofs more intuitive as compared to measurability w.r.t. σ-fields. ◊

As in (9.1) and (9.2), we define the *decision set* $\mathbb{U}_A$, and we equip it with the complete product π-field $\mathcal{U}_A$ (see Remark 3.12), called *decision field*:

$$\mathbb{U}_A := \prod_{\alpha \in A} \mathbb{U}_\alpha, \quad \mathcal{U}_A := \bigotimes_{\alpha \in A} \mathcal{U}_\alpha. \tag{10.1}$$

The *history space* $\mathbb{H}$ and its associated complete product π-field $\mathcal{H}$, called *history field*, are:

$$\mathbb{H} := \mathbb{U}_A \times \Omega, \quad \mathcal{H} := \mathcal{U}_A \otimes \mathcal{F}. \tag{10.2}$$

To each agent $\alpha \in A$ is attached an *observation function*

$$o_\alpha : \mathbb{H} \to \mathbb{Y}_\alpha. \tag{10.3}$$

Remark 10.2 Here, the information of agent α is described by a mapping $o_\alpha : \mathbb{H} \to \mathbb{Y}_\alpha$ defined over the history space $\mathbb{H} = \mathbb{U}_A \times \Omega$, whereas, in Sect. 9.2.2, it is described by an information σ-field $\mathcal{I}_\alpha \subset \mathcal{H}$. When this σ-field is a π-field (for instance, when $\mathbb{H}$ is finite), the connection between both approaches is given by

$$\mathcal{I}_\alpha = \pi(o_\alpha), \tag{10.4}$$

where the π-field generated by a mapping has been introduced in Definition 3.32. ◊

Definition 10.3 A *multi-agent stochastic input-output system* (MASIOS) is a collection consisting of agents A, states of Nature Ω and complete π-field $\mathcal{F}$, control sets and complete π-fields $\{\mathbb{U}_\alpha, \mathcal{U}_\alpha\}_{\alpha \in A}$, and observation functions $\{\mathbb{U}_\alpha, o_\alpha\}_{\alpha \in A}$.

Example 10.4 For instance, if, in the description of a sequential optimal stochastic control problem as revealed in Sect. 4.5.1, informations fields $\mathcal{I}_t$ are given by signals $\boldsymbol{Y}_t : \mathbb{H} \to \mathbb{Y}_t$, where $(\mathbb{Y}_t, \mathcal{Y}_t)$ is some measurable space (see Remark 9.8 and Eq. (10.4)), we obtain a MASIOS with agents $A = \{0, \ldots, T-1\}$. △

Example 10.5 As another example of MASIOS, consider a state model as defined in Sect. 4.4.1. We set

$$A = \{0, \ldots, T\}, \;\; \Omega = \mathbb{X}_0 \times \prod_{t=1}^{T} \mathbb{W}_t, \tag{10.5}$$

so that states of Nature are scenarios

$$\omega = \big(x_0, w(\cdot)\big) = \big(x_0, w_1, w_2, \ldots, w_T\big). \tag{10.6}$$

Identifying any $u \in \mathbb{U}_{\{0,\ldots,T\}}$ with an open-loop feedback $\gamma \equiv u$, we now define different observation functions as follows, with the help of the state map X_f (see Definition 4.6).

When the state x_t is observed at time t, this corresponds to

$$o_t(u, \omega) = X_f[0, x_0, u, w(\cdot)]_t. \tag{10.7}$$

The case when past states $x_0, \ldots, x_t$ are observed at time t is given by

$$o_t(u, \omega) = \big(X_f[0, x_0, u, w(\cdot)]_0, \ldots, X_f[0, x_0, u, w(\cdot)]_t\big). \tag{10.8}$$

△

10.2.2 Control Laws

We define the counterpart of a policy in the Witsenhausen intrinsic model in Sect. 9.2: it is a control law, that is, a random variable defined on the universe Ω. An admissible control law for a focal agent is one that is measurable w.r.t. the agent closed-loop observation after control (of all agents).

Definition 10.6 A *control law* for agent α is a random variable $\boldsymbol{U}_\alpha : \Omega \to \mathbb{U}_\alpha$, and a *collection of control laws* is a collection $\{\boldsymbol{U}_\beta\}_{\beta \in A}$ where $\boldsymbol{U}_\beta : \Omega \to \mathbb{U}_\beta$. We define the *set of collections of control laws* by:

$$\mathcal{U}_A := \prod_{\beta \in A} \mathbb{U}_\beta^\Omega = \Big\{ \boldsymbol{U} = \{\boldsymbol{U}_\beta\}_{\beta \in A} \; \Big| \; \boldsymbol{U}_\beta : \Omega \to \mathbb{U}_\beta, \; \forall \beta \in A \Big\}. \tag{10.9}$$

We warn the reader that, though typographically close, the notations $\mathcal{U}_A$ for the set of collections of control laws in (10.9) and $\mathfrak{U}_A$ for the decision field in (10.1) are distinct. In what follows, except in Remark 10.10, we do not use the decision field notation.

Remark 10.7 Both $\mathbb{U}_\alpha$ and Ω being equipped with complete π-fields, control laws are necessarily measurable. Notice that a control law for agent α is a mapping defined over the universe Ω, whereas in Sect. 9.2.2 a policy for agent α was represented by a mapping $\lambda_\alpha : \mathbb{H} \to \mathbb{U}_\alpha$. A parallel can be established between a collection $\{\boldsymbol{U}_\beta\}_{\beta \in A}$ of control laws and the mapping $M_\lambda : \Omega \to \mathbb{U}_A$ attached to a collection $\lambda \in \Lambda_A^{\mathrm{ad}}$ of admissible policies, when the solvability property holds true (see Definition 9.10). In this chapter, control laws are random variables (see Definition 3.44). ◊

Definition 10.8 For any collection $\boldsymbol{U} \in \mathcal{U}_A$ of control laws and for any agent $\alpha \in A$, the *observation of agent α after control* is the random variable $\overset{\boldsymbol{U}}{\eta_\alpha} : \Omega \to \mathbb{Y}_\alpha$ defined by

$$\overset{\boldsymbol{U}}{\eta_\alpha}(\omega) := o_\alpha\big(\boldsymbol{U}(\omega), \omega\big), \quad \forall \omega \in \Omega. \tag{10.10}$$

The collection $\{\overset{\boldsymbol{U}}{\eta_\beta}\}_{\beta \in A}$ of random variables is called *closed-loop observations*.

In general, the observation available to agent α depends, through the collection $\boldsymbol{U} = \{\boldsymbol{U}_\beta\}_{\beta \in A}$ of control laws, upon the control laws of other agents by expanding (10.10) into

$$\overset{\boldsymbol{U}}{\eta_\alpha}(\omega) = o_\alpha\big(\{\boldsymbol{U}_\beta(\omega)\}_{\beta \in A}, \omega\big). \tag{10.11}$$

A control law is said to be *admissible* for an agent if she makes her decision with no more than her observation after control.

Definition 10.9 An *admissible control law* for agent α is a control law $\boldsymbol{U}_\alpha : \Omega \to \mathbb{U}_\alpha$ such that

$$\boldsymbol{U}_\alpha \preceq \overset{\boldsymbol{U}}{\eta_\alpha}. \tag{10.12}$$

The set of *admissible (collections of) control laws* is defined by:

$$\mathcal{U}_A^{\mathrm{ad}} := \Big\{ \boldsymbol{U} = \{\boldsymbol{U}_\alpha\}_{\alpha \in A} \in \mathcal{U}_A \; \Big| \; \boldsymbol{U}_\alpha \preceq \overset{\boldsymbol{U}}{\eta_\alpha}, \; \forall \alpha \in A \Big\}. \tag{10.13}$$

The measurability constraint $\boldsymbol{U}_\alpha \overset{\boldsymbol{U}}{\preceq} \eta_\alpha$ is taken in the sense of measurability with respect to partition fields as in Definition 3.32 (see also Proposition 3.35).

Remark 10.10 Here, admissible control laws and the measurability constraint $\boldsymbol{U}_\alpha \overset{\boldsymbol{U}}{\preceq} \eta_\alpha$ are the counterparts of admissible policies $\lambda_\alpha : \mathbb{H} \to \mathbb{U}_\alpha$ measurable w.r.t. $\mathfrak{I}_\alpha$, that is, satisfying $\lambda_\alpha^{-1}(\mathfrak{U}_\alpha) \subset \mathfrak{I}_\alpha$ as in Definition 9.6. ◊

Remark 10.11 When not specified, the notation $\preceq$ is relative to mappings with common domain Ω (see Sect. 3.4.2). ◊

A special class of admissible control laws is the one made of *open-loop* or *deterministic* or *constant* control laws.

Definition 10.12 The set $\perp_A$ of *open-loop control laws*, or *deterministic control laws*, consists of the constant control laws, namely control laws measurable w.r.t. the trivial π-field $\{\emptyset, \Omega\}$ on Ω:

$$\perp_A := \left\{\boldsymbol{U} = \{\boldsymbol{U}_\alpha\}_{\alpha\in A} \in \mathcal{U}_A \;\middle|\; \boldsymbol{U}_\alpha \preceq \{\emptyset, \Omega\},\ \forall \alpha \in A\right\}. \tag{10.14}$$

Each $\boldsymbol{U}_\alpha : \Omega \to \mathbb{U}_\alpha$ in $\boldsymbol{U} = \{\boldsymbol{U}_\alpha\}_{\alpha\in A} \in \perp_A$ takes a constant value in $\mathbb{U}_\alpha$. The notation $\perp_A$ refers to the fact that the class of constant mappings is the bottom of the lattice of equivalence classes of mappings (see Proposition 3.42).

10.2.3 Precedence and Memory-Communication Relations

Thanks to the connection (10.4) between information fields and observations, we can characterize, in the MASIOS framework of Sect. 10.2, the precedence and memory-communication binary relations already introduced in Sect. 9.4.

For this purpose, we make use of the following notations. Consider $B \subset A$ a subset of agents. We set

$$u_B := \{u_\beta\}_{\beta\in B}, \tag{10.15}$$

and, for any collection $\{H_\alpha\}_{\alpha\in A}$ of mappings defined over Ω,

$$H_B := \{H_\beta\}_{\beta\in B}. \tag{10.16}$$

The *precedence* binary relation $\mathfrak{P}$ of Definition 9.15 identifies couples of agents, where the decision of the first agent indeed influences the observation of the second. By the correspondence (10.4), the subset $\langle\alpha\rangle_{\mathfrak{P}}$ of predecessors of α is (the smallest subset) such that there exist a mapping $\widetilde{o}_\alpha$ satisfying

$$o_\alpha(u, \omega) = \widetilde{o}_\alpha(u_{\langle\alpha\rangle_{\mathfrak{P}}}, \omega), \tag{10.17}$$

expressing that $o_\alpha(u, \omega)$ depends only on the components $u_{\langle\alpha\rangle_{\mathfrak{P}}} = \{u_\beta\}_{\beta\in\langle\alpha\rangle_{\mathfrak{P}}}$ of the decision u.

The *memory-communication* binary relation $\mathfrak{M}$ of Definition 9.32 identifies couples of agents, where the observation of the first one is passed on to the second one. By the correspondence (10.4), the subset $\langle\alpha\rangle_{\mathfrak{M}}$ of agents whose information is embedded within the information of agent α is (the largest subset) such that:

$$o_{\langle\alpha\rangle_{\mathfrak{M}}}(\cdot,\cdot) \preceq_{\mathbb{U}_A\times\Omega} o_\alpha(\cdot,\cdot), \quad \forall\alpha\in A. \tag{10.18}$$

Here, we specify that measurability is w.r.t. to mappings with domain $\mathbb{U}_A\times\Omega$ (see Remark 10.11).

10.2.4 A Typology of MASIOS

Thanks to the precedence and information-memory relations, we now introduce a typology of MASIOS, inspired from the discussion in Sect. 9.5.1.

Partially Nested MASIOS
We say that a MASIOS is *partially nested* when the precedence relation $\mathfrak{P}$ is included in the memory-communication relation $\mathfrak{M}$, that is, when

$$\langle\alpha\rangle_{\mathfrak{P}} \subset \langle\alpha\rangle_{\mathfrak{M}}, \quad \forall\alpha\in A, \tag{10.19}$$

or, by (10.18), when

$$o_{\langle\alpha\rangle_{\mathfrak{P}}}(\cdot,\cdot) \preceq_{\mathbb{U}_A\times\Omega} o_\alpha(\cdot,\cdot), \quad \forall\alpha\in A. \tag{10.20}$$

Remark 10.13 A consequence of Proposition 3.39 and of (10.20) is that, for all $\boldsymbol{U}\in\mathcal{U}_A$, we have that $o_{\langle\alpha\rangle_{\mathfrak{P}}}\big(\boldsymbol{U}(\cdot),\cdot\big)\preceq o_\alpha\big(\boldsymbol{U}(\cdot),\cdot\big)$ for all $\boldsymbol{U}\in\mathcal{U}_A$. This property is taken as the definition of a *partially nested information structure* in [81, 83]: it imposes conditions on the closed-loop observations (10.10), so that measurability is w.r.t. to mappings with domain Ω (see Remark 10.11). On the contrary, assumption (10.20) is an "open-loop" assumption, which does not require assumptions w.r.t. the closed-loop observations, and which makes use of measurability w.r.t. to mappings with domain $\mathbb{U}_A\times\Omega$. ◊

Sequential MASIOS
Consider the case where each agent in A is supposed to represent a time period t:

$$A=\{0,\dots,T\} \text{ where } T\in\mathbb{N}^*. \tag{10.21}$$

With the notations of Sect. 10.2.3, and especially Eq. (10.1), we have that

$$\prod_{t=0}^{T}\mathbb{U}_t=\mathbb{U}_{\{0,\dots,T\}}. \tag{10.22}$$

Following Sect. 9.5.1, the MASIOS given by the family $\{o_t\}_{t=0,\dots,T}$ of observation functions

$$o_t : \mathbb{U}_{\{0,\dots,T\}} \times \Omega \to \mathbb{Y}_t \tag{10.23}$$

is said to be a *sequential MASIOS* if it is sequential with the ordering $0, \dots, T$. By (9.50) and with the notations of Sect. 10.2.3, this is equivalent to

$$\langle 0 \rangle_{\mathfrak{P}} = \emptyset \text{ and } \langle t \rangle_{\mathfrak{P}} \subset \{0, \dots, t-1\}, \ \forall t \in \{1, \dots, T\}. \tag{10.24}$$

In other words, the observation at time t depends at most upon the past decisions $u_0, \dots, u_{t-1}$ (and the state of Nature ω). Indeed, by (10.17) and Proposition 3.38, there exist mappings $\widetilde{o}_t$, for $t = 0, \dots, T$, such that

$$o_t(u_0, \dots, u_T, \omega) = \widetilde{o}_t(u_0, \dots, u_{t-1}, \omega), \ \forall t = 1, \dots, T, \tag{10.25}$$

with the special case $o_0(u_0, \dots, u_T, \omega) = \widetilde{o}_0(\omega)$.

Example 10.14 Following Example 10.5, consider the MASIOS induced by a state model as defined in Sect. 4.4.1. It can be checked that any expression of the form (with X_f the state map of Definition 4.6)

$$o_t(u, \omega) = \widetilde{o}_t\big(X_f[0, x_0, u, w(\cdot)]_0, \dots, X_f[0, x_0, u, w(\cdot)]_t, w(\cdot)\big) \tag{10.26}$$

defines a sequential MASIOS. This includes imperfect and corrupted observations of the past states.

An important class of sequential MASIOS is given by linear state models with linear observations. More precisely, linear state models are those for which the dynamics $f_t : \mathbb{X}_t \times \mathbb{U}_t \times \mathbb{W}_{t+1} \to \mathbb{X}_{t+1}$ are linear mappings. Linear observations correspond to $o_t(u, \omega)$ being a linear expression in $w(\cdot)$, $X_f[0, x_0, u, w(\cdot)]_0, \dots, X_f[0, x_0, u, w(\cdot)]_t$. △

Quasiclassical MASIOS

As in Sect. 9.5.1, we say that a MASIOS is *quasiclassical* if it is sequential (with the ordering $0, \dots, T$) and partially nested, that is,

$$\langle 0 \rangle_{\mathfrak{P}} = \emptyset \text{ and } \langle t \rangle_{\mathfrak{P}} \subset \{0, \dots, t-1\} \cap \langle t \rangle_{\mathfrak{M}}, \ \forall t \in \{1, \dots, T\}. \tag{10.27}$$

In other words, if decisions made at time s affect the observation o_t ($s \in \langle t \rangle_{\mathfrak{P}}$), then $s \leq t-1$ and the observation o_s is embedded in the observation o_t. Indeed, by $s \in \langle t \rangle_{\mathfrak{M}}$, (10.18) and Proposition 3.38, there exists a mapping $\mathfrak{f}_{s,t}$ such that $o_s = \mathfrak{f}_{s,t}(o_t)$.

Classical MASIOS

We say that a sequential MASIOS (with the ordering $0, \dots, T$) displays *perfect memory* if

$$\{0, \dots, t\} \subset \langle t \rangle_{\mathfrak{M}}, \ \forall t \in \{0, \dots, T\}. \tag{10.28}$$

Information accumulates with time, that is, the π-fields $\mathfrak{I}_t = \pi(o_t)$ form a filtration:

$$\mathfrak{I}_0 \subset \cdots \subset \mathfrak{I}_{t-1} \subset \mathfrak{I}_t \subset \cdots \subset \mathfrak{I}_T. \tag{10.29}$$

As a consequence of (9.52), a sequential MASIOS displaying perfect memory is classical (see Sect. 9.5.1), hence quasiclassical, with the ordering $0, \dots, T$.

Remark 10.15 Define *closed-loop perfect memory* as the property that, for all $t = 0, \dots, T$,

$$o_t\big(\boldsymbol{U}(\cdot), \cdot\big) \preceq_{\Omega} o_{t+1}\big(\boldsymbol{U}(\cdot), \cdot\big), \ \forall \boldsymbol{U} \in \mathcal{U}_A^{\mathrm{ad}}. \tag{10.30}$$

As in Remark 10.13, this definition imposes conditions on the closed-loop observations (10.10), so that measurability is w.r.t. to mappings with domain Ω (see Remark 10.11). In contrast, Assumption (10.28) for the definition of perfect memory is an "open-loop" assumption, which does not require assumptions w.r.t. the closed-loop observations, and which makes use of measurability w.r.t. to mappings with domain $\mathbb{U}_{\{0,\dots,T\}} \times \Omega$. Open-loop perfect memory (10.28) implies closed-loop perfect memory (10.30). Notice that a weaker form of open-loop perfect memory, namely

$$o_t(u, \cdot) \preceq_{\Omega} o_{t+1}(u, \cdot), \ \forall u \in \mathbb{U}_{\{0,\dots,T\}},$$

does not imply closed-loop perfect memory (10.30). This can directly be seen with the following example: let $o_0(\omega) = \omega$ and $o_1(u, \omega) = u - \omega$; then $o_0(\cdot) \preceq o_1(u, \cdot)$ for all u; whereas, for $\boldsymbol{U}(\omega) = \omega$, this $\boldsymbol{U}$ is admissible since $\boldsymbol{U} \preceq o_0$, but obviously $o_0(\cdot) \not\preceq o_1\big(\boldsymbol{U}(\cdot), \cdot\big)$ since the latter is the zero mapping. ◇

10.3 No Open-Loop Dual Effect and No Dual Effect Control Laws

A collection of control laws induces a partition of the universe Ω. We say that no open-loop dual effect holds true when all constant control laws induce the same fixed partition.

10.3.1 No Open-Loop Dual Effect (NOLDE)

We now introduce the notion of *no open-loop dual effect*. For this purpose, we use the measurability equivalence $\equiv$ between mappings of Definition 3.40.

Definition 10.16 The property of *no open-loop dual effect* (NOLDE) holds true for the MASIOS discussed in Sect. 10.2 if we have that:

$$\overset{U}{\eta_\alpha} \equiv \overset{U'}{\eta_\alpha}, \quad \forall (\boldsymbol{U}, \boldsymbol{U}') \in \perp_A \times \perp_A, \quad \forall \alpha \in A. \tag{10.31}$$

In the case of NOLDE, for any agent $\alpha \in A$, all observations after open-loop control are equivalent, in the sense of measurability equivalence between mappings of Definition 3.40. Therefore, all observations after open-loop control are equivalent to a fixed mapping[1] ζ_α with domain Ω:

$$\overset{U}{\eta_\alpha} \equiv \zeta_\alpha, \quad \forall \boldsymbol{U} \in \perp_A. \tag{10.32}$$

Example 10.17 It is shown in [128] that linear state models with linear observations, as defined in Example 10.14, possess the NOLDE property if they display perfect memory as defined in Sect. 10.2.4. △

Remark 10.18 In [82], Eq. (5) expresses a similar property. ◊

The following proposition, adapted from [13], is a straightforward consequence of Proposition 3.41.

Proposition 10.19 *The property of no open-loop dual effect (NOLDE) holds true if, and only if, there exist a collection of mappings $\{\mathfrak{f}_\alpha\}_{\alpha \in A}$ where $\mathfrak{f}_\alpha : \mathbb{U}_A \times \mathbb{Z}_\alpha \to \mathbb{Y}_\alpha$ and a collection $\{\zeta_\alpha\}_{\alpha \in A}$ of random variables where $\zeta_\alpha : \Omega \to \mathbb{Z}_\alpha$ such that*

- *the partial mapping $\mathfrak{f}_\alpha(u, \cdot) : \mathbb{Z}_\alpha \to \mathbb{Y}_\alpha$ is injective, for all $u \in \mathbb{U}_A$;*
- *the observations satisfy $o_\alpha(u, \omega) = \mathfrak{f}_\alpha\big(u, \zeta_\alpha(\omega)\big)$, for all $(u, \omega) \in \mathbb{U}_A \times \Omega$.*

10.3.2 No Dual Effect Control Laws

No dual effect control laws are those control laws for which, in case of NOLDE, the closed-loop observations induce the same partitions as the constant control laws.

Definition 10.20 Assume that the NOLDE property holds true, with the fixed observations ζ as in (10.32). The *no dual effect control laws set* is made of all admissible control laws such that the closed-loop observations are equivalent to the fixed mapping ζ_α:

[1] For instance, take for ζ_α any mapping of the class of $\overset{U}{\eta_\alpha}$ for $\boldsymbol{U} \in \perp_A$.

$$\mathcal{U}_A^{\text{nde}} := \left\{ \boldsymbol{U} = \{\boldsymbol{U}_\alpha\}_{\alpha \in A} \in \mathcal{U}_A \,\middle|\, \overset{\boldsymbol{U}}{\eta_\alpha} \equiv \zeta_\alpha, \ \forall \alpha \in A \right\} \cap \mathcal{U}_A^{\text{ad}}. \tag{10.33}$$

Thus, "closing" the system with any control law belonging to the no dual effect control law set produces the same fixed closed-loop observations.

Definition 10.21 Assume that the NOLDE property holds true, with the fixed observations ζ as in (10.32). The *set of control laws measurable w.r.t. the fixed observations* $\zeta = \{\zeta_\alpha\}_{\alpha \in A}$ is defined by:

$$\mathcal{U}_A^{\zeta} := \left\{ \boldsymbol{U} = \{\boldsymbol{U}_\alpha\}_{\alpha \in A} \in \mathcal{U}_A \mid \boldsymbol{U}_\alpha \preceq \zeta_\alpha, \ \forall \alpha \in A \right\}. \tag{10.34}$$

We have the following relation between the no dual effect control laws set $\mathcal{U}_A^{\text{nde}}$ in (10.33) and the set $\mathcal{U}_A^{\zeta}$ in (10.34).

Proposition 10.22 *Assume that the NOLDE property holds true, with the fixed observations ζ as in* (10.32). *Then, no dual effect control laws are necessarily measurable w.r.t. the fixed observation ζ, that is,*

$$\mathcal{U}_A^{\text{nde}} \subset \mathcal{U}_A^{\zeta}. \tag{10.35}$$

Proof Let $\boldsymbol{U} = \{\boldsymbol{U}_\alpha\}_{\alpha \in A} \in \mathcal{U}_A^{\text{nde}}$. On the one hand, we have that $\boldsymbol{U}_\alpha \preceq \overset{\boldsymbol{U}}{\eta_\alpha}$, for all agent $\alpha \in A$, since $\boldsymbol{U} \in \mathcal{U}_A^{\text{ad}}$ by (10.33) and (10.13). On the other hand, we have that $\overset{\boldsymbol{U}}{\eta_\alpha} \equiv \zeta_\alpha$ by (10.33) and (10.32). Thus, $\boldsymbol{U}_\alpha \preceq \overset{\boldsymbol{U}}{\eta_\alpha} \equiv \zeta_\alpha$. Since this holds true for any agent α, we conclude that $\boldsymbol{U} \in \mathcal{U}_A^{\zeta}$. □

10.3.3 Characterization of No Dual Effect Control Laws

We now characterize the no dual effect control laws according to the typology discussed in Sect. 10.2.4. We use the following two lemmas.

Lemma 10.23 *Consider three mappings $H_i : \Omega \to \mathbb{Y}_i$, $i = 1, 2$ and $f : \mathbb{Y}_1 \times \Omega \to \mathbb{Y}_3$. Assume that, for all $y_1 \in \mathbb{Y}_1$, $f(y_1, \cdot) \preceq H_2(\cdot)$ and that $H_1(\cdot) \preceq H_2(\cdot)$. Then $f\big(H_1(\cdot), \cdot\big) \preceq H_2(\cdot)$.*

Proof Let $(\omega, \omega') \in \Omega^2$ be such that $H_2(\omega) = H_2(\omega')$. Since $H_1(\cdot) \preceq H_2(\cdot)$, we have that $H_1(\omega) = H_1(\omega')$ by Proposition 3.38. Putting $y_1 = H_1(\omega) = H_1(\omega')$, we thus get $f(y_1, \omega) = f(y_1, \omega')$ since $f(y_1, \cdot) \preceq H_2(\cdot)$. We conclude that

$$f\big(H_1(\omega), \omega\big) = f(y_1, \omega) = f(y_1, \omega') = f\big(H_1(\omega'), \omega'\big).$$

The proof is complete by Proposition 3.38. □

Lemma 10.24 *Let $H_i : \Omega \to \mathbb{Y}_i$, $i = 1, 2$ and $f : \mathbb{Y}_1 \times \Omega \to \mathbb{Y}_3$. Assume that, for all $y_1 \in \mathbb{Y}_1$, $H_2(\cdot) \preceq f(y_1, \cdot)$, and that $H_1(\cdot) \preceq f\big(H_1(\cdot), \cdot\big)$. Then $H_2(\cdot) \preceq f\big(H_1(\cdot), \cdot\big)$.*

Proof Let $(\omega, \omega') \in \Omega^2$ be such that $f\big(H_1(\omega), \omega\big) = f\big(H_1(\omega'), \omega'\big)$. Since $H_1(\cdot) \preceq f\big(H_1(\cdot), \cdot\big)$, we have that $H_1(\omega) = H_1(\omega')$. Putting $y_1 = H_1(\omega) = H_1(\omega')$, we thus get

$$f(y_1, \omega) = f\big(H_1(\omega), \omega\big) = f\big(H_1(\omega'), \omega'\big) = f(y_1, \omega').$$

On the other hand, we have that $H_2(\cdot) \preceq f(y_1, \cdot)$, so that $H_2(\omega) = H_2(\omega')$. The proof is complete by Proposition 3.38. □

Partially Nested MASIOS

The following main result, established in [13], provides a description of the set of no dual effect control laws for MASIOS displaying the NOLDE property.

Theorem 10.25 ([13]) *Assume that the NOLDE property holds true, with the fixed observations ζ as in (10.32). Assume that the MASIOS is partially nested as in (10.20). Then, the no dual effect control laws in (10.33) are exactly the admissible control laws which are measurable w.r.t. the fixed observation ζ:*

$$\mathcal{U}_A^{\mathrm{nde}} = \mathcal{U}_A^{\mathrm{ad}} \cap \mathcal{U}_A^{\zeta}. \tag{10.36}$$

Proof By Proposition 10.22, it suffices to show that $\mathcal{U}_A^{\mathrm{ad}} \cap \mathcal{U}_A^{\zeta} \subset \mathcal{U}_A^{\mathrm{nde}}$.

Let $\boldsymbol{U} = \{\boldsymbol{U}_\beta\}_{\beta \in A} \in \mathcal{U}_A^{\mathrm{ad}} \cap \mathcal{U}_A^{\zeta}$, that is,

$$\boldsymbol{U}_\beta \preceq \zeta_\beta \text{ and } \boldsymbol{U}_\beta \preceq \overset{\boldsymbol{U}}{\eta_\beta}, \ \forall \beta \in A. \tag{10.37}$$

Let $\alpha \in A$ be fixed: we now prove that both $\overset{\boldsymbol{U}}{\eta_\alpha} \preceq \zeta_\alpha$ and $\zeta_\alpha \preceq \overset{\boldsymbol{U}}{\eta_\alpha}$ hold true. We use the property that, by Eq. (10.17) and by abuse of notation (see (10.16)),

$$\overset{\boldsymbol{U}}{\eta_\alpha} = \overset{\boldsymbol{U}_{\langle\alpha\rangle_{\mathfrak{P}}}}{\eta_\alpha}, \ \forall \alpha \in A. \tag{10.38}$$

First, we show that $\overset{\boldsymbol{U}}{\eta_\alpha} \preceq \zeta_\alpha$. For any $u \in \mathbb{U}_A$ (identified with a constant control law), we have that:

$$\begin{aligned} \boldsymbol{U}_{\langle\alpha\rangle_{\mathfrak{P}}} &= \{\boldsymbol{U}_\beta\}_{\beta \in \langle\alpha\rangle_{\mathfrak{P}}} && \text{by (10.16)} \\ &\equiv \bigvee_{\beta \in \langle\alpha\rangle_{\mathfrak{P}}} \boldsymbol{U}_\beta && \text{by Proposition 3.42} \end{aligned}$$

$$\preceq \bigvee_{\beta \in \langle \alpha \rangle_{\mathfrak{P}}} \zeta_\beta \qquad \text{because } \boldsymbol{U}_\beta \preceq \zeta_\beta \text{ by (10.37)}$$

$$\equiv \bigvee_{\beta \in \langle \alpha \rangle_{\mathfrak{P}}} o_\beta(u, \cdot), \ \forall u \in \mathbb{U}_A \qquad \text{by (10.31), (10.32) and (10.10)}$$

$$\equiv \{o_\beta(u, \cdot)\}_{\beta \in \langle \alpha \rangle_{\mathfrak{P}}} \qquad \text{by Proposition 3.42}$$

$$\equiv o_{\langle \alpha \rangle_{\mathfrak{P}}}(u, \cdot) \qquad \text{by (10.16)}$$

$$\preceq o_\alpha(u, \cdot) \qquad \text{by (10.20)}$$

$$\equiv \zeta_\alpha(\cdot) \qquad \text{by (10.32).}$$

Therefore, we have that, on the one hand, $\boldsymbol{U}_{\langle \alpha \rangle_{\mathfrak{P}}}(\cdot) \preceq o_\alpha(u, \cdot) \equiv \zeta_\alpha(\cdot)$ and, on the other hand, $o_\alpha(u, \cdot) = \widetilde{o}_\alpha(u_{\langle \alpha \rangle_{\mathfrak{P}}}, \cdot) \equiv \zeta_\alpha(\cdot)$, for all $u \in \mathbb{U}_A$ by (10.17) and (10.32). By Lemma 10.23, we deduce that:

$$\eta_\alpha^{\boldsymbol{U}_{\langle \alpha \rangle_{\mathfrak{P}}}}(\cdot) = \widetilde{o}_\alpha\big(\boldsymbol{U}_{\langle \alpha \rangle_{\mathfrak{P}}}(\cdot), \cdot\big) \preceq \zeta_\alpha(\cdot).$$

By (10.38), we conclude that

$$\eta_\alpha^{\boldsymbol{U}} \preceq \zeta_\alpha .$$

Second, we prove that $\zeta_\alpha \preceq \eta_\alpha^{\boldsymbol{U}}$. We have that

$$\boldsymbol{U}_{\langle \alpha \rangle_{\mathfrak{P}}} = \{\boldsymbol{U}_\beta\}_{\beta \in \langle \alpha \rangle_{\mathfrak{P}}} \qquad \text{by (10.16)}$$

$$\equiv \bigvee_{\beta \in \langle \alpha \rangle_{\mathfrak{P}}} \boldsymbol{U}_\beta \qquad \text{by Proposition 3.42}$$

$$\preceq \bigvee_{\beta \in \langle \alpha \rangle_{\mathfrak{P}}} \eta_\beta^{\boldsymbol{U}} \qquad \text{because } \boldsymbol{U}_\beta \preceq \eta_\beta^{\boldsymbol{U}} \text{ by (10.37)}$$

$$\equiv \eta_{\langle \alpha \rangle_{\mathfrak{P}}}^{\boldsymbol{U}} \qquad \text{by (10.16) and Proposition 3.42}$$

$$= o_{\langle \alpha \rangle_{\mathfrak{P}}}\big(\boldsymbol{U}(\cdot), \cdot\big) \qquad \text{by (10.10)}$$

$$\preceq o_\alpha\big(\boldsymbol{U}(\cdot), \cdot\big) \qquad \text{by the partially nested property (10.20)}$$

$$= \widetilde{o}_\alpha\big(\boldsymbol{U}_{\langle \alpha \rangle_{\mathfrak{P}}}(\cdot), \cdot\big) \qquad \text{by (10.17).}$$

Therefore, on the one hand, we have just proven that

$$\boldsymbol{U}_{\langle \alpha \rangle_{\mathfrak{P}}}(\cdot) \preceq \widetilde{o}_\alpha\big(\boldsymbol{U}_{\langle \alpha \rangle_{\mathfrak{P}}}(\cdot), \cdot\big).$$

On the other hand, for all $u \in \mathbb{U}_A$, we have that

$$\begin{aligned}
\zeta_\alpha(\cdot) &\equiv \eta_\alpha^u(\cdot) && \text{by (10.32)}\\
&= o_\alpha(u,\cdot) && \text{by (10.10)}\\
&= \widetilde{o}_\alpha(u_{\langle\alpha\rangle_{\mathfrak{P}}},\cdot) && \text{by (10.17).}
\end{aligned}$$

By Lemma 10.24, we deduce that

$$\zeta_\alpha(\cdot) \preceq \widetilde{o}_\alpha\big(\boldsymbol{U}_{\langle\alpha\rangle_{\mathfrak{P}}}(\cdot),\cdot\big).$$

By (10.38), we conclude that

$$\zeta_\alpha \preceq \eta_\alpha^{\boldsymbol{U}_{\langle\alpha\rangle_{\mathfrak{P}}}} = \eta_\alpha^{\boldsymbol{U}}.$$

This completes the proof. □

Quasiclassical MASIOS

For sequential MASIOS, where each agent is supposed to represent a time period, we are able to obtain a result that is more precise than Theorem 10.25. Indeed, we now show that, for quasiclassical MASIOS displaying the NOLDE property, the no dual effect control laws are the control laws which are measurable w.r.t. the fixed observations.

Proposition 10.26 *Assume that the NOLDE property holds true, with the fixed observations ζ as in (10.32). Assume that the MASIOS is quasiclassical, as in Sect. 10.2.4 with the ordering $0,\ldots,T$ of agents. Then, the no dual effect control laws in (10.33) are the control laws which are measurable w.r.t. the fixed observation ζ, that is,*

$$\mathcal{U}^{\mathrm{nde}}_{\{0,\ldots,T\}} = \mathcal{U}^{\zeta}_{\{0,\ldots,T\}}. \tag{10.39}$$

Proof By Proposition 10.22, it suffices to show that $\mathcal{U}^{\zeta}_{\{0,\ldots,T\}} \subset \mathcal{U}^{\mathrm{nde}}_{\{0,\ldots,T\}}$. Let $\boldsymbol{U} = \{\boldsymbol{U}_t\}_{t=0,\ldots,T} \in \mathcal{U}^{\zeta}_{\{0,\ldots,T\}}$, that is,

$$\boldsymbol{U}_t \preceq \zeta_t,\ \forall t = 0,\ldots,T. \tag{10.40}$$

We prove by induction that

$$\Big(\forall t = 0,\ldots,T,\ \boldsymbol{U}_t \preceq \zeta_t\Big) \Rightarrow \Big(\forall t = 0,\ldots,T,\ \boldsymbol{U}_t(\cdot) \preceq o_t\big(\boldsymbol{U}(\cdot),\cdot\big)\Big).$$

Let the induction assumption $\mathrm{H}(t)$ be

$$\Big(\forall s = 0,\ldots,t,\ \boldsymbol{U}_s \preceq \zeta_s\Big) \Rightarrow \Big(\forall s = 0,\ldots,t,\ \boldsymbol{U}_s(\cdot) \preceq o_s\big(\boldsymbol{U}(\cdot),\cdot\big)\Big).$$

Suppose that $\boldsymbol{U}_0(\cdot) \preceq \zeta_0(\cdot)$. By (10.31), we know that $\zeta_0(\cdot) \equiv o_0(u,\cdot)$, for all $u \in \mathbb{U}$. However, o_0 is independent of u, since agent 0 has no predecessor ($[0] = \emptyset$).

Thus, we conclude that

$$\boldsymbol{U}_0(\cdot) \preceq o_0(u, \cdot) = o_0\big(\boldsymbol{U}(\cdot), \cdot\big)$$

and the induction assumption H(0) holds true.

Assume that the induction assumption H($t-1$) holds true, and suppose that

$$\boldsymbol{U}_s \preceq \zeta_s, \ \forall s = 0, \ldots, t. \tag{10.41}$$

We have that

$$\begin{aligned}
\boldsymbol{U}_{\langle t\rangle_{\mathfrak{P}}}(\cdot) &= \bigvee_{s\in\langle t\rangle_{\mathfrak{P}}} \boldsymbol{U}_s(\cdot) && \text{by (10.16)}\\
&\preceq \bigvee_{s\in\langle t\rangle_{\mathfrak{P}}} o_s\big(\boldsymbol{U}(\cdot), \cdot\big) && \text{by assumption H}(t-1)\\
& && \text{and since } \langle t\rangle_{\mathfrak{P}} \subset \{0, \ldots, t-1\} \text{ by (10.27)}\\
&\preceq \bigvee_{s\in\langle t\rangle_{\mathfrak{M}}} o_s\big(\boldsymbol{U}(\cdot), \cdot\big) && \text{by the partially nested property (10.19)}\\
&\equiv o_{\langle t\rangle_{\mathfrak{M}}}\big(\boldsymbol{U}(\cdot), \cdot\big) && \text{by (10.16)}\\
&\preceq o_t\big(\boldsymbol{U}(\cdot), \cdot\big) && \text{by (10.18)}\\
&= \widetilde{o}_t\big(\boldsymbol{U}_{\langle t\rangle_{\mathfrak{P}}}(\cdot), \cdot\big) && \text{by (10.17).}
\end{aligned}$$

Therefore, we have proved that

$$\boldsymbol{U}_{\langle t\rangle_{\mathfrak{P}}}(\cdot) \preceq \widetilde{o}_t\big(\boldsymbol{U}_{\langle t\rangle_{\mathfrak{P}}}(\cdot), \cdot\big) = o_t\big(\boldsymbol{U}(\cdot), \cdot\big).$$

Now, on the other hand, we have that

$$\boldsymbol{U}_t(\cdot) \preceq \zeta_t(\cdot) \equiv o_t(u_{\langle t\rangle_{\mathfrak{P}}}, \cdot), \ \forall \{u_s\}_{s\in\{0,\ldots,T\}} \in \mathbb{U}_{\{0,\ldots,T\}},$$

by (10.41), (10.31) and (10.17). We now use Lemma 10.24 with $H_1 = \boldsymbol{U}_t$ and $f(u, \omega) = o_t(u, \omega)$ to obtain that[2]

$$\boldsymbol{U}_t(\cdot) \preceq o_t\big(\boldsymbol{U}(\cdot), \cdot\big). \tag{10.42}$$

Thus, assumption H(t) holds true. This completes the induction. □

[2]In fact, we use a slight variation of Lemma 10.24. Indeed $\boldsymbol{U}_{\langle t\rangle_{\mathfrak{P}}}(\cdot) \preceq o_t\big(\boldsymbol{U}_{\langle t\rangle_{\mathfrak{P}}}(\cdot), \cdot\big)$ is a weaker assumption than $\boldsymbol{U}(\cdot) \preceq o_t\big(\boldsymbol{U}(\cdot), \cdot\big)$ However, thanks to (10.17), the proof of Lemma 10.24 can easily be adapted to obtain the same conclusion.

Proposition 10.26 applies to sequential MASIOS displaying perfect memory because, as a consequence of (9.52), they are classical (see Sect. 9.5.1), hence quasiclassical, with the ordering $0, \ldots, T$.

10.4 Conclusion

In this chapter, we have more deeply analyzed the "dual effect" of control previously discussed in Sects. 1.1.3, 1.2.1, 1.3.2 and 4.2.3. The specificity of sequential systems with perfect memory has been emphasized. When they display the NOLDE property, the no dual effect control laws have a simple characterization: they are the control laws which are measurable w.r.t. the fixed observations. Therefore, this chapter brings to light another element possibly explaining the importance of sequential systems with perfect memory in stochastic control.

Appendix A
Basics in Analysis and Optimization

The purpose of this appendix is to briefly review basic notions and results in Analysis and Optimization Theory. We refer the reader to the numerous textbooks on the topic (e.g. [19, 27, 46, 69, 78, 79, 102, 103, 105]) for a deeper and more rigorous treatment. Only *deterministic* problems are considered here.

A.1 Convexity, Continuity and Differentiability

We are dealing in this appendix with Hilbert spaces. A *Hilbert space* $\mathbb{U}$ is a vector space endowed with a topology deriving from a *norm*, denoted $\|\cdot\|$, which itself derives from a *scalar product*, denoted $\langle\cdot,\cdot\rangle$—that is, $\|u\|^2 = \langle u, u\rangle$—and such that $\mathbb{U}$ is *complete* for this topology. The *topological dual* of $\mathbb{U}$, denoted $\mathbb{U}^\star$, is the set of continuous linear forms (linear mappings from $\mathbb{U}$ to $\mathbb{R}$) over $\mathbb{U}$. By *Riesz representation theorem* (see [136, Theorem 4.12]), every element ℓ of $\mathbb{U}^\star$ can be represented as a mapping $u \mapsto \langle l, u\rangle$ for some l in $\mathbb{U}$. Thus, $\mathbb{U}^\star$ is identifiable to $\mathbb{U}$ by $\ell \mapsto l$, and we allow the notation $\langle \ell, u\rangle$ instead of $\ell(u)$ (in this case, $\langle\cdot,\cdot\rangle$ may be viewed as the *duality product* rather than as the scalar product). In addition to the *strong* topology of $\mathbb{U}$, the *weak* topology is also useful to consider for infinite dimensional spaces (for finite dimensional Hilbert spaces they coincide): u converges weakly towards v in the weak topology (this is denoted $u \rightharpoonup v$) if, for all $\ell \in \mathbb{U}^\star$, $\langle \ell, u\rangle \to \langle \ell, v\rangle$.

Consider a Hilbert space $\mathbb{U}$. A *cone* (with vertex at 0) is a subset of $\mathbb{U}$ such that if u belongs to it, then αu also belongs to it for all $\alpha \in \mathbb{R}^+$. Let U be a subset of $\mathbb{U}$. The *normal cone* to U at the point u_0 in U is defined by

$$N_U(u_0) := \big\{ r \in \mathbb{U} \mid \langle r, u - u_0\rangle \leq 0,\ \forall u \in U \big\}. \tag{A.1}$$

Note that $N_U(u_0)$ is actually a cone with vertex at 0 (and not at u_0).

P. Carpentier et al., *Stochastic Multi-Stage Optimization*, Probability Theory and Stochastic Modelling 75, DOI 10.1007/978-3-319-18138-7

Let J be a real-valued function over $\mathbb{U}$. The *graph* of J is the set of all points of the form $\big(u, J(u)\big)$, and the *epigraph* epi J of J is defined as the subset of points (u, y) of $\mathbb{U} \times (\mathbb{R} \cup \{+\infty\})$ such that $y \geq J(u)$.

A.1.1 Convex Functions and Subsets

Convexity theory plays a crucial role in optimization. Classical references are [64, 131]. A real-valued function J over a Hilbert space $\mathbb{U}$ is *convex* if

$$\forall u, v \in \mathbb{U},\ \forall \alpha \in [0, 1],\ J\big(\alpha u + (1-\alpha)v\big) \leq \alpha J(u) + (1-\alpha)J(v).$$

It is *strictly convex* if the inequality is strict whenever $\alpha \in (0, 1)$ and $u \neq v$; it is *strongly convex with modulus* a if there exists $a > 0$ such that, for all u and v in $\mathbb{U}$ and all α in $[0, 1]$,

$$J\big(\alpha u + (1-\alpha)v\big) \leq \alpha J(u) + (1-\alpha)J(v) - \frac{a}{2}\alpha(1-\alpha)\|u - v\|^2.$$

A function J is *concave* if $-J$ is convex (similar definition for strict and strong concavity).

As seen later on, it is useful to allow convex functions to assume the value $+\infty$. Let denote $\overline{\mathbb{R}} = \mathbb{R} \cup \{-\infty\} \cup \{+\infty\}$. The *domain* dom J of a convex function $J : \mathbb{U} \to \overline{\mathbb{R}}$ is defined as

$$\text{dom}\, J = \{u \in \mathbb{U} \mid J(u) < +\infty\}.$$

A convex function is *proper* if its domain is nonempty and if it nowhere assumes the value $-\infty$.

The notion of convexity is also relevant for subsets U of a vector space $\mathbb{U}$: U is convex if

$$\forall (u, v) \in U^2,\ \forall \alpha \in [0, 1],\ \alpha u + (1-\alpha)v \in U.$$

Indeed, the notion of convexity of functions is intimately related to that of subsets. If the former is introduced first, then the latter can be derived by means of the *characteristic function* χ_U of a subset U defined as

$$\chi_U(u) = \begin{cases} 0 & \text{if } u \in U, \\ +\infty & \text{otherwise.} \end{cases} \tag{A.2}$$

Then, the subset U is convex iff χ_U is a convex function. Conversely, if convex subsets have been defined first, a function J is convex iff epi J is a convex subset.

Convexity of functions is preserved by the following operations:

- if $J : \mathbb{U} \to \mathbb{R}$ is convex and $K : \mathbb{R} \to \mathbb{R}$ is convex *and nondecreasing*, then $K \circ J$ is convex;
- if $J_i, i = 1, \ldots, n$, is a collection of convex functions and α_i are nonnegative numbers, then $\sum_{i=1}^{n} \alpha_i J_i$ is a convex function;
- also the *upper hull* $\sup_i J_i$ is convex (in terms of epigraphs, this operation corresponds to the *intersection* of epigraphs);
- if $J : \mathbb{U} \times \mathbb{V} \to \mathbb{R}$ is convex *jointly in* u and v, then the function

$$u \mapsto \inf_v J(u, v)$$

is convex in u (but, in general, the *lower hull* of a family of convex functions is not convex).

If a subset U is not convex, there exists a *smallest* convex subset containing U: namely, this set is generated by all convex combinations of pairs of points belonging to U, that is, points of the form $\alpha u + (1 - \alpha)v$, where α ranges in $[0, 1]$ and (u, v) range in $U \times U$; it is called the *convex hull* of U and is denoted $\operatorname{co} U$. An even more useful notion is that of *closed convex hull* $\overline{\operatorname{co}}\, U$ which is the (topological) closure of $\operatorname{co} U$.

These notions transfer to functions by making use of their associated epigraphs. The convex hull $\operatorname{co} J$ of a function J is the largest convex function the graph of which lies below that of J, and the closed convex hull $\overline{\operatorname{co}}\, J$ of J is the largest convex function the graph of which lies below that of J and which has a closed epigraph: as seen below, this corresponds to the lower semicontinuity of J (Fig. A.1).

A.1.2 Continuity and Lower Semicontinuity

Let $\mathbb{U}$ and $\mathbb{V}$ be two Hilbert spaces.

Definition A.1 A function $J : \mathbb{U} \to \mathbb{V}$ is *continuous* at $u_0 \in \mathbb{U}$ if

$$\|u - u_0\| \to 0 \Rightarrow \|J(u) - J(u_0)\| \to 0.$$

A function is Lipschitz continuous with modulus L if

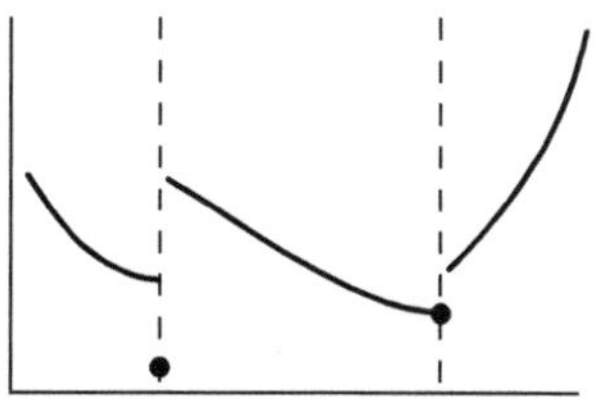

Fig. A.1 Discontinuities of l.s.c. function

$$\exists L > 0, \ \forall u, v \in \mathbb{U}, \ \|J(u) - J(v)\| \leq L \|u - v\| .$$

Of course, Lipschitz continuity is stronger than continuity. For example, the real function $u \mapsto \sqrt{u}$ over the domain $u \geq 0$ is continuous but not Lipschitz continuous.

Essentially, convex functions are continuous in the interior of their domain (see Theorem A.3 below). For convex functions, discontinuities may thus arise at the boundary of their domain (refer to the characteristic function of a convex subset as an example). In this case, *lower semicontinuity* is desirable if the function is going to be minimized (see Fig. A.2).

Definition A.2 (*lower semicontinuity*) A real-valued function J over a Hilbert space $\mathbb{U}$ is lower semicontinuous (abbreviated as l.s.c.) at u_0 if

$$\liminf_{u \to u_0} J(u) \geq J(u_0).$$

An alternative definition is that the function has a closed epigraph. A characteristic function is l.s.c. iff the corresponding subset is closed.

Theorem A.3 *A proper convex function* $J : \mathbb{U} \to \mathbb{R}$ *is continuous in every point in the interior of* dom J *in the neighborhood of which* J *is bounded from above. A proper l.s.c. convex function* J *is continuous in the interior of its domain.*

It is important to remark that an l.s.c. function is not necessarily l.s.c. when the domain space is equipped with the weak topology (instead of the strong one). But this is true for *convex* functions. Indeed, closed *convex* subsets are still closed when the space is equipped with the weak topology instead of the strong one, hence the result for convex functions when considering their epigraphs.

A.1.3 Fenchel-Legendre Transformation

Let $\mathbb{U}$ be a Hilbert space and $J : \mathbb{U} \to \mathbb{R} \cup \{+\infty\}$ be a function taking values in the extended real line. The *Fenchel conjugate* (also called *convex conjugate*) is the function $J^\star : \mathbb{U} \to \mathbb{R} \cup \{+\infty\}$ defined by:

$$J^\star(p) := \sup_{u \in \mathbb{U}} \Big(\langle p\,, u \rangle - J(u) \Big).$$

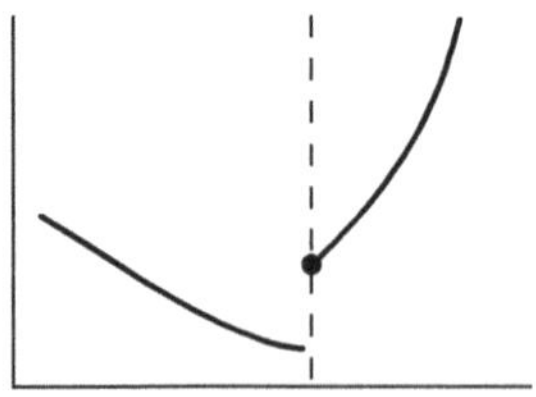

Fig. A.2 A non l.s.c. function may not reach its minimum

The Fenchel conjugate of a function J is always convex and lower semi-continuous. Let $J^{\star\star}$ be the *biconjugate* of J, that is, the Fenchel conjugate of $J^\star$:

$$J^{\star\star} = \left(J^\star\right)^\star.$$

The biconjugate $J^{\star\star}$ is the *largest* convex lower semi-continuous function less than or equal to J. By definition of the closed convex hull of J, one has that $J^{\star\star} = \overline{\text{co}}\, J$, and $J^{\star\star} \leq J$. For a proper function J, one has that $J^{\star\star} = J$ iff J is convex and lower semi-continuous (Fenchel-Moreau theorem).

For any function J and its Fenchel conjugate $J^\star$, Fenchel's inequality (also known as the Fenchel-Young inequality) holds:

$$\forall u \in \mathbb{U},\ \forall p \in \mathbb{U},\ \langle p\, , u\rangle \leq J(u) + J^\star(p).$$

A.1.4 Differentiability and Subdifferentiability

A function $J : \mathbb{U} \to \mathbb{R}$ is *Fréchet-differentiable* at u_0 if there exists an element of $\mathbb{U}^\star$, denoted $J'(u_0)$ and called the *derivative* of J at u_0, such that

$$\lim_{u\to u_0} \frac{J(u) - J(u_0) - \left\langle J'(u_0), u - u_0\right\rangle}{\|u - u_0\|} = 0.$$

A *directional derivative* of J at u_0 in the direction v, denoted $DJ(u_0;\, v)$, is defined as the following limit, if this limit exists,

$$DJ(u_0;\, v) = \lim_{\varepsilon\to 0_+} \frac{J(u_0 + \varepsilon v) - J(u_0)}{\varepsilon}.$$

If J is Fréchet-differentiable, then $DJ(u_0;\, v) = \left\langle J'(u_0), v\right\rangle$. More generally, if the directional derivative is a continuous linear function of the direction v, the function J is said *Gâteaux-differentiable*. Note that Fréchet differentiability is stronger than Gâteaux differentiability in that limits must exist when points u move to u_0 along any path and not only radially. However, the latter notion is generally sufficient in the context of optimization theory, and "differentiability" generally refers to Gâteaux-differentiability in this book.

The *gradient* of J at u_0 is the element of $\mathbb{U}$ to which $J'(u_0)$ can be identified by the Riesz theorem (see Sect. A.1), and it is denoted $\nabla J(u_0)$. If $\mathbb{U}$ is finite-dimensional, $J'(u)$ should be considered as a *row* vector whereas $\nabla J(u)$, its *transpose*, is a *column* vector of the same dimension as u. Likewise, in general, for a real-valued function G of two variables u and v, $\nabla_u G(u, v)$ denotes the primal counterpart of the partial derivative $\partial G(u, v)/\partial u$, that is, the derivative of the function $u \mapsto G(u, v)$ at point u.

Remark A.4 Consider now a mapping Θ from $\mathbb{U}$ to another Hilbert space $\mathbb{V}$. The derivative $\Theta'(u)$ is a continuous linear mapping from $\mathbb{U}$ to $\mathbb{V}$. If both spaces are finite dimensional, $\Theta'(u)$ may be represented by a matrix whose column number is the dimension of $\mathbb{U}$ and row number is the dimension of $\mathbb{V}$: this is known as the *Jacobian* matrix. Its transpose[1] is consistently denoted $\nabla\Theta(u)$. In infinite dimensions, $\nabla\Theta(u)$ is the dual or adjoint linear operator, from $\mathbb{V}$ to $\mathbb{U}$, of $\Theta'(u)$. ◊

The second-order derivative, denoted J'', is of course defined as the derivative of the first-order derivative $u \mapsto J'(u)$, provided this first-order derivative exists and is differentiable. Since J' maps $\mathbb{U}$ to $\mathbb{U}^\star$, $J''(u)$ (the second-order derivative *evaluated at u*) is a linear mapping from $\mathbb{U}$ to $\mathbb{U}^\star$ which is represented by a symmetric matrix if $\mathbb{U}$ is finite dimensional.[2] If J'' exists, we also say that J is "twice differentiable".

Even if they are continuous, convex functions are not necessarily differentiable: an example is the mapping on $\mathbb{R}$: $u \mapsto |u|$, the graph of which has a corner at 0. The notion of *subgradient* of a convex function is based on the fact that "a differentiable convex function is above all its tangents", namely,

$$\forall (u_0, u) \in \mathbb{U}^2, \;\; J(u) - J(u_0) \geq \langle \nabla J(u_0), u - u_0 \rangle \,.$$

Definition A.5 The subdifferential of a convex function $J : \mathbb{U} \to \mathbb{R}$ at u_0 is denoted $\partial J(u_0)$, if nonempty. It is the subset of $\mathbb{U}$ defined as

$$\partial J(u_0) := \big\{ r \in \mathbb{U} \,\big|\, \forall u \in \mathbb{U}, \, J(u) - J(u_0) \geq \langle r, u - u_0 \rangle \big\}.$$

An element of $\partial J(u_0)$ is called a subgradient of J at u_0, and J is subdifferentiable at u_0 if it admits at least one subgradient at this point.

At every point of subdifferentiability, the subdifferential is closed and convex. It is the set of slopes of supporting hyperplanes of the epigraph of the function at this point (see Fig. A.3).

One has the following interesting formula

$$DJ(u_0; v) = \max_{r \in \partial J(u_0)} \langle r, v \rangle \,, \tag{A.3}$$

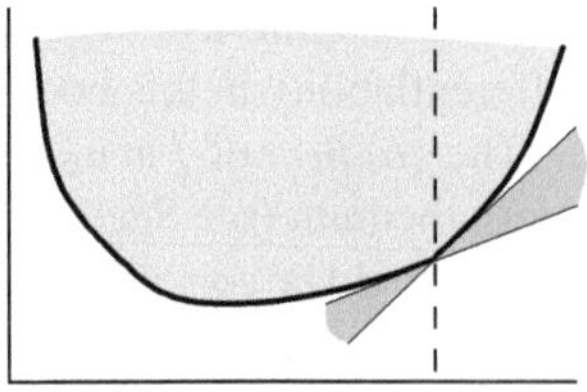

Fig. A.3 Supporting hyperplanes of the epigraph

[1] Transposition is denoted by the superscript $^\top$.

[2] In infinite dimension, it is "self-adjoint".

which shows that J is Gâteaux-differentiable iff its subdifferential is reduced to a singleton.

Example A.6 An interesting example of a subdifferential is that of the characteristic function χ_U of a closed convex subset U. If u_0 lies in the boundary[3] of U, and if u ranges in U, one has that

$$r \in \partial\chi_U(u_0) \Rightarrow \underbrace{\chi_U(u)}_{0} - \underbrace{\chi_U(u_0)}_{0} \geq \langle r, u - u_0 \rangle .$$

This inequality and Eq. (A.1) show that any r belonging to the normal cone to U at u_0 is an element of $\partial\chi_U(u_0)$ (this normal cone reduces to a half line at points where the boundary is smooth—see Fig. A.4). △

Example A.7 Another interesting example of a subdifferential is that of the upper hull of a family of convex functions

$$J(u) = \sup_v G(u, v),$$

where G is convex in u for all v. Denoting $\widehat{V}(u) = \arg\max_v G(u, v)$, and under technical assumptions (see [49]), we have that

$$\partial J(u) = \overline{\text{co}}\, \partial_u G\big(u, \widehat{V}(u)\big) = \overline{\text{co}} \bigcup_{v \in \widehat{V}(u)} \partial_u G(u, v). \qquad \triangle$$

A (possibly set-valued) mapping $A : \mathbb{U} \to 2^{\mathbb{U}}, u \mapsto A(u)$ is *monotone* if

$$\forall (u, v) \in \mathbb{U}^2,\ \forall r \in A(u),\ \forall s \in A(v),\ \langle r - s, u - v \rangle \geq 0.$$

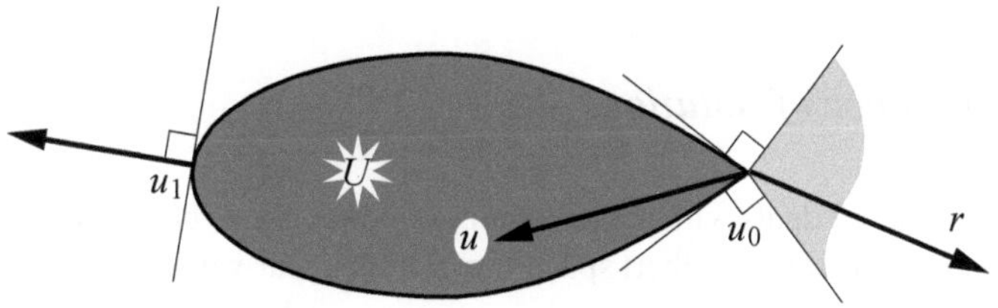

Fig. A.4 Normal cones at u_0 and u_1

[3]Otherwise, either $\partial\chi_U(u_0) = \{0\}$ if u_0 belongs to the interior of U, or no subgradients exist outside U.

It is *strongly monotone with modulus a* if

$$\exists a > 0,\ \forall (u, v) \in \mathbb{U}^2,\ \forall r \in A(u),\ \forall s \in A(v),\ \langle r - s, u - v \rangle \geq a \|u - v\|^2 .$$

For the (finite dimensional) linear operator associated with a matrix, monotonicity amounts to nonnegative definiteness whereas strong monotonicity is equivalent to positive definiteness.[4]

For a subdifferentiable convex function J, the set-valued mapping ∂J is monotone; it is strongly convex with modulus a iff ∂J is strongly monotone with modulus a. The latter property is also equivalent to the following property

$$\forall (u, v) \in \mathbb{U}^2,\ \forall r \in \partial J(u),\ J(v) - J(u) \geq \langle r, v - u \rangle + \frac{a}{2} \|v - u\|^2 ,$$

which means that a strongly convex function is bounded from below at any point by a parabola. In relation with this inequality, it is interesting to mention that a differentiable convex function J with Lipschitz continuous derivative with modulus L obeys the inequality

$$\forall (u, v) \in \mathbb{U}^2,\ J(v) - J(u) \leq \langle \nabla J(u), v - u \rangle + \frac{L}{2} \|v - u\|^2 .$$

If J is twice differentiable, then it is (strongly) convex iff $J''(u)$ is (strongly) monotone at any u.

A.2 Optimization Over a Space or an Admissible Subset

We first consider optimization problems where all the constraints are defined by means of an admissible set U^{ad}.

A.2.1 Existence of Solutions

Optimization theory deals with the problem of finding points belonging to an admissible subset U^{ad} of (say) a Hilbert space $\mathbb{U}$ at which a real-valued function J reaches its minimal (or maximal) value:

$$\min_{u \in U^{\mathrm{ad}}} J(u). \tag{A.4}$$

[4]For *linear* operators in *finite* dimension, strict and strong monotonicity cannot be distinguished.

Function J is called the *cost* or *objective function* or *criterion*. The set of points $u^\sharp$ at which this happens is denoted $\arg\min_{U^{\mathrm{ad}}} J$ (or $\arg\max_{U^{\mathrm{ad}}} J$), and each such point is called an argument of the minimum (or maximum) of J. To prevent such points from being "located at infinity", the following property of J is generally assumed: J is *coercive* over U^{ad} if

$$\left(u \in U^{\mathrm{ad}}, \|u\| \to +\infty\right) \Rightarrow J(u) \to +\infty, \tag{A.5}$$

which may be considered to hold trivially if U^{ad} is bounded.

Theorem A.8 *If J is l.s.c. and if U^{ad} is compact, then (A.4) admits solutions. The same holds true if J is convex, l.s.c. and coercive over U^{ad}, and if U^{ad} is closed and convex.*

In the latter statement, the existence of a solution of (A.4) is proved in the weak topology of $\mathbb{U}$.[5] As already explained in Sect. A.1.2, J (resp. U^{ad}) remains l.s.c. (resp. closed) when $\mathbb{U}$ is equipped with the weak topology thanks to convexity. Moreover, the coercivity of J makes it possible to reduce the admissible set to a closed and bounded one (for example $U^{\mathrm{ad}} \cap \{u \in \mathbb{U} \mid J(u) \leq J(u_0)\}$ for any given $u_0 \in U^{\mathrm{ad}}$), hence compact in the weak topology. Then the former statement applies, in the weak topology.

Convexity has another important consequence in optimization: local minima are also global minima. A *global minimum*[6] is a solution of (A.4), whereas a *local minimum* is a point $u^\sharp$ for which there exists a neighborhood in which the function J takes values which are not smaller than $J(u^\sharp)$. Moreover, if J is strictly convex, the arg min is a singleton if it exists.

A.2.2 Optimality Conditions

For smooth (not necessarily convex) functions, the following theorem can be stated.

Theorem A.9 *If U^{ad} is open and if J is (Gâteaux-)differentiable, a necessary condition for $u^\sharp$ in U^{ad} to be a local minimum is that $\nabla J(u^\sharp) = 0$ (one says that J is* stationary *at $u^\sharp$). If, moreover, J is twice differentiable, another necessary condition is that $J''(u^\sharp)$ be a monotone linear operator. A sufficient condition is that $\nabla J(u^\sharp) = 0$* and *$J''(u^\sharp)$ is strongly monotone.*

The conditions involving first-order (resp. second-order) derivatives are called *first-order* (resp. *second-order*) *optimality conditions*. The second-order optimality conditions express that J is locally convex (or strongly convex) around its minimum.

For subdifferentiable convex functions, the following condition is available.

[5] Recall that the notion of minimum does not depend on the topology.

[6] We should say "an argument of a global minimum".

Theorem A.10 *If U^{ad} is open and if J is convex and subdifferentiable, it achieves its minimum in $u^\sharp \in U^{ad}$ iff $0 \in \partial J(u^\sharp)$.*

With this result at hand, the case when U^{ad} is closed and convex can be handled by transforming (A.4) as a minimization problem over the whole space. By making use of the characteristic function of U^{ad}, (A.4) is equivalent to

$$\min_{u \in \mathbb{U}} \big(J(u) + \chi_{U^{ad}}(u)\big), \tag{A.6}$$

and the condition stated at Theorem A.10 is equivalent to any of the following three conditions:

$$u^\sharp \in U^{ad} \text{ and } \exists\, r^\sharp \in \partial J(u^\sharp),\ r^\sharp \in -\partial \chi_{U^{ad}}(u^\sharp), \tag{A.7a}$$

$$u^\sharp \in U^{ad} \text{ and } \exists\, r^\sharp \in \partial J(u^\sharp),\ \forall u \in U^{ad},\ \langle r^\sharp, u - u^\sharp \rangle \geq 0, \tag{A.7b}$$

$$u^\sharp \in U^{ad} \text{ and } \exists\, r^\sharp \in \partial J(u^\sharp),\ \forall \epsilon \geq 0,\ u^\sharp = \mathrm{proj}_{U^{ad}}\left(u^\sharp - \epsilon r^\sharp\right). \tag{A.7c}$$

Condition (A.7a) is a direct consequence of Theorem A.10 (with an additional assumption ensuring that $\partial\big(J + \chi_{U^{ad}}\big)(\cdot) = \partial J(\cdot) + \partial \chi_{U^{ad}}(\cdot)$: see e.g. [64, Chapter I, Proposition 5.6]), Condition (A.7b) arises from the expression of $\partial \chi_{U^{ad}}(u)$ obtained at Example A.6, and the optimality condition of the projection problem in (A.7c) is precisely (A.7b) (see Example A.11). Condition (A.7b) is called a *variational inequality*. It expresses the fact that there exists a subgradient $r^\sharp$ of J at $u^\sharp$ such that $-r^\sharp$ belongs to the normal cone of U^{ad} at $u^\sharp$ (see Fig. A.5 in which J is represented by its curves of constant values).

Example A.11 The projection $\mathrm{proj}_{U^{ad}}(v)$ of v over a closed convex set U^{ad} is defined as the point of U^{ad} which is the closest to v. This translates into the optimization problem

$$\min_{u \in U^{ad}} \frac{1}{2} \|u - v\|^2 ,$$

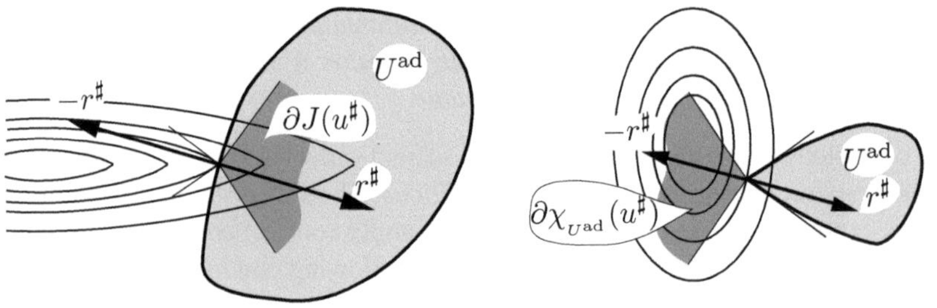

Fig. A.5 Geometric interpretation of (A.7): cases of a nondifferentiable J (*left*) and of a nonsmooth U^{ad} (*right*)

and thus by the variational inequality

$$\forall u \in U^{\mathrm{ad}},\ \big\langle \mathrm{proj}_{U^{\mathrm{ad}}}(v) - v, u - \mathrm{proj}_{U^{\mathrm{ad}}}(v) \big\rangle \geq 0,$$

which expresses that the angle of the projection direction $\mathrm{proj}_{U^{\mathrm{ad}}}(v) - v$ with any direction $u - \mathrm{proj}_{U^{\mathrm{ad}}}(v)$ pointing towards a point of U^{ad} is acute. In the particular case when U^{ad} is a cone, since one can take the values $u = 0$ and $u = 2\mathrm{proj}_{U^{\mathrm{ad}}}(v)$ in the variational inequality above, it turns out that this variational inequality decomposes into the following two:

$$\big\langle \mathrm{proj}_{U^{\mathrm{ad}}}(v) - v, \mathrm{proj}_{U^{\mathrm{ad}}}(v) \big\rangle = 0; \quad \forall u \in U^{\mathrm{ad}},\ \big\langle \mathrm{proj}_{U^{\mathrm{ad}}}(v) - v, u \big\rangle \geq 0. \qquad \triangle$$

A.3 Optimization Under Explicit Constraints

We now consider optimization problems where the constraints are expressed using equalities and inequalities.

A.3.1 First and Second-Order Stationarity Conditions

So far, constraints have been implicitly introduced by means of an admissible subset U^{ad}. This subset can be defined more explicitly via equality or inequality constraints. Therefore, we now consider (A.4) with the following definition of U^{ad} and we limit ourselves to finite dimensional problems, that is, $\mathbb{U} = \mathbb{R}^n$. In addition to the cost function, we are given two sets of functions from $\mathbb{R}^n$ to $\mathbb{R}$, namely, $\{G_i\}_{i=1,\ldots,l}$ and $\{H_j\}_{j=1,\ldots,m}$. Then,

$$U^{\mathrm{ad}} = \big\{ u \in \mathbb{U} \bigm| G_i(u) \leq 0,\ i = 1, \ldots, l \text{ and } H_j(u) = 0,\ j = 1, \ldots, m \big\}.$$

Assume that all functions J, G_i, H_j are differentiable: a fundamental result in optimization theory states that, if $u^\sharp$ is a local minimum, and under a qualification constraint condition (see below), then there must exist real values of so-called *Lagrange* or *Kuhn-Tucker multipliers* $p_i^\sharp$, $i = 1, \ldots, l$, and $q_j^\sharp$, $j = 1, \ldots, m$, such that

$$\nabla J(u^\sharp) + \sum_{i=1}^{l} p_i^\sharp \nabla G_i(u^\sharp) + \sum_{j=1}^{m} q_j^\sharp \nabla H_j(u^\sharp) = 0, \tag{A.8a}$$

$$H_j(u^\sharp) = 0,\ j = 1, \ldots, m, \tag{A.8b}$$

$$G_i(u^\sharp) \leq 0,\ p_i^\sharp \geq 0,\ i = 1, \ldots, l, \tag{A.8c}$$

$$\sum_{i=1}^{l} p_i^\sharp G_i(u^\sharp) = 0. \tag{A.8d}$$

The last equation, known as the *complementary slackness* condition, is equivalent to requiring that $p_i^\sharp G_i(u^\sharp) = 0$ for all i, due to (A.8c). Therefore, $G_i(u^\sharp) < 0$ implies that $p_i^\sharp = 0$. Another way of stating (A.8a) is to say that the *Lagrangian*

$$\mathcal{L}(u, p, q) = J(u) + \sum_{i=1}^{l} p_i G_i(u) + \sum_{j=1}^{m} q_j H_j(u), \tag{A.9}$$

is stationary at $u^\sharp$ for some "optimal" values of the multipliers. However, one must not conclude that this Lagrangian ought to be *minimal*, even if $u^\sharp$ is a true solution.

Example A.12 Consider $\mathbb{U} = \mathbb{R}$, $J(u) = -u^3$ and a single inequality constraint $u \leq 1$. Then, conditions (A.8) are satisfied for $u^\sharp = 1$ (which is the true solution) and $p^\sharp = 3$, but $u^\sharp = 1$ is a maximum (!) of $\mathcal{L}(\cdot, 3)$. △

Let $I^\sharp = \left\{i \in \{1, \ldots, l\} \mid G_i(u^\sharp) = 0\right\}$ be the set of *active* inequality constraints. Then, since $p_i^\sharp = 0$ when $i \notin I^\sharp$, (A.8a) can also be written as

$$-\nabla J(u^\sharp) = \sum_{i \in I^\sharp} p_i^\sharp \nabla G_i(u^\sharp) + \sum_{j=1}^{m} q_j^\sharp \nabla H_j(u^\sharp).$$

This expresses that (minus) the gradient $-\nabla J(u^\sharp)$ belongs to the cone generated by the gradients of active inequality constraints, namely $\{\nabla G_i\}_{i \in I^\sharp} \cup \{\nabla H_j\}_{j=1,\ldots,m} \cup \{-\nabla H_j\}_{j=1,\ldots,m}$, as long as equality constraints can be written as pairs of opposite inequality constraints. Since gradients of active constraints are orthogonal to the admissible set boundary at $u^\sharp$, (A.8a) should remind us of the situation illustrated by Fig. A.5.

However, there is a tricky point here, essentially because there might be situations where the appropriate normal cone to U^{ad} is not generated by those gradients of active constraints. Therefore, the necessary conditions (A.8) can be stated only under a *sufficient* condition of *constraint qualification*. We refer the reader to specialized books for understanding the fine points of this constraint qualification condition. Let us just quote two (nonequivalent) forms of this condition (among several others).

1. In the convex case (i.e. when all involved functions are convex and the H_j are affine), there exists some $\overline{u}$ such that

$$G_i(\overline{u}) < 0, \; i = 1, \ldots, l \; \text{ and } \; H_j(\overline{u}) = 0, \; j = 1, \ldots, m.$$

This is known as the *Slater condition*.

2. The gradients of $G_j,\ j \in I^\sharp$, and $H_j,\ j = 1, \dots, m$, at $u^\sharp$ are linearly independent.

Moreover, the conditions (A.8) are not *sufficient*, even for $u^\sharp$ to be a *local* minimum. Additional necessary and sufficient *second-order* conditions for $u^\sharp$ to be a local minimum can also be stated when the functions involved are twice differentiable.

Necessary conditions:

$$\left.\begin{array}{l}\langle \nabla G_i(u^\sharp), v\rangle = 0,\ i \in I^\sharp \\ \langle \nabla H_j(u^\sharp), v\rangle = 0,\ j = 1, \dots, m\end{array}\right\} \Rightarrow \langle v, \mathcal{L}''_{u,u}(u^\sharp, p^\sharp, q^\sharp).v\rangle \geq 0. \tag{A.10}$$

Sufficient conditions: the symbol $\geq$ is replaced by $>$ in the right-hand side above.

A.3.2 Marginal Interpretation of Multipliers

Optimal multipliers provide some information about how the optimal value of the cost function varies in the case when the right-hand side of constraints is slightly perturbed (which may be interpreted as the small change in available resources). Suppose that, for e.g. some $i_0 \in \{1, \dots, l\}$ (the same may be done for equality constraints), the problem is solved again with the slightly perturbed constraint $G_{i_0}(u) \leq \varepsilon$ (ε is a small number), whereas the other constraints remain unchanged. Let $\delta u^\sharp$ be the variation of the new solution w.r.t. the previous solution. Under the assumption that there exists an ε sufficiently small for the set of active constraints $I^\sharp$ not to change, $\delta u^\sharp$ is likely to be nonzero only if i_0 was a member of $I^\sharp$. Then, to the first-order, one has that

$$\begin{aligned}&\langle \nabla G_{i_0}(u^\sharp), \delta u^\sharp\rangle = \varepsilon,\\ &\langle \nabla G_i(u^\sharp), \delta u^\sharp\rangle = 0,\ i \in I^\sharp \setminus \{i_0\},\\ &\langle \nabla H_j(u^\sharp), \delta u^\sharp\rangle = 0,\ j = 1, \dots, m.\end{aligned}$$

From this and (A.8a), it can be concluded that the first-order variation of the optimal cost function is

$$\langle \nabla J(u^\sharp), \delta u^\sharp\rangle = -p^\sharp_{i_0}\varepsilon.$$

Hence, $-p^\sharp_{i_0}$ is the sensitivity of the optimal cost function w.r.t. a change of the right-hand side of the corresponding constraint. By the way, the fact that $p^\sharp_i = 0$ for nonactive constraints illustrates the fact that the solution is not affected by small changes in those constraints.

A.4 Optimal Control

Optimal control problems may be regarded as particular instances of general optimization problems considered in previous sections of this appendix. However, they have a special structure and, for this reason, they deserve a special attention. The global idea is to set a control problem of a dynamic system—that is, one in which the evolution in time is of interest—with the purpose of minimizing a cost function which measures the system performance over a certain time horizon, subject to certain constraints.

One distinguishes *control variables*—those practically controlled by the user—from the *state variables*—those which carry the memory of the dynamic system over time. Mathematically speaking, they may be considered as independent variables possibly involved in various equality or inequality constraints, but among these constraints, the *dynamics* of the system is typical of such problems. The dynamics may assume different forms according to whether time is continuous or discrete. In continuous time, state variable trajectories are generally related to control variable histories through (ordinary) differential equations. In discrete time, these differential equations are replaced by finite recurrent relationships.

As for the cost or objective function, it generally assumes the form of an integral—in continuous time—or of a sum—in discrete time—over the time horizon, plus possibly a final cost.

There are indeed two notions of "solution" for optimal control problems: *open-loop* and *closed-loop* or *feedback* optimal control.

- The former corresponds to expressing the optimal control variables as functions of time solely (the so-called "control history"). It means that these control values are implemented as time elapses whatever the evolution of state variables may be.
- The latter involves an *observation* function of the system state, and the optimal control—indeed the optimal *strategy* or *feedback law*—is given as a function of time *and* of past and present (but not future!) observations (for the sake of *causality*).

The reader should realize that there are no practical differences between these two implementations as long as the system is *deterministic*, that is, as long as there are no external factors influencing the evolution of the system state, out of the control itself. Indeed, in this case, for any given feedback law, one can perfectly predict the future evolution of the state, and thus of the observation and control variables, and this control history—an open-loop control solution—would obviously produce exactly the same overall behavior and lead to the same performance. Conversely, the performance that is achievable with open-loop controls can also be obtained with feedback laws since observations can simply be ignored, and time only can drive the control values implemented.

The situation is quite different in a *stochastic* environment in which unmastered external factors can cause deviations of the system state from its expected evolution—

that corresponding to zero perturbations, a fact that can be taken into account[7] through observations if closed-loop controls are used, but not with "blind" open-loop controls.

Solving deterministic optimal control problems in open-loop form essentially amounts to handling variational problems involving time functions, which leads to infinite dimensional problems if time is continuous. There are many books dealing with this subject (see e.g. [1, 32, 67, 124]), the most important result being *Pontryagin's maximum principle*.[8] In the following, we try to show how the results of previous sections can be exploited in the framework of discrete time problems in finite horizon.

For feedback optimal control, there exists a special technique called *dynamic programming* based on *Bellman's optimality principle* (see [15]). This technique can solve problems in deterministic and stochastic settings when the system state is fully and perfectly observed.[9] We only discuss the deterministic situation.

A.4.1 First-Order Optimality Conditions in Discrete Time

Consider the problem

$$\min \sum_{t=0}^{T-1} L_t(x_t, u_t) + K(x_T), \tag{A.11a}$$

$$\text{s.t.} \quad x_{t+1} = f_t(x_t, u_t), \; t = 0, \dots, T-1, \tag{A.11b}$$

$$x_0 \text{ given,}$$

where:

- t is time, $\{0, \dots, T\}$ is the time span horizon,[10]
- $x_t \in \mathbb{R}^n$ is the state vector at time t,
- $u_t \in \mathbb{R}^m$ is the control vector at time t,
- $f_t(\cdot, \cdot)$ is a $\mathbb{R}^n$-valued function representing the dynamics at time t,
- $L_t(\cdot, \cdot)$ is a $\mathbb{R}$-valued function representing the cost at time t,
- $K(\cdot)$ is the final cost.

There might be other constraints on state and control variables along the trajectory and at the final time. They would define "admissible" sets. Alternatively, these constraints, in the explicit form of equalities or inequalities, might be handled by duality in a

[7]Not only *past* deviations can be taken into account, but *future* possible deviations can also, in a way, be anticipated if some statistical knowledge of those perturbations is available.

[8]However, in this book, we prefer minimization to maximization.

[9]Indeed, more general—but not all—observation patterns can also be handled.

[10]Here, the final time T is assumed to be given, but additional constraints may define a final target in the (space × time) domain and the final time is then defined as the first time at which this target is reached; then, T becomes one of the unknowns of the problem.

way similar to what follows. For the sake of illustration, we add to the previous formulation the additional constraint

$$C(x_T) = 0, \tag{A.11c}$$

where C is a map from $\mathbb{R}^n$ to some $\mathbb{R}^l$.

Let $x = (x_0, \dots, x_T)$ and $u = (u_0, \dots, u_{T-1})$. Consider the Lagrangian

$$\mathcal{L}(x, u, p, q) = \sum_{t=0}^{T-1} L_t(x_t, u_t) + K(x_T) + \langle q, C(x_T)\rangle \\ + \sum_{t=0}^{T-1} \langle p_{t+1}, f_t(x_t, u_t) - x_{t+1}\rangle,$$

where $q \in \mathbb{R}^l$ is the multiplier associated with (A.11c); the multiplier $p = (p_1, \dots, p_T)$ associated with (A.11b) is called the *co-state* or *adjoint state*. The first-order stationary conditions (see Sect. A.3.1) read

$$\frac{\partial \mathcal{L}}{\partial x_T} = 0 \;\Rightarrow\; p_T = \left(\frac{\partial K}{\partial x}(x_T)\right)^\top + \left(\frac{\partial C}{\partial x}(x_T)\right)^\top q, \tag{A.12a}$$

and, for $t = 1, \dots, T-1$,

$$\frac{\partial \mathcal{L}}{\partial x_t} = 0 \;\Rightarrow\; p_t = \left(\frac{\partial f_t}{\partial x}(x_t, u_t)\right)^\top p_{t+1} + \left(\frac{\partial L_t}{\partial x}(x_t, u_t)\right)^\top, \tag{A.12b}$$

$$\frac{\partial \mathcal{L}}{\partial u_t} = 0 \;\Rightarrow\; 0 = \left(\frac{\partial f_t}{\partial u}(x_t, u_t)\right)^\top p_{t+1} + \left(\frac{\partial L_t}{\partial u}(x_t, u_t)\right)^\top, \tag{A.12c}$$

plus equations (A.11b) and (A.11c) as the result of the derivation with w.r.t. p and q. Note that the co-state equations (A.12b) proceed backwards in time, starting from the final condition (A.12a), whereas the state equations (A.11b) proceed forward from a given initial condition. This is known as a *two-point boundary value problem*.

Consider now a control history $\overline{u}$ that is not necessarily optimal and the corresponding state trajectory $\overline{x}$ derived from (A.11b). The final constraint (A.11c) does not need to be satisfied and is ignored in what follows. A small variation δu from the nominal $\overline{u}$ causes a deviation δx of the state trajectory and a corresponding variation δJ of the cost function J defined by (A.11a). One can express δJ in the following way. First, compute the co-state trajectory $\overline{p}$ corresponding to $\overline{u}$ and $\overline{x}$ through the backward equations (A.12b) starting from (A.12a) with $q = 0$. Then, introduce the so-called *Hamiltonian* defined by

$$\mathcal{H}_t(x_t, u_t, p_{t+1}) = L_t(x_t, u_t) + \big\langle p_{t+1}, f_t(x_t, u_t)\big\rangle. \tag{A.13}$$

Finally, one has that

$$\delta J = \sum_{t=0}^{T-1} \frac{\partial \mathcal{H}_t}{\partial u}(\overline{x}_t, \overline{u}_t, \overline{p}_{t+1})\delta u_t.$$

Note that (A.12c) can also be written as $\partial \mathcal{H}_t / \partial u = 0$; in particular, if $\mathcal{H}_t$ is convex in u, (A.12c) expresses that the optimal u_t minimizes $\mathcal{H}_t$ for the optimal values of x_t and p_{t+1}.

A.4.2 Dynamic Programming

Bellman's optimality principle states that if an optimal trajectory goes from x_0 to x_T, then necessarily the subtrajectory from x_t to x_T for any intermediate $t \in \{0, \dots, T\}$ is an optimal trajectory to go from x_t to x_T. It follows that optimal trajectories reaching x_T can be computed backwards provided that at each stage t between 0 and T, all possible intermediate values x_t be considered. This necessity of solving all subproblems issued from all "initial" states x_t at each time t explains why the method is so computationally demanding, even if the recursivity of the procedure as t moves backwards saves as much computation as possible. On the other hand, optimal controls at time t can be considered as indexed by all these intermediate states x_t, which indeed provides an optimal state feedback law.

In the same way, the optimal cost function at time t is indexed by all initial states x_t: this is the so-called *cost-to-go* or Bellman function $V_t(x_t)$. More precisely, $V_t(x_t)$ is the optimal cost function for the optimal control problem (A.11), but starting at $t \leq T$ with the initial state value x_t. Applying Bellman's optimality principle, the following (Hamilton-Jacobi-)Bellman equation is established:

$$V_t(x_t) = \min_{u_t} \Big(L_t(x_t, u_t) + V_{t+1}\big(f_t(x_t, u_t)\big)\Big) \tag{A.14a}$$

$$\text{with} \quad V_T(x_T) = \begin{cases} K(x_T) & \text{if } C(x_T) = 0; \\ +\infty & \text{otherwise.} \end{cases} \tag{A.14b}$$

The arg min in (A.14a) provides the optimal feedback law $u_t(x_t)$.

Remark A.13 Assuming enough smoothness, one can relate the Bellman function to the optimal co-state p as follows

$$(p_t)^\top = \frac{\partial V_t}{\partial x}(x_t),$$

where x_t stands for the *optimal* value of the state at time t. This means that p_t provides the sensitivity of the cost-to-go w.r.t. a change of the state x_t at t. This interpretation could have been anticipated from the fact that p_t was introduced as the multiplier of the dynamics (remember the marginal interpretation of multipliers in Sect. A.3.2). ◊

Appendix B
Basics in Probability

The results below may be found in classical books such as [30, 65, 66, 89]. We provide recalls on probability spaces, random variables, convergence of random variables, then on conditional expectation, conditional probability and stochastic kernels. We conclude with the Monte Carlo method.

B.1 Probability Space

We give the definition of a probability space, after having recalled the notions of measurable space and of measure.

B.1.1 Measurable Space

A *σ-field* on a set Ω is a collection $\mathcal{A}$ of subsets of Ω such that:

- $\emptyset \in \mathcal{A}$,
- if the sequence $\{B_n\}_{n \in N}$ is such that $B_n \in \mathcal{A}$, for $n \in N$ where N is countable, then $\bigcup_{n \in N} B_n \in \mathcal{A}$,
- if $B \in \mathcal{A}$, then the complementary set $B^{\mathrm{c}} = \Omega \backslash B \in \mathcal{A}$.

Given any collection $\mathcal{C}$ of subsets of Ω, the σ-field $\sigma(\mathcal{C})$ generated by $\mathcal{C}$ is defined to be the smallest σ-field in Ω such that $\mathcal{C} \subset \sigma(\mathcal{C})$. It is called the σ-field *generated* by $\mathcal{C}$.

A *measurable space* is a set Ω together with a σ-field $\mathcal{A}$ on Ω, and is denoted by $(\Omega, \mathcal{A})$. The elements of $\mathcal{A}$ are called *measurable sets*.

Let $(\Omega, \mathcal{A})$ and $(\mathbb{Y}, \mathcal{Y})$ be two measurable spaces. A *mapping* $\boldsymbol{Y} : \Omega \to \mathbb{Y}$ is said to be *measurable* if $\boldsymbol{Y}^{-1}(\mathcal{Y}) \subset \mathcal{A}$. The collection $\boldsymbol{Y}^{-1}(\mathcal{Y})$ of subsets of $\mathcal{A}$ is a σ-field, denoted by $\sigma(\boldsymbol{Y})$ and called the σ-field *generated* by $\boldsymbol{Y}$.

P. Carpentier et al., *Stochastic Multi-Stage Optimization*, Probability Theory and Stochastic Modelling 75, DOI 10.1007/978-3-319-18138-7

Definition B.1 For any topological space $\mathbb{Y}$, the *Borel σ-field* of Ω is the σ-field $\mathcal{B}^o_{\mathbb{Y}}$ generated by the open sets of $\mathbb{Y}$. Elements of $\mathcal{B}^o_{\mathbb{Y}}$ are called *Borel sets*. When we use measurable mappings with values in a topological space (as metric or Hilbert spaces), measurability implicitly refers to the Borel σ-field. A continuous mapping between topological spaces is measurable.

B.1.2 Measure

Let $(\Omega, \mathcal{A})$ be a measurable space. A *measure* on $(\Omega, \mathcal{A})$ is a function $\mu \colon \mathcal{A} \to \mathbb{R} \cup \{+\infty\}$ with values in the extended real numbers, such that:

1. $\mu(B) \geq 0$ for $B \in \mathcal{A}$, with equality if $B = \emptyset$;
2. if the sequence $\{B_n\}_{n \in N}$ is such that $B_n \in \mathcal{A}$, for $n \in N$ where N is countable, and the B_n are mutually disjoints, then $\mu(\bigcup_{n \in N} B_n) = \sum_{n \in N} \mu(B_n)$.

The second property is called σ-additivity, or countable additivity. The triple $(\Omega, \mathcal{A}, \mu)$ is called a *measure space*.

A measure μ is said to be *finite* if $\mu(\Omega) < +\infty$. A measure μ is said to be *σ-finite* (or a measure μ is a σ-finite measure) if there exists a countable sequence $\{B_n\}_{n \in N}$ in $\mathcal{A}$ such that $\bigcup_{n \in N} B_n = \Omega$ and $\mu(B_n) < +\infty$ for all $n \in N$.

Say that a subset $C \subset \Omega$ is *μ-negligible* in the measure space $(\Omega, \mathcal{A}, \mu)$ if there exists $B \in \mathcal{A}$ such that $C \subset B$ and $\mu(B) = 0$. The measure space $(\Omega, \mathcal{A}, \mu)$ is called *μ-complete* [26, p. 2] if every μ-negligible subset is in $\mathcal{A}$. Given a measure space $(\Omega, \mathcal{A}, \mu)$, we define a new σ-field $\mathcal{A}^{\mu}$ which consists of all the sets $B \subset \Omega$ for which there exists $B_+, B_- \in \mathcal{A}$ such that $B_- \subset B \subset B_+$ and $\mu(B_+ - B_-) = 0$. The extension of μ on $\mathcal{A}^{\mu}$ is unique and the measure space $(\Omega, \mathcal{A}, \mu)$ is complete if $\mathcal{A}^{\mu} = \mathcal{A}$. On a measurable space $(\Omega, \mathcal{A})$ without explicit reference to a measure, it is possible to define a σ-field called the σ-field of universally measurable sets over $(\Omega, \mathcal{A})$ [21, Definition 7.18]. It is the σ-field $\widehat{\mathcal{A}} := \bigcap_{\mu} \mathcal{A}^{\mu}$ obtained when the intersection is over the finite measures μ on $\mathcal{A}$. A σ-field $\mathcal{A}$ is said to be *complete* or *universally complete* if $\mathcal{A} = \widehat{\mathcal{A}}$. As an example, suppose that μ is a σ-finite measure on $\mathcal{A}$ and $\mathcal{A}$ is μ-complete. Since $\mathcal{A} \subset \widehat{\mathcal{A}} = \bigcap_{\mu'} \mathcal{A}^{\mu'} \subset \mathcal{A}^{\mu} = \mathcal{A}$ we have that $\mathcal{A} = \widehat{\mathcal{A}}$, so that the σ-field $\mathcal{A}$ is universally complete.

B.1.3 Probability Space

If $\mu(\Omega) = 1$, then $(\Omega, \mathcal{A}, \mu)$ is called a *probability space*, and the measure μ is called a *probability measure*, generally denoted by $\mathbb{P}$ and called a *probability*. Elements of $\mathcal{A}$ are called *events*

Let $(\Omega, \mathcal{A}, \mathbb{P})$ be a probability space. A condition holds *almost surely* on Ω if it holds on $\Omega \backslash N$, where N is a subset of Ω of measure 0, and abbreviated a.s. or $\mathbb{P}$-a.s..

Definition B.2 We say that a probability space $(\Omega, \mathcal{A}, \mu)$ is *non-atomic*, or alternatively call μ non-atomic, if $\mu(A) > 0$ implies the existence of $B \in \mathcal{A}$, $B \subset A$ with $0 < \mu(B) < \mu(A)$.

B.1.4 Product Probability Space

Let $(\Omega_i, \mathcal{A}_i, \mathbb{P}_i)$, $i = 1, 2$ be two probability spaces. The *product probability space* is defined as $(\Omega_1 \times \Omega_2, \mathcal{A}_1 \otimes \mathcal{A}_2, \mathbb{P}_1 \otimes \mathbb{P}_2)$, where the product σ-field $\mathcal{A}_1 \otimes \mathcal{A}_2$ has been introduced at Remark 3.25 as the one generated by the rectangles $\{G_1 \times G_2 \mid G_i \in \mathcal{A}_i,\ i = 1, 2\}$, and where the *product probability* $\mathbb{P}_1 \otimes \mathbb{P}_2$ is characterized on the rectangles by

$$\mathbb{P}_1 \otimes \mathbb{P}_2(G_1 \times G_2) = \mathbb{P}_1(G_1) \times \mathbb{P}_2(G_2). \tag{B.1}$$

It can be shown that $\mathbb{P}_1 \otimes \mathbb{P}_2$ can be extended into a probability on the product σ-field $\mathcal{A}_1 \otimes \mathcal{A}_2$. By associativity, one can define the product of a finite number of probability spaces. The case of an infinite product of probability spaces is discussed at Sect. B.7.1.

B.2 Random Variables

After having recalled the definition of a random variable, we turn to integration, with L^p-spaces and mathematical expectation. Recalls on probability image, Radon-Nikodym derivative and uniform integrability are also provided.

Definition B.3 Let $(\Omega, \mathcal{A}, \mathbb{P})$ be a probability space and $(\mathbb{Y}, \mathcal{Y})$ be a measurable space. A measurable mapping $\boldsymbol{Y} : \Omega \to \mathbb{Y}$ is called a *random variable*.

Remark B.4 When the measurable mapping $\boldsymbol{Y}$ takes values in $(\mathbb{R}, \mathcal{B}^{o}_{\mathbb{R}})$ it is called a real-valued random variable. When it takes values in $(\mathbb{R}^n, \mathcal{B}^{o}_{\mathbb{R}^n})$ it is called a random element. ♢

Two random variables $\boldsymbol{Y}$ and $\boldsymbol{Z}$ are said to be equal *almost surely*, or $\mathbb{P}$-*almost surely*, or $\mathbb{P}$-a.s., when $\mathbb{P}(\{\boldsymbol{Y} = \boldsymbol{Z}\}) = 1$. Recall that, in Probability Theory, it is customary to omit the variable ω and to write

$$\{\boldsymbol{Y} = \boldsymbol{Z}\} := \left\{\omega \in \Omega \;\middle|\; \boldsymbol{Y}(\omega) = \boldsymbol{Z}(\omega)\right\}. \tag{B.2}$$

B.2.1 L^p-Spaces

Let $0 < p \le +\infty$. The *L^p-norm* of a random variable $\boldsymbol{Y}$ with values in a Banach space is defined for $p < +\infty$ as

$$\left\| \boldsymbol{Y} \right\|_p := \left(\int_\Omega \left\| \boldsymbol{Y} \right\|^p \, \mathrm{d}\mathbb{P} \right)^{\frac{1}{p}}, \tag{B.3}$$

when the integral exists, and for $p = +\infty$ as

$$\left\| \boldsymbol{Y} \right\|_\infty = \operatorname{ess\,sup} \left\| \boldsymbol{Y} \right\| := \inf \left\{ y \in \mathbb{R} \mid \mathbb{P}\left\{ \omega \mid \left\| \boldsymbol{Y}(\omega) \right\| > y \right\} = 0 \right\}. \tag{B.4}$$

The L^∞-norm $\left\| \boldsymbol{Y} \right\|_\infty$ of $\boldsymbol{Y}$ is called the *essential supremum* of $\left\| \boldsymbol{Y} \right\|$.

The set of random variables with finite L^p-norm forms a vector space V with the usual pointwise addition and scalar multiplication of functions. Two random variables are said to be equivalent when their difference has zero L^p-norm: the *L^p-space* on Ω is the quotient space of V by this equivalence relation. Thus, random variables in L^p-space are defined up to equivalence almost surely. We use the notation $L^p(\Omega, \mathcal{A}, \mathbb{P}; \mathbb{Y})$ to specify the domain and images spaces, and the notation $L^p(\Omega, \mathcal{A}, \mathbb{P})$ for real-valued random variables. For $1 \le p \le +\infty$ the space $L^p(\Omega, \mathcal{A}, \mathbb{P})$ is a Banach space.

B.2.2 Mathematical Expectation

A real-valued random variable $\boldsymbol{Y}$ is said to be *integrable* when $\boldsymbol{Y} \in L^1(\Omega, \mathcal{A}, \mathbb{P})$ or, equivalently, when $\int_\Omega \left| \boldsymbol{Y} \right| \, \mathrm{d}\mathbb{P} < +\infty$. The mathematical *expectation* of $\boldsymbol{Y}$ is

$$\mathbb{E}(\boldsymbol{Y}) := \int_\Omega \boldsymbol{Y} \, \mathrm{d}\mathbb{P}. \tag{B.5}$$

With this notation, $\boldsymbol{Y} \in L^1(\Omega, \mathcal{A}, \mathbb{P}) \iff \mathbb{E}(\left| \boldsymbol{Y} \right|) < +\infty$.

When the dependence w.r.t. the probability $\mathbb{P}$ has to be stressed, one uses the notation $\mathbb{E}_{\mathbb{P}}(\boldsymbol{Y})$:

$$\mathbb{E}_{\mathbb{P}}(\boldsymbol{Y}) := \int_\Omega \boldsymbol{Y} \, \mathrm{d}\mathbb{P}. \tag{B.6}$$

The space $L^2(\Omega, \mathcal{A}, \mathbb{P})$ is a Hilbert space, equipped with the scalar product

$$\left\langle \boldsymbol{Y}, \boldsymbol{Z} \right\rangle := \int_\Omega \boldsymbol{Y}(\omega) \boldsymbol{Z}(\omega) \, \mathrm{d}\mathbb{P}(\omega). \tag{B.7}$$

Random variables in $L^2(\Omega, \mathcal{A}, \mathbb{P})$ are said to be *square integrable*.

B.2.3 Probability Image

Let $\boldsymbol{Y}$ be a random variable. The image measure

$$\mathbb{P}_{\boldsymbol{Y}} := \mathbb{P} \circ \boldsymbol{Y}^{-1} \tag{B.8}$$

is a probability on $(\mathbb{Y}, \mathcal{Y})$, called the *probability law of* $\boldsymbol{Y}$ or *probability distribution of* $\boldsymbol{Y}$. It is also denoted by

$$\boldsymbol{Y}_{\star}(\mathbb{P}) := \mathbb{P} \circ \boldsymbol{Y}^{-1}. \tag{B.9}$$

For any measurable bounded real-valued function φ, the mapping $\varphi(\boldsymbol{Y}) : \Omega \to \mathbb{R}$ is a random variable, and one has that

$$\mathbb{E}_{\mathbb{P}}(\varphi(\boldsymbol{Y})) = \mathbb{E}_{\mathbb{P}_{\boldsymbol{Y}}}(\varphi). \tag{B.10}$$

B.2.4 Radon-Nikodym Derivative

Let $\mathbb{P}$ and $\mathbb{Q}$ be two probabilities on $(\Omega, \mathcal{A})$. The probability $\mathbb{Q}$ is said to *have a density* w.r.t. $\mathbb{P}$ if there exists a nonnegative ($\boldsymbol{R} \geq 0$, $\mathbb{P}$-a.s.) integrable random variable $\boldsymbol{R} \in L^1(\Omega, \mathcal{A}, \mathbb{P})$ such that

$$\mathbb{E}_{\mathbb{Q}}(\boldsymbol{Z}) = \mathbb{E}_{\mathbb{P}}(\boldsymbol{R}\boldsymbol{Z}), \;\; \forall \boldsymbol{Z} \in L^1(\Omega, \mathcal{A}, \mathbb{Q}). \tag{B.11}$$

The random variable $\boldsymbol{R}$ is uniquely defined $\mathbb{P}$-a.s., is called a *density*, and is denoted by $\boldsymbol{R} = \mathrm{d}\mathbb{Q}/\mathrm{d}\mathbb{P}$.

The probabilities $\mathbb{P}$ and $\mathbb{Q}$ are said to be *equivalent* if $\mathbb{Q}$ is absolutely continuous w.r.t. $\mathbb{P}$ and $\mathbb{P}$ is absolutely continuous w.r.t. $\mathbb{Q}$. This is denoted by $\mathbb{P} \sim \mathbb{Q}$. In that case, $\boldsymbol{R} = \mathrm{d}\mathbb{Q}/\mathrm{d}\mathbb{P}$ is uniquely defined $\mathbb{P}$-a.s. and $\mathbb{Q}$-a.s., and we have that $\boldsymbol{R} > 0$ and $\mathrm{d}\mathbb{P}/\mathrm{d}\mathbb{Q} = 1/\boldsymbol{R}$:

$$\mathbb{P} \sim \mathbb{Q} \text{ and } \frac{\mathrm{d}\mathbb{Q}}{\mathrm{d}\mathbb{P}} = \boldsymbol{R} \in L^1(\Omega, \mathcal{A}, \mathbb{P}), \;\; \boldsymbol{R} > 0, \;\; \mathbb{P}\text{-a.s. or } \mathbb{Q}\text{-a.s.}. \tag{B.12}$$

B.2.5 Uniform Integrability

Consider $\{\boldsymbol{Y}_i\}_{i \in I}$ a collection of random variables with values in a Banach space. The collection $\{\boldsymbol{Y}_i\}_{i \in I}$ is said to be *uniformly continuous* if

$$\forall \epsilon > 0, \;\; \exists \alpha > 0 \text{ such that } \mathbb{P}(A) \leq \alpha \Rightarrow \sup_{i \in I} \int_A \|\boldsymbol{Y}_i\| \, \mathrm{d}\mathbb{P} \leq \epsilon,$$

and *uniformly integrable* if

$$\forall \epsilon > 0, \ \exists \alpha > 0 \text{ such that } \sup_{i \in I} \int_{\|Y_i\| > \alpha} \|Y_i\| \, \mathrm{d}\mathbb{P} \leq \epsilon. \tag{B.13}$$

B.3 Convergence of Random Variables

Let $(\Omega, \mathcal{A}, \mathbb{P})$ be a probability space. We present different notions of convergence of random variables.

B.3.1 Almost Sure Convergence

Let $\boldsymbol{Y}$ be a random variable and $\{\boldsymbol{Y}_n\}_{n \in \mathbb{N}}$ be a sequence of random variables with values in the same topological space. We say that $\{\boldsymbol{Y}_n\}_{n \in \mathbb{N}}$ converges *almost surely* towards $\boldsymbol{Y}$, denoted by $\boldsymbol{Y}_n \xrightarrow{\text{a.s.}} \boldsymbol{Y}$, if

$$\mathbb{P}\big(\boldsymbol{Y}_n \underset{n \to +\infty}{\longrightarrow} \boldsymbol{Y}\big) = 1.$$

This is denoted by $\boldsymbol{Y}_n \xrightarrow{\text{a.s.}} \boldsymbol{Y}$.

B.3.2 Convergence in L^p Norm

Let $1 \leq p \leq +\infty$. Let $\boldsymbol{Y}$ be a random variable and $\{\boldsymbol{Y}_n\}_{n \in \mathbb{N}}$ be a sequence of random variables with values in the same Banach space. The sequence $\{\boldsymbol{Y}_n\}_{n \in \mathbb{N}}$ *converges in L^p norm* towards $\boldsymbol{Y}$ if

$$\|\boldsymbol{Y}_n - \boldsymbol{Y}\|_p \underset{n \to +\infty}{\longrightarrow} 0.$$

This is denoted by $\boldsymbol{Y}_n \xrightarrow{L^p} \boldsymbol{Y}$. The L^2 convergence is called *mean square convergence*.

B.3.3 Convergence in Probability

Let $\{\boldsymbol{Y}_n\}_{n \in \mathbb{N}}$ be a sequence of random variables, taking values in a metric space $(\mathbb{Y}, d)$. The sequence $\{\boldsymbol{Y}_n\}_{n \in \mathbb{N}}$ *converges in probability* towards a random variable $\boldsymbol{Y}$ if, for every $\varepsilon > 0$,

$$\lim_{n\to\infty} \mathbb{P}(d(\boldsymbol{Y}_n, \boldsymbol{Y}) \geq \varepsilon) = 0.$$

This is denoted by $\boldsymbol{Y}_n \xrightarrow{\mathbb{P}} \boldsymbol{Y}$.

B.3.4 Convergence in Law

Let $\boldsymbol{Y}$ be a random variable and $\{\boldsymbol{Y}_n\}_{n\in\mathbb{N}}$ be a sequence of random variables with values in the same topological space. The sequence $\{\boldsymbol{Y}_n\}_{n\in\mathbb{N}}$ *converges in law* or *converges in distribution* towards a random variable $\boldsymbol{Y}$ if the sequence $\{\mathbb{P}_{\boldsymbol{Y}_n}\}_{n\in\mathbb{N}}$ of image probabilities (see Sect. B.8) narrowly converges towards $\mathbb{P}_{\boldsymbol{Y}}$, that is, if

$$\lim_{n\to+\infty} \mathbb{E}(\varphi(\boldsymbol{Y}_n)) = \mathbb{E}(\varphi(\boldsymbol{Y})), \tag{B.14}$$

for all bounded continuous function φ. This is denoted by $\boldsymbol{Y}_n \xrightarrow{\mathcal{D}} \boldsymbol{Y}$.

B.3.5 Relations Between Convergences

We have the following properties.

- Convergence almost surely implies convergence in probability.
- If a sequence converges in probability, there exists a sub-sequence which converges almost surely.
- If a sequence of random variables converges in L^2 norm, the sequence converges in probability.
- If the sequence $\{\boldsymbol{Y}_n\}_{n\in\mathbb{N}}$ converges in probability to $\boldsymbol{Y}$ then $\{\boldsymbol{Y}_n\}_{n\in\mathbb{N}}$ converges in law to $\boldsymbol{Y}$.
- When $\{\boldsymbol{Y}_n\}_{n\in\mathbb{N}}$ converges in law to a constant random variable $\boldsymbol{Y}$, then $\{\boldsymbol{Y}_n\}_{n\in\mathbb{N}}$ converges in probability to the constant $\boldsymbol{Y}$.

B.4 Conditional Expectation

Let $(\Omega, \mathcal{A}, \mathbb{P})$ be a probability space. In what follows, $\mathcal{G}$ denotes a *subfield* of $\mathcal{A}$, that is, $\mathcal{G} \subset \mathcal{A}$ and $\mathcal{G}$ is a σ-field. In this section, when not specified, a random variable is a real-valued random variable.

Let $L^2(\Omega, \mathcal{G}, \mathbb{P})$ be the closed vector subspace of square integrable $\mathcal{G}$-measurable functions in the Hilbert space $L^2(\Omega, \mathcal{A}, \mathbb{P})$. We can thus define the orthogonal projection on $L^2(\Omega, \mathcal{G}, \mathbb{P})$.

Definition B.5 If Y is a square integrable random variable, we define the *conditional expectation of* Y *knowing (the* σ*-field)* $\mathcal{G}$, and we denote by $\mathbb{E}\left(Y \mid \mathcal{G}\right)$, the orthogonal projection of Y on $L^2(\Omega, \mathcal{G}, \mathbb{P})$:

$$Z = \mathbb{E}\left(Y \mid \mathcal{G}\right) \iff Z = \underset{T \in L^2(\Omega, \mathcal{G}, \mathbb{P})}{\arg\min} \ \mathbb{E}\left(\|Y - T\|^2\right). \tag{B.15}$$

Thus, the conditional expectation solves an optimization problem under measurability constraints (among $\mathcal{G}$-measurable random variables). This was discussed in Sect. 3.5 in the case of a finite probability space.

The conditional expectation may be extended to L^1 random variables. If Y is an integrable random variable, $\mathbb{E}\left(Y \mid \mathcal{G}\right)$ is the unique $Z \in L^1(\Omega, \mathcal{G}, \mathbb{P})$ such that

$$\mathbb{E}\left(YT\right) = \mathbb{E}\left(ZT\right), \ \forall T \in L^\infty(\Omega, \mathcal{G}, \mathbb{P}). \tag{B.16}$$

In fact, the above result holds true under the weaker assumptions that Y is bounded either below or above by an integrable random variable.

The conditional expectation may be extended componentwise to L^1 random variables with values in $\mathbb{R}^d$.

Elementary Properties

Let X and Y be two integrable random variables, λ a real number. Then

$$\mathbb{E}\left(\lambda X + Y \mid \mathcal{G}\right) = \lambda \mathbb{E}\left(X \mid \mathcal{G}\right) + \mathbb{E}\left(Y \mid \mathcal{G}\right), \tag{B.17}$$

$$X \geq 0 \Rightarrow \mathbb{E}\left(X \mid \mathcal{G}\right) \geq 0, \tag{B.18}$$

$$\mathbb{E}\left(\mathbb{E}\left(X \mid \mathcal{G}\right)\right) = \mathbb{E}\left(X\right), \tag{B.19}$$

$$Y \in L^\infty(\Omega, \mathcal{G}, \mathbb{P}) \Rightarrow \mathbb{E}\left(YX \mid \mathcal{G}\right) = Y\mathbb{E}\left(X \mid \mathcal{G}\right). \tag{B.20}$$

If $\mathcal{G}^\flat$, $\mathcal{G}^\sharp$ are two subfields of $\mathcal{A}$, we have that

$$\mathcal{G}^\flat \subset \mathcal{G}^\sharp \Rightarrow \mathbb{E}\Big(\mathbb{E}\left(X \mid \mathcal{G}^\sharp\right) \Big| \mathcal{G}^\flat\Big) = \mathbb{E}(X \mid \mathcal{G}^\flat). \tag{B.21}$$

Dependence Upon the Probability Law

Proposition B.6 *Let* $\mathbb{P}$ *and* $\mathbb{Q}$ *be two equivalent probabilities on* $(\Omega, \mathcal{A})$*, with positive Radon-Nikodym derivative* R *(see Sect. B.2.4). We have that*

$$\mathbb{E}_{\mathbb{Q}}\left(X \mid \mathcal{G}\right) = \frac{\mathbb{E}_{\mathbb{P}}\left(RX \mid \mathcal{G}\right)}{\mathbb{E}_{\mathbb{P}}\left(R \mid \mathcal{G}\right)}, \ \mathbb{P}\text{-a.s. or } \mathbb{Q}\text{-a.s.}, \tag{B.22}$$

for all bounded random variable X.

Proof Let $\boldsymbol{X} \in L^{\infty}(\Omega, \mathcal{A}, \mathbb{P})$. We first note that the right hand side of Eq. (B.22) is well defined because $\boldsymbol{X}$ is bounded and $\boldsymbol{R} \in L^1(\Omega, \mathcal{A}, \mathbb{P})$. For any $G \in \mathcal{G}$, we have that

$$\begin{aligned}
\mathbb{E}_{\mathbb{Q}}\left(\boldsymbol{X}\mathbf{1}_G\right) &= \mathbb{E}_{\mathbb{P}}\left(\boldsymbol{R}\boldsymbol{X}\mathbf{1}_G\right) \text{ by (B.11)} \\
&= \mathbb{E}_{\mathbb{P}}\left(\mathbb{E}_{\mathbb{P}}\left(\boldsymbol{R}\boldsymbol{X} \mid \mathcal{G}\right)\mathbf{1}_G\right) \text{ by (B.16)} \\
&= \mathbb{E}_{\mathbb{P}}\left(\mathbb{E}_{\mathbb{P}}\left(\boldsymbol{R} \mid \mathcal{G}\right)\frac{\mathbb{E}_{\mathbb{P}}\left(\boldsymbol{R}\boldsymbol{X} \mid \mathcal{G}\right)}{\mathbb{E}_{\mathbb{P}}\left(\boldsymbol{R} \mid \mathcal{G}\right)}\mathbf{1}_G\right) \text{ since } \boldsymbol{R} > 0 \\
&= \mathbb{E}_{\mathbb{P}}\left(\boldsymbol{R}\frac{\mathbb{E}_{\mathbb{P}}\left(\boldsymbol{R}\boldsymbol{X} \mid \mathcal{G}\right)}{\mathbb{E}_{\mathbb{P}}\left(\boldsymbol{R} \mid \mathcal{G}\right)}\mathbf{1}_G\right) \text{ by (B.16)} \\
&= \mathbb{E}_{\mathbb{Q}}\left(\frac{\mathbb{E}_{\mathbb{P}}\left(\boldsymbol{R}\boldsymbol{X} \mid \mathcal{G}\right)}{\mathbb{E}_{\mathbb{P}}\left(\boldsymbol{R} \mid \mathcal{G}\right)}\mathbf{1}_G\right) \text{ by (B.11).}
\end{aligned}$$

The proof is complete by Eq. (B.16). □

Remark B.7 If $\mathrm{d}\mathbb{Q}/\mathrm{d}\mathbb{P}$ is $\mathcal{G}$-measurable, then the conditional expectation operators $\mathbb{E}_{\mathbb{Q}}(\cdot \mid \mathcal{G})$ and $\mathbb{E}_{\mathbb{P}}(\cdot \mid \mathcal{G})$ coincide ($\mathbb{Q}$-a.s. or $\mathbb{P}$-a.s.). In other words, the conditional expectation operator $\mathbb{E}_{\mathbb{P}}(\cdot \mid \mathcal{G})$ depends upon $\mathcal{G}$ and the equivalence class of $\mathbb{P}$ for the relation $\mathbb{P} \sim_{\mathcal{G}} \mathbb{Q} \iff \mathbb{P} \sim \mathbb{Q}$ and $\mathrm{d}\mathbb{Q}/\mathrm{d}\mathbb{P}$ is $\mathcal{G}$-measurable. ◊

Proposition B.8 *Let $\Phi : (\mathbb{X}, \mathcal{X}) \to (\Omega, \mathcal{A})$ be measurable, and let $\mathbb{Q}$ be a probability on $(\mathbb{X}, \mathcal{X})$. Let $\Phi_\star(\mathbb{Q})$ the probability on $(\Omega, \mathcal{A})$, image of $\mathbb{Q}$ by Φ (see Sect. B.8). For any random variable $\boldsymbol{Y}$ on Ω such that $\mathbb{E}_{\Phi_\star(\mathbb{Q})}\left(\left|\boldsymbol{Y}\right|\right) < +\infty$, we have that*

$$\mathbb{E}_{\Phi_\star(\mathbb{Q})}\left(\boldsymbol{Y} \mid \mathcal{G}\right) \circ \Phi = \mathbb{E}_{\mathbb{Q}}\left(\boldsymbol{Y} \circ \Phi \mid \Phi^{-1}(\mathcal{G})\right), \quad \mathbb{P}\text{-}a.s.. \tag{B.23}$$

Proof For any $G \in \mathcal{G}$, we have that

$$\begin{aligned}
\mathbb{E}_{\mathbb{Q}}\left(\mathbf{1}_{\Phi^{-1}(G)} \times (\boldsymbol{Y} \circ \Phi)\right) &= \mathbb{E}_{\mathbb{Q}}\left((\mathbf{1}_G \circ \Phi) \times (\boldsymbol{Y} \circ \Phi)\right) \\
&= \mathbb{E}_{\Phi_\star(\mathbb{Q})}\left(\mathbf{1}_G \times \boldsymbol{Y}\right) \text{ by (B.10)} \\
&= \mathbb{E}_{\Phi_\star(\mathbb{Q})}\left(\mathbf{1}_G \times \mathbb{E}_{\Phi_\star(\mathbb{Q})}\left(\boldsymbol{Y} \mid \mathcal{G}\right)\right) \text{ by (B.16)} \\
&= \mathbb{E}_{\mathbb{Q}}\left(\mathbf{1}_G \circ \Phi \times \mathbb{E}_{\Phi_\star(\mathbb{Q})}\left(\boldsymbol{Y} \mid \mathcal{G}\right) \circ \Phi\right) \text{ by (B.10)} \\
&= \mathbb{E}_{\mathbb{Q}}\left(\mathbf{1}_{\Phi^{-1}(G)} \times \mathbb{E}_{\Phi_\star(\mathbb{Q})}\left(\boldsymbol{Y} \mid \mathcal{G}\right) \circ \Phi\right).
\end{aligned}$$

We conclude that (B.23) holds true since $\Phi^{-1}(\mathcal{G})$ is equal to $\left\{\Phi^{-1}(G) \mid G \in \mathcal{G}\right\}$ and since the function $\mathbb{E}_{\Phi_\star(\mathbb{Q})}\left(\boldsymbol{Y} \mid \mathcal{G}\right) \circ \Phi$ is $\Phi^{-1}(\mathcal{G})$-measurable and is in $L^1(\mathbb{X}, \mathcal{X}, \mathbb{Q})$. □

Conditional Expectation w.r.t. an Atomic σ-Field

A subfield $\mathcal{G}$ of $\mathcal{A}$ is called *atomic* (see Definition 3.26) if it is generated by a countable partition $\{\Omega_n\}_{n\in N}$: $\mathcal{G} = \sigma(\Omega_n, n \in N)$, where N is countable (in bijection with a subset of $\mathbb{N}$).

Proposition B.9 *If $\mathcal{G}$ is an atomic σ-field, generated by the countable partition $\{\Omega_n\}_{n\in N}$, and if $\boldsymbol{Y}$ is an integrable random variable, then*

$$\mathbb{E}\left(\boldsymbol{Y} \mid \mathcal{G}\right) = \sum_{n,\mathbb{P}(\Omega_n)>0} \frac{\mathbb{E}\left(\mathbf{1}_{\Omega_n}\boldsymbol{Y}\right)}{\mathbb{P}(\Omega_n)} \mathbf{1}_{\Omega_n}. \tag{B.24}$$

Conditional Expectation Knowing a Random Variable

Definition B.10 If $\boldsymbol{Y}$ is an integrable random variable and $\boldsymbol{Z}$ is a random variable, we define *the conditional expectation of $\boldsymbol{Y}$ knowing (the random variable) $\boldsymbol{Z}$* as the random variable

$$\mathbb{E}\left(\boldsymbol{Y} \mid \boldsymbol{Z}\right) := \mathbb{E}\left(\boldsymbol{Y} \mid \sigma(\boldsymbol{Z})\right), \tag{B.25}$$

where $\sigma(\boldsymbol{Z})$ is the σ-field generated by the random variable $\boldsymbol{Z}$.

The following proposition results from Proposition 3.46.

Proposition B.11 *Let $\boldsymbol{Y}$ be an integrable random variable. For any random variable $\boldsymbol{Z}$, there exists a unique measurable function Ψ (unique $\mathbb{P}_{\boldsymbol{Z}}$-a.s.) such that $\mathbb{E}\left(\boldsymbol{Y} \mid \boldsymbol{Z}\right) = \Psi(\boldsymbol{Z})$. Therefore, we can define*

$$\mathbb{E}\left(\boldsymbol{Y} \mid \boldsymbol{Z} = z\right) := \Psi(z), \ \forall z \text{ such that } \mathbb{P}(Z = z) > 0. \tag{B.26}$$

Notice that Ψ depends functionally upon the random variable $(\boldsymbol{Y}, \boldsymbol{Z})$, hence, in particular, upon $\boldsymbol{Z}$. To insist upon this dependence, we rephrase the above result as follows. For any random variable $\boldsymbol{Z}$, there exists a unique measurable function $\Psi_{[\boldsymbol{Y},\boldsymbol{Z}]}$ (unique $\mathbb{P}_{\boldsymbol{Z}}$-a.s.) and $\Omega' \subset \Omega$ such that $\mathbb{P}(\Omega') = 1$ and

$$\mathbb{E}\left(\boldsymbol{Y} \mid \boldsymbol{Z}\right)(\omega) = \Psi_{[\boldsymbol{Y},\boldsymbol{Z}]}\left(\boldsymbol{Z}(\omega)\right), \ \forall \omega \in \Omega'. \tag{B.27}$$

If $\boldsymbol{Y}$ and $\boldsymbol{Z}$ are discrete random variables, by (B.24), we have that

$$\Psi_{[\boldsymbol{Y},\boldsymbol{Z}]}(z) = \sum_{y\in \boldsymbol{Y}(\Omega), z'\in \boldsymbol{Z}(\Omega)} y \frac{\mathbb{P}(\boldsymbol{Y} = y, \boldsymbol{Z} = z')}{\mathbb{P}(\boldsymbol{Z} = z')} \mathbf{1}_{\{z'\}}(z), \tag{B.28}$$

which indeed depends functionally upon $\boldsymbol{Y}$ and $\boldsymbol{Z}$, by the terms $\mathbb{P}(\boldsymbol{Y} = y, \boldsymbol{Z} = z')$ and $\mathbb{P}(\boldsymbol{Z} = z')$.

B.5 Conditional Probability

Let $(\Omega, \mathcal{A}, \mathbb{P})$ be a probability space. In what follows, $\mathcal{G}$ denotes a subfield of $\mathcal{A}$.

B.5.1 Conditional Probability w.r.t. an Event

Let $B \in \mathcal{A}$ such that $\mathbb{P}(B) > 0$, and $F \in \mathcal{A}$. We define the *conditional probability of (the event) F knowing (the event) B* as the real number

$$\mathbb{P}(F \mid B) := \frac{\mathbb{P}(F \cap B)}{\mathbb{P}(B)}. \tag{B.29}$$

Let $B \in \mathcal{A}$ such that $\mathbb{P}(B) > 0$. The *conditional probability* $\mathbb{P}_{|B} : \mathcal{A} \to [0, 1]$ defined by

$$\mathbb{P}_{|B}(F) := \mathbb{P}(F \mid B), \ \forall F \in \mathcal{A} \tag{B.30}$$

is a probability over $(\Omega, \mathcal{A})$. The probability $\mathbb{P}_{|B}$ has the density $\mathbf{1}_B / \mathbb{P}(B)$ w.r.t. $\mathbb{P}$:

$$\frac{d\mathbb{P}_{|B}}{d\mathbb{P}} = \frac{\mathbf{1}_B}{\mathbb{P}(B)}. \tag{B.31}$$

B.5.2 Conditional Expectation w.r.t. an Event

Let $B \in \mathcal{A}$ with $\mathbb{P}(B) > 0$. As $(\Omega, \mathcal{A}, \mathbb{P}_{|B})$ is a probability space, we can define the expectation under this probability, denoted $\mathbb{E}(\cdot \mid B)$. Let $\boldsymbol{Y}$ be an integrable random variable. Then *the conditional expectation of* $\boldsymbol{Y}$ *knowing the event* B is the real number

$$\mathbb{E}\left(\boldsymbol{Y} \mid B\right) := \mathbb{E}_{\mathbb{P}_{|B}}(\boldsymbol{Y}). \tag{B.32}$$

For any integrable random variable $\boldsymbol{Y}$, we have that

$$\mathbb{E}\left(\boldsymbol{Y} \mid B\right) = \frac{\mathbb{E}\left(\boldsymbol{Y}\mathbf{1}_B\right)}{\mathbb{P}(B)}. \tag{B.33}$$

Notice that, with the above notation, Eq. (B.24) can be written as:

$$\mathbb{E}\left(\boldsymbol{Y} \mid \sigma(\{\Omega_n\}_{n \in N})\right) = \sum_{n \in N, \mathbb{P}(\Omega_n) > 0} \mathbb{E}\left(\boldsymbol{Y} \mid \Omega_n\right) \mathbf{1}_{\Omega_n}. \tag{B.34}$$

B.5.3 Conditional Probability w.r.t. a σ-Field

Let $B \in \mathcal{A}$. The *conditional probability of (the event) B knowing (the subfield)* $\mathcal{G}$ is the *random variable*:

$$\mathbb{P}(B \mid \mathcal{G}) := \mathbb{E}\left(\mathbf{1}_B \mid \mathcal{G}\right). \tag{B.35}$$

If $\mathcal{G}$ is an atomic σ-field, generated by the countable partition $\{\Omega_n\}_{n \in N}$, we obtain using (B.24):

$$\mathbb{P}(B \mid \sigma(\{\Omega_n\}_{n \in N})) = \sum_{n \in N, \mathbb{P}(\Omega_n) > 0} \mathbb{P}(B \mid \Omega_n) \mathbf{1}_{\Omega_n}. \tag{B.36}$$

If $\mathbf{Z}$ is a random variable, we define *the conditional probability of the event B knowing (the random variable)* $\mathbf{Z}$ as the random variable

$$\mathbb{P}(B \mid \mathbf{Z}) := \mathbb{E}\left(\mathbf{1}_B \mid \sigma(\mathbf{Z})\right), \tag{B.37}$$

where $\sigma(\mathbf{Z})$ is the σ-field generated by the random variable $\mathbf{Z}$.

B.6 Stochastic Kernels

The main objective of this section is to formulate Proposition B.22, which states the following, widely used property. When computing a conditional expectation with respect to a σ-field $\mathcal{G}$, all the $\mathcal{G}$-measurable variables can be "frozen" during the conditional expectation evaluation. For this purpose, we review results on Borel spaces and on regular conditional laws.

B.6.1 Borel Spaces

Here we follow [21, Chap. 7].

Definition B.12 Let $\mathbb{X}$ be a topological space. We denote by $\mathcal{B}^{\mathrm{o}}_{\mathbb{X}}$ the σ-field generated by the open subsets of $\mathbb{X}$. The elements of $\mathcal{B}^{\mathrm{o}}_{\mathbb{X}}$ are called the *Borel subsets* of $\mathbb{X}$. A mapping φ between topological spaces $\mathbb{X}$ and $\mathbb{X}'$ is said to be *Borel-measurable* if $\varphi^{-1}(\mathcal{B}^{\mathrm{o}}_{\mathbb{X}'}) \subset \mathcal{B}^{\mathrm{o}}_{\mathbb{X}}$.

Definition B.13 A topological space $\mathbb{X}$ is a *Borel space* if there exists a separable[11] complete metric space $\mathbb{X}'$ as well as a Borel subset $B \in \mathcal{B}^{\mathrm{o}}_{\mathbb{X}'}$ such that $\mathbb{X}$ is

[11] A metrizable topological space is *separable* if it contains a countable dense set.

homeomorphic[12] to B. A *Borel isomorphism* φ between Borel spaces $\mathbb{X}$ and $\mathbb{X}'$ is a one-to-one Borel-measurable mapping such that φ^{-1} is Borel-measurable on $\varphi(\mathbb{X})$.

The spaces $\mathbb{R}^n$, as well as their Borel subsets, are Borel spaces. Every Borel space is metrizable and separable, and any complete separable metric space is a Borel space. Every uncountable Borel space is Borel isomorphic to $[0, 1]$, metrizable and separable. If $\mathbb{X}$ is a Borel space, then the space $\mathcal{P}(\mathbb{X})$ of probability distributions over $\mathbb{X}$ is a Borel space.

B.6.2 Stochastic Kernels

Stochastic Kernels and Parametric Disintegration

Definition B.14 ([89]) Let $(\mathbb{X}, \mathfrak{X})$ and $(\mathbb{Y}, \mathfrak{Y})$ be two measurable spaces. A *stochastic kernel* from $(\mathbb{X}, \mathfrak{X})$ to $(\mathbb{Y}, \mathfrak{Y})$ is a mapping $p : \mathbb{X} \times \mathfrak{Y} \to [0, 1]$ such that

- for any $F \in \mathfrak{Y}$, $p(\cdot, F)$ is $\mathfrak{X}$-measurable;
- for any $x \in \mathbb{X}$, $p(x, \cdot)$ is a probability on $\mathfrak{Y}$.

A *random measure* is a stochastic kernel from $(\Omega, \mathcal{A})$ to $(\Omega, \mathcal{A})$.

A stochastic kernel may equivalently be seen as a measurable mapping from $(\mathbb{X}, \mathfrak{X})$ to $\mathcal{P}(\mathbb{Y})$. Thus, as for notation and terminology, we shall speak of a *stochastic kernel* $p(x, \mathrm{d}y)$ *from* $\mathbb{X}$ *to* $\mathbb{Y}$ or of a *stochastic kernel* $p(\mathrm{d}y \mid x)$ *on* $\mathbb{Y}$ *given* $\mathbb{X}$.

Here is a composition operation on stochastic kernels.

Definition B.15 ([89]) Let $(\mathbb{X}, \mathfrak{X})$, $(\mathbb{Y}, \mathfrak{Y})$ and $(\mathbb{Z}, \mathfrak{Z})$ be three measurable spaces. Consider two stochastic kernels, $p(\mathrm{d}y \mid x)$ on $\mathbb{Y}$ given $\mathbb{X}$ and $q(\mathrm{d}z \mid y)$ on $\mathbb{Z}$ given $\mathbb{Y}$. Then, the following expression defines a stochastic kernel $p \otimes q$ on $\mathbb{Z}$ given $\mathbb{X}$:

$$(p \otimes q)(F \mid x) := \int_{\mathbb{Y}} p(\mathrm{d}y \mid x) \int_F q(\mathrm{d}z \mid y), \ \forall F \in \mathfrak{Z}. \tag{B.38}$$

The following proposition establishes that one can decompose a probability measure on a product $\mathbb{Y} \times \mathbb{Z}$ of Borel spaces as a marginal on $\mathbb{Y}$ and a stochastic kernel on $\mathbb{Z}$ given $\mathbb{Y}$. Moreover, this property remains valid when a measurable dependence w.r.t. a parameter is admitted.

Proposition B.16 (Parametric disintegration [21]) *Let $(\mathbb{X}, \mathfrak{X})$ be a measurable space, $\mathbb{Y}$ and $\mathbb{Z}$ be Borel spaces and $q(\mathrm{d}y\ \mathrm{d}z \mid x)$ be a stochastic kernel on $\mathbb{Y} \times \mathbb{Z}$ given $\mathbb{X}$. Then, there exists a stochastic kernel $r(\mathrm{d}z \mid x, y)$ on $\mathbb{Z}$ given $\mathbb{X} \times \mathbb{Y}$ and a stochastic kernel $s(\mathrm{d}y \mid x)$ on $\mathbb{Y}$ given $\mathbb{X}$ such that $q = r \otimes s$, i.e.:*

[12] A *homeomorphism* φ between topological spaces $(\mathbb{X}, \mathfrak{T})$ and $(\mathbb{X}', \mathfrak{T}')$ is one-to-one and continuous, and φ^{-1} is continuous on $\varphi(\mathbb{X})$ with the relative topology.

$$q(\mathrm{d}y\ \mathrm{d}z \mid x) = r(\mathrm{d}z \mid x, y)s(\mathrm{d}y \mid x). \tag{B.39}$$

The stochastic kernel $s(\mathrm{d}y \mid x)$ *is given by*

$$\int_F s(\mathrm{d}y \mid x) = \int_{F \times \mathbb{Z}} q(\mathrm{d}y\ \mathrm{d}z \mid x), \quad \forall F \in \mathcal{Z}. \tag{B.40}$$

Corollary B.17 *Let* $\mathbb{X}$, $\mathbb{Y}$ *and* $\mathbb{Z}$ *be Borel spaces and* $q(\mathrm{d}y\ \mathrm{d}z \mid x)$ *be a stochastic kernel over* $\mathbb{Y} \times \mathbb{Z}$ *knowing* $\mathbb{X}$. *Then, there exists a stochastic kernel* $r(\mathrm{d}z \mid x, y)$ *over* $\mathbb{Z}$ *knowing* $\mathbb{X} \times \mathbb{Y}$ *and a stochastic kernel* $s(\mathrm{d}y \mid x)$ *over* $\mathbb{Y}$ *knowing* $\mathbb{X}$ *such that* $q = r \otimes s$ *as in (B.39).*

Corollary B.18 ([21]) *Let* $\mathbb{Y}$ *and* $\mathbb{Z}$ *be Borel spaces and* $q \in \mathcal{P}(\mathbb{Y} \times \mathbb{Z})$. *Then, there exists a stochastic kernel* $r(\mathrm{d}z \mid y)$ *over* $\mathbb{Z}$ *knowing* $\mathbb{Y}$ *such that* $q = r \otimes s$*:*

$$q(\mathrm{d}y\ \mathrm{d}z) = r(\mathrm{d}z \mid y)s(\mathrm{d}y) \text{ where } s(\mathrm{d}y) = \int_{\mathbb{Z}} q(\mathrm{d}y\ \mathrm{d}z). \tag{B.41}$$

Regular Conditional Laws and Disintegration

Definition B.19 ([30, 89]) Let $(\Omega, \mathcal{A}, \mathbb{P})$ be a probability space. Let $\boldsymbol{X}$ be a random variable taking values in a measurable space $(\mathbb{X}, \mathcal{X})$.

1. Let $\mathcal{G}$ be a subfield of $\mathcal{A}$. A *regular conditional law of the random variable* $\boldsymbol{X}$ *knowing* $\mathcal{G}$ is a stochastic kernel p from $(\Omega, \mathcal{A})$ to $(\mathbb{X}, \mathcal{X})$, such that, for any $F \in \mathcal{X}$, $p(\cdot, F)$ is a version[13] of $\mathbb{P}(\boldsymbol{X} \in F \mid \mathcal{G})$.
2. Let $\boldsymbol{Y} : (\Omega, \mathcal{A}) \to (\mathbb{Y}, \mathcal{Y})$ be a random variable taking values in a measurable space $(\mathbb{Y}, \mathcal{Y})$. A *regular conditional law of the random variable* $\boldsymbol{X}$ *knowing the random variable* $\boldsymbol{Y}$ is a stochastic kernel p from $(\mathbb{Y}, \mathcal{Y})$ to $(\mathbb{X}, \mathcal{X})$, such that, for any $F \in \mathcal{X}$, $p(\boldsymbol{Y}(\cdot), F)$ is a version of $\mathbb{P}(\boldsymbol{X} \in F \mid \boldsymbol{Y})$.

When $(\mathbb{X}, \mathcal{X}) = (\Omega, \mathcal{A})$ and $\boldsymbol{X} = \mathrm{I}_\Omega : (\Omega, \mathcal{A}) \to (\Omega, \mathcal{A})$ is the identity mapping, we obtain the following particular cases. A *regular conditional law of* $\mathbb{P}$ *knowing* $\mathcal{G}$ is a random measure P such that, for any $F \in \mathcal{A}$, $P(\cdot, F)$ is a version of $\mathbb{P}(F \mid \mathcal{G})$. A *regular conditional law of* $\mathbb{P}$ *knowing* $\boldsymbol{Y}$ is a stochastic kernel p such that, for any $F \in \mathcal{A}$, $p(\boldsymbol{Y}(\cdot), F)$ is a version of $\mathbb{P}(F \mid \boldsymbol{Y})$.

As a special case, when $\boldsymbol{X}$ is $\mathcal{G}$-measurable, we may take $p(\omega, F) = \mathbf{1}_F(\boldsymbol{X}(\omega))$ as a version of $\mathbb{P}(\boldsymbol{X} \in F \mid \mathcal{G})$. This gives $p(\omega, \mathrm{d}\omega') = \delta_{\boldsymbol{X}(\omega)}(\mathrm{d}\omega')$

When a regular conditional law of $\boldsymbol{X}$ knowing $\mathcal{G}$ exists, we say that it is unique $\mathbb{P}$-a.s. if any two candidates P and Q are almost surely equal, in the sense that $\mathbb{P}(\{\omega \in \Omega \mid P(\omega, \cdot) = Q(\omega, \cdot)\}) = 1$. In that case, we denote it by $\mathbb{P}_{\boldsymbol{X}}^{\mathcal{G}}(\omega, \mathrm{d}x)$ or by $\mathbb{P}^{\mathcal{G}}(\omega, \boldsymbol{X} \in \mathrm{d}x)$. The regular conditional distribution of $\boldsymbol{X}$ knowing $\mathcal{G}$ is such that,

[13] This means that $p(\cdot, F)$ and $\mathbb{P}(\boldsymbol{X} \in F \mid \mathcal{G})$ are almost surely equal w.r.t. $\mathbb{P}$.

for any measurable function $\varphi : \mathbb{X} \to \mathbb{R}$ satisfying $\mathbb{E}\left(|\varphi(\boldsymbol{X})|\right) < +\infty$, we have that:

$$\mathbb{E}\left(\varphi(\boldsymbol{X}) \mid \mathcal{G}\right)(\cdot) = \int_{\mathbb{X}} \varphi(x)\mathbb{P}^{\mathcal{G}}_{\boldsymbol{X}}(\cdot,\, \mathrm{d}x) \quad \mathbb{P}\text{-a.s.}. \tag{B.42}$$

In the same way, *the regular conditional distribution of* $\mathbb{P}$ *knowing* $\mathcal{G}$ is denoted by $\mathbb{P}^{\mathcal{G}}(\omega,\, \mathrm{d}\omega')$. It is such that, for any integrable random variable $\boldsymbol{X} : \Omega \to \mathbb{R}$,

$$\mathbb{E}\left(\boldsymbol{X} \mid \mathcal{G}\right)(\cdot) = \int_{\Omega} \boldsymbol{X}(\omega')\mathbb{P}^{\mathcal{G}}(\cdot,\, \mathrm{d}\omega'), \;\; \mathbb{P}\text{-a.s.}. \tag{B.43}$$

In the same vein, *the regular conditional distribution of* $\boldsymbol{X}$ *knowing* $\boldsymbol{Y}$ is denoted by $\mathbb{P}^{\boldsymbol{Y}}_{\boldsymbol{X}}(y,\, \mathrm{d}x)$, by $\mathbb{P}^{\boldsymbol{Y}}(y, \boldsymbol{X} \in \mathrm{d}x)$ or by $\mathbb{P}(\boldsymbol{X} \in \mathrm{d}x \mid \boldsymbol{Y} = y)$. It is such that, for any measurable function $\varphi : \mathbb{X} \to \mathbb{R}$ satisfying $\mathbb{E}\left(|\varphi(\boldsymbol{X})|\right) < +\infty$, we have that:

$$\mathbb{E}\left(\varphi(\boldsymbol{X}) \mid \boldsymbol{Y}\right)(\cdot) = \int_{\mathbb{X}} \varphi(x)\mathbb{P}^{\boldsymbol{Y}}_{\boldsymbol{X}}(\boldsymbol{Y}(\cdot),\, \mathrm{d}x), \;\; \mathbb{P}\text{-a.s.}. \tag{B.44}$$

Example B.20 The following expressions are well known and may be easily verified. If $\mathcal{G}$ is an atomic σ-field generated by a countable partition $\{\Omega_n\}_{n \in N}$, we have that:

$$\mathbb{P}^{\mathcal{G}}(\omega,\, \mathrm{d}\omega') = \sum_{n, \mathbb{P}(\Omega_n) > 0} \mathbf{1}_{\Omega_n}(\omega)\mathbb{P}_{|\Omega_n}(\mathrm{d}\omega'). \tag{B.45}$$

If $\boldsymbol{Y}$ is a discrete random variable, we have that (see (B.29) and (B.30)):

$$\mathbb{P}^{\boldsymbol{Y}}(y,\, \mathrm{d}\omega') = \mathbb{P}_{|\{\boldsymbol{Y}=y\}}(\mathrm{d}\omega'). \tag{B.46}$$

If $\boldsymbol{X}$ and $\boldsymbol{Y}$ are discrete random variables, we have that:

$$\mathbb{P}^{\boldsymbol{Y}}_{\boldsymbol{X}}(y,\, \mathrm{d}x) = \sum_{x' \in \boldsymbol{X}(\Omega)} \mathbb{P}(\boldsymbol{X} = x' \mid \boldsymbol{Y} = y)\delta_{x'}(\mathrm{d}x). \tag{B.47}$$

If the couple $(\boldsymbol{X}, \boldsymbol{Y})$ has a density $f_{(\boldsymbol{X},\boldsymbol{Y})} > 0$ w.r.t. the Lebesgue measure on $\mathbb{R}^2$, we have that:

$$\mathbb{P}^{\boldsymbol{Y}}_{\boldsymbol{X}}(y,\, \mathrm{d}x) = \frac{f_{(\boldsymbol{X},\boldsymbol{Y})}(x, y)}{\int_{\mathbb{X}} f_{(\boldsymbol{X},\boldsymbol{Y})}(x', y)\, \mathrm{d}x'}\, \mathrm{d}x. \tag{B.48}$$

△

Proposition B.21 ([30, 89]) *Let* $\boldsymbol{X}$ *be a random variable taking values in a Borel space. If* $\boldsymbol{Y}$ *is another random variable, there exists a regular conditional distribution of* $\boldsymbol{X}$ *knowing* $\boldsymbol{Y}$*, and it is unique* $\mathbb{P}_{\boldsymbol{Y}}$*-a.s..*

Using the previous proposition with $X = \mathrm{I}_\Omega$ and when $(\Omega, \mathcal{A})$ is a Borel space we obtain as a corollary that there exists a regular conditional distribution of $\mathbb{P}$ knowing Y.

The following disintegration formula is widely used.

Proposition B.22 (Disintegration [89]) *Let $X : (\Omega, \mathcal{A}) \to (\mathbb{X}, \mathcal{X})$ be a random variable taking values in a measurable space $(\mathbb{X}, \mathcal{X})$, $\mathcal{G}$ be a subfield of $\mathcal{A}$, and $Y : (\Omega, \mathcal{A}) \to (\mathbb{Y}, \mathcal{Y})$ be a $\mathcal{G}$-measurable random variable. Let also f be a measurable function on $\mathbb{X} \times \mathbb{Y}$ such that $\mathbb{E}\left(\left|f(X, Y)\right|\right) < +\infty$. If X has a regular conditional distribution $\mathbb{P}_X^{\mathcal{G}}(\omega, \mathrm{d}x)$ knowing $\mathcal{G}$, we have that:*

$$\mathbb{E}\left(f(X, Y) \mid \mathcal{G}\right) = \int_{\mathbb{X}} \mathbb{P}_X^{\mathcal{G}}(\cdot, \mathrm{d}x) f(x, Y(\cdot)), \ \mathbb{P}\text{-a.s.}. \tag{B.49}$$

If X has a regular conditional distribution $\mathbb{P}_X^{Y}(Y(\cdot), \mathrm{d}x)$ knowing Y, we have that:

$$\mathbb{E}\left(f(X, Y) \mid Y\right) = \int_{\mathbb{X}} \mathbb{P}_X^{Y}(Y(\cdot), \mathrm{d}x) f(x, Y(\cdot)), \ \mathbb{P}\text{-a.s.}. \tag{B.50}$$

Equation (B.50) is usually written under the form

$$\mathbb{E}\left(f(X, Y) \mid \mathcal{G}\right)(\omega) = \mathbb{E}\left(f(X, y) \mid \mathcal{G}\right)(\omega)_{|y=Y(\omega)}, \ \mathbb{P}\text{-a.s.} \tag{B.51}$$

whenever Y is $\mathcal{G}$-measurable. As a corollary, we have that

$$\mathbb{E}\left(f(X, Y)\right) = \int_{\Omega} \mathbb{P}(\mathrm{d}\omega) \int_{\mathbb{X}} \mathbb{P}_X^{\mathcal{G}}(\omega, \mathrm{d}x) f(x, Y(\omega)). \tag{B.52}$$

B.7 Monte Carlo Method

The knowledge of a random phenomenon arises from experiments, which often consist of a set of independent observations. In this section, we analyze this intuitive idea. We first recall what is the product of probability spaces. We then introduce the notion of sample, and we recall the construction of the underlying probability space. We present the celebrated "Strong Law of Large Numbers" and the "Central Limit Theorem". We conclude this section by presenting numerical experiments and practical considerations.

B.7.1 Infinite-Dimensional Product of Probability Spaces

Let $\{(\mathbb{X}_n, \mathcal{X}_n, \mu_n)\}_{n\in\mathbb{N}}$ be a sequence of probability spaces. We denote by $\mathbb{X}_\infty$ the *product space*, that is the Cartesian product of $\mathbb{X}_n$

$$\mathbb{X}_\infty = \prod_{n \in \mathbb{N}} \mathbb{X}_n,$$

and we define the sequence $\{\boldsymbol{X}_n\}_{n\in\mathbb{N}}$ of *coordinate mappings*, namely

$$\begin{aligned} \boldsymbol{X}_n : \quad & \mathbb{X}_\infty && \to \mathbb{X}_n \\ & (x_0, \ldots, x_n, \ldots) && \mapsto x_n. \end{aligned}$$

Our first objective is to equip the product space with a σ-field.

Definition B.23 The σ-field $\sigma\big(\{\boldsymbol{X}_n\}_{n\in\mathbb{N}}\big)$ generated by the sequence of coordinate mappings is defined to be the smallest σ-field relative to which all $\boldsymbol{X}_n$ are measurable. It is also called the *product σ-field* with components $\mathcal{X}_n$, and is denoted

$$\mathcal{X}_\infty = \bigotimes_{n \in \mathbb{N}} \mathcal{X}_n.$$

In the product space $\mathbb{X}_\infty$, sets of the form $\prod_{n\in\mathbb{N}} A_n$ with $A_n \in \mathcal{X}_n$ are called *cylinders with finite dimensional basis* if $A_n = \mathbb{X}_n$ for all but a finite number of indices n. For such a cylinder A, the (finite) subset of $\mathbb{N}$ for which $A_n \neq \mathbb{X}_n$ is denoted $\mathbb{B}_A$.

Proposition B.24 *The product σ-field $\mathcal{X}_\infty$ is the smallest σ-field containing all cylinders with finite dimensional bases of $\mathbb{X}_\infty$.*

For a proof, see [37, Sect. 1.3].

We are now able to define a probability measure on $(\mathbb{X}_\infty, \mathcal{X}_\infty)$.

Theorem B.25 *There exists a unique probability distribution μ_∞ defined on the product σ-field $\mathcal{X}_\infty$ such that for every cylinder A with finite basis $\mathbb{B}_A$,*

$$\mu_\infty(A) = \prod_{i \in \mathbb{B}_A} \mu_i(A_i).$$

For a proof, see [37, Sect. 6.4].

The following notation is usual for the product distribution:

$$\mu_\infty = \bigotimes_{n \in \mathbb{N}} \mu_n.$$

The probability space $(\mathbb{X}_\infty, \mathcal{X}_\infty, \mu_\infty)$ is called the *infinite-dimensional product probability space* associated with the sequence $\{(\mathbb{X}_n, \mathcal{X}_n, \mu_n)\}_{n\in\mathbb{N}}$. When $(\mathbb{X}_n, \mathcal{X}_n, \mu_n) = (\mathbb{X}, \mathcal{X}, \mu)$ for all $n \in \mathbb{N}$, the following notation is used:

$$(\mathbb{X}_\infty, \mathcal{X}_\infty, \mu_\infty) = \big(\mathbb{X}^{\mathbb{N}}, \mathcal{X}^{\otimes\mathbb{N}}, \mu^{\otimes\mathbb{N}}\big).$$

B.7.2 Samples and Realizations

Let $(\Omega, \mathcal{A}, \mathbb{P})$ be a probability space and let $\boldsymbol{X}$ be a random variable, that is a measurable mapping, defined on Ω taking its values in a space $\mathbb{X}$ equipped with a σ-field $\mathcal{X}$:

$$\boldsymbol{X} : (\Omega, \mathcal{A}) \longrightarrow (\mathbb{X}, \mathcal{X}).$$

For the sake of simplicity, we restrict ourselves to the finite-dimensional case $\mathbb{X} = \mathbb{R}^p, \mathcal{X} = \mathcal{B}^{o}_{\mathbb{R}^p}$ being the associated Borel σ-field. We use the term "random variable" for $\boldsymbol{X}$ whatever the dimension p is (see Sect. B.2). We denote by μ the probability distribution of $\boldsymbol{X}$, that is the probability distribution induced by $\boldsymbol{X}$:

$$\mu(A) = \mathbb{P}(\boldsymbol{X}^{-1}(A)), \ \forall A \in \mathcal{X}.$$

Here we study problems which involve carrying out a sequence of observations of a random phenomenon. More precisely, we are interested in (possibly infinite-dimensional) sequences of independent observations of $\boldsymbol{X}$, that is, sequences of independent random variables which have the same probability distribution as $\boldsymbol{X}$. We first define the notion of sample and realization.

Definition B.26 A *n-sample* from the probability distribution μ is a sequence $(\boldsymbol{X}_1, \dots, \boldsymbol{X}_n)$ of independent random variables with the same probability distribution μ.

This definition easily extends to infinite-dimensional samples, namely sequences $\{\boldsymbol{X}_k\}_{k \in \mathbb{N}}$ of *independent identically distributed* (i.i.d.) random variables.

Before using samples, we have to know if it is always possible to get a (possibly infinite-dimensional) sample from the random variable $\boldsymbol{X}$ defined on the probability space $(\Omega, \mathcal{A}, \mathbb{P})$. The answer is usually negative, the original probability space being not "big enough" to support independent random variables.[14] There is, however, a *canonical* way to define samples. Let the infinite product of the probability spaces $(\mathbb{X}, \mathcal{X}, \mu)$ become the probability space under consideration

$$(\widetilde{\Omega}, \widetilde{\mathcal{A}}, \widetilde{\mathbb{P}}) = \left(\mathbb{X}^{\mathbb{N}}, \mathcal{X}^{\otimes \mathbb{N}}, \mu^{\otimes \mathbb{N}}\right),$$

and consider the coordinate mappings

$$\begin{aligned} \boldsymbol{X}_n \ : \qquad \mathbb{X}^{\mathbb{N}} \qquad & \rightarrow \mathbb{X} \\ (x_1, \dots, x_n, \dots) & \mapsto x_n. \end{aligned}$$

[14] Consider for example a coin toss, and let $\Omega = \{H, T\}$ and $\mathcal{A} = \{\emptyset, \{H\}, \{T\}, \Omega\}$. If the game is modeled using a real-valued random variable defined on Ω, every potential random variable representing the game can be obtained (by composition with a deterministic function) from the *unique* random variable $(H \mapsto 0, T \mapsto 1)$.

They are measurable (by definition of $\mathcal{X}^{\otimes\mathbb{N}}$), independent by construction (since $\mu^{\otimes\mathbb{N}}$ is a product probability), and their common probability distribution is μ. They thus constitute an infinite-dimensional sample of $\boldsymbol{X}$. Owing to a change of the probability space, $(\widetilde{\Omega}, \widetilde{\mathcal{A}}, \widetilde{\mathbb{P}})$ replacing $(\Omega, \mathcal{A}, \mathbb{P})$, it is possible to generate samples of arbitrary size.

Consider now the products of n probability spaces $(\mathbb{X}, \mathcal{X}, \mu)$, namely $\left(\mathbb{X}^n, \mathcal{X}^{\otimes n}, \mu^{\otimes n}\right)$. The projection mappings

$$\begin{aligned}\boldsymbol{\Pi}_n \ : \qquad \mathbb{X}^{\mathbb{N}} \qquad &\to \mathbb{X}^n \\ (x_1, \ldots, x_n, \ldots) &\mapsto (x_1, \ldots, x_n),\end{aligned}$$

are measurable. Let $\widetilde{\mathfrak{F}}_n$ be the σ-field generated by $\boldsymbol{\Pi}_n$:

$$\widetilde{\mathfrak{F}}_n = \sigma(\boldsymbol{\Pi}_n) = \sigma(\boldsymbol{X}_1, \ldots, \boldsymbol{X}_n).$$

Then $\{\widetilde{\mathfrak{F}}_n\}_{n\in\mathbb{N}}$ is the filtration associated with the sample $(\boldsymbol{X}_1, \ldots, \boldsymbol{X}_n, \ldots)$. Otherwise stated, considering $\boldsymbol{X}_n$ as the observation of $\boldsymbol{X}$ delivered at stage n, $\widetilde{\mathfrak{F}}_n$ is the σ-field generated by all observations prior to n.

B.7.3 Monte Carlo Simulation

Let $(\Omega, \mathcal{A}, \mathbb{P})$ be a probability space and let $\boldsymbol{X}$ be a random variable defined on Ω taking its values in the Borel space $\left(\mathbb{R}^p, \mathcal{B}^{\mathrm{o}}_{\mathbb{R}^p}\right)$. We denote the probability distribution of $\boldsymbol{X}$ by μ, and we consider an infinite-dimensional sample $(\boldsymbol{X}_1, \ldots, \boldsymbol{X}_n, \ldots)$ of $\boldsymbol{X}$. According to the previous paragraph, such a sequence exists up to a change of probability space. We suppose from now that $(\Omega, \mathcal{A}, \mathbb{P})$ is "big enough" for such a sequence to exist.

We first recall a classical convergence theorem (Strong Law of Large Numbers).

Theorem B.27 *Let $(\boldsymbol{X}_1, \ldots, \boldsymbol{X}_n, \ldots)$ be a sequence of i.i.d. random variables, and let $\boldsymbol{M}_n = (1/n)(\boldsymbol{X}_1 + \cdots + \boldsymbol{X}_n)$. We suppose that $\mathbb{E}\left(\|\boldsymbol{X}_1\|\right) < +\infty$. Then, the random variable $\boldsymbol{M}_n$ almost surely converges to $\mathbb{E}\left(\boldsymbol{X}_1\right)$ as n goes to infinity:*

$$\boldsymbol{M}_n \xrightarrow{a.s.} \mathbb{E}\left(\boldsymbol{X}_1\right).$$

A second classical theorem (Central Limit Theorem) gives some indication about the convergence rate of the estimator $\boldsymbol{M}_n$.

Theorem B.28 *Let $(\boldsymbol{X}_1, \ldots, \boldsymbol{X}_n, \ldots)$ be a sequence of i.i.d. random variables and let $\boldsymbol{M}_n = (1/n)(\boldsymbol{X}_1 + \cdots + \boldsymbol{X}_n)$. We suppose that $\mathbb{E}\left(\|\boldsymbol{X}_1\|^2\right) < +\infty$, and we denote $M = \mathbb{E}\left(\boldsymbol{X}_1\right)$ and $\Sigma = \mathrm{Var}\left(\boldsymbol{X}_1\right)$ the mean and the covariance matrix of $\boldsymbol{X}_1$. Then, the* *sequence of probability distributions of the random variables* $\sqrt{n}\big(\boldsymbol{M}_n - M\big)$ *narrowly converges towards the centered normal distribution with covariance matrix* Σ*:*

$$\sqrt{n}\Big(\boldsymbol{M}_n - M\Big) \xrightarrow{\mathcal{D}} \mathcal{N}_p(0, \Sigma).$$

Otherwise stated, it means that the covariance matrix of $\boldsymbol{M}_n$ is asymptotically equal to Σ/n: the convergence rate is $1/n$, and *it does not depend on the dimension of the space* $\mathbb{R}^p$.

The proof of these two celebrated theorems can be found in any textbook on Probability (e.g. [37]). Other results for the rate of convergence are available, e.g. from Large Deviations Theory [56].

B.7.4 Numerical Considerations

We are now interested in the computational point of view, that is the manipulation of random variables on a computer. As a matter of fact, if the convergence analysis of algorithms involving samples has to be carried out on random variables, their implementation on a computer is done using numerical values. The following definition is useful for numerical considerations.

Definition B.29 A *realization* $(x_1, \ldots, x_n)$ of the n-sample $(\boldsymbol{X}_1, \ldots, \boldsymbol{X}_n)$ is a value taken by the sample at some $\omega \in \Omega$:

$$(x_1, \ldots, x_n) = \big(\boldsymbol{X}_1(\omega), \ldots, \boldsymbol{X}_n(\omega)\big).$$

This definition is extended without difficulty to infinite-dimensional sample $(\boldsymbol{X}_1, \ldots, \boldsymbol{X}_n, \ldots)$. Such realizations are obtained using a *pseudo-random number generator*, that is a computational device designed to generate numbers that approximate the properties of random numbers.[15] Such a generator usually delivers the components of the realization $(x_1, \ldots, x_n)$ one by one: given a realization $(x_1, \ldots, x_{n-1})$ of a $(n-1)$-sample of $\boldsymbol{X}$, a further call to the generator produces a value x_n such that $(x_1, \ldots, x_n)$ is a realization of a n-sample of $\boldsymbol{X}$.

Let us illustrate the previous notions with help of a basic example, namely the numerical computation of the expectation of a random variable $\boldsymbol{X}$. Using a infinite dimensional sample $(\boldsymbol{X}_1, \ldots, \boldsymbol{X}_n, \ldots)$ of $\boldsymbol{X}$, we know from Theorem B.27 that $\boldsymbol{M}_n = (1/n)(\boldsymbol{X}_1 + \cdots + \boldsymbol{X}_n)$ almost surely converges to $M = \mathbb{E}\big(\boldsymbol{X}\big)$. Moreover, the last summation can be written recursively, namely

$$\boldsymbol{M}_n = \boldsymbol{M}_{n-1} - \frac{1}{n}\big(\boldsymbol{M}_{n-1} - \boldsymbol{X}_n\big).$$

[15] Sequence of numbers that lacks any pattern, in the computer science terminology.

From a realization $(x_1, \dots, x_n, \dots)$ of the sample, we deduce the realization m_n of the estimator $\boldsymbol{M}_n$:

$$m_n = \frac{1}{n}(x_1 + \cdots + x_n).$$

Note that the strong law of large numbers asserts that, except on some $\Omega_0 \subset \Omega$ such that $\mathbb{P}(\Omega_0) = 0$, each realization $(x_1, \dots, x_n, \dots)$ of the sample satisfies

$$\lim_{n \to +\infty} m_n = M.$$

Using the recursive formulation, the sequence $\{m_n\}_{n \in \mathbb{N}}$ is obtained by a computer as follows.

Algorithm B.30 (Recursive Monte Carlo Estimation)

1. Set $m_0 = 0$ and $n = 1$.
2. Draw a realization x_n of $\boldsymbol{X}$.
3. Compute $m_n = m_{n-1} - (1/n)(m_{n-1} - x_n)$.
4. Set $n = n + 1$ and go to step 2.

As already explained, the value x_n is obtained in such a way that $(x_1, \dots, x_n)$ is a realization of a n-sample of the random variable $\boldsymbol{X}$. The algorithm is stopped after a given number N of iterations (say a few thousands). Outputs of the algorithm are shown at Fig. B.1, in the specific case $M = 0$. We have represented the variation of $\|m_n\|$ over the iterations, for different values p of the dimension of the space $\mathbb{X}$.

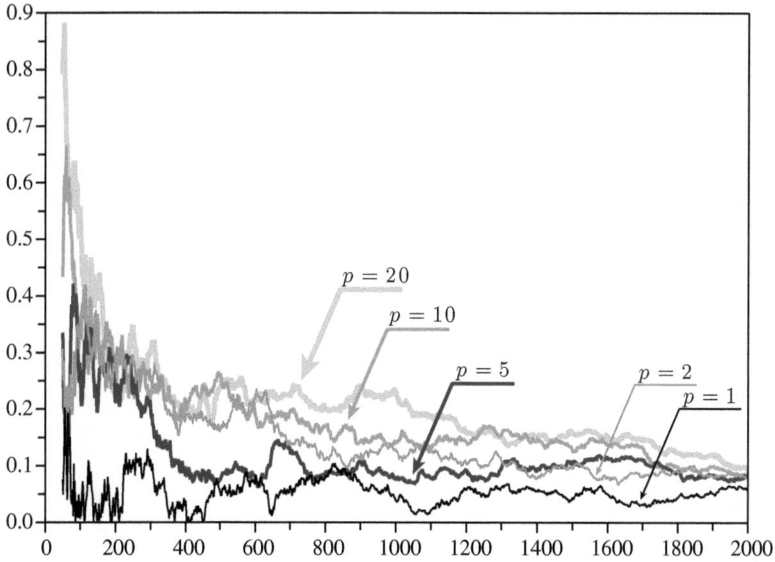

Fig. B.1 Estimation by the Monte Carlo method

Let us conclude with two remarks about this algorithm.

1. The output m_N of Algorithm B.30 is a *single* realization of the random variable $\boldsymbol{M}_N$. We have to perform multiple runs of the code in order to obtain statistical conclusions on the output.
2. It is clear from the experiments presented at Fig. B.1 that, at least asymptotically, the rate of convergence does not depend on the dimension p of the space $\mathbb{X}$.

References

1. V. Alekseev, V. Tikhomirov, and S. Fomin. *Optimal control*. Consultants Bureau, New York, 1987.
2. C. D. Aliprantis and K. C. Border. *Infinite dimensional analysis*. Springer-Verlag, Berlin, second edition, 1999.
3. M. S. Andersland and D. Teneketzis. Information structures, causality and nonsequential stochastic control I: Design-independent properties. *SIAM Journal on Control and Optimization*, 30(6):1447–1475, 1992.
4. Z. Artstein. Sensitivity to σ-fields of information in stochastic allocation. *Stochastics and Stochastics Reports*, 36(1):41–63, 1991.
5. Z. Artstein and R. J.-B. Wets. Consistency of minimizers and the SLLN for stochastic programs. *Journal of Convex Analysis*, 2(1/2):1–17, 1995.
6. H. Attouch. *Variational convergence for functions and operators*. Pitman, Boston, 1984.
7. J.-P. Aubin and H. Frankowska. *Set-valued Analysis*. Birkhäuser, Boston, 1990.
8. F. Baccelli, G. Cohen, G. Olsder, and J.-P. Quadrat. *Synchronization and Linearity – An Algebra for Discrete Event Systems*. John Wiley & Sons, New-York, 1992. This book is out of print but freely available at http://cermics.enpc.fr/~cohen-g/SED/SED1-book.html.
9. R. R. Bahadur. Statistics and subfields. *Ann. Math. Statist.*, 26(3):490–497, 1955.
10. R. R. Bahadur and E. L. Lehmann. Two comments on "sufficiency and statistical decision functions". *Ann. Math. Statist.*, 26(1):139–142, 1955.
11. M. Bansal and T. Basar. Stochastic teams with nonclassical information revisited: When is an affine law optimal? *IEEE Transactions on Automatic Control*, AC-32(6):554–559, June 1987.
12. K. Barty. *Contributions à la discrétisation des contraintes de mesurabilité, pour les problèmes d'optimisation stochastique*. PhD thesis, École Nationale des Ponts et Chaussées, 2004.
13. K. Barty, P. Carpentier, J.-P. Chancelier, G. Cohen, M. De Lara, and T. Guilbaud. Dual effect free stochastic controls. *Annals of Operation Research*, 142(1):41–62, February 2006.
14. K. Barty, J.-S. Roy, and C. Strugarek. A stochastic gradient type algorithm for closed-loop problems. *Mathematical Programming*, Series A, 119:51–78, 2009.
15. R. E. Bellman. *Dynamic Programming*. Princeton University Press, Princeton, N.J., 1957.
16. A. Ben-Tal, L. E. Ghaoui, and A. Nemirovski. *Robust Optimization*. Princeton University Press, 2009.
17. A. Benveniste, M. Métivier, and P. Priouret. *Adaptative Algorithms and Stochastic Approximation*. Springer Verlag, 1990.
18. H. Berliocchi and J.-M. Lasry. Intégrandes normales et mesures paramétrées en calcul des variations. *Bulletin de la Société Mathématique de France*, 101:129–184, 1973.

P. Carpentier et al., *Stochastic Multi-Stage Optimization*, Probability Theory and Stochastic Modelling 75, DOI 10.1007/978-3-319-18138-7

19. D. P. Bertsekas. *Nonlinear Programming*. Athena Scientific, Belmont, Massachusetts, 1999.
20. D. P. Bertsekas. *Dynamic Programming and Optimal Control*, volume 1 and 2. Athena Scientific, Belmont, Massachusetts, second edition, 2000.
21. D. P. Bertsekas and S. E. Shreve. *Stochastic Optimal Control: The Discrete-Time Case*. Athena Scientific, Belmont, Massachusetts, 1996.
22. P. Billingsley. *Probability and Measures*. John Wiley & Sons, New York, 1995.
23. J. R. Birge and R. J.-B. Wets. Designing approximation schemes for stochastic optimization problems, in particular for stochastic programs with recourse. *Mathematical Programming Studies*, 27:54–102, 1986.
24. J.-M. Bismut. An example of interaction between information and control. *IEEE Trans. Automat. Control*, 18:63–64, 1973.
25. J.-M. Bismut. An introductory approach to duality in optimal stochastic control. *SIAM Review*, 20(1):62–78, January 1978.
26. R. Blumenthal and R. Getoor. *Markov Processes and Potential Theory*. Academic Press, New York-London, 1968.
27. J. F. Bonnans, J. C. Gilbert, C. Lemarechal, and C. A. Sagastizabal. *Numerical Optimization: Theoretical and Practical Aspects*. Springer, Berlin, 2006.
28. A. Bourass and E. Giner. Kuhn-Tucker conditions and integral functionals. *Journal of Convex Analysis*, 8(2):533–553, 2001.
29. E. S. Boylan. Equi-convergence of martingales. *Annals of Mathematical Statistics*, 42:552–559, 1971.
30. L. Breiman. *Probability*. Classics in applied mathematics. SIAM, Philadelphia, second edition, 1993.
31. M. Broadie and P. Glasserman. A stochastic mesh method for pricing high dimensional american options. *The Journal of Computational Finance*, 7(4), 2004.
32. A. Bryson and Y.-C. Ho. *Applied optimal control*. Blaisdell Publishing Company, Waltham M.A., 1969.
33. P. Carpentier, G. Cohen, and J.-C. Culioli. Stochastic optimal control and decomposition-coordination methods - Part I: Theory. *In: Recent Developments in Optimization, Roland Durier and Christian Michelot (Eds.), LNEMS 429:72–87, Springer-Verlag, Berlin*, 1995.
34. C. Castaing and M. Valadier. *Convex analysis and measurable multifunctions*, volume 580. Springer, Berlin, Germany, 1977.
35. H. F. Chen, L. Guo, and A. J. Gao. Convergence and robustness of the Robbins-Monro algorithm truncated at randomly varying bounds. *Stochastic Processes and their Applications*, 27(2):217–231, 1988.
36. C. Choirat, C. Hess, and R. Seri. A functional version of the Birkhoff ergodic theorem for a normal integrand: a variational approach. *The Annals of Probability*, 31(1):63–92, 2003.
37. Y. S. Chow and H. Teicher. *Probability Theory. Independence, Interchangeability, Martingales*. Springer Verlag, 2003.
38. G. Cohen. Optimisation des grands systèmes. Unpublished lecture notes, DEA MMME, Université Paris I Panthéon-Sorbonne, 2004. http://cermics.enpc.fr/~cohen-g/documents/ParisIcours-A4-NB.pdf.
39. G. Cohen. Auxiliary problem principle and decomposition of optimization problems. *Journal of Optimization Theory and Applications*, 32(3):277–305, 1980.
40. G. Cohen and B. Miara. Optimization with an auxiliary constraint and decomposition. *SIAM Journal on Control and Optimization*, 28:137–157, 1990.
41. G. Cohen and D. Zhu. Decomposition coordination methods in large scale optimization problems. In J. Cruz, editor, *Advances in large scale systems theory and applications*, volume 1. JAI Press, Greenwich, Connecticut, 1984.
42. T. H. Cormen, C. E. Leiserson, and R. L. Rivest. *Introduction to Algorithms*. The MIT Press, Cambridge, Massachusetts, 1990.
43. K. D. Cotter. Similarity of information and behavior with a pointwise convergence topology. *Journal of Mathematical Economics*, 15(1):25–38, February 1986.

44. J. Couvreux. From the Mosco and slice convergences for convex normal integrands to that of integral functionals on L^p spaces. *SIAM Journal on Control and Optimisation*, 39(1):179–195, 2000.
45. J.-C. Culioli. *Algorithmes de décomposition/coordination en optimisation stochastique*. PhD thesis, École Nationale Supérieure des Mines de Paris, 1987.
46. J.-C. Culioli. *Introduction à l'optimisation*. Ellipses, Paris, 1994.
47. J.-C. Culioli and G. Cohen. Decomposition-coordination algorithms in stochastic optimization. *SIAM Journal on Control and Optimization*, 28(6):1372–1403, 1990.
48. A. Dallagi. *Méthodes particulaires en commande optimale stochastique*. PhD thesis, Université Paris I Panthéon-Sorbonne, 2007.
49. J. M. Danskin. *The theory of Max-Min and its application to weapons allocation problems*. Springer-Verlag, Berlin, New York, 1967.
50. G. B. Dantzig. Linear programming under uncertainty. *Management Science*, 1(3–4), April-July 1955.
51. M. De Lara and L. Doyen. *Sustainable Management of Natural Resources. Mathematical Models and Methods*. Springer-Verlag, Berlin, 2008.
52. F. Delebecque and J.-P. Quadrat. Contribution of stochastic control singular perturbation averaging and team theories to an example of large-scale system: Management of hydropower production. *IEEE transactions on Automatic Control*, AC-23(2):209–221, 1978. Special issue on Large-Scale Systems.
53. C. Dellacherie and P. Meyer. *Probabilités et potentiel*. Hermann, Paris, 1975.
54. B. Delyon. Stochastic approximation with decreasing gain: convergence and asymptotic theory. Unpublished lecture notes, Université de Rennes, 2000.
55. B. Delyon and A. Juditsky. Accelerated stochastic approximation. *SIAM Journal on Optimization*, 3(4):868–881, 1993.
56. A. Dembo and O. Zeitouni. *Large Deviations Techniques and Applications*. Springer Verlag, 1998.
57. J.-C. Dodu, M. Goursat, A. Hertz, J.-P. Quadrat, and M. Viot. Méthodes de gradient stochastique pour l'optimisation des investissements. *Bulletin de la Direction des Études et Recherches EDF*, C(2), 1981.
58. J. Dubra and F. Echenique. Information is not about measurability. *Mathematical Social Sciences*, 47(2):177–185, 2004.
59. M. Duflo. *Algorithmes stochastiques*. Springer Verlag, 1996.
60. M. Duflo. *Random Iterative Models*. Springer Verlag, 1997.
61. J. Dupacová, N. Gröwe-Kuska, and W. Römisch. Scenario reduction in stochastic programming. An approach using probability metrics. *Mathematical Programming*, 95(3):493–511, 2003.
62. J. Dupacová and R. J. Wets. Asymptotic behavior of statistical estimators and of optimal solutions of stochastic optimization problems. *The Annals of Statistics*, 16(4):1517–1549, 1988.
63. E. B. Dynkin and I. V. Evstigneev. Regular conditional expectations of correspondences. *Theory of Probability and its Applications*, XX I(2):325–338, 1976.
64. I. Ekeland and R. Temam. *Convex analysis and variational problems*. Classics in Applied Mathematics. SIAM, Philadelphia, 1999.
65. W. Feller. *An Introduction to Probability Theory and its Applications*, volume 1. John Wiley & Sons, New York, third edition, 1968.
66. W. Feller. *An Introduction to Probability Theory and its Applications*, volume 2. John Wiley & Sons, New York, third edition, 1968.
67. W. H. Fleming and R. W. Rishel. *Deterministic and Stochastic Optimal Control*. Springer-Verlag, Berlin, 1975.
68. A. Gersho and R. M. Gray. *Vector Quantization and Signal Compression*. Kluwer Academic Publishers, 1992.
69. P. E. Gill, W. Murray, and M. H. Wright. *Practical Optimization*. Academic Press, New York, 1983.

70. P. Girardeau. *Résolution de grands problèmes en optimisation stochastique dynamique et synthèse de lois de commande*. PhD thesis, Université Paris-Est, 2010.
71. H. Heitsch and W. Römisch. Scenario reduction algorithms in stochastic programming. *Computational Optimization and Applications*, 24:187–206, 2003.
72. H. Heitsch and W. Römisch. Scenario tree modeling for multistage stochastic programs. *Mathematical Programming*, 118(2):371–406, May 2009.
73. H. Heitsch and W. Römisch. Scenario tree reduction for multistage stochastic programs. *Computational Management Science*, 6:117–133, 2009.
74. C. Hess. Conditions d'optimalité pour des fonctionnelles intégrales convexes sur les espaces $\mathcal{L}^p(E)$. Cahiers de Mathématiques de la Décision No. 8203, Université Paris IX-Dauphine, 1982.
75. C. Hess. On the measurability of the conjugate and the subdifferential of a normal integrand. *Journal of Convex Analysis*, 2:153–166, 1992.
76. F. Hiai. Convergence of conditional expectations and strong law of large numbers for multivalued random variables. *Transactions of the American Mathematical Society*, 291(2):613–627, 1985.
77. F. Hiai and H. Umegaki. Integrals, conditional expectations and martingales of multivalued functions. *Journal of Multivariate Analysis*, 7:149, 1977.
78. J. B. Hiriart-Urruty and C. Lemaréchal. *Convex Analysis and Minimization Algorithms I*. Springer-Verlag, Berlin, 1993.
79. J. B. Hiriart-Urruty and C. Lemaréchal. *Convex Analysis and Minimization Algorithms II*. Springer-Verlag, Berlin, 1993.
80. Y. C. Ho. Review of the Witsenhausen problem. In *47th IEEE Conference on Decision and Control, 2008. CDC 2008*, pages 1611–1613, December 2008.
81. Y. C. Ho and K. C. Chu. Team decision theory and information structure in optimal control problems - Part I. *IEEE Trans. Automat. Control*, 17(1):15–22, February 1972.
82. Y. C. Ho and K. C. Chu. On the equivalence of information structures in static and dynamic teams. *IEEE Trans. Automat. Control*, 18(2):187–188, 1973.
83. Y. C. Ho and K. C. Chu. Information structure in dynamic multi-person control problems. *Automatica*, 10:341–351, 1974.
84. Y. C. Ho, M. P. Kastner, and E. Wong. Teams, signaling, and information theory. *IEEE Transactions on Automatic Control*, pages 305–312, April 1978.
85. J. M. Howie. *Fundamentals of Semigroup Theory*. Oxford University Press, Oxford, 1995.
86. R. Hyndman, D. M. Bashtannyk, and G. K. Grunwald. Estimating and visualizing conditional densities. *Journal of Computational and Graphical Statistics*, 5(4):315–336, 1996.
87. A. D. Ioffe. On lower semicontinuity of integral functionals. *SIAM Journal on Control and Optimisation*, 15(4):521–538, 1977.
88. J. L. Joly and F. De Thelin. Convergence of convex integrals in $\mathcal{L}^p$ spaces. *Journal of Mathematical Analysis and Applications*, 54:230–244, 1976.
89. O. Kallenberg. *Foundations of Modern Probability*. Springer-Verlag, New York, second edition, 2002.
90. S. M. Kay. *Fundamentals of Statistical Signal Processing: Estimation Theory*. Prentice Hall, 1993.
91. H. Kesten. Accelerated stochastic approximation. *Annals of Mathematical Statistics*, 29:41–59, 1958.
92. H. K. Khalil. *Nonlinear Systems*. Prentice-Hall, third edition, 2002.
93. J. Kiefer and J. Wolfowitz. Stochastic estimation of the maximum of a regression function. *Annals of Mathematical Statistics*, 23:462–466, 1952.
94. A. King and R. J.-B. Wets. Epi-consistency of convex stochastic programs. *Stochastics and Stochastics Reports*, 34:83–92, 1991.
95. H. Kudo. A note on the strong convergence of σ-algebras. *Annals of Probability*, 2(1):76–83, 1974.
96. H. J. Kushner and D. S. Clark. *Stochastic Approximation Methods for Constrained and Unconstrained Systems*. Springer Verlag, 1978.

97. H. J. Kushner and G. G. Yin. *Stochastic Approximation and Recursive Algorithms and Applications*. Springer Verlag, 2003.
98. T. L. Lai. Stochastic approximation. *The Annals of Statistics*, 31(2):391–406, 2003.
99. J. Lelong. *Étude asymptotique des algorithmes stochastiques et calcul du prix des options parisiennes*. PhD thesis, École Nationale des Ponts et Chaussées, 2007.
100. H. R. Lewis and C. H. Papadimitriou. *Elements of the Theory of Computation*. Prentice-Hall, Englewood Cliffs, 1982.
101. R. Lucchetti and R. J.-B. Wets. Convergence of minima of integral functionals, with applications to optimal control and stochastic optimization. *Statistics and Decisions*, 1(11):69–84, 1993.
102. D. G. Luenberger. *Optimization by vector space methods*. John Wiley & Sons, New York, 1997.
103. D. G. Luenberger and Y. Ye. *Linear and nonlinear programming*. Springer, New York, 2008.
104. J. Marschak and R. Radner. *Economic Theory of Teams*. Yale University Press, New Haven and London, 1972.
105. M. Minoux. *Mathematical Programming: Theory and Algorithms*. John Wiley & Sons, Chichester, 1986.
106. A. G. Mucci. Limits for martingale-like sequences. *Pacific Journal of Mathematics*, 48(1):197–202, 1973.
107. E. A. Nadaraya. On estimating regression. *Theory of Probability and its Applications*, 10:186–190, 1964.
108. A. Nemirovski, A. Juditsky, G. Lan, and A. Shapiro. Robust stochastic approximation approach to stochastic programming. *SIAM Journal on Optimization*, 19(4):1574–1609, 2009.
109. J. Neveu. *Mathematical Foundations of the Calculus of Probability*. Holden-Day, San Francisco, 1965.
110. T. Pennanen. Epi-convergent discretizations of multistage stochastic programs. *Mathematics of Operations Research*, 30(1):245–256, 2005.
111. T. Pennanen. Epi-convergent discretizations of multistage stochastic programs via integration quadratures. *Mathematical Programing, Series B*, 116:461–479, February 2009.
112. T. Pennanen and M. Koivu. Epi-convergent discretizations of stochastic programs via integration quadratures. *Numerische Mathematik*, 100(1):141–163, 2005.
113. M. Pereira and L. M. V. G. Pinto. Multi-stage stochastic optimization applied to energy planning. *Mathematical Programming*, 52:359–375, 1991.
114. G. C. Pflug. Scenario tree generation for multiperiod financial optimization by optimal discretization. *Mathematical Programming*, 89(2):251–271, 2001.
115. G. C. Pflug and A. Pichler. A distance for multistage stochastic optimization models. *SIAM Journal on Optimization*, 22(1):1–23, 2012.
116. A. Philpott and Z. Guan. On the convergence of stochastic dual dynamic programming and related methods. *Operations Research Letters*, 36:450–455, 2008.
117. L. Piccinini. Convergence of nonmonotone sub-σ-fields and convergence of associated subspaces $L^p(\mathcal{B}_n)$ ($p \in [1, +\infty]$). *Journal of Mathematical Analysis and Applications*, 225(1):73–90, September 1998.
118. L. Piccinini. A new version of the multivalued Fatou lemma. *Journal of Applied Analysis*, 4(2):231–244, 1998.
119. A. Plakhov and P. Cruz. A stochastic approximation algorithm with multiplicative step size modification. *Mathematical Methods of Statistics*, 18(2):195–200, 2009.
120. B. T. Polyak. Convergence and convergence rate of iterative stochastic algorithms. *Automation and Remote Control*, 37(12):1858–1868, 1976.
121. B. T. Polyak. New method of stochastic approximation type. *Automation and Remote Control*, 51(7):937–946, 1990.
122. B. T. Polyak and A. B. Juditsky. Acceleration of stochastic approximation by averaging. *SIAM Journal on Control and Optimization*, 30(4):838–855, 1992.
123. B. T. Polyak and Y. Z. Tsypkin. Adaptative estimation algorithms (convergence, optimality, stability). *Automation and Remote Control*, 40(3):378–389, 1979.

124. L. Pontryagin, V. Boltyanskii, R. Gamkrelidze, and E. Mishchenko. *The Mathematical Theory of Optimal Processes*. John Wiley, New York, 1962.
125. M. J. D. Powell. *Approximation Theory and Methods*. Cambridge University Press, 1981.
126. W. B. Powell. *Approximate Dynamic Programming: Solving the Curses of Dimensionality*. Wiley, second edition, 2011.
127. M. L. Puterman. *Markov Decision Processes*. John Wiley & Sons, New York, 1994.
128. J.-P. Quadrat and M. Viot. *Introduction à la commande stochastique*. École polytechnique, Palaiseau, 1999. Polycopié de cours.
129. H. Robbins and S. Monro. A stochastic approximation method. *Annals of Mathematical Statistics*, 22:400–407, 1951.
130. S. M. Robinson and R. J.-B. Wets. Stability in two-stage stochastic programming. *SIAM Journal on Control and Optimisation*, 6(25):1409–1416, 1987.
131. R. T. Rockafellar. *Convex Analysis*. Princeton University Press, Princeton, N.J., 1970.
132. R. T. Rockafellar. Convex integral functionals and duality. In *Proceedings of the University of Wisconsin, Madison, Wisconsin*, 215–236, 1971.
133. R. T. Rockafellar. Integral functionals, normal integrands and measurable selections. *Lecture Notes in Mathematics*, 543:157–207, 1976.
134. R. T. Rockafellar and R. J.-B. Wets. On the interchange of subdifferentiation and conditional expectation for convex functionals. *Stochastics*, 7:173–182, 1982.
135. R. T. Rockafellar and R. J.-B. Wets. *Variational Analysis*, volume 317 of *A series of Comprehensive Studies in Mathematics*. Springer Verlag, 1998.
136. W. Rudin. *Real and complex analysis*. McGraw-Hill Book Co., New York, 1987.
137. L. Schwartz. *Analyse IV. Applications à la théorie de la mesure*. Hermann, Paris, 1993.
138. A. Shapiro. Asymptotic properties of statistical estimators in stochastic programming. *Annals of Statistics*, 17:841–858, 1989.
139. A. Shapiro. On complexity of multistage stochastic programs. *Operations Research Letters*, 34–1:1–8, 2006.
140. A. Shapiro and T. H. de Mello. On the rate of convergence of optimal solutions of Monte Carlo approximations of stochastic programs. *SIAM Journal on Optimization*, 11(1):70–86, 2000.
141. A. Shapiro, D. Dentcheva, and A. Ruszczyński. *Lectures on Stochastic Programming: Modeling and Theory*. SIAM, Philadelphia, 2009.
142. C. Striebel. Sufficient statistics in the optimum control of stochastic systems. *J. Math. Anal. Appl*, 12:576–592, 1965.
143. C. Strugarek. Interchange of minimization and integration with measurability constraints. *Preprint* CERMICS http://cermics.enpc.fr/reports, 321, 2006.
144. G. Szász. *Théorie des treillis*. Dunod, Paris, 1971.
145. J. Thenie and J.-P. Vial. Step decision rules for multistage stochastic programming: a heuristic approach. *Automatica*, 44(6):1569–1584, June 2008.
146. L. Thibault. Espérances conditionnelles d'intégrandes semi-continus. *Annales de l'I.H.P, section B*, 17(4):337–350, 1981.
147. A. Truffert. Conditional expectation of integrands and random sets. *Annals of Operation Research*, 30:117–156, 1991.
148. P. Varaiya and R. J.-B. Wets. Stochastic dynamic optimization, approaches and computation. In M. Iri and K. Tanabe, editors, *Mathematical Programming, Recent Developments and Applications*, pages 309–332. Kluwer Academic Publisher, 1989.
149. D. Wagner. Survey of measurable selection theorems. *SIAM Journal on Control and Optimization*, 15(5):859–903, August 1977.
150. M. P. Wand and M. C. Jones. *Kernel Smoothing*. Chapman and Hall, 1995.
151. G. S. Watson. Smooth regression analysis. *Sankhya, Series A*, 26:101–116, 1964.
152. P. Whittle. *Optimization over Time: Dynamic Programming and Stochastic Control*, volume 1. John Wiley & Sons, New York, 1982.
153. P. Whittle. *Optimization over Time: Dynamic Programming and Stochastic Control*, volume 2. John Wiley & Sons, New York, 1982.

154. P. Whittle and J. F. Rudge. Simultaneous communication and control. *Advances in Applied Probability*, 8(2):365–384, June 1976.
155. H. S. Witsenhausen. A counterexample in stochastic optimal control. *SIAM Journal on Control*, 6(1):131–147, 1968.
156. H. S. Witsenhausen. On information structures, feedback and causality. *SIAM Journal on Control*, 9(2):149–160, May 1971.
157. H. S. Witsenhausen. Separation of estimation and control for discrete time systems. *Proceedings of the IEEE*, 69(11):1557–1566, November 1971.
158. H. S. Witsenhausen. A standard form for sequential stochastic control. *Mathematical Systems Theory*, 7(1):5–11, 1973.
159. H. S. Witsenhausen. The intrinsic model for discrete stochastic control: Some open problems. In A. Bensoussan and J. L. Lions, editors, *Control Theory, Numerical Methods and Computer Systems Modelling*, volume 107 of *Lecture Notes in Economics and Mathematical Systems*, pages 322–335. Springer-Verlag, 1975.
160. H. S. Witsenhausen. On policy independence of conditional expectations. *Information and Control*, 28(1):65–75, 1975.
161. T. Yoshikawa. Decomposition of dynamic team decision problems. *IEEE Trans. Automat. Control*, 23(4):627–632, 1978.
162. F. Yousefian, A. Nedic, and U. V. Shanbhag. On stochastic gradient and subgradient methods with adaptive steplength sequences. *Automatica*, 48:56–67, 2012.
163. T. V. Zandt. Information, measurability, and continuous behavior. *Journal of Mathematical Economics*, 38(3):293–309, November 2002.

Index

P. Carpentier et al., *Stochastic Multi-Stage Optimization*, Probability Theory and Stochastic Modelling 75, DOI 10.1007/978-3-319-18138-7

MIX
Papier aus verantwortungsvollen Quellen
Paper from responsible sources
FSC® C105338

If you have any concerns about our products,
you can contact us on
ProductSafety@springernature.com

In case Publisher is established outside the EU,
the EU authorized representative is:
Springer Nature Customer Service Center GmbH
Europaplatz 3, 69115 Heidelberg, Germany

Printed by Libri Plureos GmbH
in Hamburg, Germany